Computational Studies *of* New Materials II

From Ultrafast Processes and Nanostructures to Optoelectronics, Energy Storage and Nanomedicine

Computational Studies *of* New Materials II

From Ultrafast Processes and Nanostructures to Optoelectronics, Energy Storage and Nanomedicine

Editors

Thomas F George *(University of Missouri–St Louis, USA)*
Daniel Jelski *(State University of New York at New Paltz, USA)*
Renat R Letfullin *(Rose–Hulman Institute of Technology, USA)*
Guoping Zhang *(Indiana State University, USA)*

World Scientific

NEW JERSEY • LONDON • SINGAPORE • BEIJING • SHANGHAI • HONG KONG • TAIPEI • CHENNAI

Published by

World Scientific Publishing Co. Pte. Ltd.
5 Toh Tuck Link, Singapore 596224
USA office: 27 Warren Street, Suite 401-402, Hackensack, NJ 07601
UK office: 57 Shelton Street, Covent Garden, London WC2H 9HE

British Library Cataloguing-in-Publication Data
A catalogue record for this book is available from the British Library.

COMPUTATIONAL STUDIES OF NEW MATERIALS II
From Ultrafast Processes and Nanostructures to Optoelectronics, Energy Storage and Nanomedicine

ISBN-13 978-981-4287-18-0
ISBN-10 981-4287-18-0

Printed in Singapore by B & Jo Enterprise Pte Ltd

Preface

In 1999 at the request of World Scientific, Daniel Jelski and Thomas F. George edited a book entitled *Computational Studies of New Materials* consisting of fourteen chapters written by leading experts on topics including fullerenes, semiconductors, fractals, polymers and nonlinear optical processes. In 2008, Dr. Zvi Ruder, senior executive editor at World Scientific, approached us about editing a sequel, or second volume. We agreed to this venture with two additional editors — Renat R. Letfullin and Guoping Zhang — and decided upon the title *Computational Studies of New Materials II*.

While the 1999 book was quite timely when it was published, much has evolved during the past decade in the development of new materials and appropriate computational techniques, especially with the "explosion" of interest and activity in nanoscience and nanotechnology. It is worth noting that the stage was set for this from a federal perspective when in 1999 US President Bill Clinton's science advisor, Neal Lane, rated nanotechnology as one of the government's 11 inter-agency R&D priorities for the purpose of planning the FY 2001 budget.[a]

Nanomaterials, i.e., materials with dimensions on the scale of a nanometer, play a prominent role in this current 2010 book. This includes ultrafast processes stimulated by short laser pulses, such as in connection with fullerenes, and the exciting field of nanomedicine, such as selective laser cancer therapy using gold nanospheres and nanorods. Topics in addition to nanomaterials include energy storage and optoelectronics, such as in connection with polymeric light-emitting

---

[a] *OE Reports*, Number **188** (Society of Photo-Optical Instrumentation Engineers, August 1999).

diodes, semiconductor quantum wells, and tailored negative-index metamaterials and microdevices.

We ourselves have authored a number of the chapters and have invited various outstanding scientists throughout the world to serve as chapter authors. We are most impressed by the extremely high caliber of research and its presentation in the chapters by our colleague contributors. We thank the staff in the chancellor's office at UM–St. Louis for their help throughout the editorial process. We also thank Ms. Hwee Yun Tan (editor) and Ms. Jen Nie Kasim (marketing) at World Scientific for their role in producing and promoting this book.

Thomas F. George
Daniel Jelski
Renat R. Letfullin
Guoping Zhang

March 2010

Contents

List of Contributors

Doyeol Ahn
Institute of Quantum Information Processing and Systems
University of Seoul
Seoul 130-743 Korea
Tel: 82-2-2210-2551
Fax: 82-2-2210-2692
Email: dahn@uos.ac.kr

Henk F. Arnoldus
Dcpartmcnt of Physics
Mississippi State University
Mississippi State, Mississippi 39762-5167
Tel: 1-662-325-2919
Fax: 1-662-325-8898
Email: hfa1@msstate.edu

Zsolt I. Benkő
Department of Technology
University of Szeged
Boldogasszony sgt 6
H-6725 Szeged, Hungary
Tel: +36 -62-546-086
Fax: +36 -62-546-075
Email: zsolt.benko@physx.u-szeged.hu;
bzs@jgytf.u-szeged.hu

Liliana Braescu
Department of Computer Science
West University of Timisoara
Timisoara 300223, Romania
Tel: +40-256-592-221
Fax: +40-256-592-316
Email: braesculiliana@yahoo.com;
lilianabraescu@balint1.math.uvt.ro

Letizia Chiodo
Nano-Bio Spectroscopy Group
European Theoretical Spectroscopy Facility
University of the Basque Country UPV/EHU
Centro Joxe Mari Korta
E-20018 Donostia-San Sebastian, Spain
Tel: +34-943-01-8292
Fax: +34-943-01-8390
Email: lchiodo78@gmail.com; letizia.chiodo@ehu.es

Shih-I Chu
Department of Chemistry
University of Kansas
Lawrence, Kansas 66045, USA
Tel: 1-785-864-4094
Fax: 1-785-864-5396
Email: sichu@ku.edu

Thomas F. George
Office of the Chancellor and Center for Nanoscience
Departments of Chemistry & Biochemistry and Physics & Astronomy
University of Missouri–St. Louis
St. Louis, Missouri 63121, USA
Tel: 1-314-516-5252
Fax: 1-314-516-5378
Email: tfgeorge@umsl.edu

Zunnan Huang
Department of Chemistry and Biochemistry
University of Missouri–St. Louis
St. Louis, Missouri 63121, USA
Tel: 1-314-516-5318
Fax: 1-314-516-5342
Email: zn_huang@yahoo.com

Amicare Iacomino
Dipartimento di Fisica "E. Amaldi"
Università degli Studi Roma Tre
I-00146 Roma, Italy
Tel: +39-081-676433
Fax: +39-06-57337102
Email: amilcareiacomino@gmail.com

Daniel Jelski
Office of the Dean of Science & Engineering
State University of New York at New Paltz
New Paltz, New York, 12561, USA
Tel: 1-845-257-3728
Fax: 1-845-257-3730
Email: jelskid@newpaltz.edu

Tae Won Kang
Department of Physics
Dongguk University
Seoul 100-715, Korea
Tel: 82-2-2260-3205
Fax: 82-2-2278-4519
Email: twkang@dongguk.edu

Juan Maria Garciá-Lastra
Nano-Bio Spectroscopy Group
European Theoretical Spectroscopy Facility
Universidad del País Vasco
E-20018 Donostia-San Sebastián, Spain
Tel: +34-943-01-8292
Fax: +34-943-01-8390
Email: juanmaria.garcia@ehu.es

Seung Joo Lee
Quantum-Functional Semiconductor Research Center
Dongguk University
Seoul 100-715, Korea
Tel: 82-10-4744-8422 (mobile)
Fax: 82-2-2260-3945
Email: leesj@dongguk.edu

Marius Leretter
Faculty of Dentistry
Victor Babes University of Medicine and Pharmacy
Timisoara 300070, Romania
Tel: +40-256-295-257
Fax: +40-256-204-480
Email: mariusleretter@yahoo.com

Renat R. Letfullin
Department of Physics and Optical Engineering
Rose-Hulman Institute of Technology
Terre Haute, Indiana 47803, USA
Tel: 1-812-877-8570
Fax: 1-812-877-8023
Email: letfullin@rose-hulman.edu

Sheng Li
Department of Physics
Zhejiang Normal University
Jinhua, Zhejiang 321004, People's Republic of China
Tel: 86-579-8229-8952
Fax: 86-579-8229-8929
Email: sheng_lees@yahoo.com; shenglee@zjnu.cn

Xin Li
Department of Physics
Mississippi State University
Mississippi State, Mississippi 39762-5167
Tel: 1-662-312-1211
Fax: 1-662-325-8898
Email: xl121@msstate.edu

Eric Majzoub
Department of Physics and Astronomy
University of Missouri–St. Louis
St. Louis, Missouri 63121, USA
Tel: 1-314-516-5779
Fax: 1-314-516-6152
Email: majzoube@umsl.edu

Duncan John Mowbray
Nano-Bio Spectroscopy Group
European Theoretical Spectroscopy Facility
Universidad del País Vasco
E-20018 Donostia-San Sebastián, Spain
Tel: +34-943-01-8534
Fax: +34-943-01-8390
Email: dmowbray@fysik.dtu.dk

László Nánai
Department of General and Environmental Physics
University of Szeged
Boldogasszony sgt 6
H-Szeged 6725, Hungary
Tel: +36-62-544-731
Fax: +36-62-420-953
Email: nanai@physx.u-szeged.hu

Seoung-Hwan Park
Department of Electronics Engineering
Catholic University of Daegu
Kyeongbuk 712-702, Korea
Tel: 82-53-850-3226
Fax: 82-53-850-2704
Email: shpark@cu.ac.kr

Alexander K. Popov
Department of Physics and Astronomy
University of Wisconsin–Stevens Point
812 Kensington Road
Neenah, Wisconsin 54956, USA
Tel: 1-920-722-1345
Email: apopov@uwsp.edu

Angel Rubio
Nano-Bio Spectroscopy Group
European Theoretical Spectroscopy Facility
Universidad del País Vasco
E-20018 Donostia-San Sebastián, Spain
Tel: +34-943-01-8292
Fax: +34-943-01-8390
Email: arubio@ehu.es

Jie Shu
Department of Electrical and Computer Engineering
Rice University
Houston, Texas 77251-1892
Tel: 1-713-348-4692
Fax: 1-713-348-5686
Email: js35@rice.edu

Dmitry Telnov
Department of Physics
St. Petersburg State University
St. Petersburg 198504, Russian Federation
Tel: +7-812-428-4552/7200
Fax: +7-812-428-7240
Email: telnov@pcqnt1.phys.spbu.ru

Guo-Ping Tong
Department of Physics
Zhejiang Normal University
Jinhua, Zhejiang 321004, People's Republic of China
Tel: 86-579-8229-8832
Fax: 86-579-8229-8929
Email: tgp6463@zjnu.cn

Chung-In Um
Department of Physics
Korea University
Seoul 136-701, Korea
Tel: 011-9772-3013 (cell)
Fax: 02-927-3292
Email: cium@korea.ac.kr

Chung F. Wong
Department of Chemistry and Biochemistry
University of Missouri–St. Louis
St. Louis, Missouri 63121, USA
Tel: 1-314-516-5318
Fax: 1-314-516-5342
Email: wongch@umsl.edu

Sam Nyung Yi
Department of Applied Science
Korea Maritime University
Busan 606 791, Korea
Tel: 82-51-410-4448
Fax: 82-51-404-3986
Email: snyi@hhu.ac.kr

Sahng-Kyoon Yoo
Green University
Hamyang, Kyungnam 676-872, Korea
Tel: 82-55-964-0987
Fax: 82-55-962-0051
Email: susy21@hanmail.net

Guoping Zhang
Department of Physics
Indiana State University
Terre Haute, Indiana 47809, USA
Tel: 1-812-237-2044
Fax: 1-812-237-4396
Email: gpzhang@indstate.edu

Introduction

Thomas F. George

University of Missouri–St. Louis
tfgeorge@umsl.edu

Daniel Jelski

State University of New York at New Paltz
jelskid@newpaltz.edu

Renat R. Letfullin

Rose-Hulman Institute of Technology
letfullin@rose-hulman.edu

Guoping Zhang

Indiana State University
gpzhang@indstate.edu

In the world of science, ten years is a very long time!

Back in 1999, World Scientific published a volume entitled *Computational Studies of New Materials* [1] for which two of us served as the editors. A decade later, we were approached by World Scientific about editing a second volume. Adding two more editors, the four of us discussed the relationship between the first volume and this current second volume. Has the world changed so much that we need a new title? Or is this a continuation of what was done before?

This is, after all, a Rip Van Winkle experience. It's as if we had gone to sleep for ten years and awoke to find the world a completely different place. The original title was carefully chosen, where the emphasis was on the materials, and less on the computational algorithms. New materials in those days included semiconductor surfaces and doped semiconductors, fullerenes (both C_{60} and nanotubes), fractal clusters, and aperiodic crystals, among others.

In the introduction to the 1999 volume [1], we remarked at how much the world had changed since the 1970s:

Once upon a time, not too terribly long ago, materials science and solid-state physics were rougly synonymous. In those days a volume such as this would have bristled with terms such as "Brillioun zone" and "Wigner–Seitz cells." Then crystals were periodic, solids stretched infinitely in all directions, polymers consisted of monomer units, and polarizability was simply proportional to the volume. The world was simple and wonderful.

This ancient terminology, residually present in the first volume, has completely disappeared from the current text. There are several reasons for this. First, chemistry, physics and engineering are merging into a new discipline called *nanotechnology*. This is the mesoscopic size range – larger than the small molecules a chemist handles, but much smaller than the macroscopic crystals traditionally studied by physicists. In this region, solid-state terminology is no longer very helpful, but neither is the traditional language of chemistry. A new language is being invented. Biochemistry has long dealt with matter at this scale, so it should not be a surprise that our current volume includes articles of medical and biochemical interest.

Second, the state of the experimental art has changed significantly. Today we live in the era of ultrafast lasers (pulses as short as a femtosecond). Table 1 in Chapter 1 by Nánai *et al.* provides a helpful table listing the properties of modern laser. Chemists use these tools to probe molecular transition states on very short timescales. Materials science can use these tools effectively as well: it is possible to distinguish the temperature difference between electrons and phonons in the same material within the nanosecond timeframe that it takes for it all to equilibrate.

Finally, computers are much faster and software is vastly improved than 10 years ago. Of course, this is strikingly apparent to any average citizen, but it is nonetheless worth reviewing. The first volume still had some articles on the methods of quantum chemistry, e.g., detailing basis sets and algorithms. All of this is now in the background, readily available in commercial software, abbreviated by the phrase "level of theory." Quantum chemistry is no longer a task for the materials scientist except in very special circumstances. Indeed, it shows up in our volume in only one chapter by Chu and Telnov (Chapter 4) — appropriately about density functional theory approaches for probing molecules in intense, ultrashort laser fields.

These changes mean that we can now better deliver on our promise to focus on the materials, letting the algorithms live in the background. And in regard to materials, there is something new and something old. The new is, as already mentioned, the advent of nanotechnology. What is nanotechnology? It is the manipulation of matter and energy on the nanometer scale, or roughly 10 to 100 atomic radii. For structures on this scale, surfaces become very important, for either chemical, optical or physical properties.

Chapter 1 (Nánai *et al.*) is an excellent overview of laser-matter interactions at the nanoscale. After reviewing the mathematics of light-matter interactions, with special attention to the high-intensity, nonlinear case, the article considers laser-exciton-phonon interactions for nanoparticles. This includes an interesting discussion of the two-temperature model, where electrons and phonons form separate heat baths. Finally, laser-induced methods of producing nanostructures are reviewed.

An interesting contrast is Chapter 2 by Letfullin and George, which investigates ultrashort laser interactions with metal nanoparticles and argues that a one-temperature model suffices. The goal is to show how gold particles can be heated up very quickly — explosive heating.

Quantum wells, or quantum dots, are a phenomenon whose name predates the term nanotechnology, but which are indeed an excellent example of the latter. As far back as the 1980s, Louis Brus developed “quantum dots,” an early example of nanotechnology. By tuning the size and chemical composition of the quantum dot, we can create a light source of any visible frequency desired; they are becoming popular as light bulbs [2].

Chapter 11 by Park *et al.* discuss quantum wells of wurtzite ZnO-based structures with magnesium doping. As the magnesium concentration rises, the materials demonstrate high internal electric fields, leading to piezoelectric and polarizability effects.

Chapter 12 looks at how modifying the chemistry of a titanium oxide nanostructures changes the optical response. The authors computationally investigate different structures: zero-dimensional clusters, 1D rods, 2D layers, and 3D bulk. These structures are doped with nitrogen and boron. This yields a thorough description of the electronic and optical characteristics TiO_2 clusters, which is computational materials science at its finest!

One novel theme of the current volume concerns medicine — a topic unmentioned ten years ago. This, frankly, is astonishing, for until very recently medicine has been a purely experimental art. The computational study of medical materials is surely a very impressive, recent and pregnant development.

Letfullin and George contribute three, directly relevant chapters. The first (Chapter 5) is an overview of the importance of nanomaterials in medicine. This reviews the properties of nanoparticles as a function of impinging wavelength, pulse duration and single vs. multipulse modes.

The second (Chapter 6) is a model for how nanomaterials can be used to treat cancer. This builds on both Chapters 2 and 5. These nanoparticles have two features: first, on the surface are antibodies that bind specifically to cancer cells; and second, the interior of the particle, typically a gold cluster or rod, strongly absorbs a particular laser frequency. Thus large amounts of energy can be focused on cancer cells, destroying those with minimal collateral damage to the rest of the body. The third (Chapter 8) concerns the ablation of tissue by laser light — using the laser as a very fine, potentially non-invasive scalpel.

Proteins are nanoscale molecules and thus have become part of the overall discipline of materials science. While the famous protein folding problem is beyond our scope, the way proteins change shape upon binding with particular ligands is very important. A useful chapter by Huang and Chung (Chapter 9) describes methods by which this problem can be approached. There are a number of methods using classical force fields that work. Much work has been put into discovering methods for realistically modeling protein-ligand interactions that work and are inexpensive to run. Sometimes science really is indeed an art.

Chapter 7 by Braescu *et al.* investigates numerically the effect of argatroban, hirudin and melagatran on thrombin formation.

Fullerenes and other carbon structures were a topic ten years ago, and they remain of interest today. Zhang and George (Chapter 3) discuss the excitation of infrared normal modes in C_{60}. They use a tight-binding model for the electronic structure and a three-parameter harmonic force field for the phonons. These are coupled by laser light. There are two ways to excite normal modes — one is through the resonant electronic excitation that depends on the laser frequency. The second is through an off-resonant excitation that depends on the laser pulse duration.

The optical properties of light-emitting conjugated polymers are explored in Chapter 10 by Li *et al.* By contrast, Chapter 14 by Li *et al.* explores just bulk optical properties, without any specific reference to a material.

Chapter 13 by Popov and George focuses on tailoring of linear and nonlinear optical properties of a novel class of extraordinary artificial nanostructured electromagnetic materials – negative-index metamaterials. The feasibility of the design of a generation of unique nanophotonic microdevices with enhanced functionality is numerically simulated.

Chapter 15 is a second contribution by Zhang and George. This is a very interesting application of ultrafast lasers to manipulate spins on an unprecedented short time scale — laser-induced femtosecond magnetism. A magnetic field can be created by the symmetry-breaking caused by spin-orbit coupling. This phenomena, discovered in 1996, has applications to writing on magnetic media.

A fifth contribution by Letfullin and George, Chapter 16, is about metal nanoparticles in chemical lasers, where the gas medium is HF/DF. Within this are dispersed a cloud of metal nanoparticles (aluminum with a radius of 0.09 μm), which are ablated by an infrared laser. The result is the chemical creation of photons, resulting in a huge gain.

Finally, it is very appropriate that the book closes on some straight chemistry, namely the computational discovery of new hydrogen compounds (Chapter 18). An example of such a compound is $Ca(BH_4)_2$. A quantum calculation is used to determine the ground-state electronic structure of these compounds on a molecular scale. This result is then used to produce a simple classical force field that reproduces the essential properties under consideration. Obviously, charge distribution is an important component. The super-molecular structure is then modeled using this simple force field.

In conclusion, let us point out some intriguing trends. One stands out very dramatically — with a few specialized exceptions, nobody is working on quantum chemistry. The computational tools used in these papers are all decades old and are very well established. CHARMM is used to model proteins. A tight-binding calculation reproduces the electronic structure in C_{60}. Standard density functional calculations yield the charge density in borohydrides. In ten years, quantum chemistry has gone from being at the forefront of research to being a routine calculation. This is surely progress.

Second, the right tools are brought to bear on the problem at hand. Why use some computationally intensive, *ab initio* model when tight binding works just as well? CHARMM has been so well optimized for protein structure that one can use it for this application with confidence. In short, simple tools that are calibrated for the problem are found to be far superior to complex tools that attempt to be general.

This means that the problems can become much more complex and sophisticated. With today's computer power and the just mentioned ability to precisely limit the complexity, it is possible to consider complex chemistry and sophisticated structures. Of course, computational protein chemists — represented here by Huang and Wong — have been doing some of this for years. But many other articles in this volume depend on the excellent tools developed over the past decade. We do, indeed, stand on the shoulders of giants.

And as for the title? We have chosen to keep the same title. *Computational Studies of New Materials* seems even more apt today than ten years ago. We really are able to keep our promise to leave algorithms in the background. We can now use them in ever more clever and sophisticated ways in the service of materials science.

Light reading? Probably not. Depending on your background, some chapters will be more accessible than others. So don't be shy — make no effort to read the book from cover to cover. Read what catches your fancy — and enjoy.

1. D. Jelski and T. F. George, *Computational Studies of New* Materials (World Scientific, Singapore, 1999).
2. A quantum leap for lighting, *The Economist* **394** (Issue 8672), 18–19 (March 6, 2010).

Chapter 1

Laser-Matter Interactions: Nanostructures, Fabrication and Characterization

László Nánai* and Zsolt I. Benkő†
University of Szeged
**nanai@physx.u-szeged.hu*
†zsolt.benko@physx.u-szeged.hu

Renat R. Letfullin
Rose-Hulman Institute of Technology
letfullin@rose-hulman.edu

Thomas F. George
University of Missouri–St. Louis
tfgeorge@umsl.edu

The study of phenomena induced by laser radiation in both continuous (CW) and pulsed modes on solid surfaces is a widely-explored subject of modern solid-state physics and chemistry. Since the advent of the first laser in the 1960s, a huge number of scientific papers have been devoted to the investigation of different kinds of laser-matter processes, such as laser-induced damage, plasma formation, phase transitions, micro- and macroprocessing, laser-induced chemical reactions at gas–solid and liquid-solid, etc. These efforts have resulted in new laser applications in industry, e.g. cutting, welding and hardening. The laser has become a useful tool for initiating unique chemical reactions to produce advanced materials, like ultrahard ones for example. A number of technological applications, such as laser-induced deposition of metals on porous materials and semiconductor surfaces, already exist in the high-tech industry of micro- and nanoelectronics. Production of catalyzers with tailored properties of different nanostructures and components is prevalent in the chemical industry.

Many years of theoretical and experimental investigations in this area have resulted in a host of valuable physical and chemical discoveries leading to the creation of new sciences, e.g., femtochemistry. The technical development of lasers has yielded a new class of instruments — fs lasers — which are able to produce extra high (TW/cm^2) intensities within the focal spot on targets. Ultrashort lasers are becoming the source of different nonlinear events with high impact on contemporary science, such as on the study of events happening in states far from equilibrium and in nonlinear circumstances. In this chapter, we discuss fs laser-matter interactions, including the production and properties of different nanostructures.

1.1. Introduction

Theoretical and experimental investigations have been carried out to study ultrashort laser-matter interactions [1–5]. The physical phenomena associated with fs laser pulse interactions have been investigated for different target materials including dielectrics, semiconductors and metals [6–11]. Details of laser pulse impacting on solids have been analyzed in air, vacuum and different gaseous environments [12–19]. Ultrashort laser pulses have been used also for material processing and material fabrications, including production and fabrication of carbon nanotubes and other mono- and/or multicomponent nanostructures. New structures of aligned "trees" of nanotubes have been produced by using very specific laser-based techniques [20]. These new materials with their advanced mechanical, electrical, optical, magnetic and catalytic properties represent very promising targets for micro- and nanotechnology (electro-optic devices, biomedical applications, sensors and actuators). The study of the interaction of ultrashort laser pulses with different kinds of targets could lead to new basic physical and physico-chemical discoveries relating to quantum properties of matter and discoveries of new characteristic features of events on ultrashort time scales [21–26].

The chapter is organized as follows. First, we briefly provide basic classical and quantum descriptions of electromagnetic wave-matter interactions. We show that the rules of nonlinear optics must be applied for sufficiently high light intensities. Also, we briefly describe the basic principles of laser operation and design, as well as the main important beam properties making the laser a very useful tool for material processing, fabrication and characterization.

After that we give some insight into the mechanism of energy absorption and transformation in materials at the microscopic level. We show that the rapid energy impact on the electron subsystem of the bulk material is transferred to the thermal equilibrium phonon subsystem through thermalization. The subsystems time events are [27–33]:

- very short electron relaxation time (τ_{e-e}) for degeneration of the excited electron subsystem into the equilibrium electronic subsystem;
- relatively long time for electron–phonon relaxation (τ_{e-p}), during which the energy is transferred from the electronic subsystem to the phonon subsystem;
- phonon–phonon relaxation time (τ_{p-p}) during which the redistribution of the energy occurs in a phonon "battle" leading to thermal equilibrium of the material as whole.

To characterize the processes mentioned above, we give some remarkable characteristic features of the so-called two-temperature model (TTM). We also present some methods for nanomaterial fabrication by fs laser pulses (mainly by pulsed laser ablation) as well as some discussion of the very interesting properties of structures produced by laser methods [34–38].

1.2. Basic Principles of Laser-Matter Interactions

1.2.1. *Linear and nonlinear optics*

Below is given some preliminary information for light-matter interactions in the semiclassical approximation. In such an approximation, an electromagnetic wave is described classically on the basis of Maxwell's equations, while the material is described quantum mechanically, taking into account the electronic, vibrational and rotational energy levels of molecules.

From Maxwell's equations we get:

$$\overline{D} = \varepsilon_0 \overline{E} + \overline{P}(\overline{E}) \quad \text{and} \quad \overline{B} = \mu_0 \overline{H} + \mu_0 \overline{M}(\overline{H}), \qquad (1.1)$$

where $\overline{D}$ is an electric flux density, $\overline{E}$ is an electric field vector, $\overline{B}$ is a magnetic flux density, $\overline{H}$ is a magnetic field vector, ε_0 is a dielectric constant of a vacuum, which is equal to $8.85 \cdot 10^{-12}$ $C^2N^{-1}m^{-2}$ [F/m], and μ_0 is a permeability constant of free space with the value $4\pi \cdot 10^{-7}$ H/m (we use materials for which $\overline{M} = 0 \rightarrow \overline{B} = \mu_0 \overline{H}$). The polarization is defined as

$$\overline{P}(\overline{E}) = \varepsilon_0 \chi \overline{E} \ , \qquad (1.2)$$

where χ is an electric susceptibility. The wave propagation is described by the wave equation for the electric field,

$$\nabla^2 \overline{E} = \mu_0 \varepsilon_0 \frac{\partial^2 \overline{E}}{\partial t^2} + \mu_0 \varepsilon_0 \chi \frac{\partial^2 \overline{E}}{\partial t^2} = \mu_0 \varepsilon_0 \varepsilon_r \frac{\partial^2 \overline{E}}{\partial t^2} , \qquad (1.3)$$

where $\varepsilon_r = (1 + \chi)$ is a relative dielectric constant.

It is well known from optics that the frequency ν of an electromagnetic wave and the material refractive index n are related to c_0 (speed of light in vacuum) as

$$\nu^2 = \frac{1}{\mu_0 \varepsilon_0} \frac{1}{\varepsilon_r} = \frac{c_0^2}{\varepsilon_r} \quad \text{and} \quad n = \frac{c}{\nu} , \qquad (1.4)$$

where $c_0 = 3 \times 10^8$ m/s. The solution of the wave equation can be found in the form

$$E(z,t) = \mathrm{Re}\{\overline{E}(z,\omega) \exp(-ik\overline{r} + i\omega t)\} , \qquad (1.5)$$

where z is the direction of propagation, $\omega = 2\pi\nu$ is the circular frequency, and k is the wave vector ($k = 2\pi / \lambda$, with $\lambda = c / \nu$ as the wavelength).

To describe the propagation of the light in the materials we have the group velocity

$$v_g = \frac{d\omega}{dk}, \tag{1.6}$$

phase velocity

$$v_{ph} = \frac{\omega}{k} = \frac{c}{n}, \tag{1.7}$$

and the Beer–Lambert law

$$I(z) = I_0 e^{-\alpha z}, \tag{1.8}$$

where I_0 is the intensity of the light at the incidence plane, $I(z)$ is the intensity in the deep level z, and α is the absorption coefficient.

The real (ε') and imaginary (ε'') parts of the dielectric constants are related to the optical material parameters n and k through the Kramers-Kronig relations

$$n = \sqrt{\frac{\sqrt{(\varepsilon')^2 + (\varepsilon'')^2} + \varepsilon'}{2}}, \quad k = \sqrt{\frac{\sqrt{(\varepsilon')^2 + (\varepsilon'')^2} - \varepsilon'}{2}}. \tag{1.9}$$

The above equations describe the laser-matter interaction "events" for relatively low laser beam intensities where the material constants do not depend on the intensity, but on frequency. For strong laser fields comparable to the electric field strength inside atoms ($E \sim 10^9$ V/cm and $I \sim 10^{16}$ W/cm^2), we have to take into account the higher-order (nonlinear) terms for the field in the polarization $\overline{P}$ expansion,

$$\overline{P}(t) = \varepsilon_0 \left(\chi^{(1)} \overline{E}(t) + \chi^{(2)} \overline{E}^2(t) + \chi^{(3)} \overline{E}^3(t) + \ldots \right), \tag{1.10}$$

where $\chi^{(n)}$ is the nth order of the susceptibilities of the medium. As a result, we obtain different phenomena, e.g., second- and higher-order harmonic generation, double- and higher-frequency generation, parametric generation, etc.

We note that the conservation laws for photon energy and momentum (phase matching) should be applied for all processes as

$$\omega = \omega_1 + \omega_2 + \omega_3 + \ldots \quad \text{and} \quad \overline{\mathrm{k}} = \overline{\mathrm{k}}_1 + \overline{\mathrm{k}}_2 + \overline{\mathrm{k}}_3 + \ldots. \tag{1.11}$$

1.2.2. *Lasers and their characteristics*

The term 'laser' is an acronym for **l**ight **a**mplification by **s**timulated **e**mission of **r**adiation. We can obtain lasing effects in a laser-active medium if the population in a higher energetic level (N_2) is higher than in a lower energetic level (N_1) (so-called population inversion) as can be seen in Fig. 1.1. The increase of the number of atoms/molecules in the excited levels is due to pumping (by light, electric current, etc.).

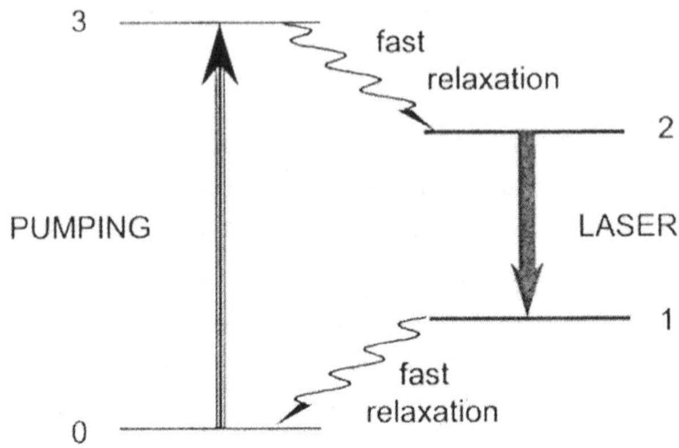

Fig. 1.1. Sample energy levels of a working laser. Here, level 2 is metastable, where the lifetime of atoms/molecules is much longer than in the excited level 3.

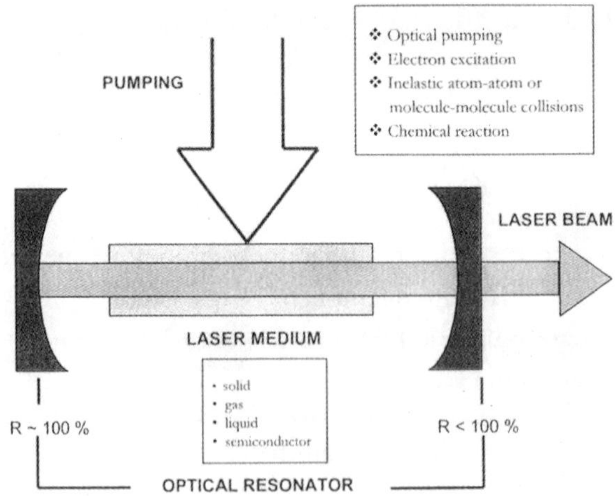

Fig. 1.2. Typical design of a laser.

The gain in laser active medium and the wavelength selection can be accomplished through an optical cavity (Fig. 1.2), which consists of two mirrors: one of them has 100% reflectivity (R) and the output mirror has a lower reflectivity. The main laser properties are summarized in Table 1.1.

Table 1.1. Main properties of a laser.

Wavelength range • 10–15 nm → 100–500 μm (100 eV → 0.01 eV) • tunable lasers: dye laser, diode laser, Ti:sapphire laser …
Monochromaticity • typically $\Delta\nu \sim$ 1 MHZ – 1 Ghz • at best $\Delta\lambda/\lambda = \Delta\nu/\nu \approx$ (1–100 Hz)/(5×10^{14} Hz) ~ 10^{-15}–10^{-12}
Directionality • $\delta\Theta = \lambda/d$, (d = beam diameter) • Typically $\delta\Theta \sim$ 1 mrad, with extra collimation → 1 μrad

Coherence
• coherence time $\Delta\tau \sim 1/\Delta\nu$
  ○ e.g., $\Delta\nu$ = 1 MHz → $\Delta\tau$ = 1 μs
• coherence length $\Delta z = c\cdot\Delta\tau$
  ○ e.g., $\Delta z = c\cdot\Delta\tau = 3\cdot10^8$ m/s · 1 μs = 300 m

Spectral brightness
• $\beta_\nu = P_\nu$ / A $\Delta\Omega$ $\Delta\nu$ [W / cm^2 str Hz]
  ○ Sun: $\beta_\nu \sim 1.5\cdot10^{-12}$ W / cm^2 str Hz
  ○ He-Ne laser (1 mW): $\beta_\nu \sim$ 25 W / cm^2 str Hz
  ○ Nd:glass laser (10 GW): $\beta_\nu \sim 2\cdot10^8$ W/cm^2 str Hz

Operation mode
• continuous wave
• pulsed operation
  ○ shortest pulses < 10 fs (10^{-14} s)
  ○ peak power at best tens of TW

1.2.3. *Femtosecond pulse lasers*

Modern investigations require laser pulses as short as possible with higher energies in order to investigate physical and chemical events on the atto- and femtosecond time scales (10^{-18}–10^{-15} s). Tunable and frequency-limited laser pulses have intensities $I \sim E\nu^3$. From the uncertainty principle, we know that the bandwidth $\Delta\nu$ is inversely proportional to the pulse duration. On the other hand, the efficiency of spontaneous emission (Einstein's coefficients) is proportional to ν^3, while the induced emission probability increases more slowly with frequency. Therefore, obtaining high-energy and short-time pulses in the UV region is possible only by using optical harmonics, first by producing the oscillations in IR spectra and then applying optical frequency conversion into the UV spectral region. It is very important to take into account the dispersion relations, i.e., the dependence of the light speed on frequency in the optical medium. Therefore the main parts of most short-pulse lasers consist of the self-mode-locking system, frequency compensation system, amplifiers, pulse stretching and compensating (CIRP), etc.

In Fig. 1.3 we demonstrate the "standard" Ti:sapphire laser architecture. High-intensity radiation can be realized based on excimer lasers with dye laser amplifiers (Fig. 1.4).

Some parameters of excimer pumped dye lasers are summarized in Table 1.2.

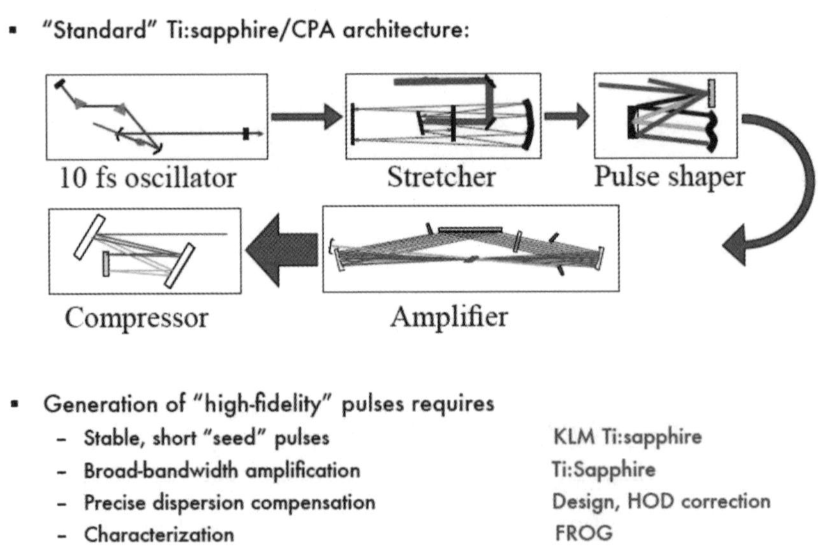

Fig. 1.3. Architectural scheme of a Ti:sapphire (fs) laser system.

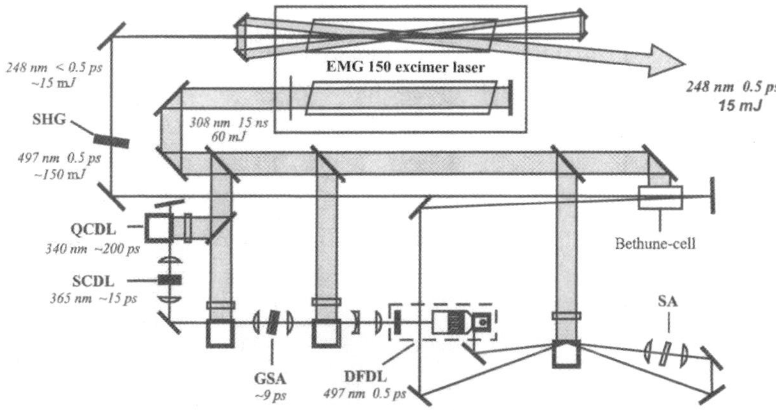

Fig. 1.4. Setup of a high-intensity excimer laser.

Table 1.2. Parameters of a few excimer lasers.

XeCl (1) 20 ps $\le \tau_p \le$ 1ns ($\Delta\lambda$ < 16 Å), $\varepsilon_{saturation} \approx$ 2.5mJ/cm^2 (2) 150 fs $\le \tau_p \le$ 5ps ($\Delta\lambda$ < 16 Å), $\varepsilon_{saturation} \approx$ 0.8–1.0 mJ/cm^2
KrF (1) 200 ps $\le \tau_p$ ($\Delta\lambda$< 15 Å), $\varepsilon_{saturation} \approx$ 2.7 mJ/cm^2, (2) 100 fs $\le \tau_p \le$ 5ps ($\Delta\lambda$ < 15 Å), $\varepsilon_{saturation} \approx$ 2.0 mJ/cm^2.

1.3. Energy Transport

1.3.1. *Impact of laser beam energy on solid matter*

Laser light can deposit a great amount of energy in a very small volume determined by the laser focal spot and penetration depth at a given wavelength. The photoexcitation of the electrons due to the large difference of electron and phonon heat capacities ($c_p >> c_e$) — especially for the case of metals and metal based nanostructures — creates non-equilibrium electron distribution leaving the lattice temperature essentially unchanged (T ~ 300 K). The rise time of the created non-degenerate electron distribution is on the order of a few fs, so that we can say that this high-temperature distribution has the same rise time as a laser pulse duration. Then, over a time scale of around 100 fs, the non-equilibrium electrons redistribute their energy among themselves. It takes time for the electron–electron Coulomb interaction to result in a local equilibrium (with temperature T_e) This process is called thermalized electron energy redistribution (with relaxation time τ_{e-e}). The excited thermalized electron gas then transfers energy through electron–phonon interactions (within the relaxation time τ_{e-p}) [11].

1.3.2. *Energy transfer between electron and phonon subsystems*

The energy transferred to the phonon bath will be redistributed among phonons during the relaxation time τ_{p-p}, leading to the equilibrium phonon temperature T_l. Therefore, we can consider the kinetic evolution of a photoexcited electron–phonon system composed of three interacting subsystems, as depicted in Fig. 1.5, where the relationship among the relaxation times depends on material parameters. For metallic and semiconducting samples, we can consider $\tau_{e-e} << \tau_{e-p}$ and $\tau_{p-p} << \tau_{e-p}$. In most of cases, $\tau_{p-p} \sim 10\tau_{e-e}$, and τ_{e-p} is as long as a few ps [12–15,38]. For disordered (nanostructured) materials (low electron density and high enough temperature), we have $\tau_{e-e} > \tau_{e-p}$ and $\tau_{e-p} >> \tau_{p-p}$. Thus, the relaxation processes occur on two timescales:

– fast, involving the non-degenerate electron–phonon interaction,
– slow, fermionic blocking related to electronic transitions.

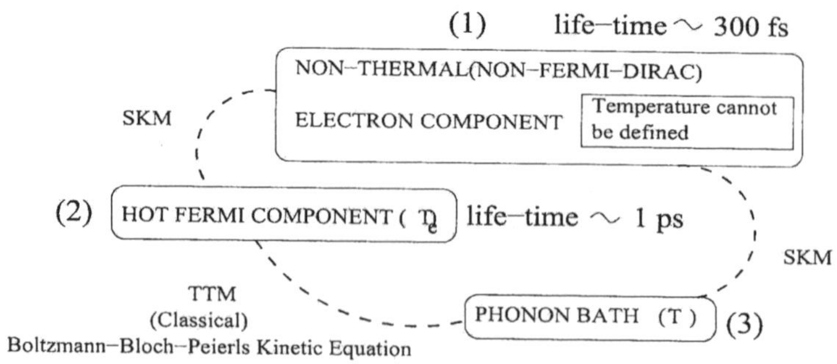

Fig. 1.5. Connections between electron and phonon energetic systems.

1.3.3. *Two-temperature model (TTM) for bulk and nanostructured materials*

The ultrafast electron dynamics on a solid surface under the action of ultrashort laser pulses have been widely studied [14,21,25,38]. The electrons located in the subsurface region of a solid are heated rapidly up to high-temperature (T_e) values by the interaction with femtosecond laser pulses, while the phonons/lattice remain in a non-perturbed room temperature state (T_l) at the time scale of a few ps. It means that for the laser pulse duration shorter than a few ps the electron gas and the lattice in the solid are in a non-thermodynamic equilibrium during the fs laser pulse irradiation [23,38].

Otherwise, during the process of the interaction of nanosecond laser pulses with a solid, the temperatures of the electron gas and lattice are characterized by a common temperature T. In the TTM, the electron and phonon subsystems are described by two separate baths with temperatures T_e and T_l. The energy transfer between two subsystems could be described by the simplest form of the TTM [38]:

$$
\begin{gathered}
C_e(T_e)\frac{\partial T_e}{\partial t} = K_0\frac{\partial^2 T_e}{\partial z^2} - G\left(T_e - T_l\right) + S(r,t) \\
C_l(T_l)\frac{\partial T_l}{\partial t} = G\left(T_e - T_l\right) \\
S(r,t) = \alpha(1-R)I_0(r,t)e^{-\alpha z}.
\end{gathered}
\tag{1.12}
$$

Here, C_e and C_l are the electronic and phonon heat capacities; K_0 is the thermal conductivity; R is the surface reflectivity; α is the absorption coefficient; G is the electron–phonon coupling coefficient; $S(r,t)$ is the energy absorbed by the solid from the laser beam with intensity I_0; and z is the direction of the laser beam propagation perpendicular to the surface. In Table 1.3, we summarize some of the above mentioned parameters for Au, Mo, Ni and Cu (technological materials). Results of our calculations based on Eq. (1.12) for Ni and Au are presented on Fig. 1.6. We should note that our calculations are in a good agreement with simulations done in Ref. 38.

Table 1.3. Thermal parameters of metals.

Metal material	G 10^{16} $Wm^{-3}K^{-1}$	C_e $Jm^{-3}K^{-2}$	K_0 $Wm^{-1}K^{-1}$	C_l 10^6 $Jm^{-3}K^{-1}$
Gold	2.1	71	318	2.5
Molybdenum	13	350	135	2.8
Nickel	36	1065	91	4.1
Copper	10	96.6	385	3.5
Silver	2.3	–	429	–

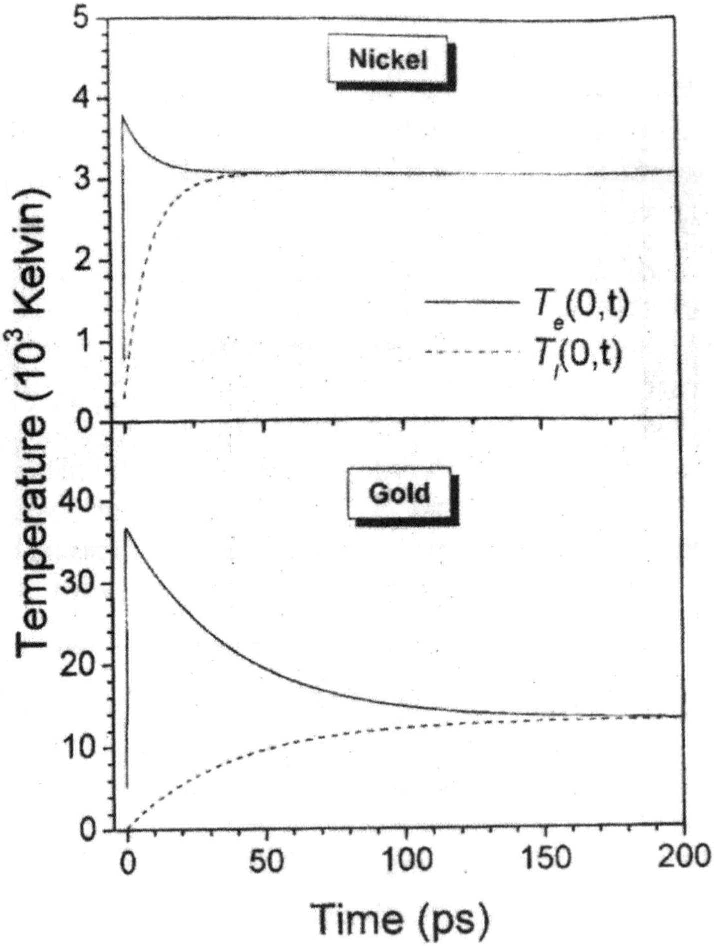

Fig. 1.6. Calculated temperature of the electron-gas (T_e) and the phonons/lattice (T_l) after ultrashort pulse laser illumination.

For the case of semiconductors, we have to take into account two types of charge carriers — electrons and holes with their different mobilities, i.e., $\mu_e \neq \mu_h$. Also, despite of free-carrier absorption and impact ionizations, multiphoton absorption should be taken into account for the situation $\hbar\omega < E_g$, where E_g is the energy gap in the semiconductor and $\hbar$ is Planck's constant. The probability of multiphoton absorption is proportional to the number of photons obeying the relation $k\hbar\omega \geq E_g$. In that case, the energy balance should be written for both electrons and holes [30–34]:

$$\frac{\partial U_e}{\partial t} + \nabla(-k_e \nabla T_e) = -G_{eo}(T_e - T_o) + \begin{bmatrix} (1-R)(\alpha + \Omega n)I(z,t) \\ +(1-R)^2 \beta^2 I^2(z,t) \end{bmatrix}$$

$$\frac{\partial U_o}{\partial t} = G_{eo}(T_e - T_o) - G_{ol}(T_o - T_a)$$

$$\frac{\partial U_a}{\partial t} + \nabla(-k_a \nabla T_a) = G_{oa}(T_o - T_a). \tag{1.13}$$

Here, Ω is the free carrier absorption coefficient; α and β are the absorption coefficients for linear and nonlinear processes, respectively; $U_0 = C_0 T_0$ and $U_a = C_a T_a$ are the optical and acoustical phonon energies; and G_{oa}, G_{eo} and G_{ol} are the coupling coefficients. For very short laser pulses ($\tau \leq 10$ fs), the ballistic transport will be dominant. After the energy distributed is among the electrons, the diffusive transport will be predominant, determining the appropriate temperature:

$$T_B(x,t) = Sech^2\left(\frac{t - \frac{x}{v}}{\tau_p}\right) e^{-\frac{x\rho}{v}} H\left(t - \frac{x}{v}\right)$$

$$T_D(x,t) = \left\{ \frac{\rho x}{v} \int_{\frac{x}{v}}^{t} Sech^2\left(\frac{t-y}{\tau_p}\right) e^{-\rho y} \frac{I_1\left[\rho\sqrt{y^2 - \left(x/v\right)^2}\right]}{\sqrt{y^2 - \left(x/v\right)^2}} dy \right\} H\left(t - \frac{x}{v}\right), \tag{1.14}$$

where τ_p is the laser pulse duration, and H is the Heaviside step function. These solutions come from the quantum heat transport equation:

$$\frac{1}{v^2}\frac{\partial^2 T}{\partial t^2}+\frac{m}{\hbar}\frac{\partial T}{\partial t}=\frac{\partial^2 T}{\partial x^2}, \tag{1.15}$$

where v is the thermal pulse propagation speed, and m is the mass of the heat carriers.

The different mobilities of electrons and holes causes a net charge density difference in the subsurface region. The appropriate "Dember" electric field related to this carrier separation can be calculated as follows

$$E(x)=\frac{e}{\varepsilon_r\varepsilon_0}\int_0^x[\rho_h(x')-\rho_e(x')]dx', \tag{1.16}$$

where ρ_e and ρ_h are the densities of electrons and holes. In Fig. 1.7, we show the results of our calculations for the charge carrier densities in GaAs. In Fig. 1.8, we demonstrate the variations in the Dember field generated in the subsurface region by 20 fs and 50 fs laser pulses.

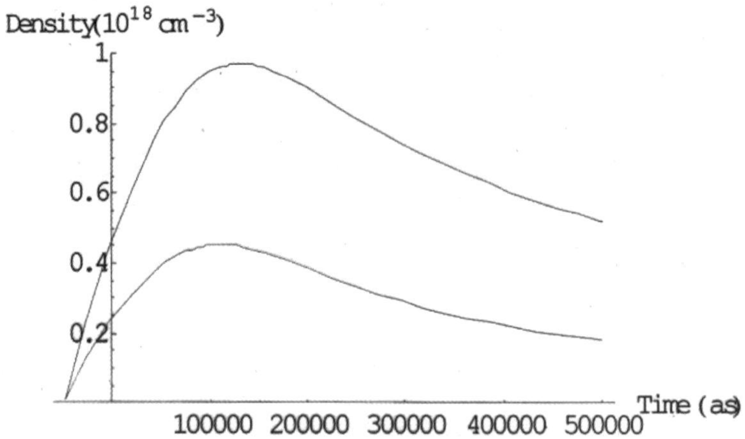

Fig. 1.7. Calculated electron and hole densities for GaAs after a short pulse laser illumination.

The changes in electric properties of the solids in the subsurface region might be registered by the so-called "pump-and-probe" experimental technique. In this technique, the excitations of electron-hole pairs by very short laser pulses can cause change in the reflectivity of the medium, which can be examined by a probe pulse delay on a pump pulse on the irradiated region. One of the possible scheme for realization of such an experiment is demonstrated on Fig. 1.9.

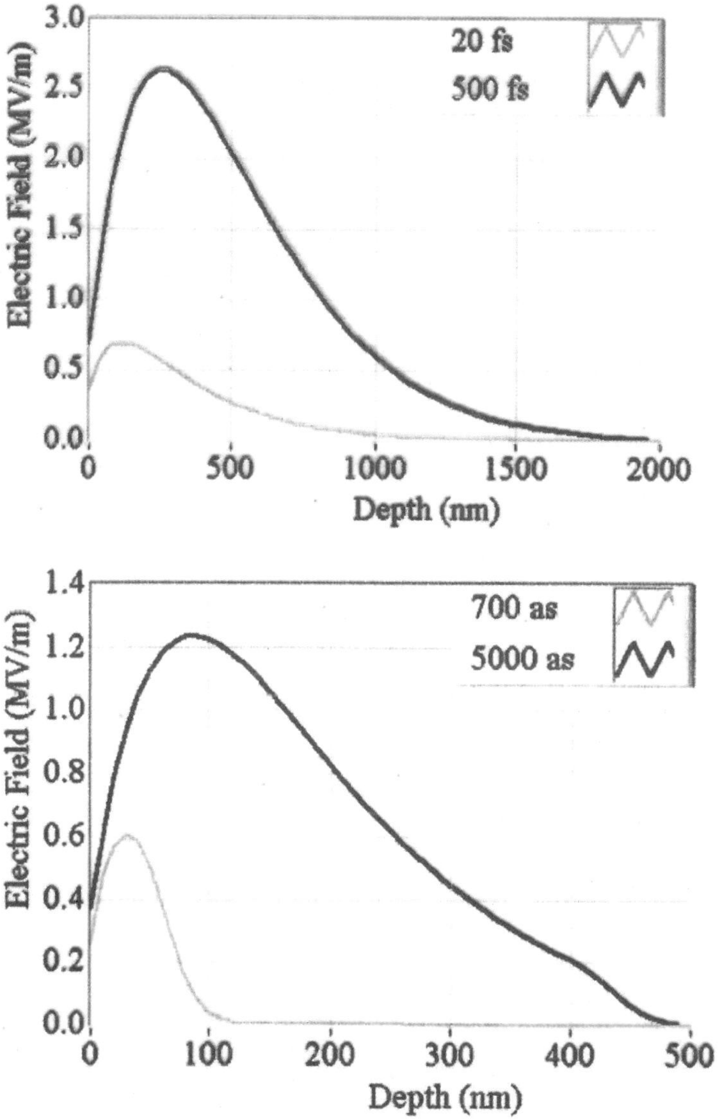

Fig. 1.8. Dember field variations.

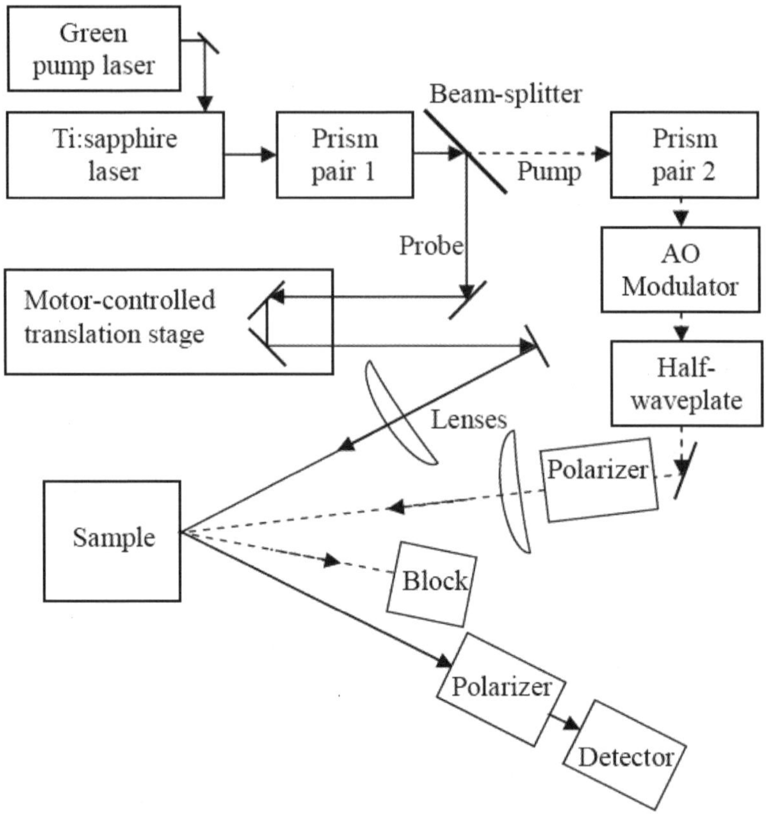

Fig. 1.9. Experimental setup for reflectivity change measurements.

A typical result demonstrating a quick reflectivity change is presented in Fig. 1.10 measured on V_2O_5 targets (semiconductors) and in Fig. 1.11 for TeO_2 (dielectric).

For the case of nanoscaled materials, there are no channels for energy relaxation through phonon interactions (generation) like in bulk materials. This means that for nanostructures, the above mentioned TTM can not be directly applied because of the non-resonant nature of the interaction of phonons and electrons. Therefore, in nanosystems, the main channel for energy transfer might be the surface phonon–electron interaction (the effective electron mean free path is on the order of the particle size). Here, along with the Fermi and Bose descriptions of electrons and phonons, we have to take into account the geometric

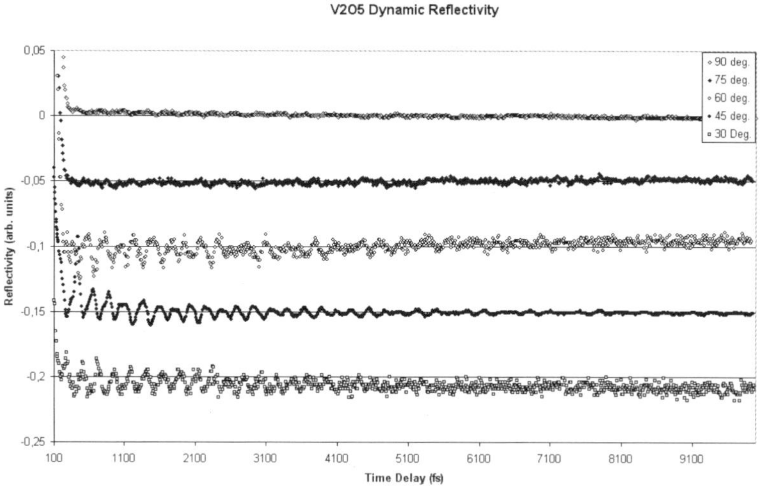

Fig. 1.10. Reflectivity change after a short laser pulse on V_2O_5.

constraints of surface phonons. We note that in the case of homogeneously photoexcited nanostructures and strong damping of phonons, there are no surface standing modes of surfaces. The interaction scheme is sketched on Fig. 1.12.

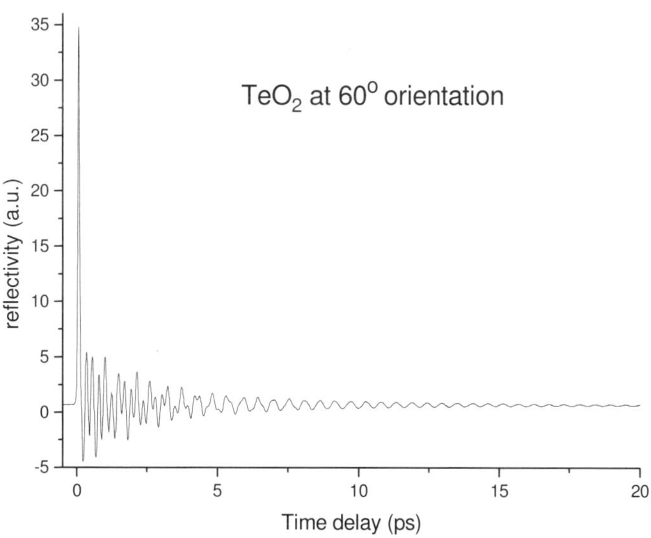

Fig. 1.11. Reflectivity change after a short laser pulse on TeO_2.

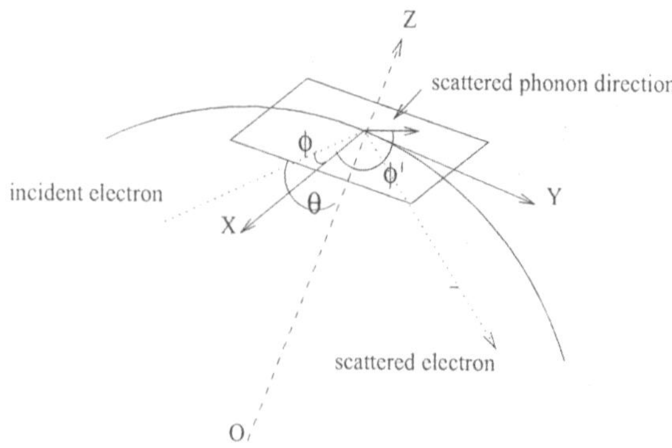

Fig. 1.12. Surface phonon-electron interaction scheme.

The probability W (according to the laws of energy and momentum conservation) that an electron from state k' will scatter into state k emitting a phonon of wave vector f is

$$W(k'-f;k')=\alpha\delta(\varepsilon_{k'}-\varepsilon_k-\hbar\omega),\quad \alpha=(\pi U_s^2/\rho VS_s^2)\,, \tag{1.17}$$

where U_s is the electron-surface phonon interaction constant: ρ and V are the material density and volume; and S_s is the surface sound speed. The time derivative of the surface phonon density is given by the Boltzmann–Bloch–Peierls equation:

$$\frac{dN_f}{dt}=\int \begin{matrix}\alpha\omega_f\{(N_f+1)N_{k'}(1-N_k)-N_fN_k(1-N_{k'})\}\\ \times\delta(\varepsilon_{k'}-\varepsilon_k-\hbar\omega)(2/(2\pi)^3)d\tau_{k'}\end{matrix}\,. \tag{1.18}$$

It can be shown that the ratio for $U_{surface}$ and U_{bulk} is:

$$\frac{U_{surface}}{U_{bulk}}=\frac{2}{3}\pi(8\pi)^{3/2}\left[\frac{U_s}{U_b}\right]^2\frac{nS_b^4}{\alpha\omega^4B_b}, \tag{1.19}$$

where T_D, and ω_D are the Debye temperature and frequency. Taking into consideration both the geometric and quantum side effects, we obtain the following for the surface phonon coupling coefficient [11]:

$$G = \frac{3\sqrt{32\pi}m^2 U_s k_s}{\pi\hbar^3 \rho v}. \tag{1.20}$$

For a gold nanoparticle (R = 10 nm, $U_s = 10^{-19}$ J, $\alpha = 4\cdot10^{-10}$ m, ρ = $19.3\cdot10^3$ kg/m^3, $n = 5.29\cdot10^{28}$ 1/m^3, T_0 = 185 K, $\omega_0 = 2.42\cdot103$ rad/s) we get $G \approx 7.1\cdot10^{13}$ J/m^{3}sK.

For the description of non-equilibrium electron kinetics, a stochastic kinetic model has have been developed. However, the microscopic regime of calculation for the non-equilibrium distribution function is complex, so that we direct the reader's attention to Refs. 11, 36–38.

1.4. Growth of Nanostructures

Nanostructures, including surfaces and single and multiwall nanotubes, might be grown by a variety of methods. The manufacturing technologies can be classified into [6–9], [16–17], [37]:

- top-down technologies,
- bottom-up technologies,
- hybrid top-down and bottom-up technologies.

The top-down process is based on adding a layer of the entire substrate (wafer) and patterning that layer with different lithographic methods (e.g., UV laser (193 nm) photolithography with phase shift masks. The bottom-up process utilizes built structures with selective addition of atoms to create structures like chemical and biological assembles. The comparison of these two classes is shown on Fig. 1.13.

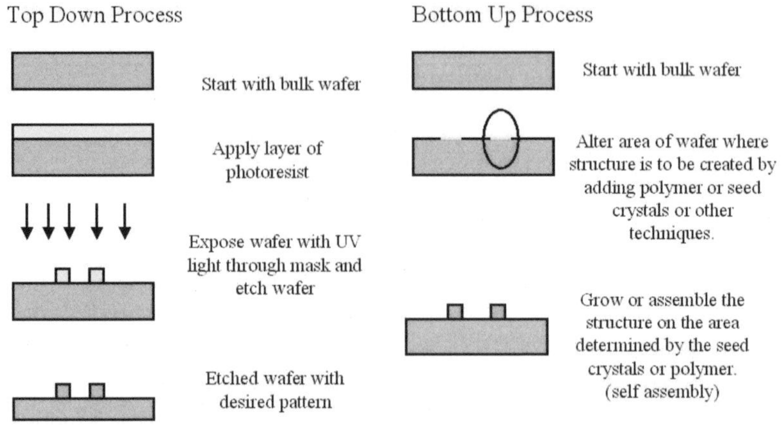

Fig. 1.13. Comparison of the top-down and bottom-up processes.

One of the most important and fruitful bottom-up methods is growing a self-assembling structure, i.e., coordinated growth from independent particles to produce larger, ordered structures (e.g., "building" of nanodots, nanowires, nanotubes, self-assembled nanolayers and different kind of interconnections).

Different physical and chemical methods of fabrication of nanostructured materials include:

- self-assembling, like surfactant systems;
- pattern formation by: photons (UV, DUV, EUV, X-rays); charged carriers (electrons, ions); physical contact methods (printing, molding, embossing); edge-based methods (near-field and different topographic approaches); reposition (liquid, gas, solid phases, MBE);
- scanning probe techniques or combined with laser methods (see Fig. 1.14).

We focus our attention on laser-based methods. The classical photonic (laser) techniques as well as laser liquid deposition, laser chemical vapor deposition, laser-induced precipitation and light induced forward transfer methods have been discussed in more detail elsewhere [2]. We now address femtosecond laser-induced ablation (micropatterning) methods.

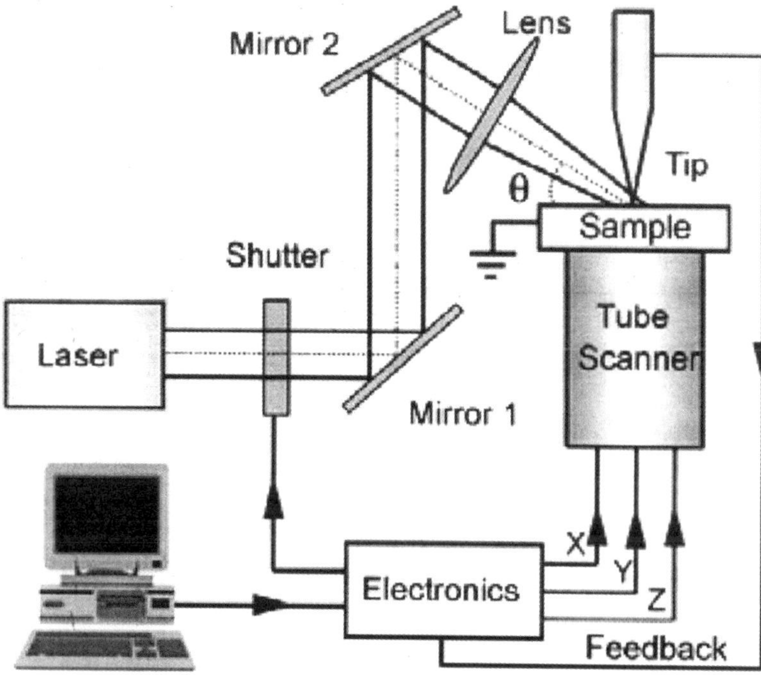

Fig. 1.14. Nanostructure fabrication setup (scanning and laser techniques).

1.4.1. *Femtosecond laser-induced growth of nanostructures*

The use of ultrashort (fs) pulses for laser ablation of materials is an effective method to obtain nanostructured materials with "tailored" properties. The ablation process itself means the collective ejection of material from a target induced by a short intensive light beam. The ablation might be realized in air, vacuum or some gas/liquid surrounding ambient medium. For the liquid case, we get clusters of various sizes dispersed in the liquid and forming a colloidal solution. The advantages of fs laser-induced ablation, as compared to longer (ps, ns) pulses, are:

- The fs beams do not interact with the ejected materials (in vacuum).
- A fs laser pulse heats a solid to higher temperature and creates a higher pressure than longer pulses at the same fluencies, since the energy delivered to the target in the time scale which is not enough for significant heat diffusion.

– The vaporization is increasing while the melting zone reducing (non density change).

Ablation in an appropriate gas environment in a reaction chamber is accompanied by stitching and aggregation of particles (ejected) in the ablated plasma flows. In Fig. 1.15 we show a typical setup for fs laser ablation. On the other hand the lasers as tools are able to provide the specific features to the processes of production of nanostructures with unique properties as:

– interaction (melting, alloying, welding, doping, size-shape aligning, optical trapping),
– characterization (electronic and vibronic properties, RAMAN photoionization properties, thermal diffusivity, mechanical strain, etc.),
– imaging (fluorescence, STM with near field (NSOM), etc.).

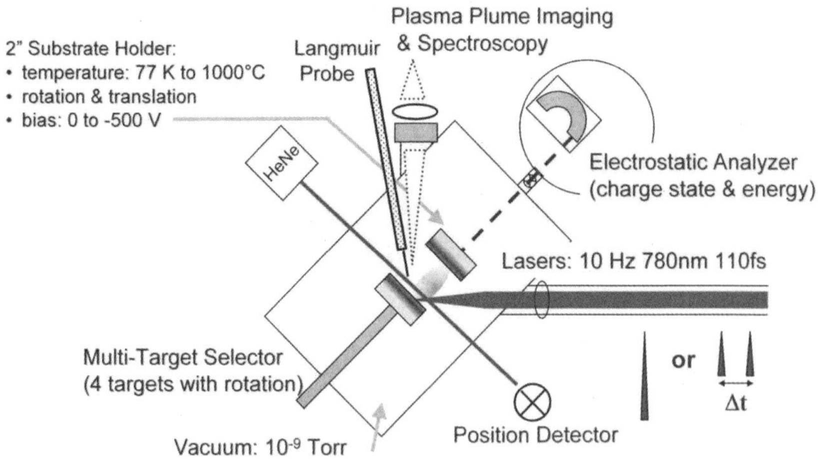

Fig. 1.15. Experimental setup for laser ablation.

In Fig. 1.16 we demonstrate the sketch of experimental setup for a near-field probe.

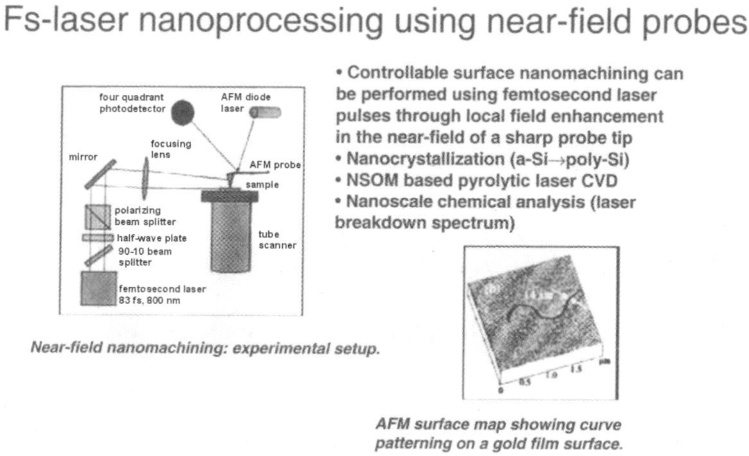

Fig. 1.16. Experimental setup for near-field probes.

The electric, optical and other properties of ablated ejects, their size distribution, morphology, etc. depend on the laser parameters (wavelength and intensity/fluence), ambient atmosphere, and geometry of the ablation experiment, as well as the composition and structure of host material. In Fig. 1.17 we demonstrate the size distribution of nanoparticles for zirconium dioxide.

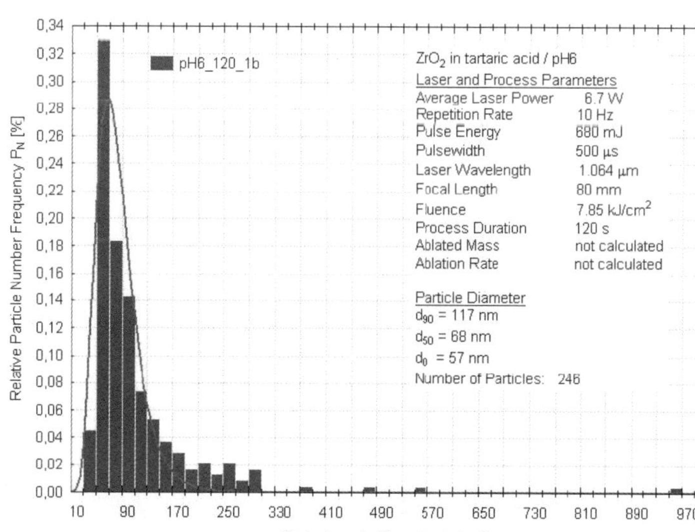

Fig. 1.17. Size distribution of ablated ZrO_2 particles.

In Table 1.4 we show some characteristic properties for a number of parameters (F is the laser pulse fluence, R_m is the nanoparticle mean radius, σ is the corresponding standard deviation, R_{max} is the maximum radius, and $\alpha = \sigma / R_m$).

The theoretical description of ablation process and nanostructure formation is suffered with many limitations because of the complexity of optical and thermo-mechanical pathway experienced by the material through the ablation process. One of the most promising computational

Table 1.4. Ablated particle sizes for some materials and laser fluences.

Material	F (J/cm^2)	I (W/cm^2)	R_m (nm)	σ (nm)	R_{max} (nm)	A
Au	0.3	2.5×10^{12}	8.0	4.7	27	0.59
Au	0.6	5.0×10^{12}	16.4	10.0	48	0.61
Au	1.2	1.0×10^{13}	14.1	10.9	55	0.77
Ag	0.3	2.5×10^{12}	11.8	8.5	46	0.72
Ag	0.6	5.0×10^{12}	13.4	8.0	45	0.60
Ag	1.2	1.0×10^{13}	15.1	11.6	52	0.77
Ni	0.3	2.5×10^{12}	19.7	13.8	60	0.70
Ni	0.6	5.0×10^{12}	17.4	11.0	55	0.63
Ni	1.2	1.0×10^{13}	24.7	19.4	100	0.78
Si	0.3	2.5×10^{12}	7.9	5.1	13	0.65
Si	0.6	5.0×10^{12}	11.6	8.6	15	0.75
Si	1.2	1.0×10^{13}	9.5	5.5	15	0.59
TbDyFe	0.6	5.0×10^{12}	44.8	24.8	80	0.55
TbDyFe	1.0	8.3×10^{12}	18.0	12.8	38	0.70

techniques is on the molecular dynamics method [16]. This involves integrating the equations of motion of an ensemble of particles whose interaction is described by a proper potential energy function (Fig. 1.18). The interaction of the sample with laser beam can be described by Lennard–Jones potential:

$$\Phi_{LJ}(r) = 4\varepsilon\left[(\sigma / r)^{12} - (\sigma / r)^{6} - (\sigma / r_c)^{12} + (\sigma / r_c)^{6}\right], \quad r \le r_c$$
$$\Phi_{LJ}(r) = 0, \quad r > r_c. \tag{1.21}$$

The thermodynamic approach, which describes the ablation mechanism in the terms of temperature and pressure allows to study the spallation, homogeneous nucleation fragmentation and vaporization phenomena accompanying the ablation. In Fig. 1.19 we give a schematic illustration of laser pulse interaction with surface layers. Fig. 1.20 demonstrates the snapshots of simulations for a laser fluence ~ 1.2 of the threshold intensity at instants in time. The time scale relates to a pulse duration τ.

Fig. 1.18. Sample surface scale for equation integration calculations.

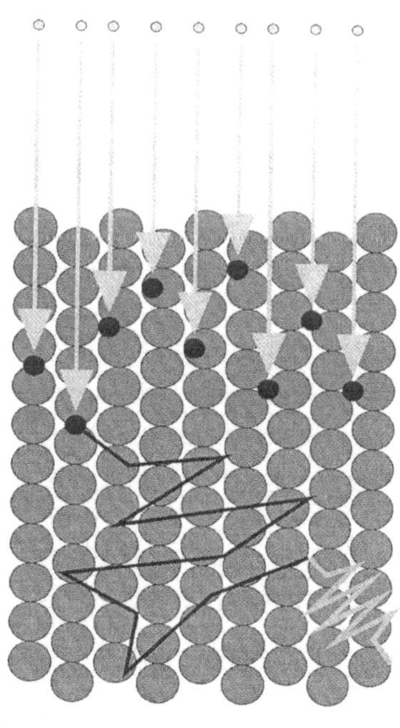

Fig. 1.19. Laser pulse – surface layers interaction scheme

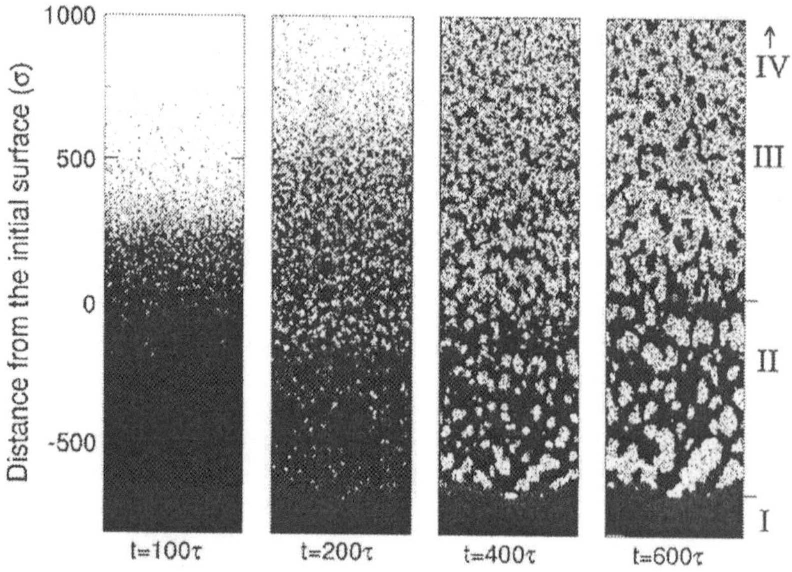

Fig. 1.20. Ablation simulation example.

1.5. Properties of Nanostructures

1.5.1. *General features*

Quantum confinement:

- In nanostructured materials, variation of the fundamental properties with size is observed when the energy spacing between the electronic levels exceeds the thermal energy (kT).
- The electronic energy levels are not continuous because of the confinement of the electronic wave function to the physical dimensions.
- The high surface area and large fraction of atoms blended in the surface layers (e.g., 30% for 1 nm and 15% for 15 nm particles) lead to strong dependence of material properties on surface effects.

We illustrate the quantum confinement for sodium in Fig. 1.21 and the energy levels for metallic and semiconductor nanoparticles in Fig. 1.22.

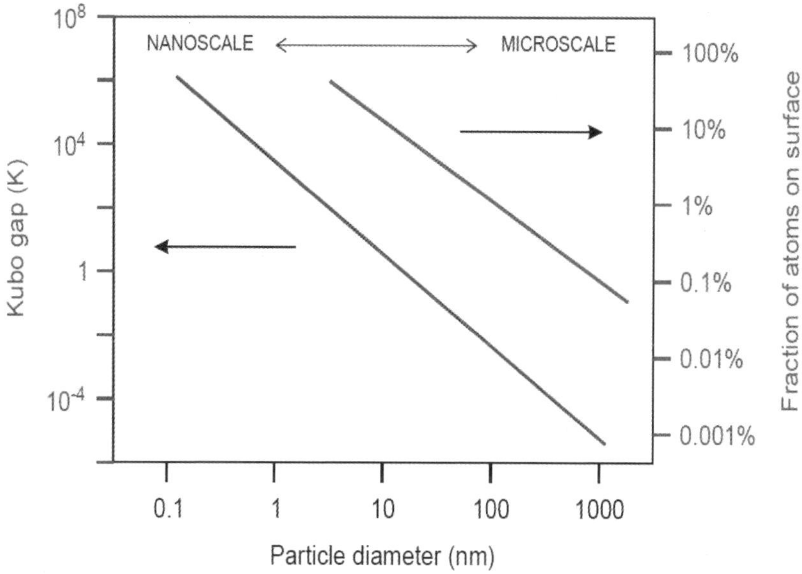

Fig. 1.21. Quantum confinement for Na.

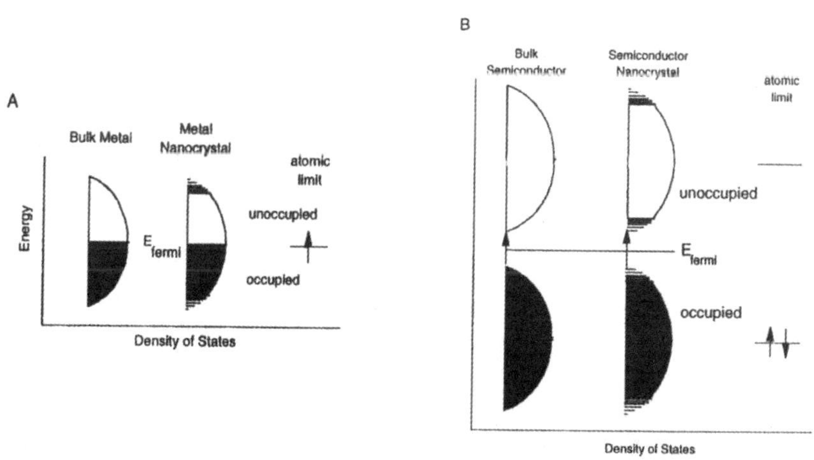

Fig. 1.22. Energy level schemes for metal and semiconductor nanoparticles.

It is seen that quantum confinement occurs for metals when the electron motion is limited by the size of the nanoparticle, and for semiconductors if the exciton Bohr radius is comparable to the nanoparticle size.

1.5.2. *Optical properties*

The dielectric properties of metal nanoparticles (spheres) are determined by the complex dielectric constant $\varepsilon = \varepsilon_1 + \varepsilon_2$. For the medium with ε_m, the effective dielectric constant is

$$\tilde{\varepsilon} = \varepsilon_m + p\frac{3\varepsilon_m}{\varepsilon + 2\varepsilon_m}(\varepsilon - \varepsilon_m), \tag{1.22}$$

where $R << \lambda$ and p (volume fraction) $<< 1$. In this case, the absorption coefficient $\alpha(\omega)$ can be written as:

$$\alpha(\omega) \approx \frac{\varepsilon_2(\omega)}{(\varepsilon_1(\omega) + 2\varepsilon_m)^2 + \varepsilon_2^2(\omega)}. \tag{1.23}$$

If $\varepsilon_1 + 2\varepsilon_m = 0$, we have a surface plasmon resonance, where the resonance frequency depends on the shape and structure of the nanoparticles and on the medium's dielectric properties. For metals, $\varepsilon(\omega) = \varepsilon^b(\omega) - \omega_p^2 / \omega(\omega + i/\tau)$, where ω^b is related to bound electrons, and $\omega_p^2 / \omega(\omega + i/\tau)$ is related to intraband transitions. In this case, the surface plasmon resonance frequency Ω_R and linewidth Γ are:

$$\begin{gathered} \Omega_R = \omega_p / \sqrt{\varepsilon_1^b(\Omega_R) + 2\varepsilon_m} \\ \Gamma = 1/\tau \end{gathered}. \tag{1.24}$$

For Ag nanoparticles with D = 13 nm and $p \sim 2\times\cdot 10^{-4}$, the related frequencies are plotted in Fig. 1.23.

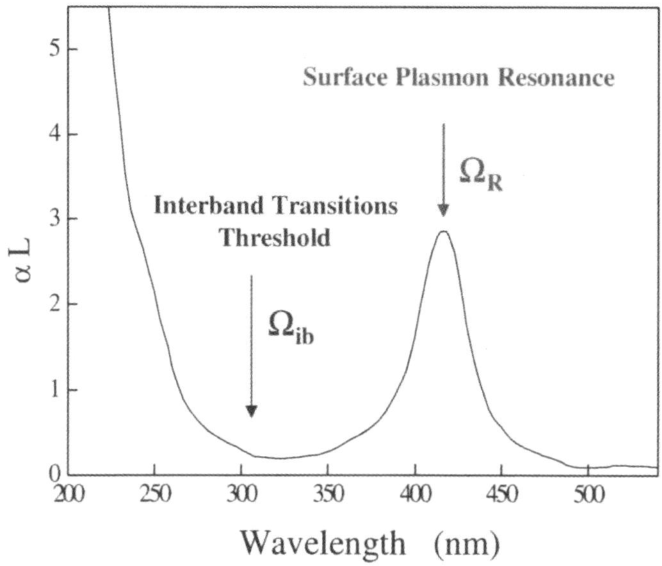

Fig. 1.23. Surface plasmon resonance for Ag nanoparticles.

1.5.3. *Mechanical properties*

For the case of nanostructures with the characteristic dimension less than 100 nm, we must take into account some characteristic differences between bulk and nanomaterials. Due to the increased surface-to-volume ratio, the contact forces are determined mainly by the surface (sub)layer atoms. It may happen that the surface forces could overcome the gravitational forces. This leads to faster dynamics and the appearance of nonlinear and non-contact forces in the form of frictions, fluids etc. Moreover, the principles of uncertainty and complexity prevail on the whole process.

The large fraction of surface atoms can modify thermomechanical properties of nanostructures, such as lowering the melting point:

$$T_b - T_m = \left[\frac{4T_b}{\Delta H \rho_p d_p}\right]\left[\sigma_p - \sigma_i\left(\frac{\rho_p}{\rho_l}\right)^{2/3}\right]. \qquad (1.25)$$

Here, T_b and T_m are the bulk and particle melting points; ΔH is the molar heat of fusion; σ, ρ, and d are the surface strength, density and diameter. The elastic (Young) modulus (E) determines the hardness and brittleness (ductility).

The hardness and strength increases with decreasing particle diameter as ~ $d^{-1/2}$ (Hall–Petch law), since the grain boundaries represent a barrier to dislocation movement.

1.5.4. *Electrical properties*

When the electronic bond structure of bulk materials changes into discrete energy levels of nanomaterials, Ohm's law is no longer valid. The Coulomb energy of an electron in a nanoparticle increases by $E_C = e^2/2C$, where C is the capacitance of the particle. At low temperatures ($kT < e^2/2C$), the electron participates in a tunneling process. Therefore, the ampere–volt characteristics (I–U) of small entities (e.g., quantum dots)

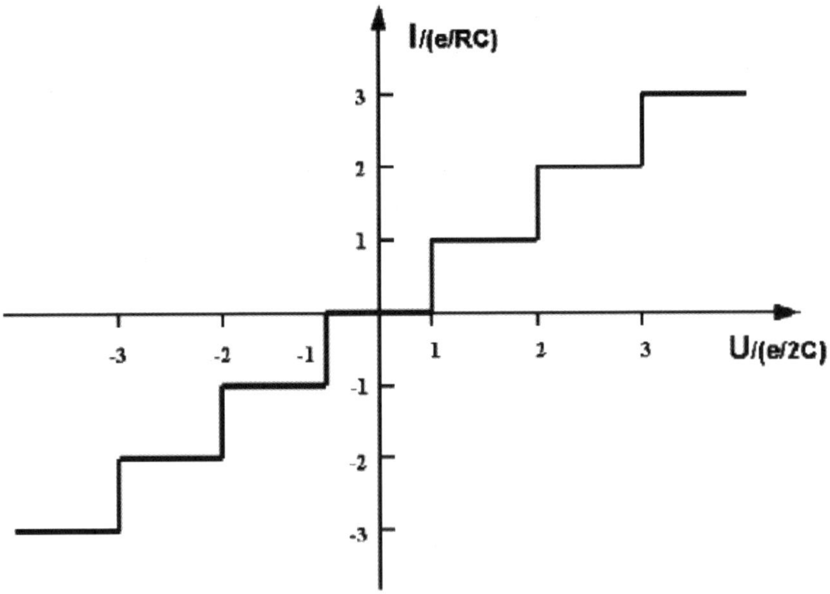

Fig. 1.24. Sample I–U characteristic of a quantum dot.

will be a staircase like shown in Fig. 1.24. Hence, no current flow is possible up to $U_C = \pm\, e/2C$ (so-called Coulomb blockade). As the Coulomb blockade is reached, electric current occurs, where the energy of the particle is compensated for by an external voltage $U = \pm\, ne/2C$. A typical measured ampere–volt characteristic plot is shown in Fig. 1.25.

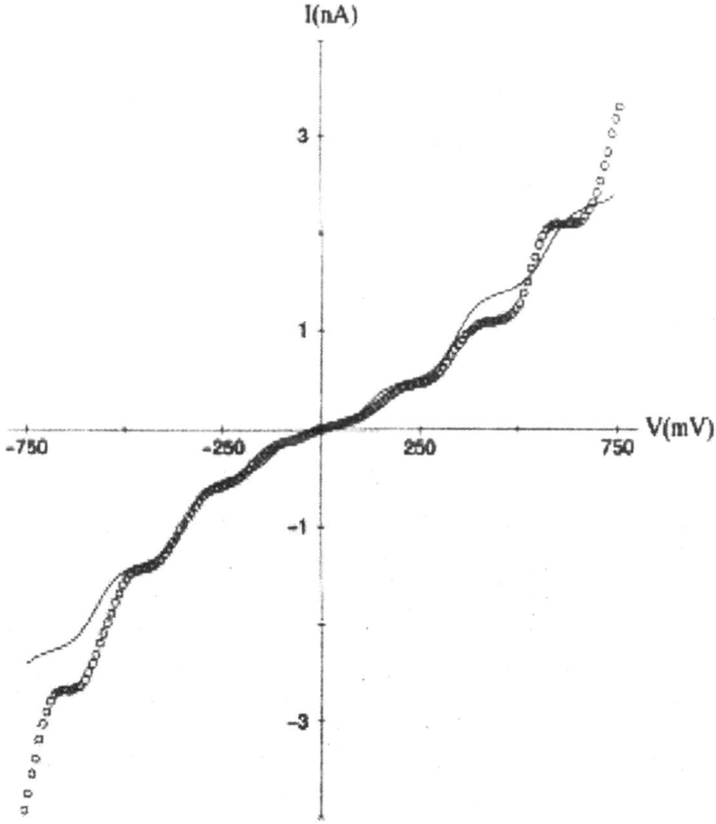

Fig. 1.25. Measured I–U characteristic of a nanostructure.

Metallic nanoparticles in such a solution behave as redox active molecules showing redox cascades.

1.5.5. *Special electric properties related to carbon nanotubes (NT)*

Carbon nanotubes (NT) are formed from C atoms built in a cylinder-shaped hexagonal-type lattice and cupped at their ends by one half of a fullerene-like molecule (Fig. 1.26).

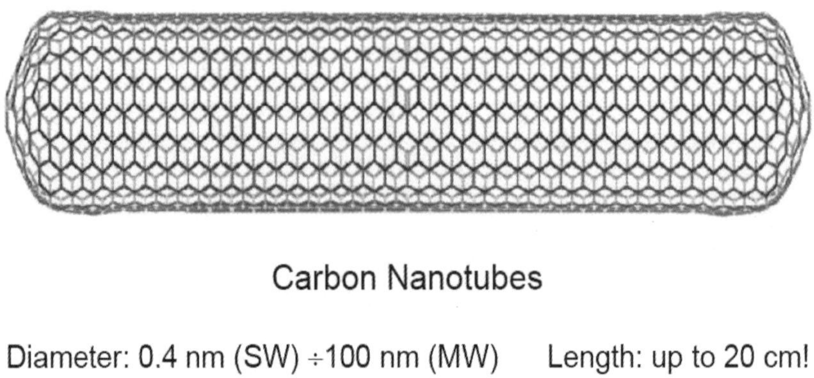

Fig. 1.26. Carbon nanotubes.

There are two types of carbon NTst: single walled nanotube (SWNT) and multiwalled nanotube (MWNT). The main guiding property for building a NT is the chirality (Fig. 1.27). The electrical properties of NTs are the most challenging features because both calculations and experiments show that they are expected to be metallic or semiconducting, depending on their structure (chirality):

- Armchair NTs are metals.
- Zig-zag and chiral NTs are
    - metals for $n - m = 3k$ (k is an integer)
    - semiconductors for $n - m \neq 3k$; $E_g \sim 1/\text{diameter}$.

Their metallic properties also are very promising because of the tensile strength, which is about 10 times larger than that of a steel wire. They have a very huge aspect ratio (length/diameter) and low density (6 times smaller than that of steel).

Chiral indices m, n
- $m = 0 \rightarrow$ zig-zag NTs ⎫ *non chiral*
- $m = n \rightarrow$ armchair NTs ⎭
- $m \neq n \rightarrow$ *chiral* NTs (two enantiomers)

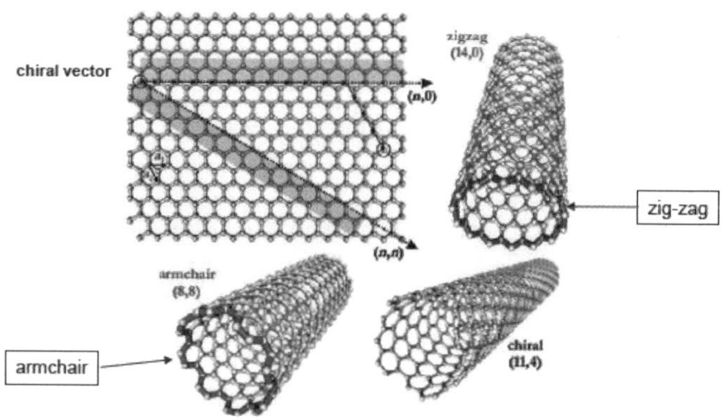

Fig. 1.27. Building of nanotubes based on chirality.

1.5.6. *Magnetic properties*

The magnetic properties of nanostructured magnetic materials are dominated by super paramagnetism, which has a very developed size dependence. The coercivity and saturation magnetization increase with a decrease in the grain size (an increase in the specific surface ratio). The most important magnetic materials are:

- magnetite (γ-Fe_2O_3)
- magnetite (Fe_3O_4)
- Co-ferrite ($CoFe_2O_4$)
- MnZn-ferrite ($MnZnFe_2O_4$).

They are in the form of:

- dry powders (size: 10–100 nm; specific surface: 8–150 m^2/g).
- surface functionalized powders, where such functionalization is essential through strong dependence of the properties on structure.
- magnetic liquids, consisting of super-paramagnetic nanoparticles with narrow particle size distribution around 10 nm covered with a thin inorganic or organic dispersant.

References

1. V. A. Markel and T. F. George, Ed., *Optics of nanostructured materials* (Wiley, New York, 2001), *Wiley series in lasers and applications*.
2. A. S. Edelstein and R. C. Commorata, *Nanomaterials: synthesis, properties and application* (Institute of Physics, Bristol and Philadelphia, 1998).
3. L. Nánai and S. Szatmári, Ultrafast processes in solids driven by fs laser pulses, *International Conference on Condensed Matter of Balkan Countries* (Mugla, Turkey, 2008), p. 27.
4. X. Xu, C. Cheng and I. H. Chowdury, Molecular dynamics study of phase change mechanisms during femtosecond laser ablation, *J. Heat Transfer* **126**, 727–33 (2004).
5. Z. W. Yan and G. Li, Electron–phonon interaction effects on the surface of states in wurtzite nitride semiconductors, *Appl. Surf. Sci.* **255**, 637–639 (2008)
6. O. Voskoboynikov, C. M. J.Wiyers, J. L. Liu and C. P. Lee, Magneto-optics of layers of semiconductor quantum dots and nano-rings, *Brasilian J. Phys.* **36**, 383–386 (2006).
7. S. Liu and Z. Tang, Nanoparticle assemblies for biological and chemical sensing, *J. Mat. Chem.* **20**, 24–35 (2010).
8. B. Teghil, L. D'Alessio, A. De Bonis, A. Galasso, N. Ibris, A. M. Salvi, A. Santogata and P. Villani, Nanoparticles and thin film formation in ultrashort pulsed laser deposition of vanadium oxide, *J. Phys. Chem. A* **113**, 14969–14974 (2009).
9. G. Miyagi, Y. Miyatani and K. Miyazaki, Nanostructuring process in femtosecond laser ablation of patterned thin film surfaces, *Rev. Laser Eng. (Suppl.)* **36**, 1210–1213 (2008).
10. S. Barcikowski, A. Hahn and A. Ostendorf, Rapid nanocomposite manufacturing – A new way of realizing multifunctional applications, in *EuroNanoForum – Nanotechnology in Industrial Applications* (Düsseldorf, 2007), pp. 51–53.
11. N. Singh, Relaxation of femtosecond photoexcited electrons in a metallic sample, *Mod. Phys. Lett. B* **18**, 979–986 (2004).
12. L. J. Lewis and D. Perez, Laser ablation with short and ultrashort laser pulses: Basic mechanism from molecular-dynamics simulations, *Appl. Surf. Sci.* **255**, 5101–5106 (2009).
13. J. Pierre, C. Boulmer-Lebourgne, R. Benserga and S. Tricot, Nanoparticle formation by femtosecond laser ablation, *J. Phys. D: Appl. Phys.* **40**, 7069–76 (2007).
14. S. Bashir, M. S. Rafique and W. Husinsky, Surface topography (nano-sized hillocks) and particle emission from metals, dielectrics and semiconductors during ultra-short laser ablation: Towards a coherent understanding of relevant processes, *Appl. Surf. Sci.* **255**, 8372–8376 (2009).
15. D. Scuderi, R. Benzerga, O. Albert, B. Reynier and J. Etchepare, Spectral and temporal characteristics of metallic nanoparticles produced by femtosecond laser pulses, *Appl. Surf. Sci.* **252**, 4360–4363 (2006).

16. J. Bernholc, C. Brabec, M. Bongiorno-Nordelli, A. Maiti, C. Roland and B. I. Yakolson, Theory of growth and mechanical properties of nanotubes, *Appl. Phys. A* **67**, 39–46 (1998).
17. S. Amoruso, G. Ausanio, R.Bruzzese, M. Vitiello and X. Wang, Femtosecond laser pulse irradiation of solid targets as a general route to nanoparticle formation in a vacuum, *Phys. Rev. B* **71**, 033406–1-4 (2005).
18. G. Saathoff, L. Miaja-Avila, M. Aeschlimann, M. M. Murname and H. C. Kapteyn, Laser-assisted photoemission from surfaces, *Phys. Rev. A* **77**, 022903–1-16 (2008).
19. M. Hong, C. S. Lim, Y. Zhou, L. S. Tan, L. Shi and T. C. Chong, Surface nano-fabrication by laser precision engineering, *Rev. Laser Eng. (Suppl.)* **36**, 1184–1187 (2008).
20. A. Thess, R. Lee, P. Nikolaev, H. Dai, P. Petit, J. Robert, C. Xu, Y. H. Lee, S. G. Kim, A. G. Rinzler, O. T. Cobbert ,G. E. Sceiseria, O. Tomanah, J. E. Fishes and R. E. Smalley, Crystalline ropes of metallic carbon nanotubes, *Science* **273**, 483–487 (1996).
21. D. Csontos and S. E. Ulloa, Modeling of transport through submicron semiconductor structures: A direct solution to the coupled Poisson-Boltzmann equations, *J Comp. Electronics* **3**, 215–219 (2004).
22. Z. H. Mai, Y. F. Lee, W. D. Song and W. K. Chim, Nano-modification on hydrogen-passivated Si surfaces by a laser assisted tunneling microscope operating in air, *Appl. Surf. Sci.* **154–5**, 360–364 (2000).
23. S. D. Brorson, A. Kazeroonian, J. S. Moodera, D. W. Face, T. K. Cheng, E. P. Ippen, M. S. Dresselhaus and G. Dresselhaus, Femtosecond room-temperature measurement of the electron–phonon coupling constant λ in metallic semiconductors, *Phys. Rev. Lett.* **64**, 2172–2175 (1990).
24. S. Gaspard, M. Oujja, R. de Nalda, C. Abrusci and F. Catalina, Nanofoaming in the surface of biopolymers by femtosecond pulsed laser irradiation, *Appl Surf. Sci.*, **254**, 1179–1184 (2007).
25. G. J. Taft, M. T. Newby, J. J. Hrebik, M. F. Onellion, T. F. George, D. Szentesi, S. Szatmári and L. Nánai, Ultrafast dynamic reflectivity of vanadiumpentoxide, *J. Mat. Res.* **23**, 308–311 (2008).
26. D. Szentesi, K. Sós, L. Nánai, S. Szatmári, X. Tang and T. F. George Terahertz-range reflectivity changes in solids induced by ultrashort laser pulses, *Int. J. Theor. Phys. Group Theory, Nonlinear Opt.* in press
27. L. Nánai, S. Szatmári, G. J. Taft and T. F. George, New trends and results in fs-laser-driven materials processing, *Nonresonant laser-matter interaction (NLMI-11)*, edited by M. N. Libenson, *Proc. Soc. Photo Opt. Instrum. Eng.* **5506**, 172–175 (2004).
28. Z. Lin and V. Zhigilei, Temperature dependence of the electron–phonon coupling, electron heat capacity and thermal conductivity in Ni under femtosecond laser irradiation, *Appl. Surf. Sci.* **253**, 6295–6300 (2007).

29. E. P. Pokatilov, V. M. Fomin, S. N. Klimin and S. N. Balaban, Characterization of nanostructures by virtue of the phenomena due to the electron–phonon interaction, *Appl. Surf. Sci.* **104/105**, 546–551 (1996).
30. D. S. Ivanov and B. Rethfeld, The effect of pulse duration on the interplay on electron heat conduction and electron–phonon interaction: Photo-mechanical versus photo-thermal damage of material targets, *Appl. Surf. Sci.* **255**, 9274–9278 (2009).
31. T. Nánai, G. Taft and T. F. George, Ultrafast phase transitions in semiconductors due to fs lasers, *Abstract Book for the 7thAnnual Conference of the Romanian Physical Society* (Timisoara, Romania, 2002), I.01.
32. L. Yun-Quan, Z. Jie and Z. L. Wen-Xi, Interaction of femtosecond laser pulses with a metal photocathode, *Chinese Physics* **14**, 1671–1675 (2005).
33. D. Szentesi, A. Juhász-Nagy, A. M. Bálint, A. Belyayev and L. Nánai, Ultrashort laser pulse induced electron–phonon dynamics in crystals, *Analele Universităţii din Vest Timişoara, Seria Fizică* (*Annals of West University of Timisoara, Physics Series, Romania*) **51**, 75–79 (2007).
34. D. Szentesi, L. Nánai, G. J. Taft, A. M. Balint and T. F. George, Laser-matter interactions on the femtosecond level: Role of the Dember effect, *Analele Universităţii din Vest Timişoara, Seria Fizică* (*Annals of West University of Timisoara, Physics Series, Romania*) **46**, 29–34 (2005).
35. S. Iijama, C. Bratec, A. Maiti and J. Bernholc, Structural flexibility of carbon nanotubes. *J. Chem. Phys.* **104**, 2089–2092 (1996).
36. L. Zhigilei, D. Ivanov , E. Levengle, S. Badigh and E. M. Bringa, Computer modeling of laser melting and spallation of metal targets, *SPIE Proc.* **5448**, 505–519 (2004).
37. D. Perez and L. J. Lewis, Ablation of solids under femtosecond laser pulses. *Phys. Rev. Lett.* **89**, 255504-1-4 (2002).
38. R. R. Letfullin, T. F. George, G. C. Duree and B. M. Bollinger, Ultrashort laser pulse heating of nanoparticles: Comparison of theoretical approaches, *Advances in Optical Technologies* **2008**, ID 251718-1-8 (2008).

Chapter 2

Nanoscale Materials in Strong Ultrashort Laser Fields

Renat R. Letfullin

Rose-Hulman Institute of Technology

letfullin@rose-hulman.edu

Thomas F. George

University of Missouri–St. Louis

tfgeorge@umsl.edu

Ultrashort laser pulses create very strong laser fields which can be used as a powerful tool to manipulate material properties. Modeling material responses in nanosize dimensions and on unprecedented short time scales imposes several challenges. In this chapter, we study two models to describe the laser pulse interaction with nanomaterials in the femtosecond, picosecond and nanosecond regimes. Also, we introduce a new mechanism of ultrashort pulse/nanomaterial interactions — laser-induced explosion of absorbing nanoparticles in strong laser fields. This mechanism is realized through fast overheating of a strongly-absorbing target during the time of an ultrashort laser pulse when the influence of heat diffusion is minimal. On the basis of simple energy balance, it is found that the threshold laser energy density for thermal explosion of different gold nanoparticles is in the range 25–40 mJ/cm^2. Explosion of nanoparticles may be accompanied by optical plasma, generation of shock waves with supersonic expansion, and particle fragmentation with fragments of high kinetic energy.

2.1. Introduction

The interaction between nanoscale materials (nanospheres, carbon nanotubes, nanorods, nanoshells, etc.) and ultrashort laser pulses (ULP) has become of great interest in condensed matter physics, nanotechnology, and laser nanomedicine [1–3]. Utilizing the strong laser fields produced by ULPs provides unique properties and functionalities for nanomaterials that have potential future applications.

Traditional knowledge of laser-nanoparticle interactions has necessitated specialized models for each case, dependent upon the laser pulse duration. Previous research has recognized the existence of these conditions, and dual-temperature models for the ultrashort laser pulse mode calculating electron and lattice subsystem temperatures are readily available (see, e.g., [4,5]). Pulses of longer duration are modeled using a uniform heating model [6–9], an appropriate approximation for pulse durations greatly exceeding the electron–phonon coupling time. On a dimensional scale, this approximation is reasonable for particle sizes not much larger than a laser wavelength, which is completely applicable for the heating of nanoparticles.

Ultrashort pulses, specifically those in the femtosecond and picosecond ranges, impose several challenges in modeling the material response. Free electrons with minimal capacity for heat are the first to absorb energy, rapidly attaining high temperatures and transferring thermal energy to the material lattice. These processes do not occur instantaneously: time must be allowed for the cooling of the electrons and the heating of the lattice. In our work, electron cooling and lattice heating have time delays on the order of femtoseconds and picoseconds, respectively. Ultrashort laser pulses end before the transfer of energy to the lattice is complete, requiring two-temperature models in order to describe the further conversion of energy from electron excitation to heat within the lattice system.

In this chapter, we demonstrate that a simpler, one-temperature model (OTM) utilizing the uniform heating approximation is appropriate for understanding ultrashort laser pulse interactions with metal nanoparticles. The approximation may be used in this situation due to the extremely-small size of nanoparticles in comparison to the wavelength of laser radiation. Using this idea, the time delay between the electron and lattice interactions will be bypassed, and the simpler model will ultimately yield results similar to the two-temperature model. Comparative simulations of the two modeling approaches are performed in this chapter to confirm OTM as an appropriate approximation for nanoparticle heating in the femtosecond, picosecond and nanosecond regimes, thus providing an effective modeling method for further laser nanotechnology research to explore.

The strongly-enhanced surface plasmon resonance (SPR) of metal nanoparticles at optical frequencies makes them excellent scatterers and absorbers of visible and near-IR light [10]. SPR creates the condition for realization of a new mechanism in the interaction of strongly-absorbing nanoparticles with ultrashort laser pulses, when their temperature rapidly reaches the thresholds for nonlinear effects and can cause an explosion of nanoparticles. Such explosion can be achieved for ultrashort laser pulses, specifically those in the picosecond and femtosecond ranges, when heat is generated within the strongly-absorbing target more rapidly than the heat can diffuse away. In this chapter, we carry out theoretical modeling to estimate the threshold laser energy densities required for the thermal explosion of nanoparticles, using the example of gold nanospheres. The explosion of nanoparticles involves the generation of a shock wave of dense vapors expanding with supersonic velocity, producing the sound waves and optical plasma.

2.2. Theoretical Background

2.2.1. *One-temperature model (OTM)*

During the interaction of a laser pulse of intensity I_0 and pulse duration τ_L with a metal nanoparticle of radius r_0, the laser energy is absorbed by free electrons due to the inverse Bremsstrahlung and then transferred from the electron gas into the lattice. In OTM, it is assumed that the electron heat transfer into the phonon subsystem is very fast, i.e., the electron and lattice temperatures are equal, $T_e = T_s$, at any instant of time. In this approximation, we can limit our description to only one lattice temperature distribution, $T_s(t,r)$, which could be found by numerical solution of a heat-mass transfer equation between the particle and the surrounding medium:

$$\frac{d}{dt}T_s(t,r)=\frac{\mu_s(T_s)}{\rho_s C_s(T_s)}\Delta T_s(t,r)+\frac{Q(t,r)}{\rho_s C_s(T_s)}-j_D(T_s)S_0+\frac{3L}{r_0 C_s(T_s)}\frac{dr_0}{dt}. \tag{2.1}$$

Here, $\Delta=\frac{\partial^2}{\partial x^2}+\frac{\partial^2}{\partial y^2}+\frac{\partial^2}{\partial z^2}$ is a Laplace operator; $\mu_s(T_s)$, $C(T_s)$, L, ρ_s and r_0 are, respectively, the heat conductivity, specific heat, evaporation heat, density and radius of the nanoparticle; $Q(t,r)$ is a heat source; $j_D(T_s)$ is the heat lost from the surface of the nanoparticle into the surrounding medium; and $S_0=4\pi r_0^2$ is the particle surface area.

The power density of energy generation in the particle $Q(t,r)$ due to radiation energy absorption is generally non-uniform throughout the particle volume, with the non-uniformity being dependent on the size and optical constants of the particle. Since for nanoparticles $2\pi r_0/\lambda<1$, we can assume that $Q(t,r)$ is uniform throughout the particle volume [6–9], and it can be described by the equation

$$Q(t,r_0)=\frac{3K_{abs}(r_0,\lambda)I_0 f(t)}{4r_0}, \tag{2.2}$$

where $K_{abs}(r_0,\lambda)$ is the absorption efficiency of the nanoparticle as a function of laser wavelength λ and particle radius r_0, and $f(t)$ is a time profile of a laser pulse.

Heat exchange between the nanoparticle surface and the surrounding medium is rapid, and heat loss becomes substantial for relatively-long laser pulses. Assuming that the heat lost from the surface of a nanoparticle occurs only due to heat diffusion into surrounding medium, the energy flux density $j_D(T_s)$ removed from the particle surface can be expressed as a nonlinear function of temperature [7–9]:

$$j_D(T_s)=\frac{\mu_\infty T_s}{(s+1)r_0^2 C_s(T_s)\rho_s}\left[\left(\frac{T_s}{T_\infty}\right)^{\alpha+1}-1\right], \tag{2.3}$$

where μ_∞ is the heat conductivity of the surrounding medium at normal temperature T_∞, and the power exponent s = const. depends on the thermal properties of the surrounding medium.

After integration over the volume for the spherically-symmetric case and transition to a uniform temperature over the particle volume, the equation which describes the kinetics of laser heating of the nanoparticle and results from Eq. (2.1) takes the following form [7–9]:

$$\frac{d}{dt}T_s = \frac{3K_{abs}(r_0,\lambda)I_0 f(t)}{4r_0\rho_s C_s(T_s)} - \frac{\mu_\infty T_s}{(s+1)r_0^2 C_s(T_s)\rho_s}\left[\left(\frac{T_s}{T_\infty}\right)^{s+1} - 1\right] + \frac{3L}{r_0 C_s(T_s)}\frac{dr_0}{dt}. \tag{2.4}$$

Here, the first term on the right side of the equation describes the heat generation into the spherical volume due to laser energy absorption by the nanoparticle. The second term describes the energy losses from the surface of the particle into the surrounding medium due to the heat diffusion process. The last term describes the energy losses due to the evaporation of the particle. This evaporation depends on the laser pulse characteristics and particle properties, and it can be realized in five different regimes (see, for instance, Refs. [8,9]): free-molecular, convective, diffusive, gas-dynamic and explosive [3] modes of evaporation. For example, within the approximation of free molecular flow, the evaporation term in Eq. (2.4) can be written as

$$4\pi r_0^2 \frac{dr_0}{dt}\rho_s = -\eta 4\pi r_0^2 V_s(T_s)\rho_v(T_s)\,, \tag{2.5}$$

where η is the accommodation coefficient, and $V_s(T_s)$ and $\rho_s(T_s)$ are, respectively, the average velocity and density of the vapor at temperature T_s. If the heating of the nanoparticle occurs below the temperature of phase transition in the particle material, the third term on the right side of Eq. (2.4) can be neglected.

2.2.2. *Two-temperature model (TTM)*

In TTM, the temperature relaxation in time and sample depth can be modeled by two coupled diffusion equations, one describing the heat conduction of electrons and the other that in the lattice. Both equations are connected by a term that is proportional to the electron–phonon coupling constant γ and to the temperature difference between electrons and lattice, originally proposed by Anisimov *et al.* [11]. We modify this original set of equations, adding a heat exchange term between the surface of the particle and the surrounding medium. Also, we are taking into account the dependence of the thermophysical parameters of electrons, particle and surrounding medium on temperature during the laser treatment, as presented in Table 2.1. The modified set of equations describing the heating of electron and lattice subsystems and the energy transfer between particle and surrounding medium has the form

$$C_e(T_e)\frac{dT_e(t)}{dt} = -\frac{dQ(z)}{dz} - \gamma(T_e - T_s) + S$$

$$C_i(T_s)\frac{dT_s(t)}{dt} = \gamma(T_e - T_s) - j_D(T_s)\,. \tag{2.6}$$

Here, $Q(z) = -k_e \frac{\partial T_e}{\partial z}$ is the heat flux; z is the direction perpendicular to the target surface; k_e is the electron thermal conductivity; $S = I_0\, f(t)\, \alpha\, exp(-\alpha\, z)$ is the laser heating source term; α is the material absorption coefficient; and $C_e(T_e)$ and $C_i(T_s)$ are the temperature-dependent heat capacities (per unit volume) of the electron and lattice subsystems. The expressions for the temperature dependence of the heat capacities, as well as the values for the electron–phonon coupling constant γ and material data, are listed in Table 2.1.

Table 2.1. Input parameters used for simulations in both models.

Parameters	Magnitude and Units	Refs.
Laser pulse shape	$f(t) = \exp[-(at-b)^2 - (at-c)^2]$ $a, b, c = const$	
Energy density	$E = 10^{-3}$ J/cm^2	
Volume density of the gold	$\rho_0 = 0.0193$ kg/cm^3	[12]
Electron heat capacity (gold)	$C_e(T_e) = aT_e^4 - bT_e^3 + cT_e^2 - dT_e + e$, J/(K^2 cm^3) $a = 6.0 \times 10^{-17}$, $b = 3.0 \times 10^{-12}$, $c = 5.07 \times 10^{-8}$, $d = 6.7589 \times 10^{-5}$, $e = 7.207 \times 10^{-2}$.	[13]
Electron–phonon coupling constant for gold	$\gamma = 2.27 \times 10^{10}$ W/(K cm^3)	[14]
Gold specific heat	$C = 129$ J/(K kg)	
Specific heat of water	$C_W(T_s) = a(1 + b(T_s - 293\text{ K}))$, J/(K kg) $a = 4182.0$. $b = 1.016 \times 10^{-4}$.	
Radius of gold nanoparticle	$r_0 = 20$ nm	
Absorption efficiency of gold nanoparticle	$K_{abs} = 4.0195$ ($r_0 = 20$ nm and $\lambda = 528$ nm)	[2,10]
Absorption coefficient of gold nanoparticle	$\alpha = 8.736 \times 10^5$ cm^{-1} ($r_0 = 20$ nm and $\lambda = 528$ nm)	[10]
Power exponent	$S = 1.0$	
Thermal conductivity of water	$\mu_0(T_s) = a(1 + b(T_s - 293\text{ K}))$, W/(K cm) $a = 0.00597$ $b = 1.78 \times 10^{-3}$.	

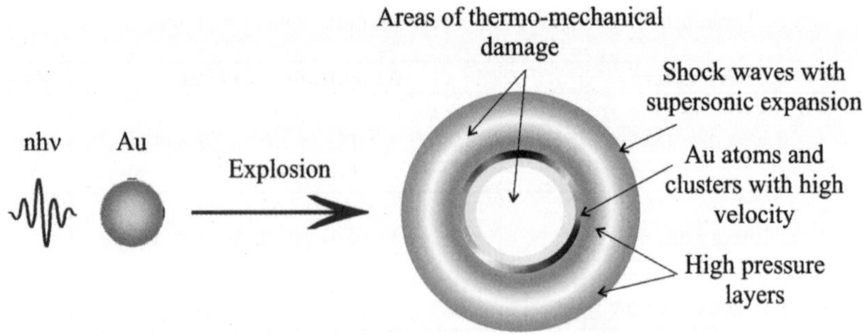

Fig. 2.1. Laser-induced thermal explosion of a gold nanoparticle.

2.3. Laser-Induced Explosion of Nanoparticles

When nanoparticles are irradiated by ultarshort laser pulses, specifically those in the picosecond and femtosecond ranges, their temperature rises very quickly to possibly reach the thresholds for nonlinear effects. Sufficient energy transfer from the laser pulses with a threshold energy density E_{expl} to the metal nanoparticles, for instance gold, results in their explosion. Laser-induced rapid explosive evaporation of gold atoms and their clusters from nanoparticles enables the generation of stress transients, shock waves and high local pressure. A phenomenological picture of these complex physical effects is shown in Fig. 2.1.

There are two main physical mechanisms that could lead to the laser-induced explosion of nanoparticles — *Coulomb explosion mode* through multiphoton ionization and *thermal explosion mode* through electron–phonon excitation-relaxation.

2.3.1. *Coulomb explosion mode*

For ultrashort laser pulses (e.g., femtosecond time scale) comparable to the time scale of electron–electron interactions and shorter than the electron–phonon interaction time, a non-photothermal mechanism of metal nanoparticle explosion can occur through multiphoton ionization, when absorbed photon energy is transferred directly to the electrons, leading to their ejection due to the Coulomb explosion mechanism. High metal nanoparticle plasmon-resonance absorption may facilitate this effect due to thermionic electron emission. An analysis of the Coulomb explosion mechanism is outside the topic of this chapter.

2.3.2. *Thermal explosion mode*

Under the action of short laser pulses in the spectral range of the surface plasmon resonances (for instance, for a solid spherical gold nanoparticle (GN) the maximum absorption is ~ 520 nm), GN atoms are excited to upper electronic states owing to absorption of many photons. Through rapid (picosecond time scale) relaxation, GN atoms decay to their ground state with effective electron–phonon conversion of the absorbed photon energy into thermal energy. Depending on the GN temperature, T_s, the following effects alone or in combination can occur:

- $T_{GNM} \leq T_s < T_{GNB}$, where T_{GNM} is the GN melting point of ~ 1,063°C [15] and T_{GNB} is the GN boiling point of ~ 2,710°C [15]. This is GN melting and fragmentation into smaller parts.
- $T_s \geq T_{GNB}$, which is GN boiling and explosion into single atoms.

The goal of the theoretical modeling is to estimate the threshold laser energy density, E_{expl}, required for realization of the thermal explosion mode of GNs, and to compare the calculated data with available experimental results. This mode is realized through the rapid overheating of a strongly-absorbing target during a short laser pulse when the influence of heat diffusion is minimal. Let us first estimate the time scale for thermal relaxation due to heat diffusion from the surface of the nanoparticle.

In the vicinity of GN $\leq 5r_0$, where r_0 is the nanoparticle's radius, the thermal relaxation time for a spherical GN can be estimated as $\tau_T = R^2/6.75k$, where k is the thermal diffusivity of the surrounding medium (water in many cases). For r_0 = 50, 100 and 200 nm, estimates of τ_T (for water, $k = 1.44 \times 10^{-3}$ cm^2/s) are approximately 2.6, 10 and 41 ns, respectively. For a laser pulse duration $\tau_L \leq \tau_T$, heat is generated within the GN more rapidly than can be diffused away, so that we can neglect heat losses from the surface of the GN due to heat diffusion into the surrounding medium. This condition for smaller GNs is valid with picosecond and femtosecond laser pulses. Nanosecond pulse durations ($\tau_L \sim$ 5–12 ns) still satisfy the condition $\tau_L \leq \tau_T$ for relatively large GNs. With this assumption, the probability of a non-PT mechanism of explosion is very low because multiphoton ionization usually occurs more effectively under femtosecond pulses.

2.3.3. *Threshold intensity for laser-induced thermal explosion*

Consider a gold nanoparticle of radius r_0 to be irradiated by a laser of intensity I_0. The power of the absorbed electromagnetic field is $P = \sigma_{abs} I_0$, where σ_{abs} is the GN's cross-section of absorption. If σ_{abs} is large enough, thermal explosion of the GN may occur at certain values of the threshold laser intensity I_{expl}, which is less than the threshold intensity for optical plasma formation in the surrounding medium [3,16]. Under the thermal explosion of GNs, we can understand the specific case for which the total energy absorbed by the GN during the time of its inertial retention in vapor state, $\tau_{expl} = R/u_s$, exceeds the energy required for the GN's complete evaporation, $\rho q V = N_{Au} q_1 V$. Here, u_s is the sound velocity in Au vapor at the critical temperature $T_{cr} \sim T_{GNB}$; ρ is the volume density of GN; N_{Au} is the number of Au atoms per unit volume; q and q_1 are the particle's specific heat of evaporation per unit mass and per particle, respectively; and V is the GN's volume. We have a laser pulse duration of $\tau_L > \tau_{expl} = R/u_s$ (~ 1 ps for $r_0 \sim$ 10 nm), where τ_{expl} is an explosive evaporation time.

To compute I_{expl}, we can write the following energy balance equation:

$$\sigma_{abs} I_{expl} \frac{r_0}{u_s} \approx \rho q V \ . \tag{2.7}$$

In terms of the absorption efficiency, $K_{abs} = \sigma_{abs} / \pi r_0^2$, the threshold intensity of the laser radiation for thermal explosion of the spherical GN can be expressed as

$$I_{expl}(r_0) \approx \frac{4\rho q u_s}{3K_{abs}(r_0)} \ . \tag{2.8}$$

Thus, the solid nanoparticle in a relatively strong laser field with intensity $I_0 \geq I_{expl}$ during the short time $\tau_{expl} \sim r_0/u_s$ transforms into a gas (vapor) sphere of radius $\sim r_0$, which has high temperature $T \sim T_{GNB}$ and high pressure $P_{vap} >> P_\infty$, where P_∞, is the ambient pressure.

We assume here that thermal explosion of a GN is accompanied by the generation of shock waves that expand with a supersonic velocity $u_s = \frac{r_0}{\tau_{expl}} = \frac{I_{expl}}{4\Delta H} \approx 10^5$ cm/s. Here, $\Delta H = \Delta H_{300-vap} + \Delta H_{vap}$, where $\Delta H_{300-vap}$ is the enthalpy change per unit volume for heating the GN from the ambient temperature to the vaporization temperature, and ΔH_{vap} is the vaporization enthalpy per unit volume. The shock waves could be waves of high acoustic or/and water vapor pressure, which spread out over long distances around an epicenter of explosion. It should be noted that the pressure produced by vaporization of a GN itself (e.g., $\sim 10^2$ atm [18]) is less than water vapor pressure around the hot GNs [17,18]. Indeed, the number of atoms vaporized per unit of time and per unit of particle surface is given by [19]

$$\frac{N_{vap}}{dt} = u_s N_{\mathrm{Au}} = \frac{I_{expl}}{4\Delta H} N_{\mathrm{Au}} \ . \tag{2.9}$$

For example, the atomic density of bulk gold is 5.9×10^{22} cm^{-3}, while each GN contains approximately 10^9 atoms, leading to an average atom density of 10^{17} cm^{-3}. Higher gold vapor pressure can only be reached on the front of the explosion, where explosion products are localized in a thin layer.

2.4. Results and Discussion

Comparative simulations using the models described above have been performed for the laser heating of a gold nanoparticle with radius

$r_0 = 20$ nm in a surrounding medium of water. The same set of input data presented in Table 2.1 was used for the calculations in both models. The temperature dependences of the electron heat capacity for the gold, specific heat and thermal conductivity for the water were obtained by interpolating the experimental data available in the literature (references are listed in Table 2.1). As can be seen from the table, the electron heat capacity is much less than the lattice heat capacity, and therefore electrons can be heated to very high transient temperatures. Then the evolution of the electron temperature involves energy transfer to the lattice and energy losses due to the electron heat transport into the target. The electron–phonon coupling process has several characteristic time scales: electron thermalization time τ_e, electron cooling time τ_c, lattice heating time τ_i, and duration of the laser pulse τ_L. The relationship between them defines three different regimes of the laser-metal interaction — femtosecond, picosecond and nanosecond modes of heating.

2.4.1. *Femtosecond pulses*

TTM and OTM have been solved numerically to predict the time dependence of the electron and lattice temperatures in the femtosecond mode when the laser pulse duration is shorter than the electron thermalization and cooling times, $\tau_L \ll \tau_e, \tau_c$. The calculations were performed for a laser pulse energy density of $E = 1.0$ mJ/cm^2 and pulse duration of $\tau_L = 60$ fs. The time profile of the femtosecond laser pulse given in Table 2.1 and shown in Fig. 2.2(b) (solid curve) corresponds to the experimentally-observed output from an amplified Ti:sapphire laser (Legend-HE from Coherent Inc.). The results of the simulations for the heating of a gold nanoparticle by a femtosecond laser pulse are shown in Figs. 2.2 and 2.3. Figure 2.2(a) displays the results obtained by TTM, and Fig. 2.2(b) presents the results of OTM simulations. A comparison of these two models is shown in Fig. 2.3.

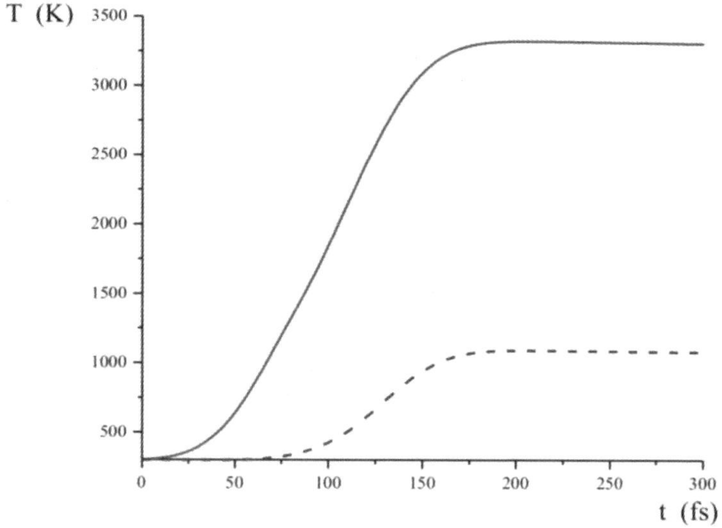

(a)

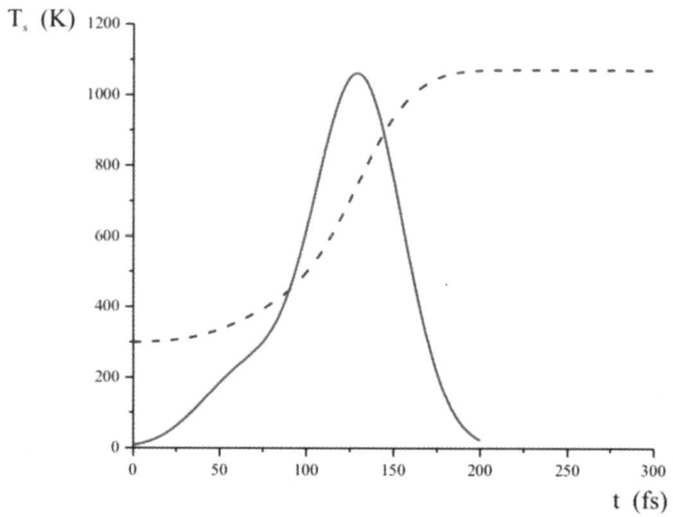

(b)

Fig. 2.2. (a) Electron (solid curve) and lattice (dashed curve) temperature evolutions on the femtosecond time scale for a gold nanoparticle predicted by TTM. (b) Evolution of the nanoparticle temperature (dashed curve) after the femtosecond laser pulse, predicted by OTM, and laser pulse shape (solid curve).

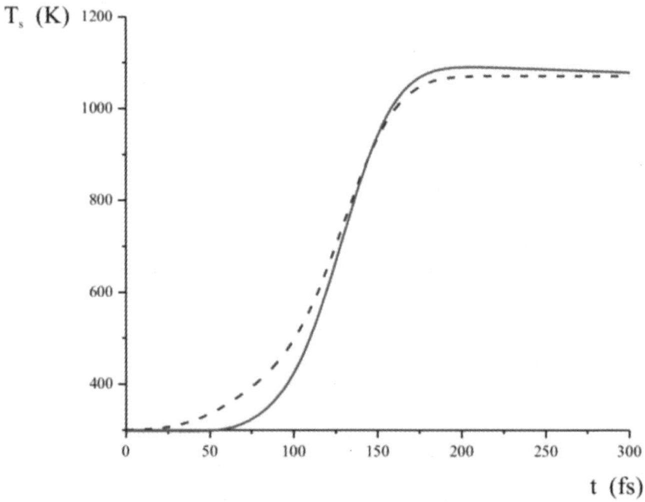

Fig. 2.3. Comparison of the TTM (solid curve) and OTM (dashed curve) predictions of the nanoparticle temperature evolutions after the 60 fs laser pulse.

As follows from Fig. 2.2(a), thermal equilibrium among the excited electrons with equilibrium temperature $T_e \approx 3300$ K for a given laser fluence is established within 175 fs. We should note that the equilibrium temperature for electrons is reached long after the end of a laser pulse, which had a duration of 60 fs. The electrons remain in the thermal equilibrium state from several hundred femtoseconds up to 1 ps (see Fig. 2.4(a)). Then the electrons cool exclusively by coupling to the lattice, resulting in a linear decay of the electron temperature during the first 10 ps (Fig. 2.4(a)). Our simulations agreed with the electron relaxation time measured in [20] for femtosecond pulse excitation of a DNA-modified gold nanoparticle.

The slow rate of electron heat diffusion into the phonon subsystem on the femtosecond time scale results in a delay of about 100 fs in the heating of the bulk sample (dashed curve in Fig. 2.2(a)). Once the electron thermal equilibrium is established, a hot electron bath raises the temperature of the cold lattice up to 1090 K for a given laser energy density $E = 1$ mJ/cm^2.

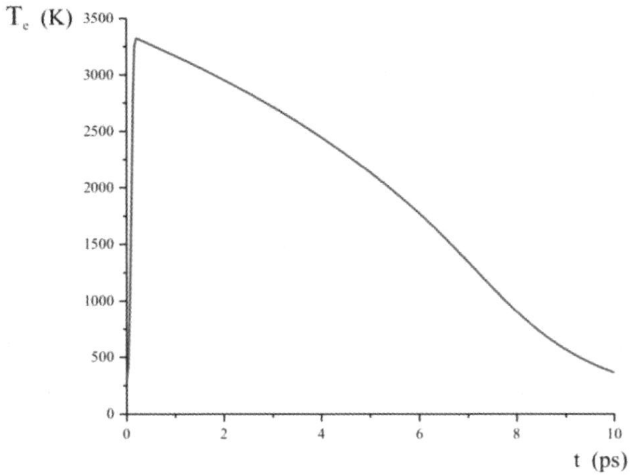

(a)

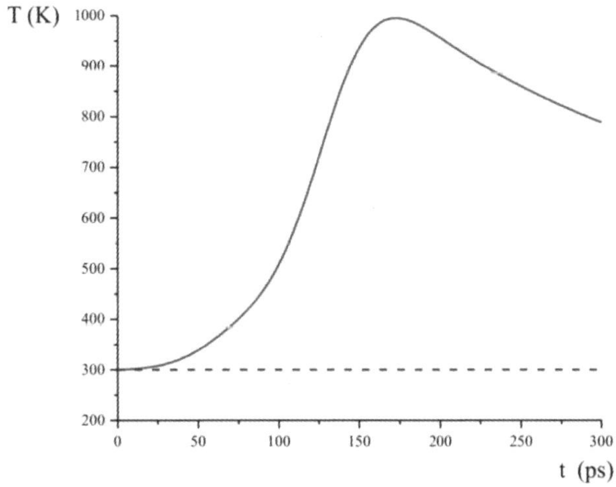

(b)

Fig. 2.4. (a) Electron temperature relaxation on the picosecond time scale. (b) Temperature-time distributions for a gold nanoparticle predicted by OTM (solid curve) and TTM (dashed curve) after the 60 ps laser pulse.

The results of heating the gold nanoparticle obtained by OTM are demonstrated in Fig. 2.2(b). This figure also shows the femtosecond laser pulse time profile used in the calculations. Comparative simulations for the evolution of the nanoparticle temperature using both models under the same conditions are presented in Fig. 2.3. It follows from these simulations that both models demonstrate the same scenario in the heating kinetics of a metal nanoparticle by a femtosecond laser pulse. Both models reveal approximately a 100 fs time delay in the heating of the particle, followed by a maximum lattice temperature of around 1090 K within 175 fs after the end of a laser pulse. Even the maximum values of the particle temperature predicted by both models are the same (see Fig. 2.3). The saturation parts in the lattice temperature curves are explained by negligibly small heat diffusion from the surface of the nanoparticle into the surrounding medium on the femtosecond time scale. A slight difference in the slopes of the temperature curves within the first 100 fs of heating occurs due to the assumption made in OTM that the electron heat transfer into the lattice subsystem is very fast. Because of this, the particle temperature in OTM promptly follows electron thermal behavior. A comparison of these two models shows that the simpler OTM gives the same results as the more precise TTM. Thus, OTM provides an adequate description of the laser heating of nanoparticles in the femtosecond regime.

2.4.2. *Picosecond pulses*

In this mode, the constants *a*, *b*, and *c* in the time profile for the laser pulse have been chosen to provide the laser pulse width of 60 ps at FWHM with the same pulse shape listed in Table 2.1 and shown in Fig. 2.2(b). As can be seen from Figs. 2.2(a) and 2.4(a), in the picosecond regime the electron thermalization time τ_e and electron cooling time τ_c are much less than the duration τ_L of the 60 ps laser pulse. Hence, during the first 10 ps the electron subsystem has already been completely cooled (Fig. 2.4(a)), and the electron temperature above the ambient no longer exists for the considered picosecond mode of heating. This is confirmed by our calculations presented in Fig. 2.4(b) (dashed curve). Thus, TTM provides a very good approximation for the femtosecond mode as soon as the electron temperature exists, but it fails to describe the laser heating of nanoparticles for longer pulse durations in the picosecond and nanosecond regimes.

Opposite to TTM, OTM describes the picosecond heating kinetics very well. A typical time evolution of the particle temperature predicted by OTM on the picosecond time scale is displayed in Fig. 2.4(b) (solid curve) for the same material constants listed in Table 2.1. The main feature is the appearance of heat lost from the surface of the nanoparticle into the surrounding medium on the picosecond time scale. After about 200 ps, cooling of the nanoparticle begins due to heat diffusion into the water. The maximum temperature reached by a 20 nm gold particle for the given laser pulse is 995 K. This temperature is sufficient to initiate any thermal killing mechanisms in cancer cells.

2.4.3. *Nanosecond pulses*

For laser heating of metal nanoparticles in the nanosecond regime, the characteristic lattice heating time τ_i is much smaller than the laser pulse duration: $\tau_L \gg \tau_i$. This means that the temperature inside the nanoparticle is nearly uniform over the whole particle at the time scale of the laser pulse duration τ_L. In this case, the electron and lattice temperatures are equal, $T_e = T_s$, so that the homogeneous heating of the particle and quasi-steady heat exchange with the surrounding medium can be described by just OTM. The characteristic lattice heating time τ_i required for the formation of a quasi-stationary temperature profile across the nanoparticle can be estimated from the formula $\tau_i = r_0^2/4\chi$, where r_0 is the particle radius and χ the thermal diffusivity of the particle material. For gold nanoparticles ($\chi = 1.18 \times 10^{-4}$ m^2/s) with radii r_0 =20–30 nm, the lattice heat diffusion time is $\tau_i \sim 2 \times 10^{-12}$ s $\ll \tau_L \sim 10^{-8}$ s.

Sample calculations have been carried out using OTM for gold nanoparticles with radii r_0 = 30–35 nm in different surrounding biomedia for an incident laser pulse of energy density E = 10 mJ/cm^2 and pulse duration τ_L = 8 ns with the time profile shown in Fig. 2.2(b). The laser pulse profile and duration 8 ns have been chosen to be close to those used in previous experiments [2]. The laser flux chosen is ten times higher than in the regimes considered above to provide approximately the same maximum nanoparticle temperature as observed for femtosecond and picosecond laser heating. We should note that 10 mJ/cm^2 is comparable to the laser fluence currently used in the photothermolysis of cancer cells [1,3]. The kinetics of heating and cooling the gold nanoparticle are demonstrated in Fig. 2.5, where (a) illustrates the time dynamics of laser heating of a 30 nm gold particle in different biological media: blood, human prostate, tumor and fat. The thermophysical characteristics of gold and biological surrounding media for different temperatures are listed in Table 2.2. Figure 2.5(b) shows results of thermal calculations for a 35 nm gold particle, which is heated and cooled in water at different heat transfer rates *s*. It follows from these calculations that during the laser pulse duration, the transfer of heat from the nanoparticle into the surrounding media is slight, and the particle rapidly reaches a high temperature. The heating rate is about 10^{12} Ks^{-1}. The temperature of the particle continues to rise even after the end of the laser pulse. The highest temperature, 770 K, for a given laser pulse fluence is observed for the heating time of 13.5 ns, when the laser pulse has already degraded (see Fig. 2.5(a)). After that time, the transfer of heat from the particle to the surrounding medium becomes increasingly important, since the energy source is no longer present in the system. The temperature of the particle and surrounding medium remains high (~ 400 K) up to 20 ns, exceeding the laser pulse duration by 2.5 times. The total time for one cycle (heating from the initial temperature 300 K to maximum temperature, followed by cooling back to the initial temperature) is about 30 ns.

2.4.4. *Laser heating of nanoparticles in biological media*

We have also examined the effect of different biological surroundings on the laser heating dynamics of 30 nm gold particles. Four biomedia were used: blood, human prostate, tumor and fat. Results of computer simulations of the time-temperature profiles of gold nanoparticles in various biological media, performed by using OTM, are plotted in Fig. 2.5(a). As follow from our calculations, the laser heating and temperature behavior of the gold nanoparticles in blood, prostate and tumor are comparable to the case of water as the surrounding medium, since the thermodynamic properties of those media are very close to each other (see Table 2.2). This means that for the thermal calculations of laser heating of biological media, the thermal properties of water can be used

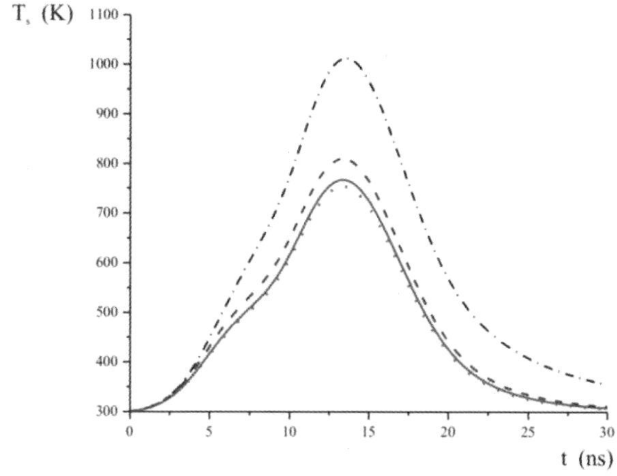

(a)

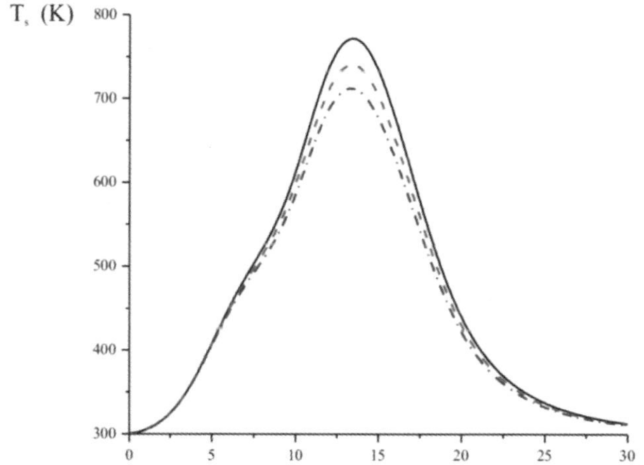

(b)

Fig. 2.5. Kinetics of heating and cooling of a gold nanoparticle by a nanosecond laser pulse of energy density 10 mJ/cm^2 and duration 8 ns. Calculations have been made by using OTM. (a) Illustration of the time dynamics of laser heating of a 30 nm gold particle in different biological media: fat (dashed-dotted curve), blood (dashed curve), tumor (solid curve) and prostate (dotted curve). (b) Results of thermal calculations for a 35 nm gold particle: heating and cooling in the water surrounding medium at different heat transfer rates s: s = 1.0 (solid curve), s = 1.25 (dashed curve) and s = 1.5 (dashed-dotted curve).

if the water content in the media is high. But the heating of a gold nanoparticle in fat is substantially different from the water case, since the fat has low thermal characteristics. Here we observe higher overheating of the particle at the same energy level and duration of the laser pulse due to the relatively low thermal conductivity of fat as compared to other biomedia.

Table 2.2. Thermophysical characteristics of the gold particle and surrounding biological tissue.

Material	**Specific heat** C (J/K kg)	**Interval of T** (K)	**Thermal conductivity** μ_0 (W/m K)	**Thermal diffusivity** χ (m^2/s)
Gold	129	273–373	318	1.18×10^{-4}
Water	4181.6–4215.6 $C(T) = 4182\ (1 + 1.016 \times 10^{-4}\ (T\text{-}293\ \text{K})$	273–373	0.597–0.682 $\mu(T) = 0.597\ (1+ 1.78 \times 10^{-3}\ (T-T_\infty)$	1.43×10^{-7}
Human prostate	3740	310	0.529	
Blood	3645–3897	273–373	0.48–0.6	1.6×10^{-7}
Fat	2975	273–373	0.185–0.233	
Tumor	3160	310	0.561	
Skin		273–373	0.210–0.410	

The temperature dynamics of the particle is sensitive to the power exponent s used in the temperature dependence of the heat lost from the surface of the nanoparticle into the surrounding medium, i.e., $j_D(T_s)$ in Eq. (2.3) (see Fig. 2.5(b)). The value of s =1 better corresponds to the real biological surrounding. The power $s > 1$ describes the medium with high thermophysical characteristics, like the cooling liquids and metals. The medium with $s < 1$ has a low thermal conductivity and can be used as a thermal isolator.

It is interesting to investigate the effect of the particle's radius on the temperature dynamics of the nanoparticle heated by the nanosecond laser radiation in the biological surroundings. There are two competitive factors here. On one hand, according to the Mie diffraction theory, the absorption efficiency of the gold nanoparticle drops with decreasing size of the particle. On the other hand, the heating rate increases for smaller particles as follows from Eq. (2.4). To find which factor has a stronger effect on the effective laser heating of a gold nanoparticle, we have calculated the maximal temperature profile for different nanoparticle radii in blood and compared it to the gold nanoparticle absorption curve. The results of these simulations are performed by using OTM and presented in Fig. 2.6. It follows from this figure that the optical effect is much stronger than the thermal effect when the radius of nanoparticle is less than 35 nm. In the radii range 1–35 nm, the overheating effect of the particle behaves according to the absorption efficiency. For a radius of 35 nm, the thermal processes dominate over the optical properties. For large radii ($\geq$ 35 nm), the maximal temperature profile is saturated with oscillations, repeating the maxima and minima of the absorption curve. The saturation of the maximal temperature curve for large particle radii is explained by the balance between heating of the particle due to absorption of laser energy and energy losses from the surface of the particle due to heat diffusion into the surrounding biological medium.

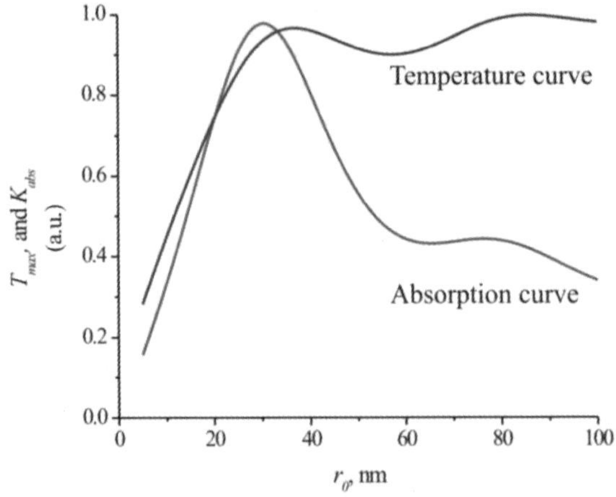

Fig. 2.6. Non-dimensional (a.u.) absorption efficiency K_{abs} and maximal temperature T_{max} curves as a function of the particle's radius for gold nanoparticles in blood heated by a laser pulse of energy density 10 mJ/cm^2 and duration 8 ns.

2.4.5. *Thermal explosion of nanoparticles*

The threshold intensity I_{expl} of a laser pulse required for thermal explosion of the nanoparticles strongly depends on the absorption efficiency of the nanoparticle. The optical properties of a sphere of arbitrary radius and dielectric constant can be calculated using Maxwell's equations. Our electrodynamics calculations for solid gold nanospheres [2,3] have shown that the maximal effect for laser heating of solid gold particles in an aqueous medium is achieved for particles with radius 35 nm at a plasmon resonance wavelength maximum of λ_{max} = 538.3 nm and absorption efficiency of K_{abs} = 4.02 (Fig. 2.7(a)). Thus, on the basis of these calculations [2,3] we can conclude that the effective laser initiation of thermal explosion of gold solid nanospheres will be achieved for a particle-size range of 10–45 nm (Fig. 2.7(b)).

The gold nanoshells show absorption efficiency which is comparable to K_{abs} of solid gold nanospheres [10]. However, the nanoshell optical resonance lies in the near-infrared (NIR) region (λ_{max} = 892 nm for a 70 nm gold shell [10]), where the biological tissue transmissivity is the highest. The optical characteristics of nanoparticles can be significantly

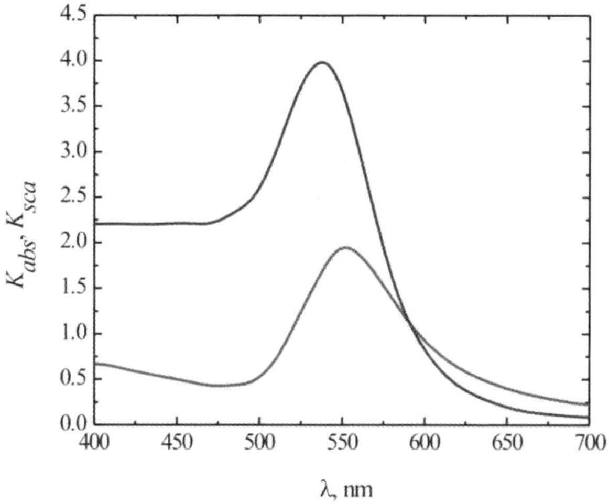

(a)

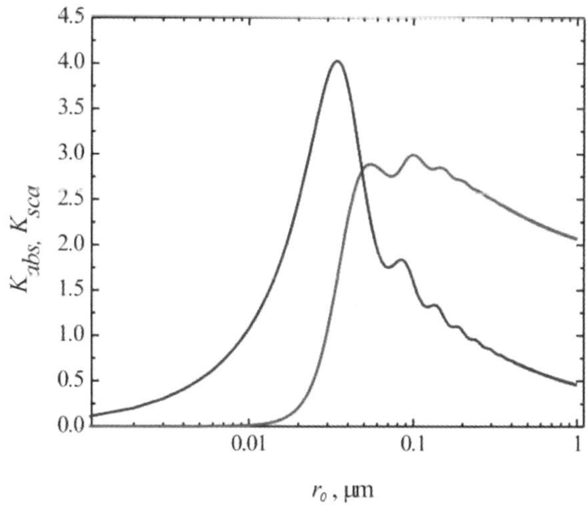

(b)

Fig. 2.7. Computer calculations of the absorption, K_{abs} (upper curves on each graph), and scattering, K_{sca} (lower curves), coefficients for gold nanoparticles in an aqueous suspension: (a) absorption and scattering spectrum of the gold particle in the visible range λ = 400–700 nm and (b) illustrations of the dependences of K_{abs} and K_{sca} on particle size for the optimal wavelength λ = 538.3 nm.

changed by altering their shapes. For example, gold nanorods have an NIR plasmon resonance (corresponding to the mode with the electric field parallel to the long axis of the nanorod) at a much smaller size. The magnitude of their NIR resonance absorption efficiency for an effective radius r_{eff} = 11.43 nm is about 14 [10]. This radius r_{eff} for a particle of arbitrary shape defines the volume V of the nanorod:

$$r_{eff} = (3V/4\pi)^{1/3}.$$

Thus, all three gold nanoparticle types (solid nanospheres, nanoshells and nanorods) have high absorption efficiency K_{abs} from 4 to 14 in the visible and NIR spectral ranges.

Using Eq. (2.8) we can estimate the lower level of the threshold energy density of a laser pulse required for thermal explosion of the GN using $E_{expl}(R) \approx I_{expl}(R) \times \tau_{expl}$, where the threshold intensity, I_{expl}, depends on the absorption efficiency of the nanoparticle. For the particular laser radiation (λ = 532 nm) used in the experiments [2], the threshold energy density for thermal explosion of a solid gold nanosphere of size r_0 = 35 nm (absorption efficiency of K_{abs} = 4.02 (Fig. 2.7)) is E_{expl} = 38.5 mJ/cm^2. The laser threshold energy density, E_{expl}, strongly depends on the types of nanoparticles (e.g., gold solid nanospheres, nanoshells and nanorods) [3,16]. For example, gold nanorods have a NIR resonance absorption efficiency of approximately 14 for an effective radius of 11.43 nm [10]. Due to the higher plasmon-resonance absorption efficiency of nanorods, the threshold energy density can be reduced by using gold nanorods up to E_{expl} = 25 mJ/cm^2.

The estimated values for E_{expl} are in good agreement with some available experimental results listed in Table 2.3. Indeed, for spherical GNs with average size 45 nm irradiated with second-harmonic Nd:YAG laser pulses (532 nm), the laser fluence threshold for fragmentation associated with GN melting and boiling phenomena is 30 mJ/cm^2 [15]. For a 30-ps pulse, 25-nm GN fragmentation has been observed at 23 mJ/cm^2, with a slight effect on changing the GN shape even at 2–5 mJ/cm^2 [21,22].

Table 2.3. Theoretical and experimental values for energy density thresholds.

Theoretical thresholds E_{expl} (mJ/cm^2)	**Experimental thresholds** E_{fragm} (mJ/cm^2)
Gold solid sphere, nanoshell of r_0 = 35–40 nm: E_{expl} = 38.5 mJ/cm^2	Gold solid sphere of r_0 = 45 nm: E_{fragm} = 30 mJ/cm^2
Gold nanorod of $r_{eff} \approx 11$ nm: E_{expl} = 11.2 mJ/cm^2	Gold solid sphere of r_0 = 25 nm: E_{fragm} = 23 mJ/cm^2

2.5. Conclusions and Summary

The comparative analysis of OTM and TTM for heating of a metal nanoparticle in the femtosecond, picosecond and nanosecond regimes has shown that:

- In the femtosecond mode, the thermal equilibrium among the excited electrons is established within the first 175 fs, long after the end of the laser pulse duration.
- The electrons remain in the thermal equilibrium state up to 1 ps.
- Both models demonstrate the same scenario in the heating kinetics of a metal nanoparticle by a femtosecond laser pulse: about a 100 fs time delay in the heating of the particle is observed, until reaching a maximum lattice temperature and saturation in temperature curves after 175 fs.
- The electron cooling time due to coupling to the lattice is about 10 ps, which imposes an upper time limit for TTM application.
- TTM gives a very good approximation for the femtosecomd mode while an electron temperature exists, but it fails to describe the laser heating of nanoparticles for longer pulse durations in the picosecond and nanosecond regimes.
- OTM shows that the heat lost from the surface of the nanoparticle into the surrounding medium becomes noticeable after 200 ps.

- The heating of a metal nanoparticle by a nanosecond laser pulse in fat provides higher particle overheating than in blood, prostate and water as the surrounding media due to the thermally-isolating property of the fat.
- The optical properties of the nanoparticle have a much stronger effect on the heating dynamics in the nanosecond mode than the thermal effects when the radius of the particle is less than 35 nm. For larger particles, the thermal processes dominate the optical properties, and the temperature curve is determined by the balance between heating of the nanoparticle and energy losses from the surface of the particle due to heat diffusion into the surrounding biological medium.

Thus, the comparison of the two models shows that OTM provides an adequate description of the laser heating of nanoparticles in the femtosecond, picosecond and nanosecond regimes.

We have also studied a new mechanism of laser/nanoparticle interaction — laser-induced thermal explosion of nanoparticles by ultrashort pulses. Thermal explosion is realized when heat is generated within a strongly-absorbing target more rapidly than the heat can diffuse away. On the basis of simple energy balance, it is found that the threshold energy density of a single laser pulse required for the thermal explosion of a solid gold nanosphere is approximately 40 mJ/cm^2. The nanoparticle's explosion threshold energy density can be reduced further (up to 11 mJ/cm^2) by using large nanorods (and probably nanoshells). The explosion of nanoparticles is accompanied by the generation of a shock wave of dense vapors expanding with supersonic velocity, producing the sound waves and optical plasma. It is important that most of these phenomena can explain some published experimental results whose interpretation was performed without taking into account this effect.

References

1. C. M. Pitsillides, E. K. Joe, X. Wei, R. R. Anderson and C. P. Lin, Selective cell targeting with light-absorbing microparticles and nanoparticles, *Biophys. J.* **84**, 4023–4032 (2003).

2. V. P. Zharov, R. R. Letfullin and E. Galitovskaya, Microbubbles-overlapping mode for laser killing of cancer cells with absorbing nanoparticle clusters, *J. Phys. D: Appl. Phys.* **38**, 2571–2581 (2005).
3. R. R. Letfullin, C. Joenathan, T. F. George and V. P. Zharov, Cancer cell killing by laser-induced thermal explosion of nanoparticles, *J. Nanomedicine* **1**, 473–480 (2006).
4. J. L. Wu, C. M. Wang and G. M. Zhang, Ultrafast optical response of the Au-BaO thin film stimulated by a femtosecond pulsed laser," *J. Appl. Phys.* **83**, 7855–7859 (1998).
5. B. N. Chichkov, C. Momma, S. Nolte, F. Alvensleben and A. Tünnermann, Femtosecond, picosecond and nanosecond laser ablation of solids, *Appl. Phys. A* **63**,109–115 (1996).
6. V. K. Pustovalov, Theoretical study of heating of spherical nanoparticles in media by short laser pulses, *Chem. Phys.* **308**, 103–108 (2005).
7. R. R. Letfullin and V. I. Igoshin, Multipass optical reactor for laser processing of disperse materials, *Quantum Electron.* **25**, 684–689 (1995).
8. R. R. Letfullin, Solid aerosols into the strong laser fields, *Bulletin of the Samara State Technical University: Physical–Mathematical Sciences* **4**, 243–263 (1996).
9. R. R. Letfullin and V. I. Igoshin, Theoretical modeling of plasma formation and generation of electromagnetic fields in the gas dispersed media under the action of laser radiation, *Trudy FIAN* **217**, 112–135 (1993).
10. P. K. Jain, K. S. Lee, I. H. El-Sayed and M. A. El-Sayed, Calculated absorption and scattering properties of gold nanoparticles of different size, shape and composition: Applications in biological imaging and biomedicine, *J. Phys. Chem. B* **110**, 7238–7248 (2006).
11. S. I. Anisimov, B. L. Kapeliovich and T. L. Perel'man, Electron emission from metal surfaces exposed to ultrashort laser pulses, *Sov. Phys. JETP* **39**, 375 (1974).
12. R. B. Ross, *Metallic Materials Specification Handbook*, 4th edn. (Chapman & Hall, London, 1992).
13. Z. Lin and L. V. Zhigilei, Electron–phonon coupling and electron heat capacity of metals under conditions of strong electron–phonon nonequilibrium, *Phys. Rev. B* **77**, 075133–075150 (2008).
14. J. Hohlfeld, S. Wellershoff, J. Güdde, U. Conrad, V. Jähnke and E. Matthias, Electron and lattice dynamics following optical excitation of metals, *Chem. Phys.* **251**, 237–258 (2000).
15. A. Takami, H. Kurita and S. Koda, Laser-induced size reduction of noble particles, *J. Phys. Chem. B*, **103**, 1226–1232 (1999).
16. R. R. Letfullin, V. P. Zharov, C. Joenathan and T. F. George, Nano-photothermolysis of cancer cells, *SPIE Newsroom (Society of Photo-Optical Instrumentation Engineers)*, DOI: 10.1117/2.1200701.0634–1-2 (2007).

17. S. Inasawa, M. Sugiyama and Y. Yamaguchi, Bimodal size distribution of gold nanoparticles under picosecond laser pulses, *J. Phys. Chem. B* **109**, 9404–9410 (2005).

18. A. Vogel and V. Venugolapan, Mechanism of pulsed laser ablation of biological tissue, *Chem. Rev.* **103**, 577–644 (2003).

19. L. Boufendi, A. Bouchoule, B. Dubreuil, E. Stoffels, W. W. Stoffels and M. L. deGiorgi, Study of initial dust formation in an Ar-SiH_4 discharge by laser-induced particle explosive evaporation, *J. Appl. Phys.* **76**, 148–153 (1994).

20. P. K. Jain, W. Qian and M. A. El-Sayed, Ultrafast cooling of photoexcited electrons in gold nanoparticle-thiolated DNA conjugates involves the dissociation of the gold-thiol bond, *J. Am. Chem. Soc.* **128**, 2426–2433 (2006).

21. Z. Peng, T. Walther and K. Kleinermanns, Photofragmentation of phase-transferred gold nanoparticles by intense pulsed laser light, *J Phys Chem B* **109**, 15735–40 (2005).

22. Z. Peng, T. Walther and K. Kleinermanns, Influence of intense pulsed laser irradiation on optical and morphological properties of gold nanoparticle aggregates produced by surface acid-base reactions, *Langmuir* **21**, 4249–4253 (2005).

Chapter 3

Exciting Infrared Normal Modes in C_{60} by an Ultrafast Laser

G. P. Zhang
Indiana State University
gpzhang@indstate.edu

Thomas F. George
University of Missouri–St. Louis
tfgeorge@umsl.edu

Selectively exciting a normal mode in a large molecule by an ultrafast laser is important to many laser engineered processes such as chemical reactions, but with so many degrees of freedom such selection is often challenging. It is shown here that the normal-mode selectivity can be achieved to some degree in C_{60} by carefully tuning laser parameters. Two methods are identified to selectively excite infrared modes. The first is to directly excite them by resonant excitation, and the second relies on the laser pulse duration with a slightly off-resonant laser frequency. These methods can be directly tested experimentally through time-resolved absorption spectra.

3.1. Introduction

Ultrafast infrared spectra play an increasingly important role in probing dynamics in many different systems ranging from small atoms to large biological molecules [1]. The infrared (IR) spectra, with their energy comparable to the energy scale of many physical, chemical and biological processes [2], can time-resolve IR-mode vibrations. As shown previously, this technique enables one to probe the dynamics in superconductors [3] and strongly-correlated transition metal oxides. Consequently, the dynamics of Cooper pairing and band-gap opening can be monitored in the time domain, which is essential to understanding the underlying mechanism of high-temperature superconductivity. The key to the success of such an investigation depends on the realization of normal-mode selectivity. Similarly, for a laser-controlled chemical reaction, one wants to know whether such a selection is possible. With more than two decades of intensive investigations, it is clear that such selection is possible in small molecules but often difficult in larger molecules. Therefore, a theoretical approach is very helpful in this regard, but the challenge is huge, since one has to include not only the complicated molecular states but also the laser field simultaneously. Fortunately, C_{60} provides an excellent venue for exploring the infrared mode selectivity in a system that is large enough to exhibit many features that larger molecules have, but it is still simple enough to simulate theoretically and probe experimentally.

Experimental and theoretical realization of selective Raman excitation in C_{60} is truly encouraging [4–8]. In this chapter, we are motivated to investigate how to selectively excite IR modes in C_{60}. We show that the normal-mode selectivity can be achieved to some degree in C_{60} by carefully tuning laser parameters. The infrared active modes can be selectively excited either by resonant excitation or slightly off-resonant excitation with an optimal laser pulse duration. Comparing with the selective Raman excitation which solely relies on the pulse duration, the infrared mode is easier to selectively excite. Finally, we compute the time-resolved absorption spectrum to demonstrate the possibility to observe our predictions experimentally. The experimental realization of normal-mode control may pave the way to probing Cooper pair breaking in superconductors such as K_3C_{60}.

The rest of the paper is arranged as follows. In Sec. 3.2, we present our theoretical scheme. In Sec. 3.3, the results will be presented, followed by the time-resolved pump-probe absorption spectrum in Sec. 3.4. Section 3.5 is the conclusion.

3.2. Theory

We use a tight-binding model to simulate C_{60} [9,10],

$$H_0 = -\sum_{ij,\sigma} t_{ij}(c^\dagger_{i,\sigma}c_{j,\sigma} + h.c.) + \frac{K_1}{2}\sum_{i,j}(r_{ij} - d_0)^2 + \frac{K_2}{2}\sum_i d\theta^2_{i,5} + \frac{K_3}{2}\sum_i (d\theta^2_{i,6,1} + d\theta^2_{i,6,2}) , \tag{3.1}$$

where the first term is the electronic Hamiltonian with the hopping integral, $t_{ij} = t^0 - \alpha(|\mathbf{r}_i - \mathbf{r}_j| - d_0)$, between nearest neighbor atoms at $\mathbf{r}_i$ and $\mathbf{r}_j$, and $r_{ij} = |\mathbf{r}_i - \mathbf{r}_j|$. Here t^0 is the average hopping constant, α is the electron-lattice coupling constant, and d_0 is the nearest-neighbor distance in diamond. In Eq. (3.1), $c^\dagger_{i,\sigma}$ is the electron creation operator at site i with spin $\sigma(=\uparrow\downarrow)$ [8]. The next three terms on the RHS are the lattice stretching, pentagon–hexagon and hexagon–hexagon bending energies, respectively. By fitting the energy gap, bond lengths of two kinds and 174 normal-mode frequencies, You *et al.* [11] have determined the above parameters as $t^0 = 1.91$ eV, $\alpha = 5.0$ eV/Å, $K_1 = 42$ eV/Å^2, $K_2 = 8$ eV/rad^2, $K_3 = 7$ eV/rad^2 and $d = 1.5532$ Å. These parameters will be fixed in our calculation.

In order to simulate the dynamical properties, we include the laser field as described by

$$H_I = -e \sum_{i\sigma} \mathbf{E}(t) \cdot \mathbf{r}_i \; n_{i\sigma} \ , \tag{3.2}$$

where $\mathbf{r}_i$ is the electron position, $n_{i\sigma}$ is the electron number operator, and $|\mathbf{E}(t)| = A\cos[\omega(t-t_0)]\exp[-(t-t_0)^2/\tau^2]$ [12]. Here A, ω, τ, e, t and t_0 are the field strength, laser frequency, pulse duration or width, electron charge, time and time delay, respectively. We numerically integrate the Liouville equation for the electron density matrices [8,12–14],

$$-i\hbar \frac{\partial \langle \rho^\sigma_{ij} \rangle}{\partial t} = \langle [\rho^\sigma_{ij}, H] \rangle \ , \tag{3.3}$$

where $H = H_0 + H_I$, $\rho^\sigma_{ij} = c^\dagger_{i\sigma} c_{j\sigma}$ is the density matrix operator, and $\langle\ \rangle$ represents the expectation value. Information about the lattice dynamics is obtained from the displacements and velocities of each carbon atom.

To accurately measure how one mode is excited, we employ a kinetic energy-based normal-mode analysis [15]. While the details are presented already [15], the main idea is that one expands the kinetic energy in terms of normal-mode kinetic energies. By measuring the normal-mode kinetic energy, we can tell whether a mode is highly excited or not, which is extremely convenient for nonlinear processes. This technique is exact even for a highly-nonlinear process, because the kinetic energy always depends on the velocity squared, irrespective of whether or not the process is nonlinear. This is different from the potential energy where, for a highly-nonlinear process, higher-order terms will appear so that a normal-mode analysis breaks down. In practice, the kinetic-energy-based normal-mode analysis is also very convenient to use.

3.3. Results and Discussion

In C_{60} there are 12 infrared active normal modes (4 distinct 3-fold degenerate modes with frequencies at 525, 575, 1182 and 1428 cm^{-1}). While there is plenty of experimental research on these infrared modes using continuum or ultrafast lasers, their emphases are different from ours [16] and are not on selectively exciting a few infrared active modes [1]. Sension *et al.* [17] used an infrared laser to excite C_{60}, but their emphasis was on charge transfer to C_{60}. Fleicher *et al.* employed an infrared laser to investigate C_{60} fragmentation [18]. Lee *et al.* investigated photoinduced electron transfer in conducting-polymer-C_{60} composites by infrared photoexcitation spectroscopy [19]. Meskers *et al.* measured time-resolved infrared absorption in a study of photoinduced charge transfer in a polythiophene-methanofullerene composite film [20]. We anticipate that our theoretical results will motivate further experimental investigations in C_{60}, as similar experimental studies have been done in molecular systems for a long time [1].

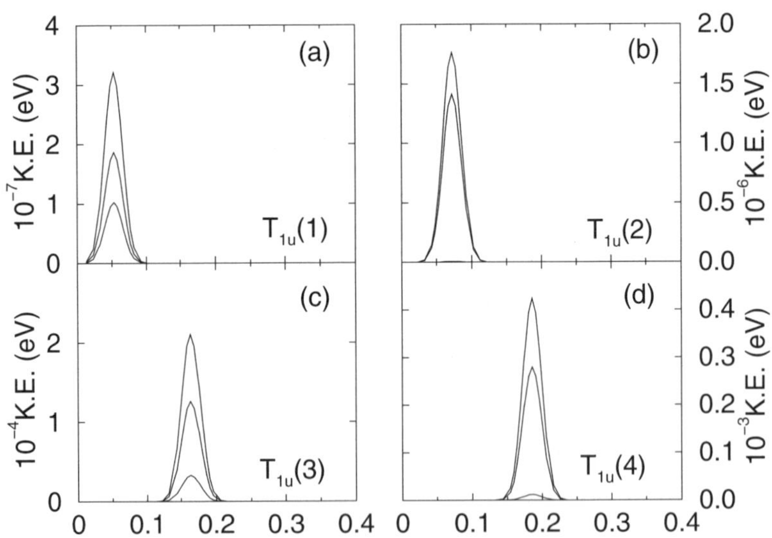

Fig. 3.1. Selective excitation of infrared modes by a laser. Four IR-active modes from $T_{1u}(1)$ to $T_{1u}(4)$ are shown in (a)–(d). The resonant peaks occur accurately at their respective eigenfrequencies. The laser pulse duration is 50 fs.

In the following, we will explore two different selection processes, where one is based on the laser frequency and the other on the pulse duration. IR excitation is a first-order process and is very sensitive to the incident laser frequency. This motivates us to directly use the laser frequency to selectively excite a few normal modes. The laser field strength is chosen as 0.01 V/Å and the pulse duration as 50 fs. Figure 3.1 shows the kinetic energy for each mode of T_{1u} as a function of the laser energy for four distinctive modes. We find that $T_{1u}(1)$, $T_{1u}(2)$, $T_{1u}(3)$and $T_{1u}(4)$ all are resonantly enhanced at their respective eigenfrequencies. For instance, $T_{1u}(1)$ resonates at 0.0529 eV (see Fig. 3.1(a)). This reveals an important method, namely, that as far as the mode frequencies are different, the resonant excitation is an effective way to selectively target a few specific IR modes.

While the effect of laser frequency is straightforward, it is unknown how the laser pulse duration affects the IR mode excitation. For the Raman modes, the laser pulse duration is almost the sole parameter that one can employ to select a few modes [15]. We choose the incident photon energy as 0.0529 eV, which is exactly at the $T_{1u}(1)$'s eigenfrequency. The results are shown in Fig. 3.2. It is evident that for shorter pulse durations, the $T_{1u}(3)$ and $T_{1u}(4)$ modes dominate (see Figs. 3.2(c) and 3.2(d)). This offers an opportunity to selectively excite those two modes. But as the pulse duration becomes longer, they lose their dominance quickly (see Figs. 3.2(c) and 3.2(d)). On the other hand, the resonant mode $T_{1u}(1)$ still increases superlinearly. The most unique mode is $T_{1u}(2)$, where its excitation first increases and then reaches its maximum around 50 fs, after which the kinetic energy for this mode decreases. The reason is because the $T_{1u}(1)$ and $T_{1u}(2)$ modes are very close in energy, and with a finite pulse duration, the Fourier-transformed field has frequency components close to the eigenfrequency of the $T_{1u}(2)$ mode. But as the pulse duration becomes longer, then the component decreases, since in the continuous-wave limit one must recover Fermi's golden rule.

We find that the uniqueness $T_{1u}(2)$ is directly related to the excitation wavelength. If we excite at the $T_{1u}(2)$'s eigenfrequency, then the $T_{1u}(1)$ and $T_{1u}(2)$ modes exchange their roles. If we increase the incident photon energy to the $T_{1u}(3)$ or $T_{1u}(4)$ mode, then the $T_{1u}(1)$ and $T_{1u}(2)$ modes show a decay in the same fashion as the $T_{1u}(3)$ mode. In other words, whenever the incident energy is tuned to a specific mode, this mode's kinetic energy increases superlinearly, while the modes close to this mode's eigenfrequency have an optimal pulse duration, and all the other modes de-

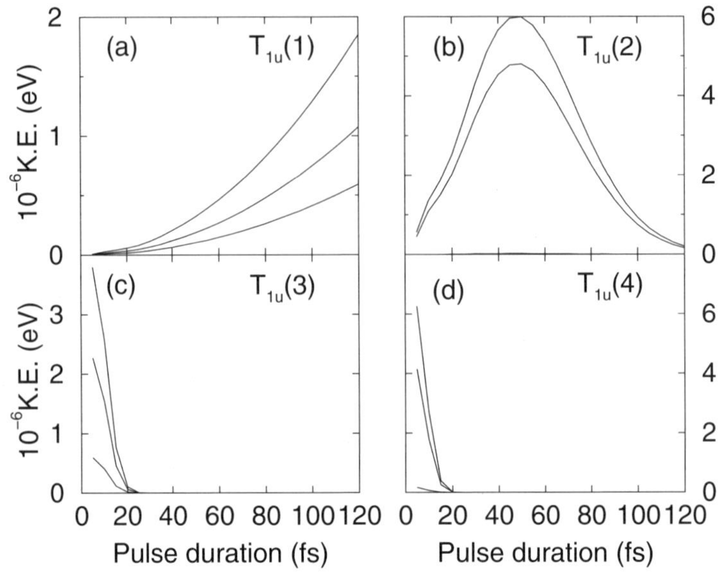

Fig. 3.2. Additional control of IR-mode excitation by a laser. The laser energy is tuned at the eigenfrequency of the $T_{1u}(1)$ mode. The laser field strength is 0.01 V/Å. (a) Superlinear increase in the kinetic energy of the $T_{1u}(1)$ mode, which is very different from the Raman mode (see fig. 3.3). (b) The $T_{1u}(2)$ mode has a local optimal laser pulse duration since its eigenenergy is close to the incident laser frequency. (c) and (d) are for the $T_{1u}(3)$ and $T_{1u}(4)$ modes and do not show any optimal pulse duration.

cay with the pulse duration. The optimal pulse duration provides another means to selectively excite a few IR modes, provided the mode frequencies are not degenerate or clustered together. Therefore, one has two ways to selectively excite IR modes. Note that since the superconducting energy gap is normally in the IR region, our results may have some applications in Cooper pair-breaking dynamics.

3.4. Time-Resolved Pump-Probe Absorption

In order to make a connection with future experiments, we compute the time-dependent absorption spectrum. For a typical pump-probe experiment, the pump pulse first excites the system, and after a time delay T, a probe pulse probes the electronic and vibrational excitations introduced by the pump. The total field is a sum of the pump and probe fields [21],

$$\mathbf{E}(t, T) = \mathbf{E}_{\text{pump}}(t) + \mathbf{E}_{\text{probe}}(t + T) . \tag{3.4}$$

Note that these fields are vectors. To avoid artifacts from the interference between these two pulses, their polarizations are chosen to be perpendicular to each other. Both pulses have a Gaussian shape with pulse duration of 40 fs and frequency of 0.25 eV. From the above analysis, we should be able to directly target the infrared modes.

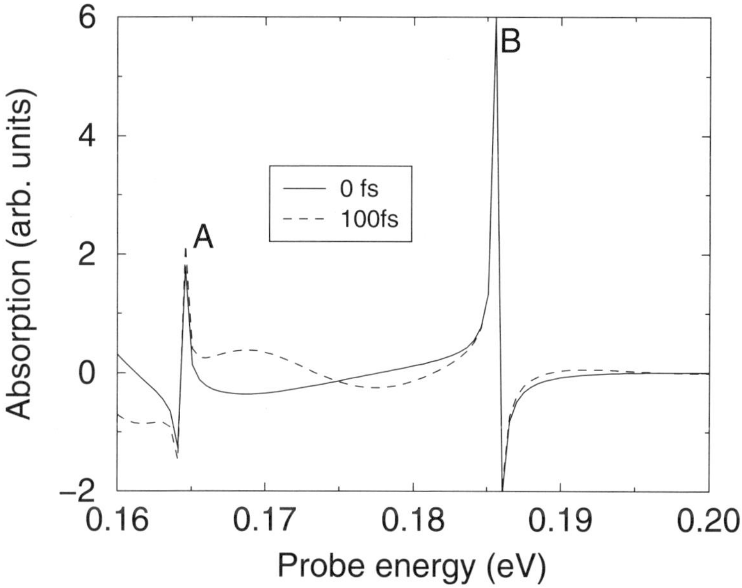

Fig. 3.3. Time-resolved absorption spectrum for two delays between pump and probe pulses. Both laser energies are centered at 0.25 eV. The pulse duration is 40 fs for each. The field strength is 0.05 V/Å for the pump and 0.01 V/Å for the probe. The solid line denotes the spectrum at $T = 0$ fs, while the dashed line represents the results at the time delay $T = 100$ fs.

The pump pulse has a strong field strength of 0.05 V/Å, and the probe has a weak field strength of 0.01 V/Å. The time-resolved absorption spectrum is computed from

$$\alpha(\omega, T) \propto \omega \mathrm{Im}(P_{\mathrm{probe}}(\omega)/E_{\mathrm{probe}}(\omega)) \ , \tag{3.5}$$

where $P_{\mathrm{probe}}(\omega)$ is the Fourier-transformed polarization $P_{\mathrm{probe}}(t) = \sum_i r_i n_i(t)$ along the probe pulse direction, and $E_{\mathrm{probe}}(\omega)$ is the Fourier transform of $E_{\mathrm{probe}}(t)$. Here $n_i(t)$ is computed from the Liouville equation (Eq. (3.3)). To properly take into account the environment effect, we introduce two relaxation times, T_1 and T_2, which are commonly called longitudinal and transverse times. From the experimental relaxation times [6], we choose $T_1 = 200$ fs and $T_2 = 100$ fs.

Figure 3.3 shows the results for two delays. The solid line represents the results at $T = 0$. We notice there are two prominent peaks, A and B. Peak A is situated at 0.164 eV, while peak B is at 0.185 eV. If we compare these two peaks with Figs. 3.1(c) and 3.1(d), we find that peaks A and B correspond to the $T_{1u}(3)$ and $T_{1u}(4)$ modes, respectively. This demonstrates that the kinetic-energy-based analysis reveals the same information as the pump-probe technique, but from a different perspective. The latter can be detected experimentally, while insights can be gained from the kinetic-energy-based analysis.

If we delay the probe by 100 fs, we see that the main peak positions stay the same, which indicates the vibrational excitation is in the ground electronic state. A new absorption appears around 0.17 eV, which is related to the system excitation due to the laser pumping. We expect that infrared ultrafast laser experiments can easily detect these features, which provides an opportunity to verify our theoretical predictions experimentally. Such experimental capability has been available for some time [1].

3.5. Conclusions

We have investigated the possibility to selectively excite infrared-active vibrational modes in C_{60}. Such modes can be selectively excited by either resonant pumping or off-resonant excitation with an optimal pulse duration. Overall, the IR modes have better control than Raman modes. We expect that these findings can be detected experimentally, as also supported by our time-resolved pump-probe absorption results. Pump-probe experiments with a short pulse of 12 fs have already probed the rapid oscillations of the A_g modes [7]. Even shorter pulses are now routinely available. Therefore, our study may motivate further experimental investigations.

Acknowledgments

This work was supported by the U. S. Department of Energy under Contract No. DE-FG02-06ER46304 and also by a Promising Scholars grant from Indiana State University. We acknowledge part of the work as done on Indiana State University's high-performance computers. This research used resources of the National Energy Research Scientific Computing Center at Lawrence Berkeley National Laboratory, which is supported by the

Office of Science of the U.S. Department of Energy under Contract No. DE-AC02-05CH11231. Initial studies used resources of the Argonne Leadership Computing Facility at Argonne National Laboratory, which is supported by the Office of Science of the U.S. Department of Energy under Contract No. DE-AC02-06CH11357.

References

1. M. D. Fayer, *Ultrafast Infrared and Raman Spectroscopy* (Marcel Dekker, New York, 2001); C. Ferrante, A. Tokmakoff, C. Taiti, A. S. Kwok, R. S. Francis, K. D. Rector and M. D. Fayer, Vibrational population dynamics in liquids and glasses: IR pump-probe experiments from 10 K to 300 K, in *Ultrafast Processes in Spectroscopy*, edited by O. Svelto, S. De Silvestri, and G. Denardo (Plenum, New York, 1996), p. 115*ff*.
2. A. H. Zewail, Femtochemistry: Recent progress in studies of dynamics and control of reactions and their transition states, *J. Phys. Chem.* **100**, 12701–12724 (1996).
3. S. B. Fleischer, E. P. Ippen, G. Dresselhaus, M. S. Dresselhaus, A. M. Rao, P. Zhou and P. C. Eklund, Femtosecond optical dynamics of C_{60} and M_3C_{60}, *Appl. Phys. Lett.* **62**, 3241–3243 (1993); S. B. Fleischer, B. Pevzner, D. J. Dougherty, H. J. Zeiger, G. Dresselhaus, M. S. Dresselhaus, E. P. Ippen and A. F. Hebard, Coherent phonons in alkali metal-doped C_{60}, *Appl. Phys. Lett.* **71**, 2734–2736 (1997).
4. M. Boyle, M. Heden, C. P. Schulz, E. E. B. Campbell and I. V. Hertel, Two-color pump-probe study and internal-energy dependence of Rydberg-state excitation in C_{60}, *Phys. Rev. A* **70**, 051201–51204 (2004).
5. H. Hohmann, C. Callegari, S. Furrer, D. Grosenick, E. E. B. Campbell and I. V. Hertel, Photoionization and fragmentation dynamics of C_{60}, *Phys. Rev. Lett.* **73**, 1919–1922 (1994).
6. T. N. Thomas, J. F. Ryan, R. A. Taylor, D Mihailovic and R. Zamboni, Time-resolved optical studies of photoexcited states in C_{60}, *Int. J. Mod. Phys. B* **6**, 3931–3934 (1992); T. N. Thomas, R. A. Taylor, J. F. Ryan, D. Mihailovic and R. Zamboni, Ultrafast dynamics of photoexcited states in C_{60}, *Europhys. Lett.* **25**, 403–408 (1994).
7. S. L. Dexheimer, D. M. Mittlemann, R. W. Schoenlien, W. Vareka, X.-D. Xiang, A. Zettl and C. V. Shank, Ultrafast dynamics of solid C_{60}, in *Ultrafast Phenomena VIII*, edited by J. -L. Martin, A. Migus, G. A. Mourou and A. H. Zewail (Springer-Verlag, Heidelberg, 1993), pp. 81–86; *ibid.*, *SPIE Proc.* **1861**, 328–332 (1991).
8. G. P. Zhang and T. F. George, Controlling vibrational excitations in C_{60} by laser pulse duration, *Phys. Rev. Lett.* **93**, 147401-1/147401-4 (2004); G. P. Zhang, Optical high harmonic generations in C_{60}, *Phys. Rev. Lett.* **95**, 047401-1/047401-4 (2005).

9. G. P. Zhang, X. Sun and T. F. George, Nonlinear optical response and ultrafast dynamics in C_{60}, *J. Phys. Chem. A* **113**, 1175-1188 (2009).
10. G. P. Zhang, R. T. Fu, X. Sun, D. L. Lin and T. F. George, Relaxation process of charge transfer in C_{60}, *Phys. Rev. B* **50**, 11976–11980 (1994); X. Sun, G. P. Zhang, Y. S. Ma, R. L. Fu, X. C. Shen, K. H. Lee, T. Y. Park, T. F. George and L. N. Pandey, Relaxation process of self-trapping exciton in C_{60}, *Phys. Rev. B* **53**, 15481–15484 (1996); G. P. Zhang, X. Sun, T. F. George and L. N. Pandey, Normal mode analysis for a comparative study of relaxation processes of charge transfer and photo-excitation in C_{60}, *J. Chem. Phys.* **106**, 6398–6403 (1997).
11. W. M. You, C. L. Wang, F. C. Zhang and Z. B. Su, Application of a Su-Schrieffer-Heeger-like model to the intramolecular electron-phonon coupling in C_{60} clusters, *Phys. Rev. B* **47**, 4765–4770 (1993).
12. G. P. Zhang and W. Hübner, Laser-induced ultrafast demagnetization in ferromagnetic metals, *Phys. Rev. Lett.* **85**, 3025–3028 (2000).
13. G. P. Zhang, Hartree-Fock dynamical electron-correlation effects in C_{60} after laser excitation, *Phys. Rev. Lett.* **91**, 176801-1/176801-4 (2003).
14. G. P. Zhang, X. Sun and T. F. George,Laser-induced ultrafast dynamics in C_{60}, *Phys. Rev. B* **68**, 165401-1/165401-5 (2003).
15. G. P. Zhang and T. F. George, Normal-mode selectivity in ultrafast Raman excitations in C_{60}, *Phys. Rev. B* **73**, 035422-1/035422-6 (2006).
16. M. S. Dresselhaus, G. Dresselhaus and P. C. Eklund, *Science of Fullerenes and Carbon Nanotubes* (Academic Press, San Diego, 1996); G. von Helden, I. Holleman, A. J. A. van Roij, G. M. H. Knippels, A. F. G. van der Meer and G. Meijer, Shedding new light on thermionic electron emission of fullerenes *Phys. Rev. Lett.* **81**, 1825–1828 (1998).
17. R. J. Sension, A. Z. Szarka, G. R. Smith and R. M. Hochstrasser, Ultrafast photoinduced electron transfer to C_{60}, *Chem. Phys. Lett.* **185**, 179–183 (1991).
18. S. B. Fleischer, B. Pevzner, D. J. Dougherty, E. P. Ippen, M. S. Dresselhaus and A. F. Hebard, Phototransformation in visible and near-IR femtosecond pump-probe studies of C_{60}, *Appl. Phys. Lett.* **69**, 296 (1996).
19. K. Lee, R. A. J. Janssen, N. S. Sariciftci and A. J. Heeger, Direct evidence of photoinduced electron transfer in conducting-polymer-C_{60} composites by infrared photoexcitation spectroscopy, *Phys. Rev. B* **49**, 5781–5784 (1994).
20. S. C. J. Meskers, P. A. van Hal, A. J. H. Spering, J. C. Hummelen, A. F. G. van de Meer and R. A. J. Janssen, Time-resolved infrared-absorption study of photoinduced charge transfer in a polythiophene-methanofullerene composite film, *Phys. Rev. B* **61**, 9917–9920 (2000).
21. W. T. Pollard, S. L. Dexheimer, Q. Wang, L. A. Peteanu, C. V. Shank and R. A. Mathies, Theory of dynamic absorption spectroscopy of nonstationary states. 4. Application to 12-fs resonant impulsive Ramon spectroscopy of bacteriorhodopsin, *J. Phys. Chem.* **96**, 6147–6158 (1992).

Chapter 4

Self-Interaction-Free Time-Dependent Density Functional Theoretical Approaches for Probing Atomic and Molecular Multiphoton Processes in Intense Ultrashort Laser Fields

Shih-I Chu

University of Kansas
and
National Taiwan University
sichu@ku.edu

Dmitry A. Telnov

St. Petersburg State University
telnov@pcqnt1.phys.spbu.ru

In this chapter, we present a short account of some recent development of *self-interaction-free* density functional theory (DFT) and time-dependent density functional theory (TDDFT) for accurate and efficient treatment of the electronic structure, and time-dependent quantum dynamics of many-electron atomic and molecular systems. In addition we present several advanced numerical techniques for efficient and high-precision treatment of the self-interaction-free DFT/TDDFT equations. The usefulness of these procedures is illustrated by a few case studies of atomic and molecular processes in intense ultrashort laser fields, a subject of much current interest in strong-field atomic and molecular physics as well as attosecond science and technology.

4.1. Introduction

In the last couple of decades, the density-functional theory (DFT) has been a widely used formalism for electron structure calculations of atoms, molecules, and solids [1–3]. The DFT is based on the earlier fundamental work of Hohenberg and Kohn [4] and Kohn and Sham [5]. In the Kohn–Sham DFT formalism [5], the electron density is decomposed into a set of orbitals, leading to a set of one-electron Schrödinger-like equations to be solved self-consistently. The Kohn–Sham equations are structurally similar to the Hartree–Fock equations but include in principle exactly the *many-*

body effects through a *local* exchange-correlation (xc) potential. Thus DFT is computationally much less expensive than the traditional *ab initio* many-electron wave function approaches and this accounts for its great success for large systems. However, the DFT is well developed mainly for the *ground-state* properties only. The treatment of the *excited states* within the DFT is a more recent development [6–10].

The essential element of DFT is the input of the *exchange-correlation* (xc) energy functional whose exact form is unknown. The simplest approximation for the xc energy functional is through the *local spin-density approximation* (LSDA) [1,11] of homogeneous electronic gas. A deficiency of the LSDA is that the xc potential decays exponentially and does not follow the correct *long-range* asymptotic Coulombic ($-1/r$) behavior. As a result, the LSDA electrons are too weakly bound and for negative ions even unbound. More accurate forms of the xc energy functionals are available from the *generalized gradient approximation* (GGA) [12–15], which takes into account the *gradient* of electron density. However, the xc potentials derived from these GGA energy functionals suffer similar problems like in LSDA and do not have the proper long-range asymptotic potential behavior either. Thus while the total energies of the ground states predicted by these GGA density functionals [12–15] are reasonably accurate, the excited-state energies and the ionization potentials obtained from the highest occupied orbital energies of atoms and molecules are not satisfactory, typically 30% to 50% too low [1,16]. The problem of the incorrect long-range behavior of the LSDA and GGA energy functionals can be attributed to the existence of the *self-interaction energy* [1,3,16,17]. For proper treatment of atomic and molecular dynamics such as collisions or multiphoton ionization processes *etc.*, it is necessary that both the ionization potential and the excited-state properties be described more accurately. In addition, the treatment of time-dependent processes will require the use of time-dependent density functional theory (TDDFT).

The TDDFT extends the concept of stationary DFT to time-dependent domain. For any interacting many-particle quantum system subject to a given time-dependent potential, all physical observables are uniquely determined by knowledge of the time-dependent density and the state of the system at any instant in time [18,19]. In particular, if the time-dependent potential is turned on at some time t_0 and the system has been in its ground state until t_0, all observables are unique functionals of only the density. In this case, the initial state of the system at time t_0 will be a unique functional of the ground state density itself, i.e. of the density at

t_0. This unique relationship allows one to derive a computational scheme in which the effect of the particle–particle interaction is represented by a density-dependent single-particle potential, so that the time evolution of an interacting system can be investigated by solving a time-dependent auxiliary single-particle problem. Additional simplifications can be obtained in the linear response regime [18–20]. In the last several years there is considerable effort and success in the extension of the (weak-field) TDDFT and the use of linear response theory to the study of electronic excited state energies [21,22], frequency-dependent multipole polarizabilities [22,23], optical spectra of molecules, clusters, and nanocrystals [22,25], autoionizing resonances [16], etc.

The primary focus of this article is to discuss some of the recent developments and applications of *self-interaction-free* TDDFT for the study of atomic and molecular multiphoton processes in intense ultrashort laser fields. The strong-field atomic and molecular physics is one of the most active fields of forefront research in attosecond science and technology [26,27]. The rapid advent of high-power and ultrashort-pulse laser technology in the last decade has facilitated the experimental exploration of multiphoton and very high-order ($>$ 300th order) nonlinear optical processes, leading to the discovery of a host of novel strong-field phenomena, such as multiphoton and above-threshold ionization (MPI/ATI) of atoms, multiphoton and above-threshold dissociation (MPD/ATD) of molecules, multiple high-order harmonic generation (HHG), chemical bond softening and hardening, Coulomb explosion, and coherent control of chemical and physical processes, etc. For the treatment of these strong-field processes, the conventional high-order perturbation approach is generally not adequate. On the other hand, nonperturbative approach using *ab initio* wave functions requires the solution of $(3N+1)$th-order time-dependent Schrödinger equation in space and time, where N is the number of electrons. But this is well beyond the capability of current computer technology for $N > 2$. Even for the case of $N = 2$, fully *ab initio* time-dependent study is still at the beginning stage. The single-active-electron (SAE) model [28,29] with frozen core is thus commonly used for describing the strong-field processes. However, within the SAE model, important physical phenomena such as excited state resonances, dynamical response from individual valence spin-orbital, inner core excitation, nonsequential ionization, and dynamical electron correlations, etc., cannot be treated. Clearly, a more complete formalism beyond the SAE and other phenomenological models is very desirable at this time for more comprehensive and accurate treatment of atomic and molecular physics and chemical physics in strong fields.

We note, however, that the conventional (weak-field) TDDFT is not adequate for the treatment of strong field processes. Similar to the stationary DFT case, due to the existence of the self-interaction energy, TDDFT calculations using *adiabatic* LDA or GGA energy functionals do not have the correct long-range asymptotic Coulombic ($-1/r$) potential. Moreover, nonperturbative framework for TDDFT will be required for the treatment of strong field processes. The recent development of *self-interaction-free* DFT and TDDFT removes some of these problems and provide powerful and practical nonperturbative frameworks for quantitative treatment of highly excited states and strong-field processes of many-electron quantum systems.

In the following, we shall present a short account of some of the recent development of the self-interaction-free TDDFT and their applications to the nonperturbative treatment of multiphoton dynamics and very-high-order nonlinear optical processes of atomic and molecular systems in intense ultrashort laser fields.

4.2. Recent Development of Self-Interaction-Free TDDFT for Nonperturbative Treatment of Atomic Multiphoton Processes in Intense Laser Fields

The central result of modern TDDFT is a set of time-dependent Kohn–Sham (TDKS) equations which are structurally similar to the time-dependent Hartree–Fock (TDHF) equations but include (in principle exactly) all many-body effects through a *local* time-dependent exchange-correlation (xc) potential [18,19]. To date, most applications of TDDFT fall in the regime of *weak-field* linear or nonlinear response and the *adiabatic* LSDA energy functional is often used [18,20]. Applications of the time-dependent LSDA approach has been made to the photo-response of atoms, molecules, clusters, nanocrystals, semiconductor surfaces, and bulk semiconductor in the weak-field perturbative regime [20,24,25].

As indicated in the introduction section, the conventional (weak-field) TDKS formalism cannot be directly applied to the study of multiphoton processes in *intense* laser fields. In this section, we discuss a TDDFT with *optimized effective potential* (OEP) and *self-interaction correction* (SIC) for *nonperturbative* treatment of many-electron quantum systems in intense laser fields [30]. It is based on the extension of the corresponding *steady-state* procedure [16] to the time domain and makes use of the semianalytical approximation of OEP by Krieger, Li, and Iafrate (KLI) [31]. We note that a related TDOEP-KLI method without the use of SIC was proposed by Ullrich and Gross [32]. The latter method provides an accurate procedure for the calculation of the exchange part of the time-dependent potential. But computationally it can be more time consuming than the conventional TDKS approach since the TDOEP-KLI procedure requires the construction of Hartree-Fock-like *nonlocal* potential at each time step. The advantage of the TDOEP/KLI-SIC approach [30] is that it allows the construction of self-interaction-free time-dependent *local* OEP which is also orbital *independent*. This greatly facilitates the study of time-dependent processes of many-electron quantum systems in strong fields.

4.2.1. *TDDFT with OEP/KLI-SIC for atomic multiphoton processes in intense pulsed laser fields*

The quantum mechanical *action* of a many-electron system interacting with an external field can be expressed as [30,32]

$$\begin{aligned}A[\{\psi_{n\sigma}\}] &= \sum_{\sigma}\sum_{n=1}^{N_\sigma}\int_{-\infty}^{t_1} dt \int \psi_{n\sigma}^*(\mathbf{r},t)\left[i\frac{\partial}{\partial t}+\frac{1}{2}\nabla^2\right]\psi_{n\sigma}(\mathbf{r},t)d\mathbf{r}\\ &- \sum_{\sigma}\int_{-\infty}^{t_1} dt\int \rho_\sigma(\mathbf{r},t)[v_{\mathrm{n}}(\mathbf{r})+v_{\mathrm{ext}}(\mathbf{r},t)]d\mathbf{r}\\ &- \frac{1}{2}\int_{-\infty}^{t_1} dt\int\int \frac{\rho(\mathbf{r},t)\rho(\mathbf{r}',t)}{|\mathbf{r}-\mathbf{r}'|}d\mathbf{r}d\mathbf{r}'\\ &- A_{\mathrm{xc}}[\{\psi_{n\sigma}\}],\end{aligned} \tag{4.1}$$

where $\psi_{n\sigma}(\mathbf{r},t)$ are the time-dependent spin-orbitals, $N=\sum_\sigma N_\sigma$ is the total number of electrons, $v_{\mathrm{n}}(\mathbf{r})$ is the electron-nucleus Coulomb interaction, $v_{\mathrm{ext}}(\mathbf{r},t)$ describes the coupling of the electron to the external laser fields, and $A_{\mathrm{xc}}[\{\psi_{n\sigma}\}]$ is the exchange-correlation (xc) action functional. The electron spin-densities $\rho_\sigma(\mathbf{r},t)$ are defined as follows:

$$\rho_\sigma(\mathbf{r},t)=\sum_{n=1}^{N_\sigma}|\psi_{n\sigma}(\mathbf{r},t)|^2, \tag{4.2}$$

and the total electron density $\rho(\mathbf{r},t)$ is obtained by summation of the spin-densities:

$$\rho(\mathbf{r},t)=\sum_{\sigma}\rho_\sigma(\mathbf{r},t). \tag{4.3}$$

The spin orbitals satisfy the one-electron Schrödinger-like equation:

$$i\frac{\partial}{\partial t}\psi_{n\sigma}(\mathbf{r},t) = \left(-\frac{1}{2}\nabla^2 + V_\sigma(\mathbf{r},t)\right)\psi_{n\sigma}(\mathbf{r},t), \tag{4.4}$$

where $V_\sigma(\mathbf{r},t)$ will be the TDOEP if we choose the set of spin-orbitals $\{\psi_{n\sigma}\}$ which render the total action functional $A[\{\psi_{n\sigma}\}]$ *stationary*:

$$\frac{\delta A[\{\psi_{n\sigma}\}]}{\delta V_\sigma(\mathbf{r},t)} = 0. \tag{4.5}$$

Generally, $V_\sigma(\mathbf{r},t)$ contains the memory effect. It depends not only on the density at the time moment t but also on the densities at preceding times. However, if we use the following explicit SIC expression for the exchange-correlation (xc) action functional [30],

$$\begin{aligned} A_{\mathrm{xc}}[\{\psi_{n\sigma}\}] = &\int_{-\infty}^{t_1} dt E_{\mathrm{xc}}[\rho_\uparrow(\mathbf{r},t), \rho_\downarrow(\mathbf{r},t)] \\ &- \sum_\sigma \sum_{n=1}^{N_\sigma} \int_{-\infty}^{t_1} dt\,\{J[\rho_{n\sigma}] + E_{\mathrm{xc}}[\rho_{n\sigma}, 0]\}, \end{aligned} \tag{4.6}$$

then the memory term vanishes identically. Similar results are obtained as long as one uses an explicit E_{xc} form (such as that in LSDA or GGA) of energy functional and the adiabatic approximation.

The use of the SIC form in Eq. (4.6) removes the spurious self-interaction terms in conventional TDDFT and results in a proper *long-range* asymptotic potential. Another major advantage of this procedure is that only *local* potential is required to construct the *orbital-independent* OEP. This facilitates considerably the numerical computation.

By extending the steady-state OEP/KLI-SIC procedure [16] to the time-dependent domain, we obtain the time-dependent (TD) OEP as:

$$V_\sigma(\mathbf{r},t) = v_{\mathrm{n}}(\mathbf{r}) + v_{\mathrm{ext}}(\mathbf{r},t) + \frac{\delta J[\rho]}{\delta\rho_\sigma(\mathbf{r},t)} + V_\sigma^{\mathrm{SIC}}(\mathbf{r},t), \tag{4.7}$$

where

$$\begin{aligned} V_\sigma^{\mathrm{SIC}}(\mathbf{r},t) = &\sum_n \frac{\rho_{n\sigma}(\mathbf{r},t)}{\rho_\sigma(\mathbf{r},t)}\{v_{n\sigma}(\mathbf{r},t) \\ &+ \left[\overline{V}_{n\sigma}^{\mathrm{SIC}}(t) - \overline{v}_{n\sigma}(t)\right]\}, \end{aligned} \tag{4.8}$$

$$\begin{aligned} v_{n\sigma}(\mathbf{r},t) = &\frac{\delta E_{\mathrm{xc}}[\rho_\uparrow(\mathbf{r},t), \rho_\downarrow(\mathbf{r},t)]}{\delta\rho_\sigma(\mathbf{r},t)} - \frac{\delta J[\rho_{n\sigma}(\mathbf{r},t)]}{\delta\rho_{n\sigma}(\mathbf{r},t)} \\ &- \frac{\delta E_{xc}[\rho_{n\sigma}(\mathbf{r},t), 0]}{\delta\rho_{n\sigma}(\mathbf{r},t)}, \end{aligned} \tag{4.9}$$

and

$$\overline{V}_{n\sigma}^{\text{SIC}}(t) = \langle\psi_{n\sigma}|V_{\sigma}^{\text{SIC}}(\mathbf{r},t)|\psi_{n\sigma}\rangle, \tag{4.10}$$

$$\overline{v}_{n\sigma}(t) = \langle\psi_{n\sigma}|v_{n\sigma}(\mathbf{r},t)|\psi_{n\sigma}\rangle. \tag{4.11}$$

Equations (4.4) and (4.7) are to be solved self-consistently. Note that since the exact form of $V_{\text{xc},\sigma}(\mathbf{r},t)$ is unknown, the *adiabatic* approximation is often used in the TDDFT calculations:

$$V_{\text{xc},\sigma}(\mathbf{r},t) = V_{\text{xc},\sigma}[\rho_\sigma]|_{\rho_\sigma=\rho_\sigma(\mathbf{r},t)}\,. \tag{4.12}$$

Finally Eq. (4.4) is an initial value problem and the initial wave function can be determined by

$$\psi_{n\sigma}(\mathbf{r},t)_{t=0} = \phi_{n\sigma}(\mathbf{r})\exp(-i\epsilon_{n\sigma}t)|_{t=0}, \tag{4.13}$$

where, $\phi_{n\sigma}(\mathbf{r})$ and $\epsilon_{n\sigma}$ are the eigenfunction and eigenvalue of the time-independent Kohn–Sham equation (with OEP/KLI-SIC) for the static case [16].

4.2.2. *Time-dependent generalized pseudospectral method for numerical solution of self-interaction-free TDDFT equations*

In this section we briefly describe a numerical procedure recently developed for accurate and efficient solution of the time-dependent OEP/SIC equation, Eq. (4.4). The commonly used procedures for the time propagation of the Schrödinger or TDDFT equation employ *equal-spacing* spatial grid discretization [33,34]. For processes such as HHG, accurate time-dependent wave functions are required to achieve convergence since the intensity of various harmonic peaks can span a range of many (10 to 20) orders of magnitude. High-precision wavefunctions are, however, more difficult to achieve by the conventional equal-spacing spatial-grid-discretization time-dependent techniques, due to the Coulomb singularity at the origin and the long-range behavior of the Coulomb potential. To achieve more accurate wave function propagation, a numerical procedure, the *time-dependent generalized pseudospectral* (TDGPS) method [35] has been recently introduced. The TDGPS procedure consists of the following two basic elements: (i) The GPS technique [36,37] is used for *nonuniform* optimal grid discretization of the radial coordinates and the Hamiltonian. It has been shown that the number of grid points required in the GPS procedure can be orders of magnitude smaller than those used by the conventional equal-spacing discretization methods. (ii) A split-operator technique in the *energy* representation is introduced for efficient time propagation of the wave functions. More detailed discussion of the TDGPS method is given in [35] and in a later molecular section.

4.2.3. *Multiphoton Quantum Dynamics of Atomic Systems in Intense Laser Fields*

In this section, we discuss an application of the TDOEP/KLI-SIC formalism to the nonperturbative study of *multiple high-order harmonic generation* of atoms in intense laser fields. The study of the HHG phenomena is one of the most rapidly developing topics in strong-field atomic and molecular physics [28,38–40]. The generation of harmonic orders well in excess of 100 from noble gas, diatomic and polyatomic molecules, and cluster targets has been demonstrated by several recent experiments [38–40]. Thus the HHG mechanism provides a simple and powerful new route to generate coherent x-ray laser source which is technically much less demanding and less energy intensive than current plasma based x-ray schemes. The availability of such a *compact* laboratory (*table-top*) system for the generation of coherent x-rays holds promise as a source for biological holography and nonlinear optics in the x-ray regime. Another potential new application of HHG processes is the possibility of generating laser pulses of *ultrashort* duration (tens of attoseconds) in the near future, leading a way to perform *attosecond* spectroscopy and study new dynamical phenomena with *attosecond* time resolution.

To study HHG, we start from the calculation of the total induced dipole moment of N-electron systems which can be expressed in terms of electron density ρ as follows:

$$d(t) = \int \rho(\mathbf{r},t) z d\mathbf{r} = \sum_{i\sigma} \langle \psi_{i\sigma}(\mathbf{r},t)|z|\psi_{i\sigma}(\mathbf{r},t)\rangle. \tag{4.14}$$

The corresponding HHG power spectrum can now be obtained by the Fourier transformation of the respective time-dependent dipole moment:

$$P(\omega) = \left| \frac{1}{t_f - t_i} \int_{t_i}^{t_f} d(t) \exp(-i\omega t)\, dt \right|^2 \equiv |d(\omega)|^2. \tag{4.15}$$

One recent application of the TDOEP/KLI-SIC formalism is to study the role of *dynamical electron correlation* on HHG of He atoms in intense linearly polarized (LP) laser pulses [30]. Of particular interest is the study of the mechanism responsible for the production of the "higher" harmonics observed in the experiment [41] which cannot be explained by the SAE

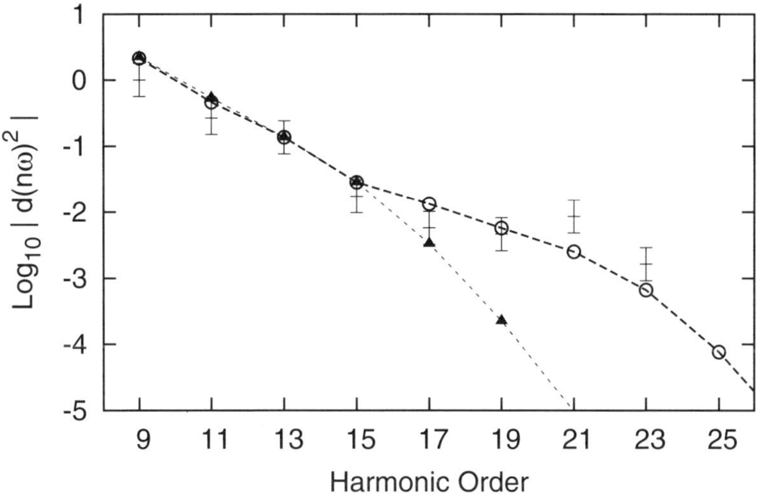

Fig. 4.1. The HHG spectrum of He obtained from the all-electron calculation (open circle) and from the SAE model (filled triangle). The experimental data (with error bar) are also shown for comparison. The HHG yields are normalized to the 13th harmonic peak. The laser peak intensity used in the calculation is $I = 3.5 \times 10^{15}$ W/cm^2 and wavelength $\lambda = 248.6$ nm.

model [28,29]. Figure 4.1 shows that while the SAE model fails to produce the higher harmonics, the TDDFT/KLI-SIC results agree well with the experimental data in both lower and higher HHG regimes, indicating the important role played by the dynamical electron correlation [30]. More detailed study of the HHG processes of rare gas (He, Ne, Ar) atoms has been recently reported [42].

4.3. Self-Interaction-Free TDDFT for Molecular Multiphoton Processes in Intense Laser Fields

4.3.1. *Time-dependent generalized pseudospectral method for numerical solution of self-interaction-free TDDFT equations in two-center systems*

In the following, we extend the time-dependent generalized pseudospectral (TDGPS) procedure to the numerical solution of the time-dependent equations in two-center multielectron systems. Consider the solution of the time-dependent one-electron Kohn–Sham equations for spin orbitals $\psi_{n\sigma}(\mathbf{r}, t)$:

$$i\frac{\partial}{\partial t}\psi_{n\sigma}(\mathbf{r},t) = \left[-\frac{1}{2}\nabla^2 + v_{\text{eff},\sigma}(\mathbf{r},t)\right]\psi_{n\sigma}(\mathbf{r},t), \qquad n = 1, 2, ..., N_\sigma. \tag{4.16}$$

The time-dependent effective potential $v_{\text{eff},\sigma}(\mathbf{r},t)$ is a functional of the electron spin densities $\rho_\sigma(\mathbf{r},t)$. The potential $v_{\text{eff},\sigma}(\mathbf{r},t)$ can be written in the general form

$$v_{\text{eff},\sigma}(\mathbf{r},t) = v_{\text{n}}(\mathbf{r}) + v_{\text{H}}(\mathbf{r},t) + v_{\text{xc},\sigma}(\mathbf{r},t) + v_{\text{ext}}(\mathbf{r},t) \tag{4.17}$$

where $v_{\text{n}}(\mathbf{r})$ is the electron interaction with the nuclei,

$$v_{\text{n}}(\mathbf{r}) = -\frac{Z}{|\mathbf{R}_1 - \mathbf{r}|} - \frac{Z}{|\mathbf{R}_2 - \mathbf{r}|} \tag{4.18}$$

with Z being the nuclear charge (we consider homonuclear diatomic molecules only), and $\mathbf{R}_1$ and $\mathbf{R}_2$ being the positions of the nuclei (which are assumed to be fixed at their equilibrium positions); $v_{\text{H}}(\mathbf{r},t)$ is the Hartree potential due to electron–electron Coulomb interaction,

$$v_{\text{H}}(\mathbf{r},t) = \int \frac{\rho(\mathbf{r}',t)d^3r'}{|\mathbf{r} - \mathbf{r}'|}. \tag{4.19}$$

The potential $v_{\text{ext}}(\mathbf{r},t)$ in Eq. (4.17) describes the interaction with the laser field. Using the dipole approximation and the length gauge, it can be expressed as follows:

$$v_{\text{ext}}(\mathbf{r},t) = (\mathbf{F}(t) \cdot \mathbf{r}). \tag{4.20}$$

Here $\mathbf{F}(t)$ is the electric field strength of the laser field, and the linear polarization is assumed. For the laser pulses with the sine-squared envelope, one has:

$$\mathbf{F}(t) = \mathbf{F}_0 \sin^2\frac{\pi t}{T}\sin\omega_0 t \tag{4.21}$$

where T and ω_0 denote the pulse duration and the carrier frequency, respectively; F_0 is the peak field strength.

The wave functions and operators are discretized with the help of the generalized pseudospectral (GPS) method in prolate spheroidal coordinates [43–46]. The prolate spheroidal coordinates ξ, η, and φ are related to the Cartesian coordinates x, y, and z as follows [47]:

$$\begin{aligned} x &= a\sqrt{(\xi^2-1)(1-\eta^2)}\cos\varphi, \\ y &= a\sqrt{(\xi^2-1)(1-\eta^2)}\sin\varphi, \\ z &= a\xi\eta \quad (1 \le \xi < \infty, \quad -1 \le \eta \le 1). \end{aligned} \tag{4.22}$$

In Eq. (4.22), we assume that the molecular axis is directed along the z axis, and the nuclei are located on this axis at the positions $-a$ and a, so the internuclear separation $R = 2a$. For the unperturbed molecule, the projection m of the angular momentum onto the molecular axis is conserved, and the *exact* spin orbitals have factors $(\xi^2 - 1)^{|m|/2}(1 - \eta^2)^{|m|/2}$ which are non-analytical at nuclei for odd $|m|$. Straightforward numerical differentiation of such functions could result in significant loss of accuracy. Therefore different forms of the kinetic energy operators have been suggested for even and odd m [44,48]. However, for the molecules in the linearly polarized laser field with arbitrary orientations of the molecular axis, the projection of the electron angular momentum onto the molecular axis is not conserved any longer. In this case, we apply a full 3D discretization with respect to the coordinates ξ, η, and φ. For ξ and η, we use the GPS discretization with non-uniform distribution of the grid points; for φ, the Fourier grid (FG) method [49] with uniform spacing of the grid points is more appropriate. To take care of the possible singularities at the nuclei, we use special mapping transformations of the coordinates ξ and η [50] which make the wave functions analytic at the nuclei for both even and odd projections of the angular momentum. The discretized kinetic energy operator takes the form of the matrix $T_{ijk;i'j'k'}$:

$$T_{ijk;i'j'k'} = \frac{1}{2a^2}\left[\frac{T^{(\xi)}_{ii'}\delta_{jj'} + T^{(\eta)}_{jj'}\delta_{ii'}}{\sqrt{(\xi_i^2 - \eta_j^2)(\xi_{i'}^2 - \eta_{j'}^2)}}\delta_{kk'} + \frac{T^{(\varphi)}_{kk'}\delta_{ii'}\delta_{jj'}}{(\xi_i^2 - 1)(1 - \eta_j^2)}\right] \tag{4.23}$$

where the partial matrices $T^{(\xi)}_{ii'}$, $T^{(\eta)}_{jj'}$, and $T^{(\varphi)}_{kk'}$ related to the coordinates ξ, η, and φ, respectively, have quite simple expressions [50].

The time-dependent Kohn–Sham equations (4.16) are solved by means of the split-operator method with spectral expansion of the propagator matrices [43–45,51]. We employ the following split-operator, second-order short-time propagation formula:

$$\begin{aligned}\psi_{n\sigma}(\mathbf{r}, t + \Delta t) &= \exp\left(-\frac{i}{2}\Delta t\,\widehat{H}_0\right) \\ &\times \exp\left(-i\Delta t\,V(\mathbf{r}, t + \frac{1}{2}\Delta t)\right) \\ &\times \exp\left(-\frac{i}{2}\Delta t\,\widehat{H}_0\right)\psi_{n\sigma}(\mathbf{r}, t) + O((\Delta t)^3).\end{aligned} \tag{4.24}$$

Here Δt is the time propagation step, $\widehat{H}_0$ is the unperturbed electronic Hamiltonian which includes the kinetic energy and the the effective potential before the laser field switched on,

$$\widehat{H}_0 = -\frac{1}{2}\nabla^2 + v_{\mathrm{eff},\sigma}(\mathbf{r},0). \tag{4.25}$$

The potential $V(\mathbf{r},t)$ describes the interaction with the laser field and can be expressed as follows:

$$V(\mathbf{r},t) = v_{\mathrm{eff},\sigma}(\mathbf{r},t) - v_{\mathrm{eff},\sigma}(\mathbf{r},0). \tag{4.26}$$

It contains the direct interaction with the field $v_{\mathrm{ext}}(\mathbf{r},t)$ (4.20) as well as terms due to the variation of the density. For the field polarized under the angle γ with respect to the molecular axis, the direct interaction can be expressed as follows, using the prolate spheroidal coordinates:

$$\begin{aligned} v_{\mathrm{ext}}(\xi,\eta,\varphi,t) = aF(t)\Big(&\xi\eta\cos\gamma + \\ &\sqrt{(\xi^2-1)(1-\eta^2)}\cos\varphi\sin\gamma\Big)\,. \end{aligned} \tag{4.27}$$

Note that Eq. (4.24) is different from the conventional split-operator techniques [33,52], where $\widehat{H}_0$ is usually chosen to be the kinetic energy operator and $\hat{V}$ the remaining Hamiltonian depending on the spatial coordinates only. The use of the *energy*-representation in Eq. (4.24) allows the explicit *elimination* of the undesirable fast-oscillating *high-energy* components and speeds up considerably the time propagation [35,43,51]. For the given Δt, the propagator matrix $\exp\left(-\frac{i}{2}\Delta t\,\widehat{H}_0\right)$ is time-independent and constructed only once from the spectral expansion of the unperturbed Hamiltonian $\widehat{H}_0$ before the propagation process starts. The matrix $\exp\left(-i\Delta t\,V(\mathbf{r},t+\frac{1}{2}\Delta t)\right)$ is time-dependent and must be calculated at each time step. However, for interaction with the laser field in the length gauge, this matrix is diagonal (as any multiplication by the function of the coordinates in the GPS and FG methods), and its calculation is not time-consuming.

4.3.2. *Exploration of the underlying mechanisms for high harmonic generation of H_2 in intense laser fields*

In this section we show an application of the TDOEP/KLI-SIC procedure to the study of high-order harmonic generation of H_2 in intense pulsed laser fields. First we discuss the field-free electronic structure calculations using the steady-state OEP/KLI-SIC procedure [16,43], and the GPS procedure [43,51] is extended to discretize the molecular Hamiltonian in the prolate spheroidal coordinates. For H_2, the calculated ground-state energy is -1.1336 a.u. (using LSDA exchange energy functional only) and -1.1828 a.u. (including both LSDA exchange and correlation energy functionals), the latter is within 1% of the exact value of -1.174448 a.u. If the GGA energy functional such as that of BLYP [1] is used, the calculated ground-state energy is improved to -1.17444 a.u.

Consider now H_2 molecules subject to an intense laser field with the wavelength 1064 nm, $\sin^2$ pulse shape, and 20 optical cycles in pulse length, linearly polarized along the molecular axis. The time-dependent xc potential is constructed by means of the time-dependent OEP/KLI-SIC procedure using the adiabatic LSDA exchange and correlation energy functional. In the following, we shall focus our discussion on the HHG process of H_2 molecules from the ground vibrational state with the internuclear separation R fixed at the equilibrium distance ($R = 1.4$ a.u.).

The solution of the TDOEP/KLI-SIC equation is performed by means of the TDGPS method described above. To explore the detailed spectral and temporal structure of HHG and the underlying mechanisms in different energy regimes, one can perform the time-frequency analysis by means of the wavelet transform [53,54] of the induced dipole,

$$A_W(t_0, \omega) = \int d(t) W_{t_0,\omega}(t) dt \equiv d_\omega(t), \tag{4.28}$$

with the wavelet kernel $W_{t_0,\omega}(t) = \sqrt{\omega} W(\omega(t - t_0))$. For the harmonic emission, a natural choice of the mother wavelet is given by the Morlet wavelet [54]

$$W(x) = (1/\sqrt{\tau}) e^{ix} e^{-x^2/2\tau^2}. \tag{4.29}$$

Figure 4.2 shows the modulus of the time-frequency profiles of H_2 (at $R = 1.4a_0$) in (1064 nm, 20 o.c., $\sin^2$ pulse shape, and 10^{14} W/cm^2) laser fields, revealing striking and vivid details of the spectral and temporal structures. Several salient features are noticed. (a) First, for the lowest few harmonics, the time profile (at a given frequency) shows a smooth function of the driving laser pulse. This is an indication that the *multiphoton* mechanism dominates this lower harmonic regime. In this regime, the probability of absorbing N-photons is roughly proportional to I^N, and I (laser intensity) is proportional to $E(t)^2$. (b) Second, the smooth time profile is getting shorter (in time duration) and broadened (in frequency) as the

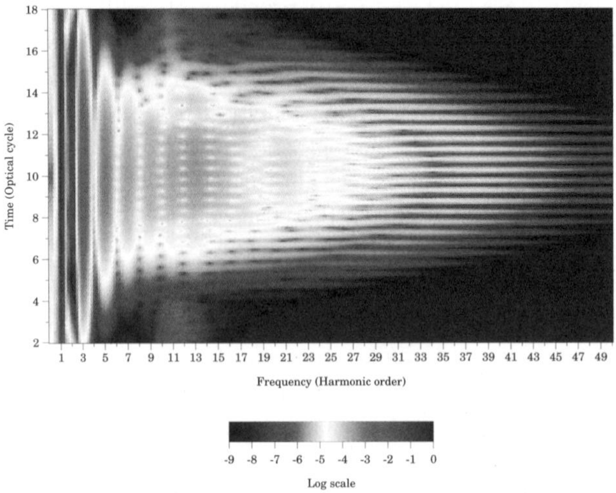

Fig. 4.2. The time-frequency spectra (modulus) of H_2 (at $R = 1.4a_0$) in (1064 nm, 20 o.c., $\sin^2$ pulse shape) laser fields with peak intensity 10^{14} W/cm^2. The plot density is in logarithmic scale (in the powers of 10). For color image, see Fig. 5 of Ref. [43].

harmonic order is increased, as is evident in Fig. 4.2 from the 1st to the 7th harmonics. As the harmonic order is further increased, the time profiles (see particularly the 11th harmonic in Fig. 4.2) develop extended fine structures. This can be attributed to the effect of excited states and the onset of the ionization threshold. (c) Third, for those high harmonics in the plateau regime well above the ionization threshold, the most prominent feature is the development of fast burst time profiles. At a given time, we see that such bursts actually form a *continuous frequency profile* in Fig. 4.2. This is a clear evidence of the existence of the bremsstrahlung radiation emitted by each recollision of the electron wavepacket with the parent ionic core(s). In contrast, we find that the (multiphoton-dominant) lowest-order harmonics form a *continuous time profile* at a given frequency. In the intermediate energy regime where both multiphoton and tunneling mechanisms contribute, the time-frequency profiles show a net-like structure.

4.3.3. *Multiphoton ionization and high-order harmonic generation of N_2*

The exact form of the exchange-correlation (xc) potential $v_{\mathrm{xc},\sigma}(\mathbf{r}, t)$ is unknown. However, high-quality approximations to the xc potential are becoming available. When these potentials, determined by time-independent ground-state DFT, are used along with TDDFT in the electronic structure calculations, both inner shell and excited states can be calculated rather accurately [56]. In the time-dependent calculations, we adopt the commonly used adiabatic approximation, where the xc potential is calculated with the time-dependent density. The adiabatic approximation had recently many successful applications to atomic and molecular processes in intense external fields [55,56]. For the studies of the diatomic molecules [50,57], we utilize the LBα (van Leeuwen–Baerends) xc potential [58]:

$$v_{\mathrm{xc},\sigma}^{\mathrm{LB}\alpha}(\mathbf{r},t) = \alpha v_{\mathrm{x},\sigma}^{\mathrm{LSDA}}(\mathbf{r},t) + v_{\mathrm{c},\sigma}^{\mathrm{LSDA}}(\mathbf{r},t) - \frac{\beta x_\sigma^2(\mathbf{r},t)\rho_\sigma^{1/3}(\mathbf{r},t)}{1+3\beta x_\sigma(\mathbf{r},t)\ln\{x_\sigma(\mathbf{r},t)+[x_\sigma^2(\mathbf{r},t)+1]^{1/2}\}}. \tag{4.30}$$

The LBα potential contains two parameters, α and β, which have been adjusted in time-independent DFT calculations of several molecular systems and have the values $\alpha = 1.19$ and $\beta = 0.01$ [58]. The first two terms in Eq. (4.30), $v_{\mathrm{x},\sigma}^{\mathrm{LSDA}}$ and $v_{\mathrm{c},\sigma}^{\mathrm{LSDA}}$ are the exchange and correlation potentials within the local spin density approximation (LSDA). The last term in Eq. (4.30) is the gradient correction with $x_\sigma(\mathbf{r}) = |\nabla\rho_\sigma(\mathbf{r})|/\rho_\sigma^{4/3}(\mathbf{r})$, which ensures the proper long-range asymptotic behavior $v_{\mathrm{xc},\sigma}^{\mathrm{LB}\alpha} \to -1/r$ as $r \to \infty$. The potential (4.30) has proved to be reliable in molecular TDDFT studies [59,60]. The correct long-range asymptotic behavior of the LBα potential is crucial in photoionization problems since it allows to reproduce accurate molecular orbital energies, and the proper treatment of the molecular continuum.

In our calculations of multiphoton processes in N_2, we used the laser wavelength 800 nm ($\omega_0 = 0.056954$ a.u.) and the sine-squared envelope with 20 optical cycles. The propagation procedure based on Eq. (4.24) is applied sequentially starting at $t = 0$ and ending at $t = T$. As a result, the spin orbitals $\psi_{n\sigma}(\xi,\eta,\varphi,t)$ are obtained on a uniform time grid within the interval $[0,T]$. The space domain is finite with the linear dimension restricted by the end point R_b. We choose $R_b = 40$ a.u.; the corresponding space volume contains all relevant physics for the laser field parameters used in the calculations. Between 20 a.u. and 40 a.u., we apply an absorber which smoothly brings down the wave function for each spin orbital without spurious reflections. Absorbed parts of the wave packet localized beyond 20 a.u. describe unbound states populated during the ionization process. Because of the absorber, the normalization integrals of the wave functions $\psi_{n\sigma}(\mathbf{r},t)$ decrease in time. Calculated after the pulse, they determine the

Table 4.1. Absolute values of spin orbital energies of N_2 and Ar. (A) Present DFT calculations (eV). (B) Experimental ionization energies (eV).

Molecule	Spin-orbital	A	B	
N_2	$2\sigma_u$	18.5	18.7	(Ref. [62])
	$1\pi_u$	16.9	17.2	(Ref. [62])
	$3\sigma_g$ (HOMO)	15.5	15.6	(Ref. [62])
Ar	$3s$	29.0	29.3	(Ref. [63])
	$3p$	15.3	15.8	(Ref. [63])

ionization probabilities $\mathcal{P}_{n\sigma}^{(\mathrm{i})}$ for each spin orbital:

$$\mathcal{P}_{n\sigma}^{(\mathrm{i})} = 1 - \int d^3r |\psi_{n\sigma}(\mathbf{r}, T)|^2. \tag{4.31}$$

We note that the quantities $\mathcal{P}_{n\sigma}^{(\mathrm{i})}$ represent the ionization probabilities for the electron occupying the unperturbed $\psi_{n\sigma}(\mathbf{r}, t = 0)$ spin orbital *before* the laser pulse. The total ionization probability $\mathcal{P}^{(\mathrm{i})}$ can be calculated as follows:

$$\mathcal{P}^{(\mathrm{i})} = 1 - \prod_{n\sigma} \left(1 - \mathcal{P}_{n\sigma}^{(\mathrm{i})}\right). \tag{4.32}$$

The total ionization probability as defined by Eq. (4.32) reduces to the sum of the spin orbital probabilities only in the limit of the weak laser field (small $\mathcal{P}_{n\sigma}^{(\mathrm{i})}$).

In the calculations, we used the experimental value of the equilibrium internuclear separation in N_2 (2.074 a.u. [61]). In Table 4.1, we summarize the energies for the spin orbitals that have a significant contribution to MPI and HHG and the corresponding experimental vertical ionization energies. Also presented are the data for the companion Ar atom which has an ionization potential close to that of N_2 and is expected to manifest close ionization probabilities as well. The agreement between the calculated and experimental values is fairly good for all three systems indicating a good quality of the LBα exchange-correlation potential.

4.3.3.1. *Multiphoton ionization*

We present the orientation-dependent MPI probabilities for N_2 molecule at the peak intensity 2×10^{14} W/cm^2 (Fig. 4.3). The orientation dependence of the total MPI probability is in a good accord with the experimental observations [64,65] for this molecule and reflects the symmetry of its

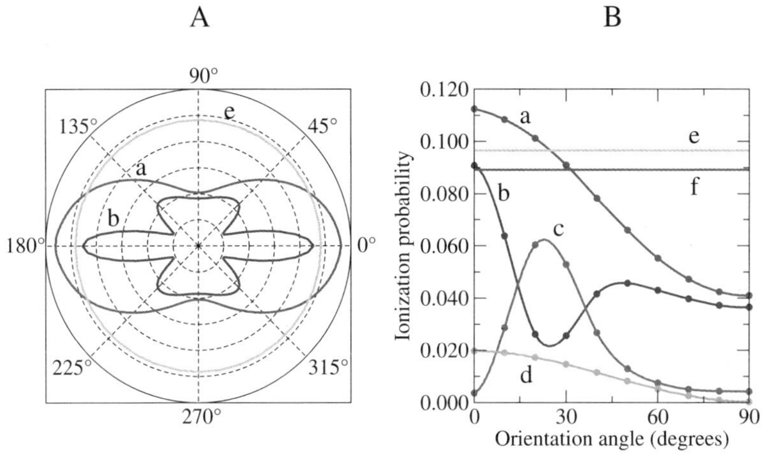

Fig. 4.3. MPI probabilities of N_2 molecule and Ar atom for the peak intensity 2×10^{14} W/cm^2 in polar (panel A) and Cartesian (panel B) coordinates: (a) total probability for N_2, (b) $3\sigma_g$ (HOMO) probability for N_2, (c) $1\pi_u$ (HOMO−1) probability for N_2, (d) $2\sigma_u$ probability for N_2, (e) total probability for Ar, and (f) $3p_0$ probability for Ar.

highest-occupied molecular orbital (HOMO): the maximum corresponds to the parallel orientation. However, multielectron effects are quite important for N_2, particularly at intermediate orientation angles. In the angle range around 30°, the orbital probability of HOMO−1 ($1\pi_u$) is larger than that of HOMO ($3\sigma_g$). Despite the orbital probabilities have local minima and maxima, the total probability shows monotonous dependence on the orientation angle. With increasing the peak intensity of the laser field, the orientation angle distribution of the total ionization probability becomes more isotropic. For comparison, we show also the ionization probability of the Ar atom. As one can see from Fig. 4.3, the absolute values of the ionization probabilities of N_2 and Ar are close to each other. However, the inner shell contributions are less important for Ar: the total probability is dominated by the highest-occupied ($3p$) shell contribution. An analysis of the spin orbital energies (Table 4.1) can help to understand the relative importance of MPI from the inner shells in N_2 compared to that in Ar. The smaller the ionization potential of the electronic shell, the easier it can be ionized. That is why HOMO is generally expected to give the main contribution to the MPI probability. However, in N_2 the ionization potential of HOMO−1 is quite close to that of HOMO (the difference between the calculated values is 1.4 eV), and in the strong enough laser field both shells show comparable ionization probabilities (a possible resonance between HOMO and HOMO−1 in the 800 nm laser field also favors that; see discussion of HHG in Sec. 4.3.3.2 below). At the same time, the gap between the $3p$ and $3s$ spin orbital energies in Ar is much larger (our calculation gives the value 13.7 eV), and the $3p$ contribution to the MPI probability remains dominant for all three laser intensities.

4.3.3.2. *High-order harmonic generation*

For non-monochromatic fields, the spectral density of the radiation energy emitted for all the time is given by the following expression [66]:

$$S(\omega) = \frac{4\omega^4}{6\pi c^3} |\widetilde{\mathbf{D}}(\omega)|^2. \tag{4.33}$$

Here ω is the frequency of radiation, c is the velocity of light, and $\widetilde{\mathbf{D}}(\omega)$ is a Fourier transform of the time-dependent dipole moment:

$$\widetilde{\mathbf{D}}(\omega) = \int_{-\infty}^{\infty} dt \mathbf{D}(t) \exp(i\omega t). \tag{4.34}$$

The dipole moment is evaluated as an expectation value of the electron radius-vector with the time-dependent total electron density $\rho(\mathbf{r}, t)$:

$$\mathbf{D}(t) = \int d^3r \; \mathbf{r}\rho(\mathbf{r}, t). \tag{4.35}$$

For a long enough laser pulse, the radiation energy spectrum (4.33) contains peaks corresponding to odd harmonics of the carrier frequency ω_0. We define the energy $\mathcal{E}(N_\mathrm{h})$ emitted in the N_hth harmonic (N_h is an odd integer number) as follows:

$$\mathcal{E}(N_\mathrm{h}) = \int_{(N_\mathrm{h}-1)\omega_0}^{(N_\mathrm{h}+1)\omega_0} d\omega S(\omega). \tag{4.36}$$

In Fig. 4.4 we present the HHG data for N_2 at the peak intensity 2×10^{14} W/cm^2. The cutoff position in the HHG spectrum for this intensity is expected at the harmonic order 35, in fair agreement with the computed data. To show the orientation dependence of the HHG spectra, we choose three values of the orientation angle γ: 0°, 40°, and 90° which represent the limiting cases of the parallel and perpendicular orientation as well as the intermediate angle case. The orientation dependence of HHG resembles that of MPI: HHG is more intense for the orientations where MPI reaches its maximum. It is clearly seen that the HHG signal at 0° is dominant in the low-order and central parts of the spectrum whereas for higher harmonics a stronger signal is observed at 40°. One can also see that the emission of the harmonic radiation at the perpendicular orientation ($\gamma = 90°$) is suppressed

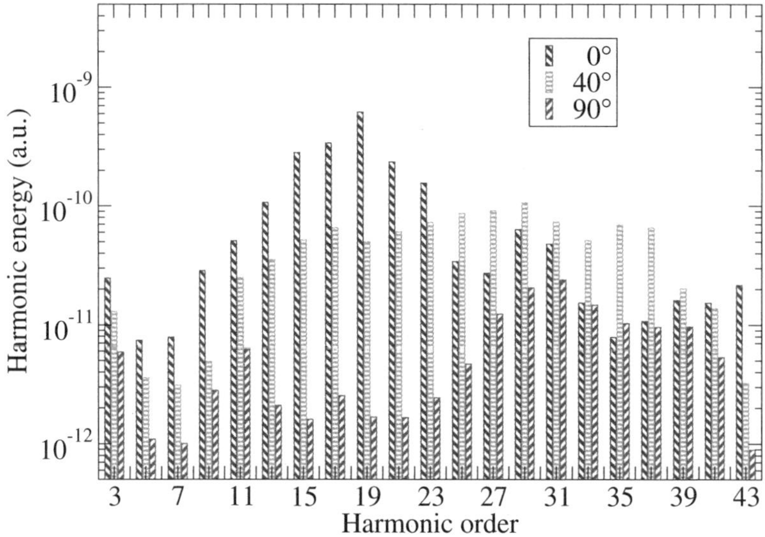

Fig. 4.4. Energy emitted in harmonic radiation by N_2 molecule for the peak intensity 2×10^{14} W/cm^2: left bar, orientation angle $\gamma = 0°$; middle bar, orientation angle $\gamma = 40°$; right bar, orientation angle $\gamma = 90°$.

in the low-order and central parts of the HHG spectrum. The maximum in the harmonic energy distribution at 90° is shifted to higher orders. This result is in a good accord with the recent experimental measurements on N_2 [67].

4.3.4. *Multielectron effects on the orientation dependence of multiphoton ionization of CO_2*

In this section, we present all-electron TDDFT calculations of the orientation-dependent MPI of the three-center CO_2 molecule [68]. The electronic structure of CO_2 is solved with the help of the Voronoi-cell finite difference (VFD) method [69]. In contrast to the ordinary finite difference method with regular uniform grids, the VFD method can accommodate any type of grid distributions, so-called unstructured grids, with the help of geometrical flexibility of the Voronoi diagram. To attack multicenter Coulombic singularity in all-electron calculations of polyatomic molecules, highly adaptive molecular grids are used [68] in this study.

Table 4.2 compares experimental vertical ionization potentials [70] of CO_2 and absolute values of orbital binding energies computed with the LBα

Table 4.2. Absolute values of spin-orbital energies of CO_2. (A) Present DFT calculations (eV). (B) Experimental vertical ionization energies (eV).

Spin-orbital	A	B
$1\pi_g$ (HOMO)	13.9	13.8 (Ref. [70])
$1\pi_u$	17.5	17.6 (Ref. [70])
$3\sigma_u$	17.2	18.1 (Ref. [70])
$4\sigma_g$	18.5	19.4 (Ref. [70])

potential. Molecular grids are constructed by a combination of spherical atomic grids covering large distances ($r_{\max}$ ~20 Å). The C–O bond length is fixed at 1.162 Å [71]. As one can see from Table 4.2, the calculated orbital binding energies are in fairly good agreement with the experimental data, particularly those for HOMO ($1\pi_g$) and HOMO−1 ($1\pi_u$).

Figure 4.5 shows the orientation dependence of the total ionization probability. The laser parameters used are 20-optical-cycle $\sin^2$-envelope laser pulses with two different sets of the wavelength and the peak intensity: (a) 820 nm and 1.1×10^{14} W/cm^2, and (b) 800 nm and 5×10^{13} W/cm^2. For comparison, Fig. 4.5 includes experimental measurements [64,65] and MO–ADK results [64,72]. All data sets are normalized to their maximum value. In Fig. 4.5(a), two dashed lines of experiment are due to uncertainty of the measured alignment distribution [64]. The total ionization probability computed by TDDFT manifests the center-fat propeller shape with the peak at 40°. The position of the peak agrees well with both experiments [64,65] which give it at 45°. As for the broadness of the central pattern, our results agree well with the data of Ref. [65] but are different from that of Ref. [64], the latter showing a narrower pattern. This discrepancy may be related to the experimental uncertainty in the molecular alignment processes.

We now examine contributions of the individual orbitals to the total ionization probability. As the calculations at 820 nm and 1.1×10^{14} W/cm^2 reveal [68], HOMO ($1\pi_g$) is dominant in the total ionization and other contributions ($1\pi_u$, $3\sigma_u$, and $4\sigma_g$) are about 10 times smaller. Contributions of $2\sigma_u$ and $3\sigma_g$ are negligible. In fact, the unperturbed π orbitals are degenerate: one lies on the xz-plane ($1\pi_{g,x}$ and $1\pi_{u,x}$) and the other lies on the yz-plane ($1\pi_{g,y}$ and $1\pi_{u,y}$). As the field whose polarization vector varies in the xz-plane is applied to the molecule, $1\pi_{g,x}$ provides the most dominant contribution to the orientation dependence of the total ionization probability, which has the maximum at 45°; $1\pi_{g,y}$ shows a dumbbell shape

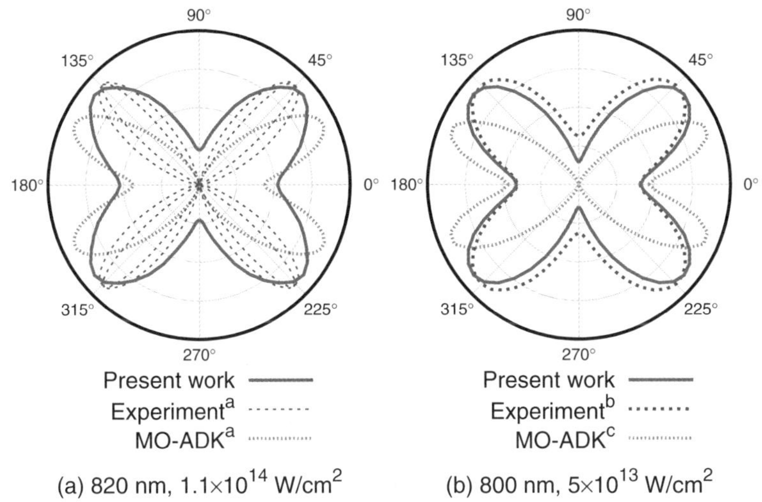

Fig. 4.5. Orientation dependence of total ionization probability of CO_2.
[a]Ref. [64];
[b]Ref. [65];
[c]Ref. [72].

with the same probability as $1\pi_{g,x}$ at 0°. Thus the center-fat propeller shape of the total ionization probability in Fig. 4.5 is mostly reflected by contributions of the two HOMOs ($1\pi_{g,x}$ and $1\pi_{g,y}$). On the other side, the MO–ADK model predicts the butterfly shape with the peak at 25° [64, 72], in large deviation from the experimental data. In contrast with MO–ADK model, the self-interaction-free TDDFT approach [68] incorporates multi-electron correlation and multiple orbital effects, and the results are in excellent agreement with experimental observation. The TDDFT [68] results show the significance of the electron correlations and suggest that all the valence orbitals should be taken into account, even when HOMO dominates the ionization process.

4.4. Conclusions

In this article, we have presented *self-interaction-free* TDDFT approaches recently developed for accurate and efficient treatment of the time-dependent dynamics of many-electron quantum systems. They allow the construction of orbital-independent single-particle local potential which is self-interaction free and possesses the correct $(-1/r)$ long-range asymptotic behavior. With the self-interaction-free potential, the energy of the highest occupied spin-orbital provides a good approximation to the ionization potential. The generalized pseudospectral (GPS) technique allows the construction of non-uniform and optimal spatial grids, denser mesh nearby each nucleus and sparser mesh at longer range, leading to high-precision solution of both electronic structure and time-dependent quantum dynamics with the use of only a modest number of spatial grid points. The self-interaction-free TDDFT formalism along with the use of the time-dependent GPS numerical technique provides a powerful new nonperturbative time-dependent approach for in-depth and quantitative exploration of the effects of electron correlation and multiple orbitals on strong field multiphoton processes at an unprecedented detail. We note that an alternative and complementary approach not discussed in this article, is the generalized Floquet formulation of TDDFT and TD-current DFT [73–76]. This is a *time-independent* approach for nonperturbative treatment of many-electron quantum systems in *periodic* or *quasi-periodic* time-dependent fields. Further extension of these approaches to large molecular systems is currently in progress.

Acknowledgments

This work was partially supported by the Chemical Sciences, Geosciences and Biosciences Division of the Office of Basic Energy Sciences, Office of Sciences, Department of Energy, and by the National Science Foundation. We also would like to acknowledge the partial support of National Science Council of Taiwan (Grant No. 97-2112-M-002-003-MY3) and National Taiwan University (Grants Nos. 98R0045 and 99R80870).

References

1. R. G. Parr and W. T. Yang, *Density-Function Theory of Atoms and Molecules* (Oxford University Press, New York, 1989).
2. N. H. March, *Electron Density Theory of Atoms and Molecules* (Academic Press, San Diego, 1992).
3. E. K. U. Gross, F. J. Dobson and M. Petersilka, in *Density Functional Theory* (Springer, New York, 1996), p.81.
4. P. Hohenberg and W. Kohn, Inhomogeneous electron gas, *Phys. Rev.* **136** (3B), B864–B871 (1964).
5. W. Kohn and L. J. Sham, Self-consistent equations including exchange and correlation effects, *Phys. Rev.* **140** (4A), A1133–A1138 (1965).

6. A. Görling, Density-functional theory for excited states, *Phys. Rev. A.* **54** (5), 3912–3915 (1996).
7. Z. Y. Zhou and S. I. Chu, Spin-dependent localized Hartree–Fock density-functional approach for the accurate treatment of inner-shell excitation of closed-shell atoms, *Phys. Rev. A.* **75** (1), 014501 (4) (2007).
8. M. K. Harbola, Density-functional approach to obtaining excited states: study of some open-shell atomic systems, *Phys. Rev. A.* **65** (5), 052504 (6) (2002).
9. A. K. Roy and S. I. Chu, Density-functional calculations on singly and doubly excited Rydberg states of many-electron atoms, *Phys. Rev. A.* **65** (5), 052508 (9) (2002).
10. M. Slamet, R. Singh, L. Massa and V. Sahni, Quantal density-functional theory of excited states: The state arbitrariness of the model noninteracting system, *Phys. Rev. A.* **68** (4), 042504 (9) (2003).
11. S. J. Vosko, L. Wilk and M. Nusair, Accurate spin-dependent electron liquid correlation energies for local spin density calculations: a critical analysis, *Can. J. Phys.* **58** (8), 1200–1211 (1980).
12. A. D. Becke, Density-functional exchange-energy approximation with correct asymptotic behavior, *Phys. Rev. A.* **38** (6), 3098–3100 (1988).
13. C. Lee, W. Yang and R. G. Parr, Development of the Colle-Salvetti correlation-energy formula into a functional of the electron density, *Phys. Rev. B.* **37** (2), 785–789 (1988).
14. J. P. Perdew and Y. Wang, Accurate and simple density functional for the electronic exchange energy: generalized gradient approximation, *Phys. Rev. B.* **33** (12), 8800–8802 (1986).
15. Q. Zhao and R. G. Parr, Local exchange-correlation functional: numerical test for atoms and ions, *Phys. Rev. A.* **46** (9), R5320–R5323 (1992).
16. X. M. Tong and S. I. Chu, Density-functional theory with optimized effective potential and self-interaction correction for ground states and autoionizing resonances, *Phys. Rev. A.* **55** (5), 3406–3416 (1997).
17. J. P. Perdew and A. Zunger, Self-interaction correction to density-functional approximations for many-electron systems, *Phys. Rev. B.* **23** (10), 5048–5079 (1981).
18. E. Runge and E. K. U. Gross, Density-functional theory for time-dependent systems, *Phys. Rev. Lett.* **52** (12), 997–1000 (1984).
19. E. K. U. Gross and W. Kohn, Local density-functional theory of frequency-dependent linear response, *Phys. Rev. Lett.* **55** (26), 2850–2852 (1985).
20. A. Zangwill and P. Soven, Density-functional approach to local-field effects in finite systems: photoabsorption in the rare gases, *Phys. Rev. A.* **21** (5), 1561–1572 (1980).
21. G. Onida, L. Reining and A. Rubio, Electronic excitations: density-functional versus many-body Green's-function approaches, *Rev. Mod. Phys.* **74** (2), 601–659 (2002).
22. For a recent review, see, K. Burke, J. Werschnik and E. K. U. Grosset, Time-dependent density functional theory: past, present, and future, *J. Chem. Phys.* **123** (6), 062206 (9) (2005), and references therein.

23. V. P. Osinga, S. J. A. van Gisbergen, J. G. Snijders and E. J. Baerends, Density functional results for isotropic and anisotropic multipole polarizabilities and C_6, C_7, and C_8 Van der Waals dispersion coefficients for molecules, *J. Chem. Phys.* **106** (12), 5091–5101 (1997).
24. K. Yabana and G. F. Bertsch, Application of the time-dependent local density approximation to optical activity, *Phys. Rev. A.* **60** (2), 1271–1279 (1999).
25. J. R. Chelikowsky, L. Kronik and I. Vasiliev, Time-dependent density-functional calculations for the optical spectra of molecules, clusters, and nanocrystals, *J. Phys. Condens. Matter.* **15** (35), R1517–R1547 (2003).
26. F. Krausz and M. Ivanov, Attosecond physics, *Rev. Mod. Phys.* **81** (1), 163–234 (2009).
27. P. B. Corkum and F. Krausz, Attosecond science, *Nature Phys.* **3**, 381–387 (2007).
28. A. L'Huillier, K. J. Schafer and K. C. Kulander, Theoretical aspects of intense field harmonic generation, *J. Phys. B.* **24** (15), 3315–3341 (1991).
29. J. L. Krause, K. J. Schafer and K. C. Kulander, Calculation of photoemission from atoms subject to intense laser fields, *Phys. Rev. A.* **45** (7), 4998–5010 (1992).
30. X. M. Tong and S. I. Chu, Time-dependent density-functional theory for strong-field multiphoton processes: application to the study of the role of dynamical electron correlation in multiple high-order harmonic generation, *Phys. Rev. A*, **57** (1), 452–461 (1998).
31. J. B. Krieger, Y. Li and G. F. Iafrate, Construction and application of an accurate local spin-polarized Kohn–Sham potential with integer discontinuity: exchange-only theory, *Phys. Rev. A.* **45** (1), 101–126 (1992).
32. C. A. Ullrich, U. J. Gossmann and E. K. U. Gross, Time-dependent optimized effective potential, *Phys. Rev. Lett.* **74** (6), 872–875 (1995).
33. M. R. Hermann and J. A. Fleck, Jr., Split-operator spectral method for solving the time-dependent Schrödinger equation in spherical coordinates, *Phys. Rev. A.* **38** (12), 6000–6012 (1988).
34. K. C. Kulander, Time-dependent Hartree–Fock theory of multiphoton ionization: helium, *Phys. Rev. A.* **36** (6), 2726–2738 (1987).
35. X. M. Tong and S. I. Chu, Theoretical intense study of multiple high-order harmonic generation by ultrashort pulsed laser fields: a new generalized pseudospectral time-dependent method, *Chem. Phys.* **217** (2,3), 119–130 (1997).
36. G. Yao and S. I. Chu, Generalized pseudospectral methods with mappings for bound and resonance state problems, *Chem. Phys. Lett.* **204** (3,4), 381–388 (1993).
37. J. Wang, S. I. Chu and C. Laughlin, Multiphoton detachment of H^-. II. Intensity-dependent photodetachment rates and threshold behavior — complex-scaling generalized pseudospectral method, *Phys. Rev. A.* **50** (4), 3208–3215 (1994).
38. I. P. Christov, J. Zhou, J. Peatross, A. Rundquist, M. M. Murnane and H. C. Kapteyn, Nonadiabatic effects in high-harmonic generation with ultrashort pulses, *Phys. Rev. Lett.* **77** (9), 1743–1746 (1996).

39. Z. Chang, A. Rundquist, H. Wang, M. M. Murnane and H. C. Kapteyn, Generation of coherent soft X rays at 2.7 nm using high harmonics, *Phys. Rev. Lett.* **79** (16), 2967–2970 (1997).
40. M. Schnürer, Ch. Spielmann, P. Wobrauschek, C. Streli, N. H. Burnett, C. Kan, K. Ferencz, R. Koppitsch, Z. Cheng, T. Brabec and F. Krausz, Coherent 0.5-keV X-ray emission from helium driven by a sub-10-fs laser, *Phys. Rev. Lett.* **80** (15), 3236–3239 (1998).
41. N. Sarukura, K. Hata, T. Adachi, R. Nodomi, M. Watanabe and S. Watanabe, Coherent soft-x-ray generation by the harmonics of an ultrahigh-power KrF laser, *Phys. Rev. A.* **43** (3), 1669–1672 (1991).
42. X. M. Tong and S. I. Chu, Multiphoton ionization and high-order harmonic generation of He, Ne, and Ar atoms in intense pulsed laser fields: self-interaction-free time-dependent density-functional theoretical approach, *Phys. Rev. A.* **64** (1), 013417 (8) (2001).
43. X. Chu and S. I. Chu, Self-interaction-free time-dependent density-functional theory for molecular processes in strong fields: high-order harmonic generation of H_2 in intense laser fields, *Phys. Rev. A.* **63** (2), 023411 (10) (2001).
44. D. A. Telnov and S. I. Chu, *Ab initio* study of the orientation effects in multiphoton ionization and high-order harmonic generation from the ground and excited electronic states of H_2^+, *Phys. Rev. A.* **76** (4), 043412 (10) (2007).
45. D. A. Telnov and S. I. Chu, *Ab initio* study of high-order harmonic generation of H_2^+ in intense laser fields: time-dependent non-Hermitian Floquet approach, *Phys. Rev. A.* **71** (1), 013408 (10) (2005).
46. X. Chu and S. I. Chu, Time-dependent density-functional theory for molecular processes in strong fields: study of multiphoton processes and dynamical response of individual valence electrons of N_2 in intense laser fields, *Phys. Rev. A.* **64** (6), 063404 (9) (2001).
47. M. Abramowitz and I. Stegun (eds.), *Handbook of Mathematical Functions* (Dover, New York, 1965).
48. L. Tao, C. W. McCurdy and T. N. Rescigno, Grid-based methods for diatomic quantum scattering problems: a finite-element discrete-variable representation in prolate spheroidal coordinates, *Phys. Rev. A.* **79** (1), 012719 (9) (2009).
49. C. C. Marston and G. G. Balint-Kurti, The Fourier grid Hamiltonian method for bound state eigenvalues and eigenfunctions, *J. Chem. Phys.* **91** (6), 3571–3576 (1989).
50. D. A. Telnov and S. I. Chu, Effects of multiple electronic shells on strong-field multiphoton ionization and high-order harmonic generation of diatomic molecules with arbitrary orientation: an all-electron time-dependent density-functional approach, *Phys. Rev. A.* **80** (4), 043412 (12) (2009).
51. X. Chu and S. I. Chu, Complex-scaling generalized pseudospectral method for quasienergy resonance states in two-center systems: application to the Floquet study of charge resonance enhanced multiphoton ionization of molecular ions in intense low-frequency laser fields, *Phys. Rev. A.* **63** (1), 013414 (10) (2001).
52. M. D. Feit, J. A. Fleck, Jr. and A. Steiger, Solution of the Schrödinger equation by a spectral method, *J. Comput. Phys.* **47** (3), 412–433 (1982).

53. X. M. Tong and S. I. Chu, Probing the spectral and temporal structures of high-order harmonic generation in intense laser pulses, *Phys. Rev. A.* **61** (2), 021802 (4) (2000).
54. C. K. Chui, *An Introduction to Wavelets* (Academic Press, New York, 1992).
55. S. I. Chu, Recent development of self-interaction-free time-dependent density-functional theory for nonperturbative treatment of atomic and molecular multiphoton processes in intense laser fields, *J. Chem. Phys.* **123** (6), 062207 (16) (2005).
56. M. A. L. Marques, C. A. Ullrich, F. Nogueira, A. Rubio, K. Burke and E. K. U. Gross (eds.), *Time-Dependent Density Functional Theory* (Springer, Berlin, 2006).
57. D. A. Telnov and S. I. Chu, Effects of electron structure and multielectron dynamical response on strong-field multiphoton ionization of diatomic molecules with arbitrary orientation: an all-electron time-dependent density-functional-theory approach, *Phys. Rev. A.* **79** (4), 041401(R) (4) (2009).
58. P. R. T. Schipper, O. V. Gritsenko, S. J. A. van Gisbergen and E. J. Baerends, Molecular calculations of excitation energies and (hyper)polarizabilities with a statistical average of orbital model exchange-correlation potentials, *J. Chem. Phys.* **112** (3), 1344–1352 (2000).
59. X. Chu and S. I. Chu, Role of the electronic structure and multielectron responses in ionization mechanisms of diatomic molecules in intense short-pulse lasers: an all-electron *ab initio* study, Phys. Rev. A **70** (6), 061402(R) (4) (2004).
60. J. Heslar, J. Carrera, D. Telnov and S. I. Chu, High-order harmonic generation of heteronuclear diatomic molecules in intense ultrashort laser fields: an all-electron TDDFT study, *Int. J. Quant. Chem.* **107** (15), 3159–3168 (2007).
61. K. P. Huber and G. Herzberg, *Molecular spectra and molecular structure. IV. Constants of diatomic molecules* (Van Nostrand Reinhold, New York, 1979).
62. A. Lofthus and P. H. Krupenie, The spectrum of molecular nitrogen, *J. Phys. Chem. Ref. Data* **6** (1), 113–307 (1977).
63. K. Siegbahn, C. Nordling, G. Johansson, J. Hedman, P. F. Hedén, K. Hamrin, U. Gelius, T. Bergmark, L. O. Werme, R. Manne and Y. Baer, *ESCA Applied to Free Molecules* (North Holland, Amsterdam, 1969).
64. D. Pavičić, K. F. Lee, D. M. Rayner, P. B. Corkum and D. M. Villeneuve, Direct measurement of the angular dependence of ionization for N_2, O_2, and CO_2 in intense laser fields, *Phys. Rev. Lett.* **98** (24), 243001 (4) (2007).
65. I. Thomann, R. Lock, V. Sharma, E. Gagnon, S. T. Pratt, H. C. Kapteyn, M. M. Murnane and W. Li, Direct measurement of the angular dependence of the single-photon ionization of aligned N_2 and CO_2, *J. Phys. Chem. A.* **112** (39), 9382–9386 (2008).
66. L. D. Landau and E. M. Lifshitz, *The classical theory of fields.* (Pergamon Press, Oxford, 1975).
67. B. K. McFarland, J. P. Farrell, P. H. Bucksbaum and M. Gühr, High harmonic generation from multiple orbitals in N_2, *Science.* **322**, 1232–1235 (2008).

68. S. K. Son and S. I. Chu, Multielectron effects on the orientation dependence and photoelectron angular distribution of multiphoton ionization of CO_2 in strong laser fields, *Phys. Rev. A* **80** (1), 011403(R) (4) (2009).
69. N. Sukumar, Voronoi cell finite difference method for the diffusion operator on arbitrary unstructured grids, *Int. J. Numer. Methods Eng.* **57** (1), 1–34 (2003).
70. D. W. Turner, C. Baker, A. D. Baker and C. R. Brundle, *Molecular photoelectron spectroscopy* (Wiley, London, 1970).
71. G. Herzberg, *Molecular spectra and molecular structure, III. Electronic spectra and electronic structure of polyatomic molecules* (Van Nostrand, New York, 1966).
72. A. T. Le, X. M. Tong and C. D. Lin, Alignment dependence of high-order harmonic generation from CO_2, *J. Mod. Opt.* **54** (7), 967–980 (2007).
73. D. A. Telnov and S. I. Chu, Floquet formulation of time-dependent density functional theory, *Chem. Phys. Lett.* **264**, 466–476 (1997).
74. D. A. Telnov and S. I. Chu, Generalized Floquet theoretical formulation of time-dependent density functional theory for many-electron systems in multicolor laser fields, *Int. J. Quant. Chem.* **69** (3), 305–315 (1998).
75. D. A. Telnov and S. I. Chu, Generalized Floquet formulation of time-dependent current-density-functional theory, *Phys. Rev. A.* **58** (6), 4749–4756 (1998).
76. D. A. Telnov and S. I. Chu, Multiphoton above-threshold detachment of Li^-: exterior-complex-scaling – generalized-pseudospectral method for calculations of complex-quasienergy resonances in Floquet formulation of time-dependent density-functional theory, *Phys. Rev. A.* **66** (4), 043417 (13) (2002).

Chapter 5

Nanomaterials in Nanomedicine

Renat R. Letfullin
Rose-Hulman Institute of Technology
letfullin@rose-hulman.edu

Thomas F. George
University of Missouri–St. Louis
tfgeorge@umsl.edu

Nanoparticles are being researched as a noninvasive method for selectively killing cancer cells. With particular antibody coatings on nanoparticles, they attach to the abnormal cells of interest (cancer or otherwise). Once attached, nanoparticles can be heated with UV/visible/IR or RF pulses, heating the surrounding area of the cell to the point of death. Researchers often use single-pulse or multipulse lasers when conducting nanoparticle ablation research. The laser heating of nanoparticles is very sensitive to the time structure of the incident pulsed laser radiation, the time interval between the pulses, and the number of pulses used in the experiments. In this chapter, time-dependent simulations and detailed analyses are carried out for different nonstationary pulsed laser-nanoparticle interaction modes, and the advantages and disadvantages of multipulse (set of short pulses) and single-pulse laser heating of nanoparticles are shown. A comparative analysis for both radiation modes (single-pulse and multipulse) are discussed for laser heating of metal nanotargets on nanosecond, picosecond and femtosecond time scales to make recommendations for efficient laser heating of nanomaterials in nanomedicine experiments.

5.1. Introduction

In recent years, there has been a tremendous increase in research at the nanoscale for materials. One particular area is the application of nanoparticles to enhance the diagnostic and treatment methods available for cancer [1–7]. The application of nanotechnology for laser thermal-based killing of abnormal cells (e.g., cancer cells) targeted with absorbing nanoparticles (e.g., gold solid nanospheres, nanoshells or nanorods) is becoming an extensive area of research. Studies have shown that by coating the surface of nanoparticles with a specific protein (a 'targeting agent,' normally an antibody), the nanoparticles will bind to a complementary protein such as found on a cancer cell [2,6–11]. After the nanoparticles are bound to the cancerous cells, they can be heated with electromagnetic radiation, inducing a variety of effects around the particles [6,8,10,11]. The heated particle can cause the cell to experience hyperthermia, resulting in surface-protein denaturing [12] and changing membrane permeability [13]. Alternatively, the nanoparticle itself can heat to the point of melting, evaporation or explosion [4,5], causing further damage to cells. These effects can be used to increase the sensitivity of photoacoustic diagnosis or aid in therapy, such as selective photothermolysis, by selective thermal killing of tumor cells into which absorbing nanoparticles have been incorporated. The potential advantages of these new photothermal sensitizers heated with short laser pulses may include:

- selective cancer-cell targeting by means of conjugation of absorbing particles (e.g., gold nanospheres, nanoshells or nanorods) with specific antibodies;
- localized tumor damage without harmful effects on surrounding healthy tissue;
- absorption at longer wavelengths in the transparency window of most biotissues;
- no undesired side effects (e.g., cytotoxicity or cutaneous photosensitivity); and

- relatively fast treatment involving potentially just one or several laser pulses.

Progress towards the development of selective nano-photothermolysis technology requires the investigation of new physical concepts and new approaches to the study of short/ultrashort laser pulse interactions with biological systems containing nanostructures. The extent of the particle heating depends on many factors, divided primarily into three categories: (1) particle material, size and shape; (2) laser pulse wavelength, energy density, duration, pulse shape and generation frequency; and (3) properties of the surrounding medium.

Even with the many factors involved in an experimental setup that contribute to the individual heating characteristics of a nanoparticle, several factors stay relatively constant across many experimental situations. Specifically, spherical gold nanoparticles are commonly used because they are relatively easy to fabricate, nontoxic, easily conjugated to antibodies, and strong absorbers [14]. Nanosecond pulse-width lasers with pulse firing frequencies of 10 Hz are often used because they are widely available and cheaper than pico- and femtosecond lasers or lasers with higher firing frequencies. Additionally, water or a phosphate buffered saline (PBS) solution is used as a surrounding medium due to large similarities with a biological cell. Several experiments that use these factors are studied in our analysis [6,8,11,15–18].

In experiments, Zharov *et al.* [6] used nanoparticle heating for bacterial killing. Spherical gold nanoparticles of diameter 10, 20, and 40 nm were heated with laser light, and it was determined that at 3–5 J/cm^2, only 1–3 pulses are required for harmful effects on bacteria, while at 0.5–1 J/cm^2, at least 100 pulses are required to produce harmful effects. Pitsillides *et al.* [8] explored micro- and nanoparticles for selective cell therapy, finding that for 20 nm diameter gold nanospheres irradiated by 20 ns, 532 nm pulses at 0.5 J/cm^2, there is a much greater correlation between the number of particles attached to the cell than on the number of pulses when considering cellular damage. Zharov *et al.* [11] investigated bubble formation from nanoparticles. Using a smooth distribution of 40 nm diameter gold nanospheres on a cell, they found that 30 pulses of 0.5 J/cm^2 or one pulse of ~ 2 J/cm^2 results in complete cell death. Alternatively, with cluster formations, cell damage occurs after 100 laser pulses at 80 mJ/cm^2. Peng *et al.* [15] focused on the effects of laser irradiation on particle size and peak absorption wavelength. They showed that using a Nd:YAG laser and 21 nm spherical gold nanoparticles, irradiation causes the particles to fragment to an average diameter of 4.9 nm, changing the peak absorption wavelengths. Kalambur *et al.* [16] used iron-oxide nanoparticles to compare RF heating with multipulse laser heating and determined that multipulse lasers produce a higher cellular uptake of particles, enabling greater cellular damage. Hleb and Lapotko [17] used gold nanorods to determine high-energy effects of lasers on nanorods and nanospheres, such as how long the particles maintain photothermal properties (before deterioration) and the effects of multiple pulses in the form of a 'pulse train,' determining that very rapid pulse repetition leads to increased bubble effects. Lastly, Takahashi *et al.* [18] used gold nanorods irradiated with multipulse near-IR lasers to find that selective cell damage is achievable.

One of the greatest potential benefits from multipulse lasers is accumulative heating in the target. If multipulse lasers could be used to quickly accumulate heat in nanoparticles with many consecutive low-energy pulses, then the laser energy density would not be a prohibitive aspect of the treatment. Multiple low-energy density pulses could be used rather than a single high-energy density pulse to achieve the same nanoparticle temperature, sparing healthy cells from excessive heating as well as assuring that the energy density required for treatment is below the medical standard of 100 mJ/cm^2 [11].

The focus of this chapter is to compare and contrast the effects of multipulse (set of short pulses) versus single-pulse laser heating of nanoparticles. The laser heating of nanoparticles is very sensitive to the time structure of the incident pulsed laser radiation: pulse shape and duration, and the number of pulses per unit time. We limit the maximum temperature of the theoretical calculations to the melting point of the material (T_M ~ 1336 K for bulk gold material and remaining above 1100 K for gold particles larger than 5 nm in diameter [19]), so that the particles will not undergo advanced phenomena of heating (evaporation, melting or explosion), but require that the nanoparticles surpass 433 K as required for protein denaturing [14]. Based on theoretical calculations of the time dynamics of nanoparticle heating in various experimental circumstances, we will demonstrate that the 'standard' experimental setup does not benefit from a multipulse over single-pulse laser heating mode of metal nanoparticles for short and ultrashort pulses.

5.2. Nano-optics: Lorentz–Mie formalism

The optimal wavelength of laser radiation and optimal size range of nanoparticles for effective laser killing of cancer cells can be found by using an extended Lorentz-Mie diffraction theory, taking into account the plasmon-resonance absorption effect in metal nanoparticles. In the most general case, calculations based on the Mie theory are reduced to searching for the scattering matrix of j particles, $S^j(\theta, \varphi)$, consisting of four complex functions, $S_i^j(\theta, \varphi)$ (I = 1, . . ,4), describing the amplitude and phase of a scattered scalar wave in any direction. Forward scattering (θ = 0°) contains the attenuation process of an electromagnetic wave, and for the case of spherical particles, $S_3^j = S_4^j = 0$. We can limit the description to a single scattering amplitude function:

$$S^j(0) = S_1^j(0) = S_2^j(0) = \frac{1}{2}\sum_{l=1}^{\infty}(2l+1)(a_l^j + b_l^j). \tag{5.1}$$

The Mie coefficients a_l and b_l contain the characteristics of the dispersal medium and are calculated through the cylindrical Bessel function of the first kind, $\psi_l(y)$, and the Hankel function of the second kind, $\xi_l(\rho)$, both with half-integral indexes:

$$a_l = \frac{\psi_l'(y)\psi_l(\rho) - \tilde{m}\psi_l(y)\psi_l'(\rho)}{\psi_l'(y)\xi_l(\rho) - \tilde{m}\psi_l(y)\xi_l'(\rho)}, \tag{5.2}$$

$$b_l = \frac{\tilde{m}\psi_l'(y)\psi_l(\rho) - \psi_l(y)\psi_l'(\rho)}{\tilde{m}\psi_l'(y)\xi_l(\rho) - \psi_l(y)\xi_l'(\rho)}.$$

Here, $\tilde{m} = m_0/m_1$ is the relative value of the refractive index of the medium; $m_0 = n_0 - i\chi_0$ and $m_1 = n_1 - i\chi_1$ are the complex refractive indices of the particle material and the aqueous suspension, respectively; $\rho = 2\pi r_0/\lambda$ is the Mie parameter; and $y = 2\pi r_0 n_0/\lambda$, $\psi_l(u) = (\pi u/2)^{1/2} J^{(1)}_{l+1/2}$, $\xi_l(u) = (\pi u/2)^{1/2} H^{(2)}_{l+1/2}$, and $\psi_l' = d\psi_l(u)/du$. With knowledge of $S^j(0)$, it is possible to calculate the integrated optical performance of the particles (i.e., the dimensionless efficiency coefficients of scattering, $K^j_{sca}(\rho,\tilde{m}) = \sigma^j_{sca}(\rho,\tilde{m})/\sigma_0$, absorption, $K^j_{abs}(\rho,\tilde{m}) = \sigma^j_{abs}(\rho,\tilde{m})/\sigma_0$, and attenuation, $K^j_{att}(\rho,\tilde{m}) = \sigma^j_{att}(\rho,\tilde{m})/\sigma_0$, of the radiation at a given wavelength) as

$$K^j_{att}(\rho,\tilde{m}) = \frac{4\pi}{k^2}\mathrm{Re}\{S^j(0)\},$$

$$K^j_{sca}(\rho,\tilde{m}) = \frac{2}{\rho^2}\sum_{l=1}^{\infty}(2l+1)\{|a_l^j|^2 + |b_l^j|^2\}, \tag{5.3}$$

$$K^j_{abs}(\rho,\tilde{m}) = K_{ext}(\rho,\tilde{m}) - K_{sca}(\rho,\tilde{m}),$$

where $k = 2\pi/\lambda$ is the wave number, and $\sigma^j_{sca}(\rho,\tilde{m})$, $\sigma^j_{abs}(\rho,\tilde{m})$, $\sigma^j_{att}(\rho,\tilde{m})$ and σ_0 are the scattering, absorption, attenuation and geometric cross-sections of *j* particles, respectively.

In a simulation of electromagnetic wave propagation in a dispersed medium, our previously developed effective algorithm [20] was used. Here, the cylindrical functions of real or imaginary arguments and their derivatives, which occur in the expressions for the Mie coefficients a_l and b_l (see Eq. (5.5)), are calculated as the ratio of the function and its derivative by using the recurrent relationships for direct and inverse recursions. Such an approach allows us to make an effective and accurate determination of the optical properties of a dispersal medium over a broad range of Mie parameters, $\rho = 2\pi r_0 / \lambda$ = 0.001–1500, below the diffraction limit simultaneously with small and large values of the real n_0 and the imaginary χ_0 parts of the refractive index of the particle's and surrounding medium substance.

5.2.1. *Absorption and scattering spectrum*

The Lorentz-Mie formalism (Eqs. (5.1)-(5.3)) requires the use of two dimensionless input parameters, $\rho = 2\pi r_0 / \lambda$ and $\delta = \rho\tilde{m}$, where $\tilde{m}$ is the relative value of the complex refractive index of the nanoparticles in the surrounding medium at the wavelength λ. Computer calculations of the absorption, K_{abs}, and scattering, K_{sca}, coefficients for gold nanoparticles in an aqueous suspension are plotted in Fig. 5.1. Figure 5.1(a) shows the absorption and scattering spectrum of the gold particle in the visible range λ = 400–700 nm, and Fig. 5.1(b) illustrates the dependences of K_{abs} and K_{sca} on particle size for the optimal wavelength. It is evident from Fig. 5.1 that the absorption coefficient K_{abs} has a strong maximum at the wavelength λ = 538.3 nm and for a gold particle radius of 35 nm. Gold nanoparticles are effective absorbers of the laser radiation ($K_{abs} > 1, K_{sca}$) over a wide range of the spectrum, 400–580 nm, with a maximum coefficient K_{abs} = 4.02 at the wavelength λ = 538.3 nm (Fig. 5.1(a)).

5.2.2. *Absorption and scattering vs. particle size*

High absorbance at optimal wavelength is observed also for a relatively large size range of gold nanoparticles (Fig. 5.1(b)). We see that the absorption curve at the level $K_{abs} \geq 1$ has a wide particle radius range of r_0 = 10–210 nm. This means that in this size range the absorption cross-section of gold particles σ_{abs} at the given λ = 538.3 nm exceeds the particle geometric cross-section σ_0.

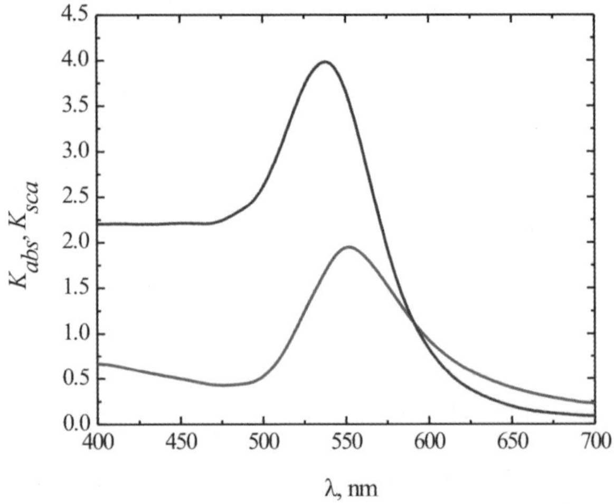

(a)

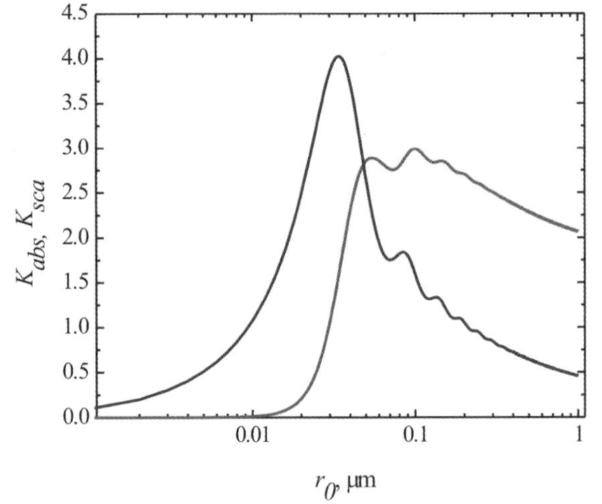

(b)

Fig. 5.1. Computer calculations of the absorption, K_{abs}, and scattering, K_{sca}, coefficients for gold nanoparticles in an aqueous suspension: (a) absorption (upper curve) and scattering (lower curve) spectrum of the gold particle in the visible range λ = 400–700 nm; (b) dependences of K_{abs}, (upper curve) and K_{sca} (lower curve) on particle size for the optimal wavelength.

Light scattering at $\lambda = 538.3$ nm by gold nanoparticles in an aqueous suspension becomes apparent for particles with a radius > 30 nm. For large particle sizes, the scattering coefficient K_{sca} rapidly increases and reaches the value of the absorption coefficient at a particle radius of 50 nm. For particles with a radius ≥ 50 nm, scattering of laser radiation predominates over absorption, and the suspension under consideration containing the nanoparticles becomes a strongly scattering medium at the given wavelength.

For particles with radii (r_0) in the range of 1–45 nm, the absorption coefficient, K_{abs}, is considerably greater than the scattering coefficient, K_{sca} (i.e., the efficiency of laser heating of nanoparticles in this size range is high). Thus, the optimal nanoparticle-size range for effective laser initiation of thermal explosion in tumor cells is 10–45 nm, where $K_{abs} \geq 1$ and $K_{abs} > K_{sca}$. The maximal effect of laser heating of gold particles in an aqueous medium can be achieved for particles with a radius of 35 nm at the wavelength $\lambda = 538.3$ nm ($K_{abs} = 4.02$).

5.2.3. *Optical properties of gold nanoparticle in low-absorbing media*

We examine the effects of the medium refractive index on the light-absorbing properties of gold particles in suspension as well as attached to a surface. Four media are used, namely water, cytoplasm, cell membrane and collagen. The refractive indices of the media at the different wavelengths and the results of the Mie theory calculations of the maximum light — absorption factor (K_{abs}), wavelength (λ_{max}) and radius (r_{max}) for gold particles in media of different refractive indexes — are listed in Table 5.1. Results of computer simulations of the absorption coefficient, K_{abs}, as a function of the wavelength and radius of gold nanoparticles in various biological media are plotted in Fig. 5.2. Figure 5.2(a) shows the absorption spectrum of the gold particle over the visible range λ = 400–700 nm, and Fig. 5.2(b) illustrates the dependences of K_{abs} on particle size for different surrounding media.

Table 5.1. Mie theory results for the absorption factor (K_{abs}), wavelength (λ_{max}) and radius (r_{max}) for gold particles in media of different refractive indexes used in the simulations.

λ (nm)	Refractive index, N	K_{abs} (max)	λ_{max} (nm)	r_{max} (nm)
	Water			
400	1.339			
500	1.334	4.0195	538.3	35
600	1.332			
700	1.331			
	Cytoplasm			
400	1.35			
500	1.36	4.145	540	33
600	1.365			
700	1.367			
	Blood			
400	1.354			
500	1.4	4.376	542	30
600	1.39			
700	1.383			
	Cell Membrane			
400	1.54			
500	1.5	4.743	547	28
600	1.46			
700	1.4			
	Collagen			
400	1.6			
500	1.55	4.954	549	26
600	1.5			
700	1.45			

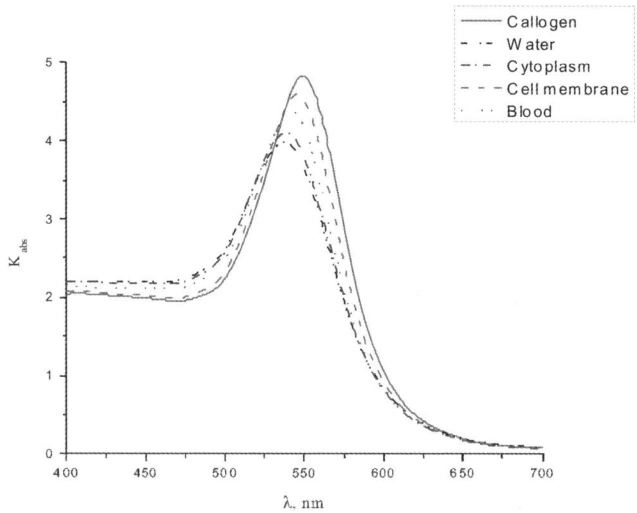

(a)

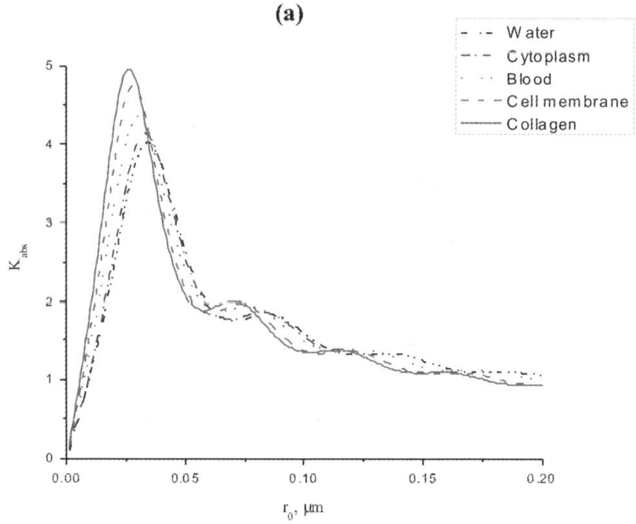

(b)

Fig. 5.2. (a) Absorption spectrum of the gold nanoparticle over the visible range $\lambda = 400$–700 nm. (b) Absorption efficiency K_{abs}, as a function of particle size for different surrounding media.

As follows from our calculations, the optical properties of the gold nanoparticles in low-absorbing biological media are substantially different from the case of the surrounding water medium. The maximal peak of the absorption factor increases continuously with increase in the refractive index of the surrounding media, and the color of the incident light changes from green to red, i.e., the red-shift effect of the absorption maximum with refractive index (see Fig. 5.2(a)). Maximum absorption by gold nanoparticles is observed for the surrounding collagen medium, where the absorption cross-section of gold nanoparticle exceeds its geometric cross-section by a factor of five for the optimal wavelength 549 nm. The presence of the low-absorbing biological surroundings also shifts the optimal radius of the gold nanoparticle for effective absorption in a smaller size region from 35 nm for the water medium to 26 nm for the collagen case (see Table 5.1). Thus, the absorption efficiency of gold nanoparticles is considerably higher in the low-absorbing biological media in comparison to the surrounding water medium.

5.3. Kinetics of Heating and Cooling of Nanoparticles in Biological Environment

5.3.1. *Time dynamics of the nanoparticle's temperature*

In the above section on nano-optics, we used the generalized Lorenz–Mie diffraction theory to determine the absorption characteristics of spherical particles in various biological media. Using the complex indexes of refraction for the medium and particle, as well as the particle size, we determined the absorption efficiency of the spherical particle at a given wavelength. The wavelength of interest in each case depends on the particular laser used in the experiments discussed above.

In this section, we use the heat transfer equation with several simplifying assumptions to calculate the particle temperature as a function of time. It has been shown previously that a one-temperature model (OTM) is appropriate for metal nanoparticle heating for lasers of pulse durations in the femto-, pico- and nanosecond ranges [21]. The first approximation used in this model is that the electron and lattice temperatures are equal (due to fast transfer of electron heat to the phonon subsystem). This provides the following heat-mass transfer rate equation for a single lattice temperature, $T_s(t,r)$, that describes the laser interaction with the particle and surrounding medium:

$$\frac{d}{dt}T_s(t,r) = \frac{\mu_s(T_s)}{\rho_s C_s(T_s)}\Delta T_s(t,r) + \frac{Q(t,r)}{\rho_s C_s(T_s)} - j_D(T_s)S_0 + \frac{3L}{r_0 C_s(T_s)}\frac{dr_0}{dt} \quad . \tag{5.4}$$

Here, $\mu_s(T_s)$, $C(T_s)$, L, ρ_s and r_0 are, respectively, the heat conductivity of the surrounding medium and the specific heat, evaporation heat, density and radius of the nanoparticle. The second assumption is that the particle is heated uniformly, provided $\tau_D = \frac{r_0^2}{4\chi} < \tau_L$, where τ_D is the time to heat the particle given its radius (r_0) and thermal diffusivity (χ), and laser pulse duration (τ_L). This is appropriate for the heating of metal nanoparticles by nanosecond laser pulses in our analysis and allows the use of a simplified relationship for the power density of the energy generated, $Q(t,r_0)$, in the particle:

$$Q(t,r_0) = \frac{3K_{abs}(r_0,\lambda)I_0 f(t)}{4r_0} \quad . \tag{5.5}$$

Here, $K_{abs}(r_0,\lambda)$ is the absorption efficiency of the nanoparticle, $f(t)$ is the time profile of the laser pulse, and I_0 is the incident intensity of the laser pulse. The experiments analyzed utilize Gaussian beam profiles in the form of $f(t) = e^{-(a*t-b)^2}$; the coefficients are adjusted to create a pulse profile of the pulse duration of the laser used in the experiments, as shown in Fig. 5.3. Another approximation limits heat loss to diffusion from the particle surface into the surrounding medium, defining the energy flux density, $j_D(T_s)$, as

$$j_D(T_s) = \frac{\mu_\infty T_s}{(\alpha+1)r_0^2 C_s(T_s)\rho_s}\left[\left(\frac{T_s}{T_\infty}\right)^{\alpha+1} - 1\right], \tag{5.6}$$

where μ_∞ and T_∞ are, respectively, the heat conductivity and temperature of the surrounding medium at equilibrium. α is an exponent set to make

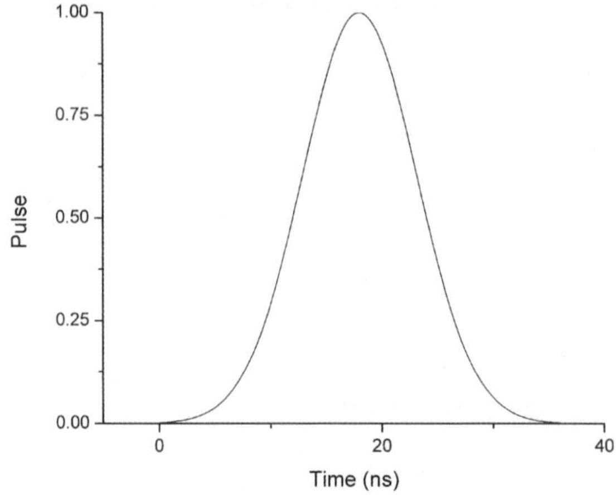

Fig. 5.3. 12 ns Gaussian laser pulse profile from Ref. [6] with coefficients a = 1.388E8 and $b = 2.5$.

the dynamics more realistic between the particle and medium. $\alpha > 1$ relates to high thermophysical characteristics such as metals, while $\alpha < 1$ relates to low thermophysical characteristics such as insulators; $\alpha = 1$ is a good approximation for biological media, which are of interest in the present paper. The last assumption is that the particle temperature remains below the temperature of evaporation. The result, Eq. (5.7), determines the particle temperature in terms of the energy accumulation from the incident laser radiation in the first term and energy loss due to heat diffusion into the medium in the second term:

$$\frac{d}{dt}T_s = \frac{3K_{abs}(r_0,\lambda)I_0 f(t)}{4r_0\rho_s C_s(T_s)} - \frac{\mu_\infty T_s}{(\alpha+1)r_0^2 C_s(T_s)\rho_s}\left[\left(\frac{T_s}{T_\infty}\right)^{\alpha+1} - 1\right]. \qquad (5.7)$$

5.3.2. *Comparison of single-pulse and multipulse modes of heating*

To assure confidence of our model, we compare the absorption efficiency as well as the maximum particle temperature calculated by Pitsillides *et al.* [8] with our own calculated values. For a 30 nm diameter spherical gold particle irradiated by a 20 ns, 532 nm pulse at 0.5 J/cm^2, Pitsillides determine K_{abs} = 2 and a maximum particle temperature of 2500 K. Using the Lorenz–Mie diffraction theory, we calculate K_{abs} = 1.94 (Fig. 5.4), and with the OTM (Eq. (5.7)) we compute the maximum temperature to be on the order of 2350 K (Fig. 5.5). Using a similar approach, we determine the absorption efficiency and time-temperature profile for each of the situations intended for the analysis.

The summary data obtained from experimental papers [6,8,11,15–18] can be found in Table 5.2, while the additional input data we use for our calculations are listed in Table 5.3. We have made modifications to the energy density when the particle temperature exceeds the melting point of the material to reduce the particle temperature into our range of interest. Peng *et al.* [15] use an incident energy density (0.13 J/cm^2) at a wavelength of 532 nm that causes particles to exceed the melting point resulting in fragmentation. To make a case for reducing the incident energy density, we perform calculations to determine how changing the energy density would change the time-temperature profile of the nanoparticle heating and cooling. Our calculations show that changing the energy density has no effect on the time it takes for the particle to heat to the maximum temperature. Figure 5.6 demonstrates this with three different time-temperature profiles: a 21 nm diameter gold sphere irradiated by a 0.13 J/cm^2 pulse (solid) and 0.0325 J/cm^2 pulse (dash), and a 4.9 nm diameter gold sphere irradiated by a 0.13 J/cm^2 pulse (dot). Changing the energy density from a 0.13 J/cm^2 pulse to a 0.0325 J/cm^2 pulse has no effect on the time required for the particle to reach its maximum temperature, which occurs 9 ns after the incident pulse in both cases.

The cooling time for the particle is slightly affected by the magnitude of the maximum temperature and is thus linked to the energy density. Heating the particles to 1500 K and 800 K, respectively, for 0.13 J/cm^2 and 0.0325 J/cm^2 pulses, we find that the cooling time is only slightly longer for the higher energy density pulse. For the 0.13 J/cm^2 pulse, the particle cools back to the ambient temperature 11.4 ns after the peak, while for the 0.0325 J/cm^2 pulse the particle takes 10.6 ns after the maximum to cool.

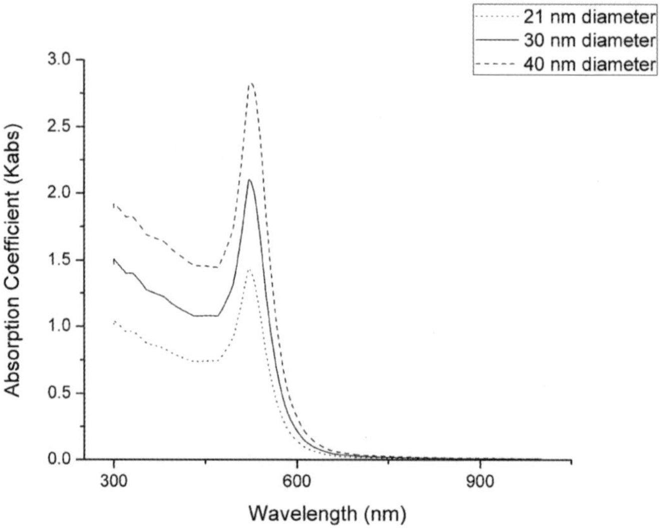

Fig. 5.4. Gold nanosphere absorption efficiency curves in water determined with the Lorenz–Mie diffraction theory.

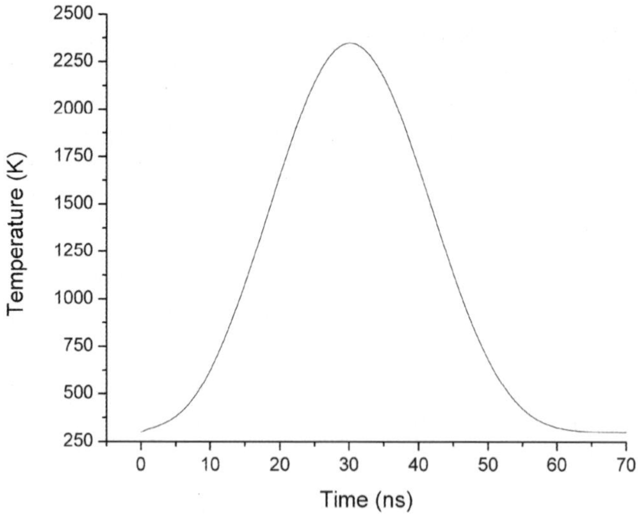

Fig. 5.5. Time-temperature profile of a 30 nm diameter gold nanoparticle in water irradiated by a 20 ns, 0.5 J/cm^2 pulse at 532 nm.

Table 5.2. Experimental data for multipulse laser nanoparticle heating.

Reference	[6]	[8]	[11]	[15]	[16]	[17]	[18]
Laser	Nd:YAG	Nd:YAG	Nd:YAG	Nd:YAG	Nd:YAG	Ti:sapphire	Nd:YAG
Wavelength (nm)	532	565	532	532	532	750	1064
Pulse Energy (J/cm^2)	0.5	0.5	0.5	0.13	0.03	0.95	.002
Pulse duration (ns)	12	20	12	6	7	10	7
Frequency of generation (Hz)	10[a]	10[a]	10[a]	10	20	11.8 MHz / 50 MHz	10
Particle							
Material	Gold	Gold	Gold	Gold	Iron oxide	Gold	Gold
Structure	Sphere	Sphere	Sphere	Sphere	Sphere	Rod	Rod
Diameter (nm)	40	30	40	21/4.9	10	23.6[b]	22.76[b]
Aspect ratio	-	-	-	-	-	3.2	5.9
Other							
Number of pulses	100	100	30	3000	600	2	10
Surrounding medium	PBS	PBS	Water	Water	Water	Water	Water

[a] HOYA ConBio MedLite C Series and IV Series Nd:YAG lasers have a maximum generating frequency of 10 Hz, so the generation frequency for lasers when not provided is assumed to be 10 Hz.

[b] It is common to use an effective radius (r_{eff}) when the particle is of nonstandard shape; this is the radius of a sphere that provides the equivalent volume as an arbitrary shape.

Table 5.3. Additional input parameters for our theoretical calculations.

Reference	[6]	[8]	[11]	[15]	[16]	[17]	[18]
Particle							
Calculated absorption efficiency	2.7225[a] (λ = 532 nm)	0.72[a] (λ = 565 nm)	2.72250 (λ = 532 nm)	1.29367 (λ = 532 nm)	1.68663 (λ = 532 nm)	11.5[b] (λ = 725 nm)	15.75[b] (λ = 860 nm)
Specific heat (J/kg-K)	129	129	129	129	937	129	129
Density (kg/cm^3)	0.0193	0.019	0.0193	0.0193	0.00524	0.0193	0.0193
Initial temperature (K)	300	300	300	300	300	300	300
Laser							
Energy density (J/cm^2) (original / revised)	0.5/ 0.01	0.5 / 0.1	0.5 / 0.01	0.13/ 0.0325	0.03	0.95 / 0.01	0.002
Medium							
Thermal conductivity (W/cm-K)	0.0075	0.0075	0.0060	0.0060	0.0060	0.0060	0.00604

[a] Absorption efficiency values calculated for nanoparticles in water due to difficulty in obtaining refractive index data for a PBS solution.
[b] Estimated absorption efficiency from Jain *et al.* [22] based on the effective radius and aspect ratio of the nanorods, using the closest values given based on wavelength, r_{eff}, and aspect ratio.

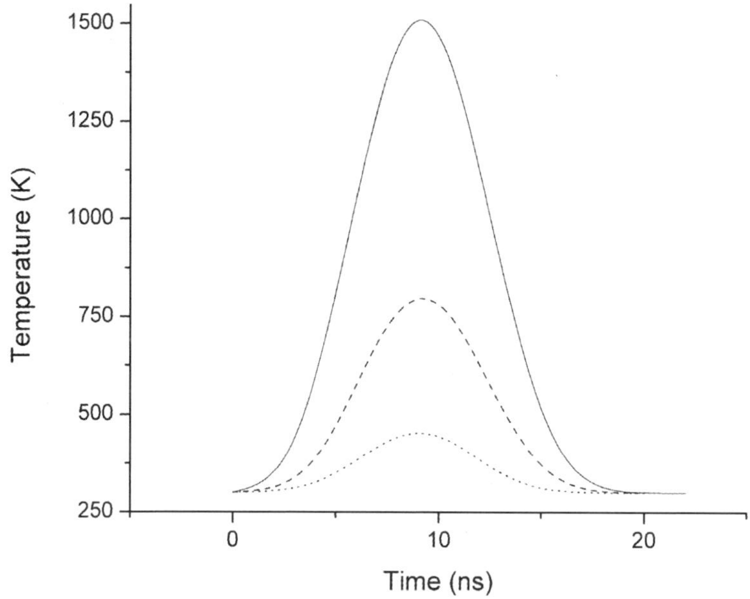

Fig. 5.6. Time-temperature profiles for 21 nm diameter (solid) and 4.9 nm diameter (dot) gold particles in water irradiated by a 6 ns, 0.13 J/cm^2 pulse at 532 nm and a 21 nm diameter gold particle irradiated by the same pulse at 0.0325 J/cm^2 (dash).

5.3.3. *Single-pulse mode*

To analyze the effects due to varying characteristics of the experiments, in this section we compare theoretical calculations for those experiments that have contrasting characteristics, such as pulse duration, particle size and shape, and medium in the single-pulse mode of heating. Considering particle shape — between gold nanorods and gold nanospheres — we extrapolate from Jain *et al.* [22] that changing a particle shape significantly alters the peak absorption wavelengths and the respective coefficients of absorption, but these can be equated to spherical particles with an effective radius and aspect ratio $\left(AR = \frac{Length}{Diameter}\right)$. In Ref. [17], the gold nanorods with r_{eff} = 11.8 nm and AR = 3.2 have an absorption efficiency around 11.5 when irradiated with 725 nm light [22]. Contrasted with gold nanospheres in Ref. [15] (r_0 = 11.5 nm, K_{abs} = 1.29 at λ = 532 nm), a

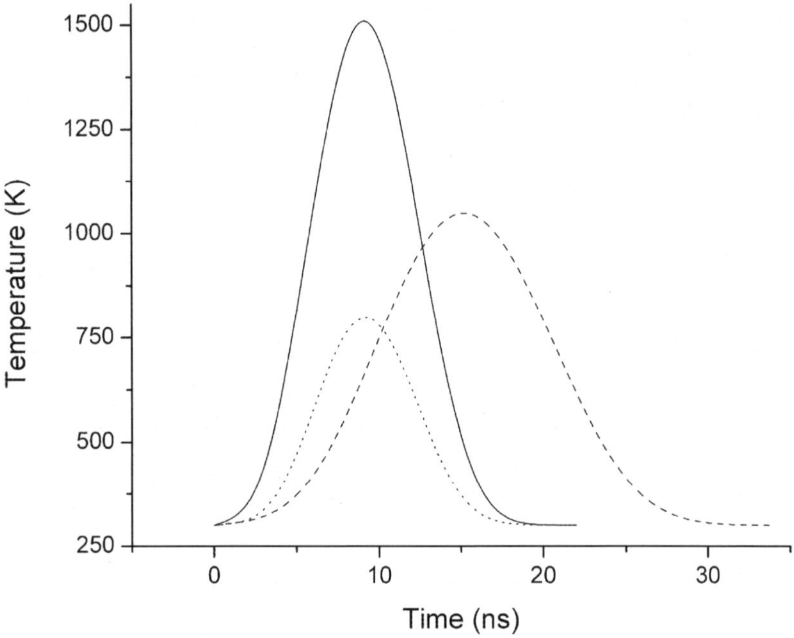

Fig. 5.7. Time-temperature profiles for 21 nm diameter nanospheres with incident laser pulse duration of 6 ns at an energy density of 0.13 J/cm^2 (solid) and 0.0325 J/cm^2 (dot) and nanorods with an effective diameter of 23.6 nm irradiated by a 10 ns, 0.01 J/cm^2 pulse (dash).

much lower energy density can be used to reach the same ablation temperature for the nanorods due to the extremely high absorption efficiency. Figure 5.7 shows the time-temperature profiles for Ref. [15], nanospheres (d = 21 nm) under 0.13 J/cm^2 and 0.0325 J/cm^2 pulses at 532 nm, as well as the temperature profile for the nanorods (r_{eff} = 11.8 nm, AR = 3.2) from Ref. [17] irradiated by 0.01 J/cm^2. This figure demonstrates how a much lower energy density can result in higher nanorod temperatures when compared to nanospheres of equivalent radius. The tunable characteristics of nanorods allows greater adjustment for which wavelength will result in a maximum absorption efficiency and how high the absorption efficiency will be, thus making nanorods potentially more effective for ablative treatments than nanospheres. As we will show in the next section, in terms of accumulative heating with a

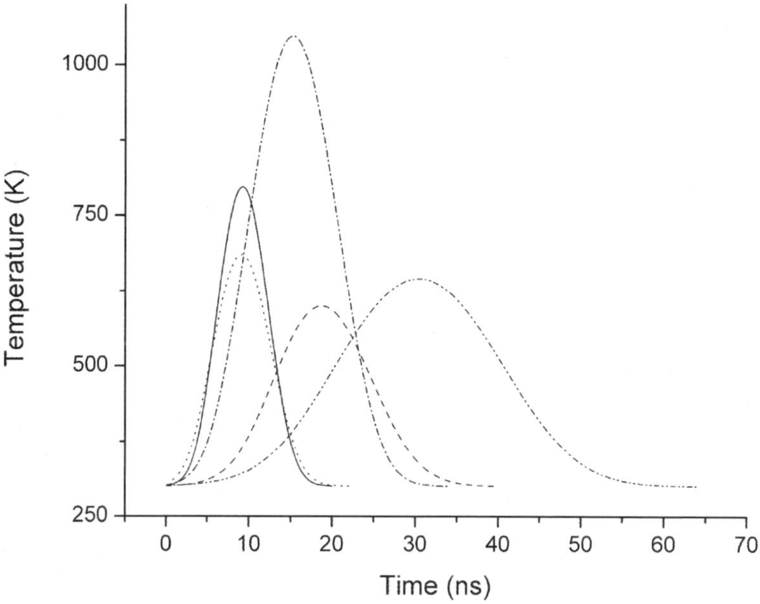

Fig. 5.8. Time-temperature profiles for five different pulse durations used in single-pulse mode heating. Three gold nanosphere cases: Ref. [6] — 12 ns pulse (dash); Ref. [8] — 20 ns (dash-dot-dot); and Ref. [15] — 6 ns (solid). Two gold nanorod cases: Ref. [17] — 10 ns (dash-dot) and Ref. [18] — 7 ns (dot).

multipulse mode of irradiation, the difference between nanorods and nanospheres is not significant.

Another characteristic of the experimental setups [6,8,11,15–18] is a laser pulse duration of roughly 10 ns. Figure 5.8 shows time-temperature profiles for five different pulse durations ranging from 6 ns to 20 ns. It follows from our calculations that during the laser pulse, the transfer of heat from the nanoparticle into the surrounding media is slight, and the particle rapidly reaches its maximum temperature. The heating rate is about 10^{12} Ks^{-1}, depending on the pulse duration and incident energy density. The temperature of the particle continues to rise even after the end of the laser pulse. Using Ref. [17] as an example, the highest temperature of 1048.6 K for a 10 ns pulse is observed at a heating time of 15.2 ns, when the laser pulse has already degraded. After then, the transfer of heat from the particle to the surrounding medium becomes critical, since the energy source is no longer present in the system. The temperature of the particle and surrounding medium remains high (above 425 K) up to 24.6 ns after the pulse is fired, exceeding the laser pulse duration by 2.5 times. The total time for one cycle (heating from the initial temperature of 300 K to a maximum temperature and then cooling back to the initial temperature) is 32.8 ns, over triple the pulse duration. Varying the pulse duration changes the maximum temperature of the particle as well as the time span of heating and cooling.

Due to the nature of the experiments and the intended application of the method for treatment of cancer inside the human body, the medium characteristics cannot be changed. Furthermore, the ability to change the material of the nanoparticle is also limited due to toxicity, but in the case of metal nanoparticles, they should all show similar heating and cooling kinetics to those seen in gold and iron (III) oxide (Fe_3O_4). As we have shown above, the particle size, particle shape and laser energy density do not significantly change the temporal behavior of nanoparticle heating and cooling, but just the magnitude of the maximum temperature reached. The laser pulse duration changes the maximum particle temperature as well as the temporal span of heating and cooling, following relatively predictable multiplicative values with respect to the pulse duration: the maximum temperature is at 1.46 times the pulse duration; the temperature remains above 425 K until 2.2 times the pulse duration; and the entire heating and cooling cycle takes about 3.23 times the pulse duration. These aspects could be useful for further single-pulse investigations, but they do not come into play when considering multipulse heating of nanoparticles discussed below.

5.3.4. *Multipulse mode*

Using the scaled down energy density values determined in previous sections and a function to simulate multiple pulses, we have calculated the time profile of the nanoparticle temperature in a multipulse heating mode to determine what would enable accumulative heating of nanoparticles. To simulate multiple pulses, we use the same function as in the single-pulse simulations, but with the addition of several values of b to shift each consecutive pulse: $f(t) = e^{-(a*t-b_1)^2} + e^{-(a*t-b_2)^2} + e^{-(a*t-b_3)^2}$. Three

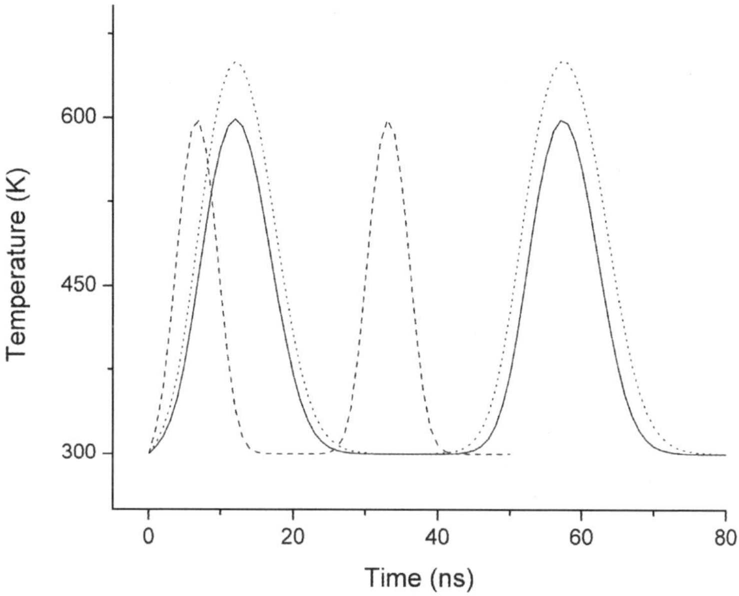

Fig. 5.9. Multipulse time-temperature profiles for Ref. [6] (solid), [11] (dot) and [16] (dash) with modified pulse generation frequencies.

multipulse cases are shown in Fig. 5.9, demonstrating the particle heating and cooling over time for the conditions of Refs. [6,11,16], listed in Table 5.2. These simulations utilize a 37 MHz and 62.5 MHz generation frequency due to the large time gap (relative to the pulse width) between each incident pulse when using the 10 Hz generation frequency specified in the references. These profiles demonstrate visually that even at relatively high generation frequencies, multiple pulses do not produce any accumulative heating effect over time. The particle reaches the peak temperature and falls to equilibrium (300 K) in roughly 3 times the pulse duration, but the time delay between the pulses in experiments exceeds the pulse duration by five to ten million times.

Several of the experimental papers show effects from multipulse laser heating that are not present in the single-pulse laser mode, even though the pulse generation frequency for the multipulse mode is well below the MHz generation frequency. Zharov *et al.* [6] discuss bacterial killing and find that at 3–5 J/cm^2, only 1–3 pulses are required for bacterial damage, while at 0.5–1 J/cm^2 more than 100 pulses are required. We feel the reason for these effects are that at 3–5 J/cm^2 the particles undergo phenomena beyond simple heating and ablation of the surrounding cell due to the extremely high temperature of the nanoparticle, and thus have a large area of effect. In the case of 0.5–1 J/cm^2 pulses, the nanoparticles can still surpass the melting threshold, but without extreme particle phenomena (such as explosion) being observed. In either case, each pulse is incident on ambient temperature particles, raising their temperature to the critical temperature for cell death. Additionally, the energy densities involved are well above the medical standard of 100 mJ/cm^2, so such results are helpful to show the high-energy phenomena and potential treatment options. However, the methods of implementing treatment require refinement before direct medical applications can be developed. Our current investigation to determine the potential of accumulative heating effects in metal nanoparticles is an attempt at localizing damage to cancer cells by using low-energy density lasers while still killing the cancer cells.

After determining most other experimental characteristics (such as pulse duration, particle size, etc.) has no ability to create an accumulative heating phenomenon of interest in metal nanoparticles, the pulse generation frequency used in multipulse scenarios was tested with a high-pulse generation frequency experiment by Hleb *et al.* [17]. This experiment uses two pulses of 10 ns duration that are separated by 20 ns and 150 ns time intervals — 50 MHz and 11.8 MHz, respectively. Figure 5.10 shows three different multipulse calculations based on the experimental setup used in Ref. [17] for gold nanorods (r_{eff} = 11.8 nm, AR = 3.2, K_{abs} = 11.5 at 725 nm) in water irradiated by 0.01 J/cm^2, 10 ns pulses with a generation frequency of 50 MHz, 66.6 MHz and 80 MHz. We exclude the case with pulse separation of 150 ns due to the previously shown rapid cooling of metal nanoparticles (32.8 ns for a 10 ns pulse in Ref. [17]) as well as the experimental result by Hleb *et al.* [17] that there are no additional effects from two pulses separated by 150 ns over a single pulse. Even at the high-pulse generation frequencies shown, the 10 ns pulse does not create an accumulative heating effect in the metal nanoparticles. Furthermore, changing the pulse duration concurrently with high-frequency pulse generation simulations only changes the magnitude of the maximum particle temperature and does

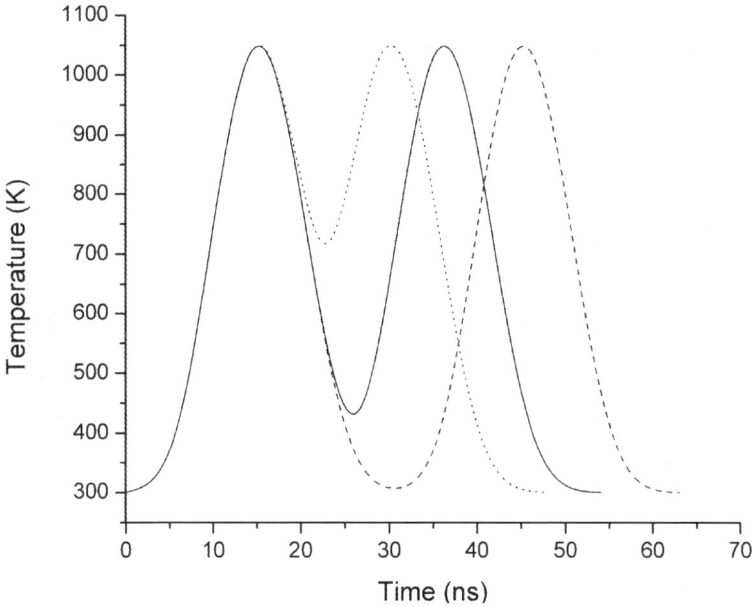

Fig. 5.10. Time-temperature profile for multipulse laser heating of gold nanorods (r_{eff} = 11.8 nm, AR = 3.2, K_{abs} = 11.5 at 725 nm) irradiated by 0.01 J/cm^2, 10 ns pulses at a repetition rate of 50 MHz (dash), 66.6 MHz (solid) and 80 MHz (dot).

not enable the accumulative heating effect in the metal nanoparticles of interest to our investigation.

5.4. Summary

In this chapter, we have performed time-dependent simulations and detailed analyses of different non-stationary laser-nanoparticle interaction modes to determine the accumulative heating potential of metal nanoparticles irradiated by multipulse lasers. Our analysis of the single-pulse mode of heating of metal nanoparticles in a biological cell environment has shown that alterations to the particle size, particle shape and material (metal) all have effects on the magnitude of the maximum particle temperature. Additionally, modifications to the laser energy density or wavelength can also be used to change the maximum particle temperature. The most substantial change within single-pulse heating is the effect of pulse duration on the time dynamics of particle heating and cooling. By increasing the duration, the nanoparticle maximum temperature decreases, but heating and cooling both take longer, allowing a longer time period for the nanoparticle to be above the denaturing temperature of 433 K. This increased duration at a high temperature could prove beneficial to assuring cellular damage. Alternatively, decreasing the pulse duration increases the maximum temperature reached by the nanoparticle while lowering the time for heating and cooling. Such laser parameter modifications could allow for an easier and more refined treatment than modification of particle size or shape, while still having dramatic implications on the effectiveness of the treatment. From our analysis of multipulse metal nanoparticle heating, we have determined that the multipulse mode of heating does not create accumulative heating effects in metal nanospheres or nanorods (diameters of 5–40 nm) due to the heating and cooling kinetics of metal nanoparticles in an aqueous environment.

References

1. T. M. Fahmy, P. M. Fong, A. Goyal and W. M. Saltzman, Targeted for drug delivery, *Nanotoday* (August 18–26, 2005).
2. Y. Fukumori and H. Ichikawa, Nanoparticles for cancer therapy and diagnosis, *Advanced Powder Technol.* **17**, 1–28 (2006).
3. S. Nie, Y. Xing, G. J. Kim and J. W. Simons, Nanotechnology applications in cancer, *Annu. Rev. Biomed. Eng.* **9**, 257–88 (2007).
4. R. R. Letfullin, C. Joenathan, T. F. George and V. P. Zharov, Laser-induced explosion of gold nanoparticles: potential role for nanophotothermolysis of cancer, *Nanomedicine* **1**, 473–480 (2006).
5. R. R. Letfullin, V. P. Zharov, C. Joenathan and T. F. George, Nano-photothermolysis of cancer cells, *SPIE Newsroom (Society of Photo-Optical Instrumentation Engineers)* DOI: 10.1117/2.1200701.0634–1–2 (2007).
6. V. P. Zharov, K. E. Mercer, E. N. Galitovskaya and M. S. Smeltzer, Photothermal nanotherapeutics and nanodiagnostics for selective killing of bacteria targeted with gold nanoparticles, *Biophys. J.* **90**, 619–627 (2006).
7. G. A. Mansoori, P. Mohazzabi, P. McCormack and S. Jabbari, Nanotechnology in cancer prevention, detection, and treatment: bright future lies ahead, *World Review of Science, Technology, and Sustainable Development* **4**, 226–257 (2007).

8. C. M. Pitsillides, E. K. Joe, X. Wei, R. R. Anderson and C. P. Lin, Selective cell targeting with light-absorbing microparticles and nanoparticles, *Biophys. J.* **84**, 4023–4032 (2003).
9. M. J. Vicent, Polymer-drug conjugates as modulators of cellular apoptosis, *AAPS Journal* **9**, E200-E207 (2007).
10. J. Khandare and T. Minko, Polymer-drug conjugates: progress in polymeric prodrugs, *Prog. Polym. Sci.* **31**, 359–397 (2006).
11. V. P. Zharov, R. R. Letfullin and E. N. Galitovskaya, Microbubbles-overlapping mode for laser killing of cancer cells with absorbing nanoparticle clusters, *J. Phys. D: Appl. Phys.* **38**, 2571–2581 (2005).
12. J. R. Lepock, H. E. Frey and K. P. Ritchie, Protein denaturation in intact hepatocytes and isolated cellular organelles during heat shock, *J. Cell Biol.* **122**, 1267–1276 (1993).
13. C. Yao, R. Rahmanzadeh, E. Endl, Z. Zhang, J. Gerdes and G. Hüttmann, Elevation of plasma membrane permeability by laser irradiation of selectively bound nanoparticles, *J. Biomed. Opt.* **10**, 064012–1–8 (2005).
14. V. K. Pustovalov, A. S. Smetannikov and V. P. Zharov, Photothermal and accompanied phenomena of selective nanophotothermolysis with gold nanoparticles and laser pulses, *Laser Phys. Lett.* **5**, 775–792 (2008).
15. Z. Peng, T. Walther and K. Kleinermanns, Influence of intense pulsed laser irradiation on optical and morphological properties of gold nanoparticle aggregates produced by surface acid-base reactions, *Langmuir* **21**, 4249–4253 (2005).
16. V. S. Kalambur, E. K. Longmire and J. C. Bischof, Cellular level loading and heating of superparamagnetic iron oxide nanoparticles, *Langmuir* **23**, 12329–12336 (2007).
17. E. Y. Hleb and D. O. Lapotko, Photothermal properties of gold nanoparticles under exposure to high optical energies, *Nanotechnology* **19**, 1–10 (2008).
18. H. Takahashi, T. Niidome, A. Nariai, Y. Niidome and S. Yamada, Gold nanorod-sensitized cell death: microscopic observation of single living cells irradiated by pulsed near-infrared laser light in the presence of gold nanorods, *Chem. Lett.* **35**, 500–501 (2006).
19. P. Buffat and J. P. Borel, Size effect on the melting temperature of gold particles, *Phys. Rev. A* **13**, 2287–2298 (1976).
20. R. R. Letfullin and V. I. Igoshin, Multipass optical reactor for laser processing of disperse materials, *Quantum Electronics* **25**, 684–689 (1995).
21. R. R. Letfullin, T. F. George, G. C. Duree and B. M. Bollinger, Ultrashort laser pulse heating of nanoparticles: comparison of theoretical approaches, *Advances in Optical Technologies* **2008**, ID 251718–1–8 (2008).
22. P. K. Jain, K. S. Lee, I. H. El-Sayed and M. A. El-Sayed, Calculated absorption and scattering properties of gold nanoparticles of different size, shape, and composition: applications in biological imaging and biomedicine, *J. Phys. Chem. B* **110**, 7238–7248 (2006).

Chapter 6

New Dynamic Modes for Selective Laser Cancer Nanotherapy

Renat R. Letfullin

Rose-Hulman Institute of Technology

letfullin@rose-hulman.edu

Thomas F. George

University of Missouri–St. Louis

tfgeorge@umsl.edu

The application of nanotechnology for laser thermal-based killing of abnormal cells (e.g., cancer cells) targeted with absorbing nanoparticles (e.g., gold solid nanospheres, nanoshells or nanorods) is becoming an extensive area of research. In this chapter, we develop a theory for selective laser nanophotothermolysis of abnormal biological cells with gold nanoparticles and self-assembled nanoclusters. The theory takes into account laser-induced linear and nonlinear synergistic effects in biological cells containing nanostructures, with focus on optical, thermal, bubble formation and nanoparticle explosion phenomena. On the basis of the developed models, we discuss new ideas and new dynamic modes for cancer treatment by laser activated nanoheaters, involving nanoclusters aggregation in living cells, microbubbles overlapping around laser heated intracellular nanoparticles/clusters, and laser thermal explosion mode of single nanoparticles — "*nanobombs*"— delivered to the cells.

6.1. Introduction

Nanophotothermolysis using pulsed lasers and absorbing nanoparticles (e.g., gold solid nanospheres, nanoshells or nanorods) attached to specific targets has recently demonstrated great potential for selective damage to cancer cells [1–19], bacteria [2,10], viruses [10] and DNA [10]. The temperature of gold nanoparticles irradiated by short/ultrashort laser pulses rises very quickly, reaching thresholds for various nonlinear synergetic effects in a cancer cell volume, such as thermal coagulation, vaporization, sound production, bubbles and shock wave generations. These effects can be used to increase the sensitivity of photoacoustic diagnosis or aid in therapy, such as selective nanophotothermolysis, by selective thermal killing of tumor cells into which absorbing nanoparticles have been incorporated.

Under certain delivery and particle-accumulation conditions, nanoparticles can form clusters into the cancer cell volume or cell membrane surface as shown on Fig. 6.1. A cluster is a group of closely-located nanoparticles (10–30 nm each) separated by the thickness of antibodies with the cluster's total size of 200–400 nm, which depends on the radii of the nanoparticles and duration of incubation time. This is the time required to collect the appropriate number of nanoparticles in the cell or cell membrane. The presence of nanoclusters may provide further synergistic enhancement of selective nanophotothermolysis due to:

- increase of the nanocluster's average local absorption;
- red shifting of absorption to the near-IR range (window of transparency of most biotissue [4]) that may be achieved with conventional gold solid nanospheres, which are simpler to prepare, stable and non-toxic, and cheaper compared to nanoparticles with special design (e.g. rods, shells, etc);
- better heating efficiency achieved; and
- decrease in threshold energy for the bubble formation and cell damage.

In this chapter, we make a first attempt to describe theoretically the damage of abnormal cells produced by the nanoclusters, and then compare this to the single-particle case. Two basic mechanisms of the killing of cancer cells in the cluster aggregation mode (CAM) corresponding to a nanosecond laser pulse duration are studied here – thermal damage

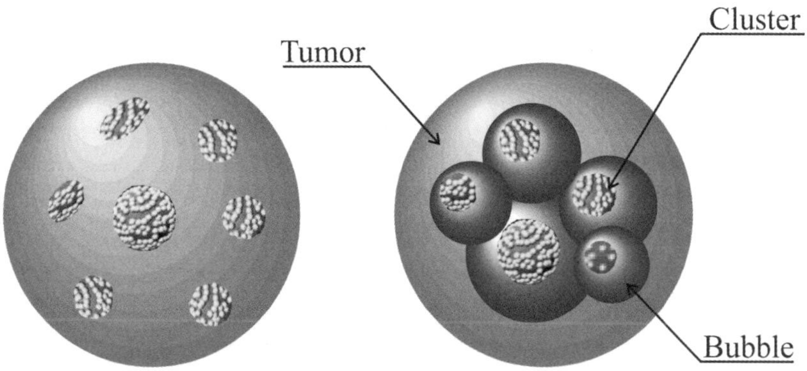

Fig. 6.1. Principle of the nanocluster aggregation mode in selective nanophotothermolysis of cancer.

of abnormal cells by laser-heated cluster, and bubble generation and their overlapping around the nanoparticles/nanocluster as shown on Fig. 6.1.

Utilizing the shorter pulses on the pico- and femtosecond time scales, a new alternative technique for selective killing of cancer cells by thermal explosion of a single or a few nanoparticles can be achieved. When nanoparticles are irradiated by ultrshort laser pulses, their temperature rapidly reaches a critical value for nanoparticle thermal explosion, which we refer to as the "nanobomb" mode of heating, as schematically shown on the Fig. 6.2. This involves the generation of strong shock waves of vapor or acoustic pressure expanding in the cell volume with supersonic velocity. The phenomena accompanying thermal explosion of the nanoparticles can produce a large damage area of the abnormal cell, providing an effective therapeutic effect for cancer cell killing.

In this chapter, we develop a theory for laser-induced linear and nonlinear synergistic effects in biological cells containing nanostructures with focus on cluster aggregation, bubble formation and nanoparticle explosion phenomena. The theory is based on our experience in theoretical studies relevant to the nanoparticles [3,11–18].

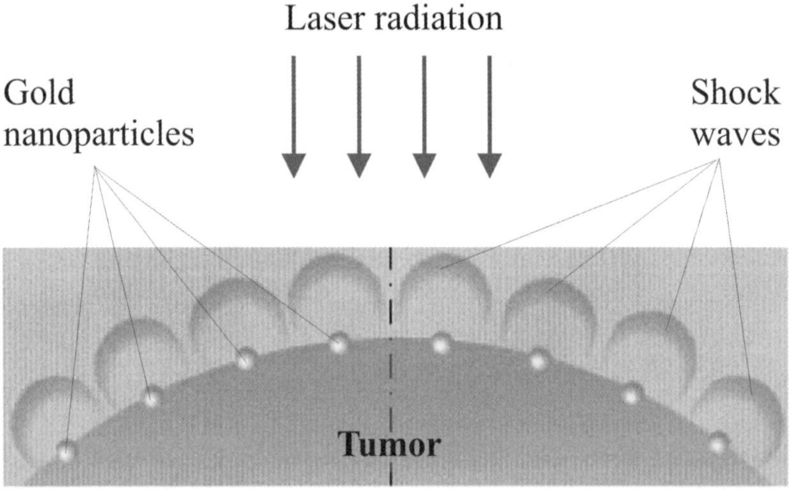

Fig. 6.2. Principle of the thermal explosion mode of nanoparticles –"nanobombs", in selective nanophotothermolysis of cancer.

6.2. Nanocluster Aggregation Mode

Our experiments [3] show that a relatively smooth distribution of single particles is observed for a standard incubation time (15 min) in cells, antibodies and nanoparticle mixtures. The average distance between particles in this case is about 0.5 μm. With increasing incubation time, the number of particles in a cell increases, and the distances between them decrease to 100–300 nm. Further increase in the incubation time leads to the aggregation of particles into clusters (see Fig. 6.1) consisting of closely-located three to seven particles on the cell membrane or inside the cell volume (interparticle spacing becomes about 10–30 nm). A long incubation time leads to the creation of clusters of more significant size (up to 200–300 nm as experimentally observed) consisting of 10–20 particles. Similar results are obtained in model experiments with silver nanoparticles [6,7]. Here, the effective therapeutic effect for cancer cell killing is achieved due to a large damage area at the relatively low energy density of the incident laser pulse.

6.2.1. *Space distribution of temperature fields around the nanoparticle*

For simulations of a space distribution of the temperature in- and outside the nanoparticle, we use here the heat transfer model developed by Goldenberg and Tranter [19] for a uniformly-heated homogeneous sphere embedded in an infinite homogeneous medium. Consider a homogeneous sphere of radius r_0 surrounded by an infinite homogeneous medium, with heat produced in the sphere for time $t > 0$ at the constant rate A per unit time per unit volume. Since the surrounding medium is transparent for the chosen wavelength, we can neglect here thermally-induced changes in the cellular refractive index during the action of nanosecond laser pulse. Both the sphere and medium are initially at zero temperature. Let the suffix 1 refer to the sphere and 2 to the medium, and T, μ, and χ denote the temperature, thermal conductivity and diffusivity respectively. Then the heat transfer equations with boundary conditions are

$$\left.\begin{aligned} \frac{1}{\chi_1}\frac{\partial T_1}{\partial t} &= \frac{1}{r^2}\frac{\partial}{\partial r}\left(r^2\frac{\partial T_1}{\partial r}\right)+\frac{A}{\mu_1}, \quad 0 \le r < r_0, \\ \frac{1}{\chi_2}\frac{\partial T_2}{\partial t} &= \frac{1}{r^2}\frac{\partial}{\partial r}\left(r^2\frac{\partial T_2}{\partial r}\right), \qquad r > r_0, \end{aligned}\right\} \quad t > 0 \tag{6.1}$$

$$T_1 = T_2 = 0, \quad \text{when } t = 0 \tag{6.2}$$

$$\left.\begin{aligned} T_1 &= T_2 \\ \mu_1\frac{\partial T_1}{\partial r} &= \mu_2\frac{\partial T_2}{\partial r} \end{aligned}\right\} \quad \text{for } r = r_0 \tag{6.3}$$

$$T_1 \text{ finite as r} \to 0 \text{ and } T_2 \text{ finite as } r \to \infty. \tag{6.4}$$

Using the Laplace transformation for T, defined by

$$\overline{T} = \int_0^\infty e^{-pt}T(t)dt = L[T(t)],$$

Equations. (6.1)–(6.4) can be reduced to the stationary heat transfer equations:

$$\left.\begin{aligned} &\frac{\partial^2}{\partial r^2}\left(r\overline{T_1}\right) - rq_1^2\overline{T_1} = -\frac{rA}{\mu_1 p}, \; 0 \le r < r_0, \\ &\frac{\partial^2}{\partial r^2}\left(r\overline{T_2}\right) - rq_{21}^2\overline{T_2} = 0, \qquad r > r_0, \end{aligned}\right\} \tag{6.5}$$

$$\left.\begin{aligned} &\overline{T_1} = \overline{T_2} \\ &\mu_1 \frac{\partial \overline{T_1}}{\partial r} = \mu_2 \frac{\partial \overline{T_2}}{\partial r} \end{aligned}\right\} \text{ for } r = r_0 \tag{6.6}$$

$$\overline{T_1} \text{ finite as } r \to 0 \text{ and } \overline{T_2} \text{ finite as } r \to \infty, \tag{6.7}$$

where $q_1^2 = p/\chi_1$, $q_2^2 = p/\chi_2$.

The solution of Eqs. (6.5)–(6.7) for the temperature distribution inside the heated sphere is [19]:

$$\overline{T_1} = \frac{r_0^2 A}{\mu_1}\left\{\begin{aligned} &\frac{1}{3}\frac{\mu_1}{\mu_2} + \frac{1}{6}\left(1 - \frac{r^2}{r_0^2}\right) - \\ &-\frac{2r_0 b}{\pi r}\int_0^\infty \frac{\exp(-y^2 t/\gamma_1)}{y^2}\frac{(\sin y - y\cos y)\sin(ry/r_0)}{[(c\sin y - y\cos y)^2 + b^2 y^2 \sin^2 y]}dy \end{aligned}\right\}, \tag{6.8}$$

$$\text{where } b = \frac{\mu_2}{\mu_1}\sqrt{\frac{\chi_1}{\chi_2}}, \quad c = 1 - \frac{\mu_2}{\mu_1} \text{ and } \gamma_1 = \frac{r_0^2}{\chi_1}. \tag{6.9}$$

The temperature at the center of the sphere is

$$(\overline{T_1})_{r=0} = \frac{r_0^2 A}{\mu_1}\left\{\begin{aligned} &\frac{1}{3}\frac{\mu_1}{\mu_2} + \frac{1}{6} - \\ &-\frac{2b}{\pi}\int_0^\infty \frac{\exp(-y^2 t/\gamma_1)}{y}\frac{(\sin y - y\cos y)\sin(ry/r_0)}{[(c\sin y - y\cos y)^2 + b^2 y^2 \sin^2 y]}dy \end{aligned}\right\}, \tag{6.10}$$

The temperature outside the sphere is

$$\overline{T_2} = \frac{r_0^3 A}{r\mu_1}\left\{\begin{array}{l}\frac{1}{3}\frac{\mu_1}{\mu_2} - \\ -\frac{2}{\pi}\int_0^\infty \frac{\exp(-y^2 t/\gamma_1)}{y^3}\frac{(\sin y - y\cos y)[by\sin y\cos\sigma y - (c\sin y - y\cos y)\sin\sigma y]}{[(c\sin y - y\cos y)^2 + b^2 y^2\sin^2 y]}dy\end{array}\right\}, \tag{6.11}$$

with b, c, γ_l as in Eq. (6.9) and

$$\sigma = \left(\frac{r}{r_0} - 1\right)\sqrt{\frac{\chi_1}{\chi_2}}. \tag{6.12}$$

The heat source is characterized here by constant rate A per unit time per unit volume as

$$A = \frac{E_{abs}}{\tau_{rel}V} = \frac{K_{abs}\varepsilon_L S}{\tau_{rel}V}, \tag{6.13}$$

where E_{abs} is the energy absorbed by the nanoparticle; ε_L is the laser pulse energy density; K_{abs} is the absorption efficiency of the nanoparticle at the given wavelength of laser radiation; S and V are the area and volume of the absorbed center; and τ_{rel} is the thermal relaxation time in the biological medium. By using a spherical symmetry for the absorbing center of radius r_0, we can reduce (6.13) to

$$A = \frac{3K_{abs}\varepsilon_L}{\tau_{rel}r_0}. \tag{6.14}$$

The model described above can be used for simulating the space distribution of the temperature for single or many heat sources in cell volume.

6.2.2. *Laser heating of single nanoparticle in bio-media*

Let us first calculate the laser heating of single gold nanoparticles with radii r_0 = 30–35 nm in the water and surrounding blood bio-media when the incident laser pulse has an energy density of E = 5 J/cm^2 and pulse duration of τ_L = 8 ns. The laser flux is chosen at the level of 5 J/cm^2 to provide cell lethality due to particle heating during one single laser pulse shot, and it is comparable to the laser fluence currently used in clinical treatments of pigmented skin lesions. The results of these simulations are presented in Figs. 6.3 and 6.4. Figure 6.3 illustrates the space behavior of the temperature inside the 30 and 35 nm gold particles heated and cooled in the surrounding water and blood media. Figure 6.4 shows the space distribution of the temperature outside these nanoparticles.

The events that take place after absorption of the laser pulse energy by small particles depend on the size of the locally-heated region and the duration of laser exposure. Long pulses that exceed the thermal relaxation time in biological tissue, $\tau_r = l^2/4\chi_t$ (where χ_t is the thermal diffusivity of the biological tissue, and l is the minimum size of the locally heated region), cause heating of both the particle and the surrounding media. If the size of the locally-heated region l is chosen to be equal to the size of the laser focal spot, which is about 1 μm, the thermal relaxation time for water is 1.75 μs and for blood is 1.56 μs which is much longer than the laser pulse duration (~ 10 ns) used in our research. For ultrashort laser pulses when $\tau_L << \tau_r$, the absorbed energy can be thermally confined within the target, causing rapid heating of the absorber itself. The extreme temperature rise in the absorber can induce explosive vaporization of the nanoparticle, as described in last section of this chapter.

6.2.3. *Internal temperature distribution in single nanoparticle*

The integration of Eqs. (6.8)–(6.10) gives the temperature distribution inside the heated nanoparticle surrounded by the bio-media. Our numerical simulations of the internal 30 nm gold particle's temperature distribution heated by a 5 J/cm^2 laser pulse of 8 ns width in the surrounding water and blood media show that the particle reaches the maximum temperature T_{max} ~ 5789.5 K in water and T_{max} = 6870 K in blood at the center of the sphere. Then we observe a small decrease of the inside temperature by 6 K to the particle's boundaries, establishing the stationary surface temperature of 5783.3 K in water and 6863.6 K in blood (see Fig. 6.3). Since the temperature variation inside the nanoparticle occurs over a small range of 6 K, the internal particle's temperature can be assumed to be constant (Fig. 6.3(b)). Thus, the quasi-stationary and homogeneous approach can be applied to modeling the laser heating and evaporation of the nanoparticles in the biological media.

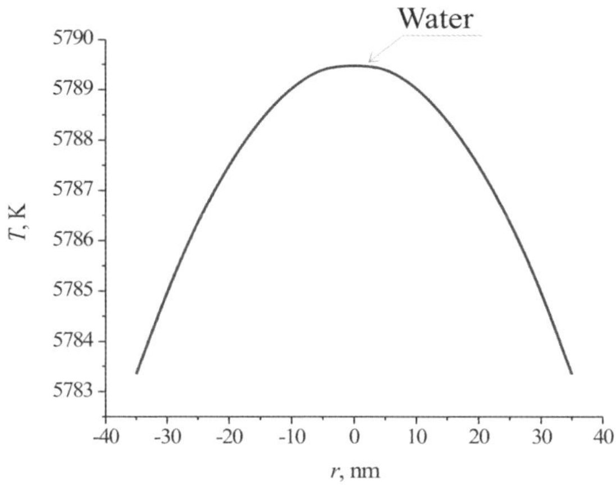

(a)

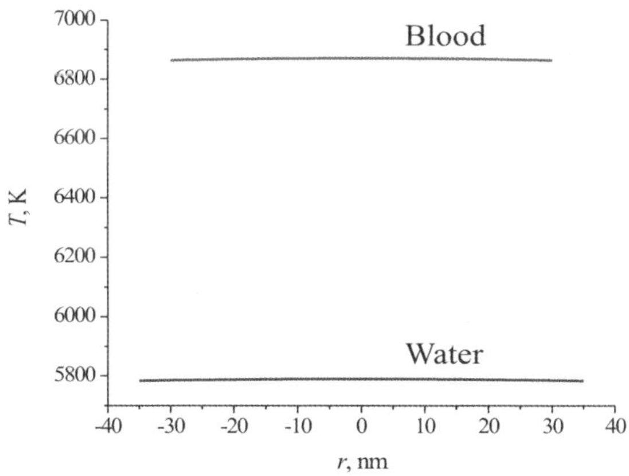

(b)

Fig. 6.3. One-dimensional distribution of the temperature inside the gold nanoparticle in the surrounding water and blood media irradiated by a single laser pulse of the energy density 5 J/cm^2 and duration 8 ns.

6.2.4. *Outside temperature distribution around a single nanoparticle*

As follows from Fig. 6.4, there is a significant heat loss from the surface of a nanoparticle to the surrounding medium, suggesting that even lower energy pulses are enough to achieve true and large thermal damage of the biological surroundings. If the temperature of the surrounding medium exceeds some thermal thresholds, secondary phenomena of cell killing effects may develop, such as photothermal ablation, sound generation, bubble formation and so on. For most cancer cell killing effects the threshold temperature is about 150°C (423 K). Our calculations show that at this level of the threshold temperature, the damage area produced by a heated 30 nm gold particle in blood has a size of ~ 1 μm in diameter, which exceeds by 16 times the size of the nanoparticle (see Fig. 6.4(a)). Thus, under chosen conditions, the damage area produced by one 30 nm particle heated by single laser pulse of 5 J/cm^2 is comparable to the size of the laser focal spot.

To find the lowest energy level of the laser pulse required to achieve thermal thresholds for cell death, we have conducted calculations of the temperature space distribution of gold nanoparticles irradiated by different laser fluxes. The results of these simulations are illustrated in the Fig. 6.4(b)). As follows from these calculations, a single 8 ns laser pulse of energy density 2.5 J/cm^2 provides heating of the 30 nm gold particle in the surrounding blood up to the boiling point of the gold, ~ 3400 K. The melting point of gold ~ 1400 K is reached for a lower energy pulse density of 1 J/cm^2 for 8 ns laser pulse duration. Decreasing the energy density of the incident single laser pulse leads to smaller damage volume around the nanoparticle. Using the threshold temperature for cell death, 423 K, we can determine the size of cell damage produced by the heated 30 nm gold particle. For the 8 ns single laser pulse with energy density 2.5 J/cm^2, the damage volume around the nanoparticle has 0.5 μm diameter, which is smaller but still comparable to the focal spot of incident laser pulse. In the case of a 1 J/cm^2 single laser pulse, the damage area is 180 nm in diameter, which is only thrice the diameter of the nanoparticle. Thus, for the experimental realization of selective nanophotothermolysis of cancer cells by a single 30 nm gold particle we

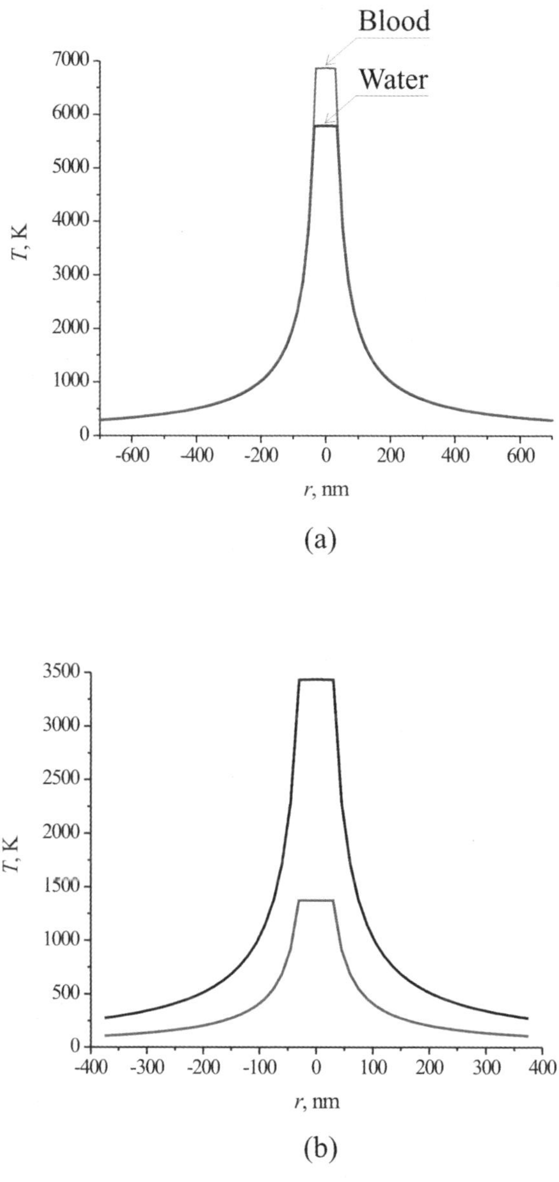

Fig. 6.4. Space distribution of the temperature outside the 30 nm gold particle (a) in the water and surrounding blood media irradiated by a single laser pulse of energy density 5 J/cm^2 and duration 8 ns and (b) in a blood medium irradiated by a single laser pulse of the energy density 2.5 J/cm^2 (upper curve) and 1 J/cm^2 (lower curve).

recommend the use of a laser pulse with energy density of 2.5 J/cm^2 for 8 ns width.

6.2.5. *Temperature simulations for many heat sources*

The model of Eqs. (6.5)–(6.13) can be used for simulations of the space distribution of the temperature produced by many heat sources. Figure 6.5(a) illustrates the temperature fields around the closely-located nanoparticles in the surrounding blood medium irradiated by single laser pulse with energy density of 2.5 J/cm^2 and pulse duration 8 ns. As follows from these calculations, to provide the threshold temperature 423 K between the particles everywhere in the cell volume (i.e., thermal damage in the cell), the interparticle distance l (i.e., between the centers of neighboring nanoparticles) should not exceed $l \leq 480$ nm.

The chosen above criteria can be reached by varying l and the energy density of the incident laser pulse. In order to find the optimal distance l, where the minimal temperature between the nanoparticles is equal to 423 K, we have conducted calculations of the interparticle temperature distribution for different values of the laser energy density. Results of these simulations presented in the Fig. 6.6 show that an optimum l does not exist in the framework of the thermal calculations: l linearly increases with the energy density of the incident laser pulse.

The interparticle distance l sets the lower limit for the nanoparticle concentration n required for cell death (i.e., the number of particles per unit volume) as

$$n = \left(\frac{1}{l^3}\right). \qquad (6.15)$$

Let us estimate n and the absolute number N of nanoparticles to provide the high-temperature distribution inside the cancer cell volume. For l = 480 nm, the concentration of n = 9.0×10^{12} cm^{-3} is homogeneously distributed throughout the cell volume. N, can be found as $N = nV_{\text{cell}}$, where V_{cell} is the cell volume. In the case of the 10 μm diameter cells of the MDA-MB-231 breast cancer line, the number of gold 30 nm particles

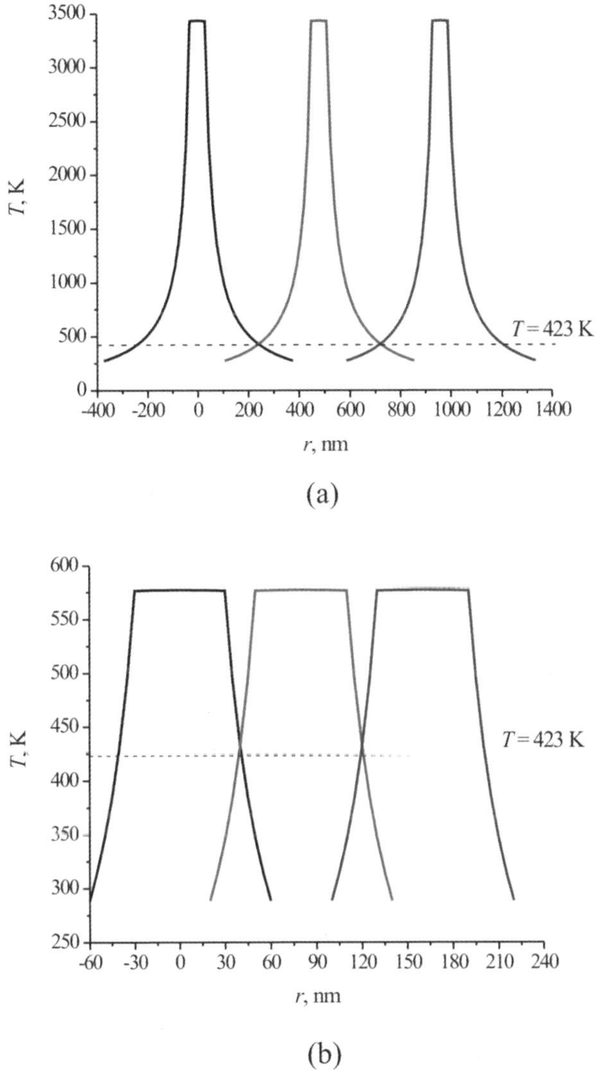

Fig. 6.5. Space distribution of the temperature fields produced in the blood medium (a) by many heat sources irradiated by a single laser pulse with energy density of 2.5 J/cm^2 and pulse duration 8 ns and (b) inside the cluster consisting of 30 nm gold particles irradiated by a 8 ns single laser pulse of energy density 0.42 J/cm^2.

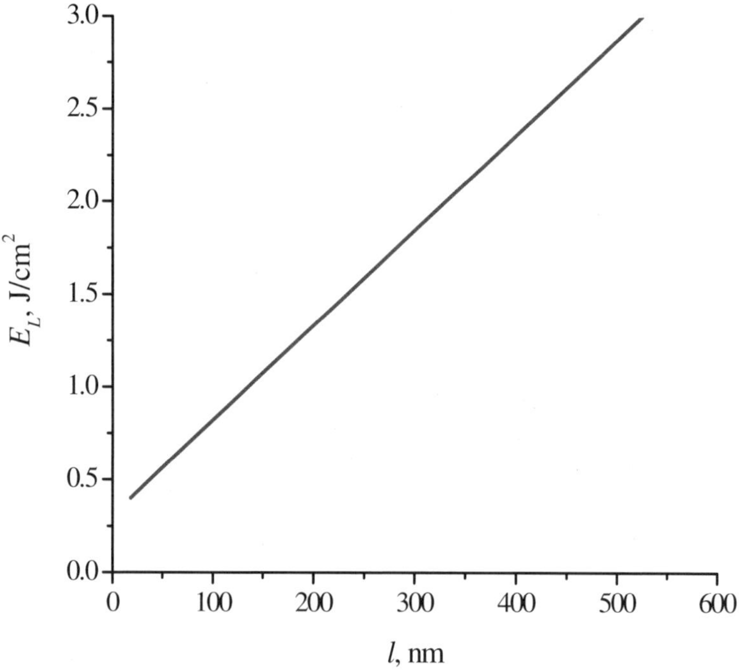

Fig. 6.6. Interparticle distance for different values of the laser energy density.

needed to provide the threshold temperature everywhere in the cell volume (thermal damage) is $N \approx 4{,}710$. This is a relatively large number of particles per cell, requiring a long incubation time to collect them in the cell or cell membrane. As shown our experiments [3], during the relatively long incubation time, the nanoparticles collide with each other, stick together and form nanoparticle clusters. The temperature distribution inside a cluster is expected to be significantly higher than the threshold temperature 423 K, since the particles in the cluster are compacted much closer to each other. Let we calculate the space distribution of the temperature field inside and outside the cluster formed by 30 nm particles conjugated with antibodies.

6.2.6. *Temperature distribution inside the nanocluster*

Our experiments [3] show that the distance between the centers of neighboring particles in a cluster is 80 nm, i.e., 20 nm between the surfaces of the 30 nm particles. The interparticle distance in the cluster could not be less than 20 nm, since the particles are covered by antibodies with a thickness of about 10 nm. The model described in Sec. 6.2.1 is used to calculate the temperature distribution inside the cluster consisting of separate gold nanoparticles with radius r_0 = 30 nm in the surrounding blood medium irradiated by a single laser pulse of energy density E = 2.5 J/cm^2 and pulse duration τ_L = 8 ns. As follows from these calculations, the 2.5 J/cm^2 single laser pulse provides an extremely high minimal temperature of 2574 K inside the cluster (minimal temperature between the nanoparticles). The surface temperature of the nanoheaters is 3432 K. So, at the same energy level of the incident laser radiation use for laser heating of the single nanoparticle, the minimal temperature inside the cluster exceeds by 6 times the threshold temperature for cell killing. Thus, the aggregation of the nanoparticles in clusters gives us a unique possibility to dramatically decrease the incident laser pulse energy with the same efficiency of cell death.

To find the lowest energy level of the incident single laser pulse, we have conducted calculations of the space distribution of the temperature inside the cluster consisting of 30 nm gold particles for different laser fluxes. The size of the cluster is chosen to be smaller than that of the laser focal spot. The results of these simulations for the lowest energy providing the threshold temperature not less than 423 K inside the cluster (between nanoparticles) are illustrated in Fig. 6.5(b)). As follows from these calculations, the 8 ns single laser pulse of the energy density 0.42 J/cm^2 is enough to heat the 30 nm gold particles inside the cluster up to 577 K and to reach the minimal temperature of 432 K in the cluster volume. There is evidence that specific properties of gold nanoparticles may be the reason for the remarkably low melting temperature (573 K) [20]. Thus, clusters aggregated from small nanoparticles can provide thermal damage of cancer cells at the significantly low energy level of 0.42 J/cm^2 of the incident single laser pulse.

6.2.7 Temperature distribution outside the nanocluster

We assume that the cluster is aggregated from 30 nm gold particles with 20 nm interparticle distances homogeneously compacted in the cylinder of length 380 nm. The cluster is located along the r-axis and surrounded by the blood. It is irradiated by the single laser pulse of energy density 1 J/cm^2 and pulse duration of 8 ns at FWHM. Results of the simulations for space distribution of the temperature field inside and outside the cluster in the surrounding blood medium are demonstrated in Fig. 6.7(a)). As follows from these calculations, the temperature inside the cluster oscillates in the range of 1374–1056 K along its length. So, the cluster heated by a single laser pulse of energy density 1 J/cm^2 has a high inside temperature which exceeds by thrice the threshold temperature for cell death. Outside temperature distribution shows us that the cluster provides a large thermal damage area of ~ 0.5 μm in cell volume at the threshold temperature level 423 K. To reach the same thermal damage area by a single nanoparticle, a laser pulse with energy density 2.5 times higher than for the cluster case is required.

We should note that since the cluster is treated here as a discrete medium (each particle considered separately), the obtained above results for cluster temperature and thermal damage area have lowest values. The upper limit for these values can be found for a homogeneous (continuous) distribution of the medium in the cluster. To keep our notations here the same, we will label such a continuously distributed medium as a homogeneous cluster. There are two reasons in support the homogeneous distribution in the cluster.

The first reason is the optics area. The size of nanoparticles in a cluster is much smaller than the wavelength of laser radiation, $r_0 << \lambda$. As shown in our "Nanomaterials in Nanomedicine" chapter of this book, according to the Lorentz-Mie scattering theory the cross-section of absorption exceeds by four times the geometric cross-section of the gold nanoparticle, $\sigma_{abs} > \sigma_0$. Hence, cross-sections of absorption of each nanoparticle are overlapping in the cluster. Because of this the light doesn't distinguish the separate nanoparticles inside cluster, and the laser energy is distributed homogeneously among the particles.

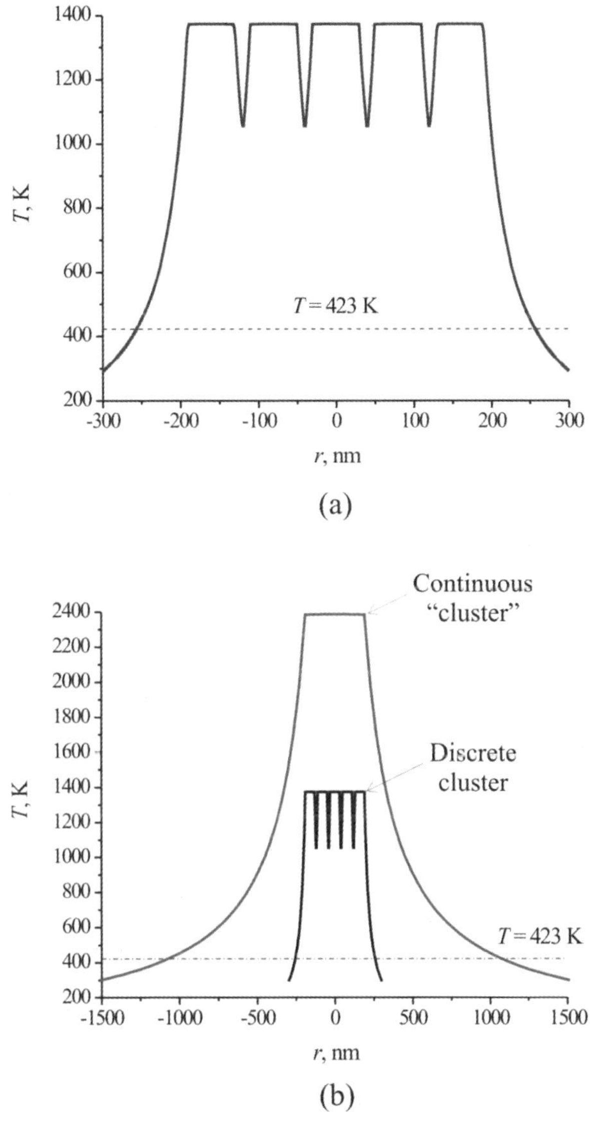

Fig. 6.7. One-dimensional distribution of the temperature field inside and outside the discrete cluster (a) and continuous "cluster" (b) consisting of 30 nm gold particles in the surrounding blood medium irradiated by a single laser pulse of energy density 1 J/cm^2.

The second reason is the heat transfer frame. Since the temperature variation inside nanocluster occurs in the relatively small temperature range of 318 K (45°C), we can assume that the internal cluster temperature is homogeneous and constant. This means that from a thermal calculations point of view, the cluster can be treated as homogeneous (continuous) gold particles with the same size.

Let us evaluate the damage area produced by continuous gold "clusters" in the blood medium. The temperature distribution for gold particles of radius 190 nm (cluster diameter 380 nm) in blood irradiated by a single laser pulse of energy density 1 J/cm^2 is illustrated on Fig. 6.7(b)). As follows from these simulations, the upper level of temperature inside the homogeneous cluster is 2384 K, and the thermal damage area is 2.13 μm in diameter at the threshold temperature level 423 K. Thus, the 380 nm cluster provides a huge thermal damage area with minimum size of 0.5 μm (for a discrete cluster) up to a maximal size of 2 μm (for a homogeneous cluster) in a cell volume irradiated by a single laser pulse of energy density 1 J/cm^2.

On the basis of our theoretical modeling, we can conclude that the aggregation of nanoparticles in a cluster mode of nano-photothermolysis raises the efficiency of cancer cell treatment due to the large damage area at the relatively low energy density of the incident single laser pulse. This conclusion is supported by the following results:

- The same laser energy density used for heating a single nanoparticle to the minimal temperature for thermal cell damage provides a temperature 6 times greater inside the cluster.
- Clusters can produce thermal damage of cancer cells at a significantly lower energy density level 0.42 J/cm^2 of the incident single laser pulse in comparison to a single particle.
- Clusters provides a larger thermal damage area with minimal size of 0.5 μm (for the discrete cluster case) up to maximal size of 2 μm (for continuous "cluster") in a cell volume irradiated by a single laser pulse of energy density 1 J/cm^2 and duration of 8 ns. To reach the same thermal damage area of 0.5 μm by a single nanoparticle, a laser pulse with energy density 2.5 times higher than for the cluster case is required.

We should note that the damage area is estimated here by using the threshold temperature required for the initiation of cell killing effects, i.e., thermal damage. However, the actual damage area is expected to be significantly larger since the damage can occur on the cellular level, for example, due to bubble formation around the nano-absorbers and its expansion in the cell volume. The next section is devoted to the investigation of the microbubble dynamics around the nanoparticles/nanoclusters in biological media.

6.3. Microbubble Overlapping Mode

As shown in the previous section, the significant heat conduction away from a heated nanoparticle requires timescales on the order of microseconds (1.75 μs for water and 1.56 μs for blood). Because of this, for nanoseconds laser pulses, most of the pulse energy remains localized at the absorbing nanoparticles/cluster whose temperature rises high enough to cause vaporization of the immediate surrounding medium. This will create a bubble that then expends outward from the nanoparticle. We now investigate bubble dynamics resulting from laser pulses with duration of 8 ns at FWHM.

The experiments [3] show that it is enough to heat the strongly absorbing nanoparticles above the boiling temperature of water only or above the critical temperature of the phase transition in water to generate the bubbles around the particles. So, for a relatively low laser flux, the nanoparticle will be surrounded by a rapidly expanding layer of vapor. Let us calculate the threshold energy density of laser pulse deposited into the nanoparticle, which is required to raise its surface temperature to the level of critical temperature for the bubble generation.

An experimental nucleation temperature of T_{thr} = 150°C (423 K) for heterogeneous bubble generation on the melanosome's surface was determined by Kelly [21]. Our calculations (Fig. 6.8) show that a threshold energy density of 0.33 J/cm^2 for the single 8 ns laser pulse is required to reach a surface temperature of 450 K for the 30 nm gold nanoparticles in blood medium. Thus, the bubble mode of the photothermolysis can provide larger damage area at the relatively low

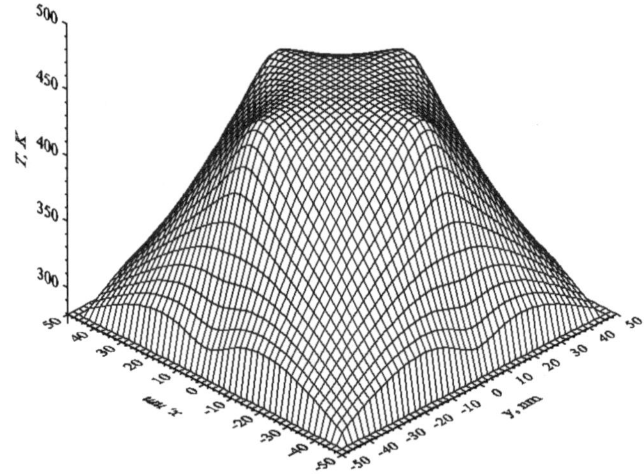

Fig. 6.8. Two-dimensional distribution of the temperature inside and outside a 30 nm gold nanoparticle in the surrounding water and blood medium irradiated by a single laser pulse of energy density 0.33 J/cm^2 and duration 8 ns.

optical flux of 0.33 J/cm^2 (τ_L = 8 ns) due to rapid expanding nanobubbles around the nanoparticles.

6.3.1. *Spherical Bubble Dynamics*

Consider a spherical bubble of radius $R(t)$ (where t is time) in an infinite domain of liquid whose temperature and pressure far from the bubble are T_∞ and $p_\infty(t)$ respectively. We will assume that the liquid density ρ_L and dynamic viscosity η_L are constant and uniform. It will also be assumed that the contents of the bubble are homogeneous, and that its temperature $T(t)$ and pressure $p(t)$ within the bubble are always uniform. The radius of the bubble will be one of the primary results of the analysis. In the absence of mass transport across the boundary (evaporation or condensation) and with the above assumptions, the bubble dynamics can be described by the generalized Rayleigh–Plesset equation [22]:

$$\frac{p_V(T_\infty)-p_\infty(t)}{\rho_L}+\frac{p_V(T)-p_\infty(T_\infty)}{\rho_L}+\frac{p_G}{\rho_L}= \\ =R\frac{d^2R}{dt^2}+\frac{3}{2}\left(\frac{dR}{dt}\right)^2+\frac{4\eta_L}{\rho_L R}\frac{dR}{dt}+\frac{2S}{\rho_L R}, \tag{6.16}$$

where $p_V(T)$ is the saturated vapor density; and p_G is the partial pressure of the contaminant gas contained in the bubble. The first term on the left side of this equation is the instantaneous tension or driving term determined by the conditions far from the bubble. The second term will be referred to as the thermal term. It will be assumed here that the behavior of the gas in the bubble is polytropic so that

$$p_G = p_{G_0}\left(\frac{r_0}{R}\right)^{3k}, \tag{6.17}$$

where p_{G_0} is the partial pressure of the contaminant gas contained in the bubble at some reference size $R = r_0$, k is approximately constant. Clearly, $k = 1$ implies a constant bubble temperature, and $k = \gamma$ would model adiabatic behavior, where γ is the ratio of the specific heat of the vapor at constant pressure to the specific heat at constant volume.

Equation (6.16) can be readily integrated numerically to find $R(t)$ and the rate of bubble growth $v = \frac{dR}{dt}$ given the input $p_\infty(t)$, the temperature T_∞, and the other constants. Initial conditions are also required and, in the context of cavitating flows, it is appropriate to assume that a nanobubble of radius $R = r_0$ is in equilibrium at $t = 0$ in the fluid at a pressure $p_\infty(0)$, so that

$$p_{G_0} = p_\infty(0) - p_V(T_\infty) + \frac{2S}{r_0} \tag{6.18}$$

and $(dR/dt)_{t=0} = 0$.

6.3.2. *Microbubbles overlapping*

As discussed in Sec. 6.2, under certain delivery and particle accumulation conditions, the particles can locate on the membrane surface very close to each other, as shown in Fig. 6.1. In this mode we can expect that the optical, thermal, acoustic, and bubble formation phenomena initiated by laser radiation around each particle to overlap and create some synergistic effects [3]. For example, the characteristic radii for these overlapping effects can be defined by an optical radius $R_{opt} = (\sigma_{abs}/\pi)^{1/2}$, thermal diffusion length $R_T = (kt)^{1/2}$, sound transfer distance $R_{ac} = c_s t$, and microbuble radius $R_{bubble} = v_b t$, where σ_{abs} is the cross-section of absorption, k is the heat diffusion coefficient, c_s is the speed of sound, and v_b is the bubble growth velocity. Since for a certain range of gold nanoparticle sizes the optical radius can exceed the geometrical particle radius r_o at the given wavelength of laser radiation, a plasmon-plasmon resonance at $R_{opt} > L$ (L is the distance between particles) can be achieved. This resonance leads not only to the known "red-shifting" of maximum absorption, [23] but also increases the integrated absorption coefficient of the particle's cluster. The overlapping thermal fields at $R_T > L$ (during or after the laser pulse t_p) lead to a dramatic increase in thermal and accompaning effects (e.g., to sudden appearance of different nonlinear effects such as phase transitions). The interaction of acoustic waves at $R_{ac} > L/2$ significantly changes the local refractive index, which might be crucial for diagnostic purposes. Finally, the interaction of growing bubbles (initially of nanoscale sizes) at $R_{bubble} > L/2$ leads to a substantial increase in the average bubble size, with a decrease in the bubble-formation threshold (which, for single nanoparticles, is relatively high), and probably an increase in the bubble growth velocity. All of these phenomena are very crucial in changing the distance between nanoparticles, and they are important for both high sensitive diagnostics and efficient therapy.

We focus here on therapeutic application of this technology. Indeed, laser-induced overheating effects around nanoparticles may create many therapeutic actions through the microbubble formation phenomena, accompanied by acoustic and cavitation effects, mechanical stress, and laser-induced hydrodynamic pressure arising from the bubble rapid expansion and collapse. In this section, we discuss a dynamic mode for selective cancer treatment by laser activated nanoheaters, involving the situation where the bubbles are overlapping inside the cell volume. The bubble overlapping mode (BOM) may dramatically increase the efficiency of the cancer treatment by laser-heated nanoparticles as the result of a large damage range.

Since bubbles do not all appear simultaneously, there are few possibilities for bubbles overlapping shown in Fig. 6.1. The bubbles can overlap at different nucleation times and spreading velocities, or simultaneously with the same spreading velocity when they reach each other midway between the neighboring gold particles. In our calculations, we choose stronger criterion than overlapping at midpoints. Effective overlapping with large damage of tumor cells can be achieved if each bubble reaches the neighboring nanoparticles themselves, not just the midpoints between nanoparticles.

6.3.3. *Condition for nanoparticles concentration*

The criteria for BOM introduced above impose the requirement on an interparticle distance *L*, the distance between two neighboring nanoparticles. Since the bubble must reach the neighboring particles during the growing time τ_{growth}, the integration of the Rayleigh–Plesset equation (6.16) [22] for bubble dynamics at the nucleation temperature $T_{\rm nuc}$ gives the requirement for interparticle distance *L* as

$$L \le \sqrt{\frac{2}{3}\frac{P_{sat}(T_{nuc}) - P_{\infty}}{\rho(T_{\infty})}}\tau_{growth} \tag{6.19}$$

where $P_{sat}(T_{nuc})$ is the saturated vapor pressure at the nucleation temperature, P_{∞} is the ambient pressure, and $\rho(T_{\infty})$ is the water density. This sets the lower limit to the required for BOM concentration of nanoparticles, the number of particles per unit volume, as

$$n = \left(1/L^3\right). \tag{6.20}$$

Let us now estimate the interparticle distance, the nanoparticle's concentration, and their absolute number collected into the tumor cell to provide the BOM by laser activated nanoheaters. The interparticle distance L calculated by Eq. (6.19) for the experimentally measured nucleation temperature of $T_{nuc} = 150$ °C is

$$L \approx 15.8\tau_{growth} \,. \tag{6.21}$$

The bubble growing time τ_{growth} is limited by the bubble nucleation time t_{nuc} and bubble lifetime τ_{buble}, i.e., $t_{nuc} \leq \tau_{growth} < \tau_{buble}$. In our experiments [3], the nucleation time of the bubbles around isolated gold nanoparticles in aqueous suspension and irradiated by nanosecond laser pulses is $t_{nuc} \sim 100$ ns. The bubble lifetime depends on the laser energy and ranges from 100 ns to 2 μs. In the overwhelming majority of experiments, including ours, the bubble lifetime is in the range τ_{bubble} = 200–400 ns. So, we can conclude that the bubble growing time is close to the nucleation time, i.e., $\tau_{growth} \sim t_{nuc}$. Then, from the rate equation (6.21), we can estimate the distance between nanoparticles (microbubble radius) required for the BOM as $L = R_{bubble} \approx 1.6$ μm, which corresponds to the results of our experiments. Thus, Eq. (6.20) gives the lower limit of the concentration of nanoparticles to provide the BOM as $n \approx 2.44 \times 10^{11}$ cm^{-3}. The absolute number of particles homogeneously distributed inside a tumor cell for this mode can be found as $N = nV_{cell}$, where V_{cell} is the cell volume. For the case of the breast cancer line MDA-MB-231 of 15 μm size, the number of gold nanoparticles providing BOM is $N \approx 430$.

6.3.4. *Condition for nanoparticle sizes*

The optimal range of nanoparticle sizes for effective laser initiation of BOM in tumor cells is governed by the nanoparticle optics described in our "Nanomaterials in Nanomedicine" chapter of this book. The optical characteristics of spherical nanoparticles dispersed in a biological medium at a given radiation wavelength λ can be calculated on the basis of Lorentz-Mie diffraction theory at the single-scattering approximation. The Mie formalism requires the use of two dimensionless input parameters, $\rho = 2\pi r_0 / \lambda$ and $\delta = \rho\tilde{m}$, where $\tilde{m}$ is the relative value of the complex refractive index of the nanoparticles in the surrounding medium at the wavelength λ. Computer calculations of the dependences of the absorption K_{abs} and scattering K_{sca} coefficients at the given laser wavelength on the size of gold nanoparticles in aqueous suspension are plotted in Fig. 6.9. Figure 6.9(a) shows results for gold particles in water heated by the laser radiation of wavelength λ = 450 nm ($\tilde{m}$ = 1.122–1.416i), and Fig. 6.9(b) is for the case of wavelength λ = 633 nm ($\tilde{m}$ = 0.1126–2.5754i). It is evident from Fig. 6.9(a) that the absorption coefficient K_{abs} has a strong maximum at the radius of 50 nm for the gold particle at the given wavelength (for 532 nm wavelength, the maximum radius is 30 nm).

The width of the absorption maximum at the level $K_{abs} \geq 1$ corresponds to the particle radii range of r_0 = 20–600 nm. This means that in this size range, the absorption cross-section of gold particles at the given wavelength λ = 450 nm exceeds the particle geometric cross-section. Light scattering at λ = 450 nm by gold nanoparticles in aqueous suspension becomes apparent for particle radii larger than 50 nm. For large particle sizes, the scattering coefficient K_{sca} rapidly increases and acquires the value of absorption coefficient for the particle radius of 150 nm. For particles of radii $r_0 \geq$ 150 nm, the scattering of laser radiation predominates over the absorption, and the considered suspension containing the nanoparticles becomes a strongly scattered medium at the given wavelength.

For particle radius range r_0 = 1–50 nm, the absorption coefficient K_{abs} is considerably greater than the scattering coefficient K_{sca}, i.e., the efficiency of laser heating of nanoparticles in this size range is high. Thus, the optimal range of nanoparticle sizes for effective laser initiation of BOM in tumor cells is 1–50 nm. The maximal effect of laser heating of gold particles can be achieved for particles of radii 20–50 nm, where $K_{abs} \geq 1$ and $K_{sca} < 1$.

Figure 6.9(b) demonstrates the $K_{sca}(r_0)$ and $K_{abs}(r_0)$ curves for the wavelength of λ = 633 nm. We observe very weak absorption and a strong scattering of the light at this wavelength by the gold nanoparticles. The absorption slightly predominates over the scattering for a small range of nanoparticle radii when $r_0 \leq$ 35 nm. It follows that laser light with wavelength λ = 633 nm can be recommended for diagnostic purposes only. However, creation of nanoclusters on a cell surface may

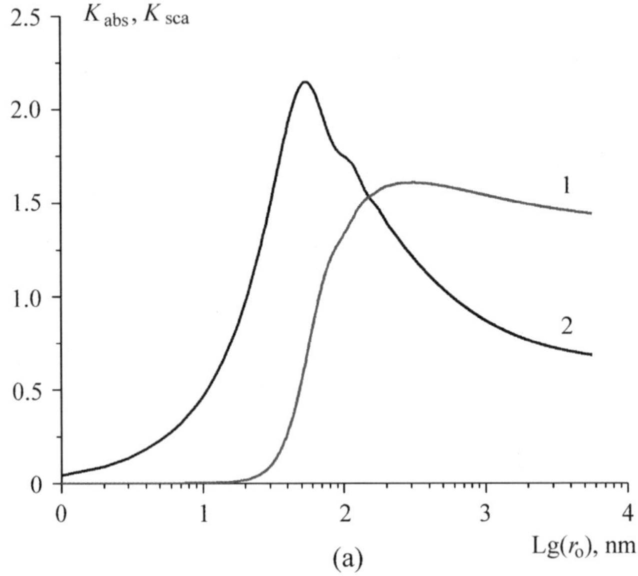

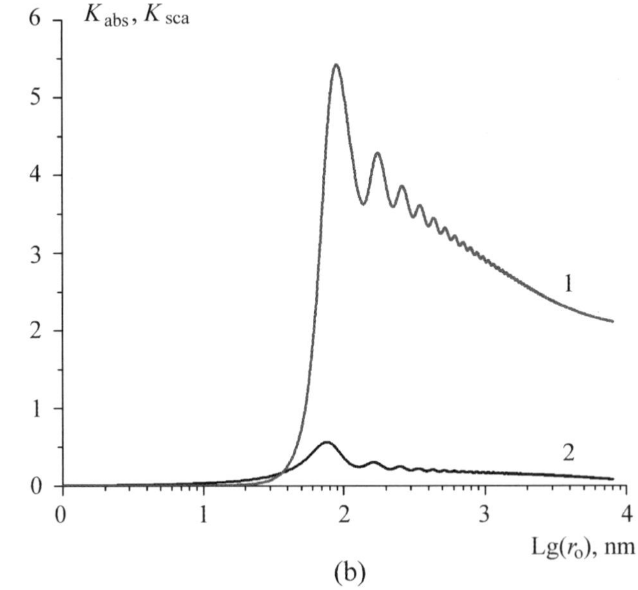

Fig. 6.9. Dependencies of the scattering K_{sca} (curve 1) and absorption K_{abs} (curve 2) coefficients at the laser wavelength λ = 450 nm on the size r_0 of gold particles in water medium ($\tilde{m}$ = 1.122–1.416i) (a) and for the case λ = 633 nm ($\tilde{m}$ = 0.1126–2.5754i) (b).

lead to shifting the absorption maximum to a near-IR range [23] that can also be used for therapeutic purposes.

Thus, the proposed theoretical model for the nanoparticle optics along with bubble dynamics allows us to find the following conditions for BOM realization:

- The maximal effect of laser heating of gold nanoparticles over the wavelength range λ = 450–550 nm can be achieved for the particle radii of 20–50 nm;
- The optimal nanoparticle concentration is $n \approx 2.44 \times 10^{11}$ cm^{-3}, which corresponds to an absolute number of nanoparticles in a breast cancer cell volume of 430 (88 on the cell surface).
- The bubble radius providing BOM is $R_{bubble} \approx 1.6$ μm.

These theoretical predictions of the BOM conditions have been confirmed by our experimental results [3], which include the study of nanocluster-related phenomena as discussed in Sec. 6.2 and below.

6.3.5. *Microbubble generation around nanoclusters*

The model described above has been used to calculate the bubble growth with time around the 30 nm gold particles and homogeneous cluster with radius 190 nm surrounded by the blood bio-medium when the incident laser pulse has an energy density of E = 0.33 J/cm^2 and pulse duration of τ_L = 8 ns. The physical properties of blood used in the calculations are listed in Table 6.1. The results of the simulations are presented in Figs. 6.10 and 6.11. Figure 6.10 illustrates the time behavior of the radius and its rate for the bubble generated around the 30 nm particles heated and cooled in the blood. Figure 6.11 demonstrates the difference in the time dynamics of bubble growth caused by the single 30 nm particle and 190 nm homogeneous cluster.

Calculations are performed for both cases at constant bubble temperature (k = 1) and for adiabatic conditions ($k = \gamma$= 4/3). Where under such conditions, the time dependence of the bubble radius $R(t)$ is a better fit to our experimental results [3]. As follows from our calculations,

Table 6.1. Physical properties of blood at 1 atm pressure used in the simulations.

Properties	Notation used in the model	Values	Units
Density	ρ_L	998.2	kg/m^3
Viscosity (water)	η_L	1.022×10^{-4}	kgs/m^2
Surface tension (water – saturated vapor)	S	7.2×10^{-3}	kg/m
Thermal conductivity at T = 273–373 K	μ_0	0.48 – 0.6	W/mK
Thermal diffusivity	χ	1.6×10^{-7}	m^2/s
Specific heat at T = 273–373 K	C	3645 – 3897	J/Kkg
Ratio of specific heats (water/vapor)	γ	1.33	

the adiabatic expansion of the bubble provides a final radius of 3.5 μm for the bubble lifetime 200 ns, which agrees well agreed with our experimental observations in [3].

The general feature of the simulation results is linear and smooth growth of the bubbles' radius with time when the bubbles' lifetime exceeds 30 ns. The time dynamics of the temperature calculated in the "Nanomaterials in Nanomedicine" chapter of this book for 30 nm nanoparticles heated by a 8 ns laser pulse, show that the total time for one cycle (heating of the nanoparticles from initial temperature 300 K to maximum temperature, and then cooling back to the initial temperature) is about 30 ns. Thus, linear growth of the bubble radius with time after 30 ns can be explained by the absence of an active heat source in the medium. This result is confirmed also by the rate dynamics $v(t) = dR/dt$ for the bubble radius shown in Fig. 6.10(b). In the time range 0–30 ns, the both the curves $R(t)$ and $v(t)$ have strong nonlinearity due to active heat transfer from nanoparticle to the surrounding medium. During the first several nanoseconds, the bubble rapidly expands with a high velocity of 40 m/s. Then, the expansion rate drops exponentially and reaches a constant saturation value of 16 m/s for $t > 30$ ns. Since there is no heating source in the system after 30 ns, the microbubble expands uniformly and linearly at a constant rate.

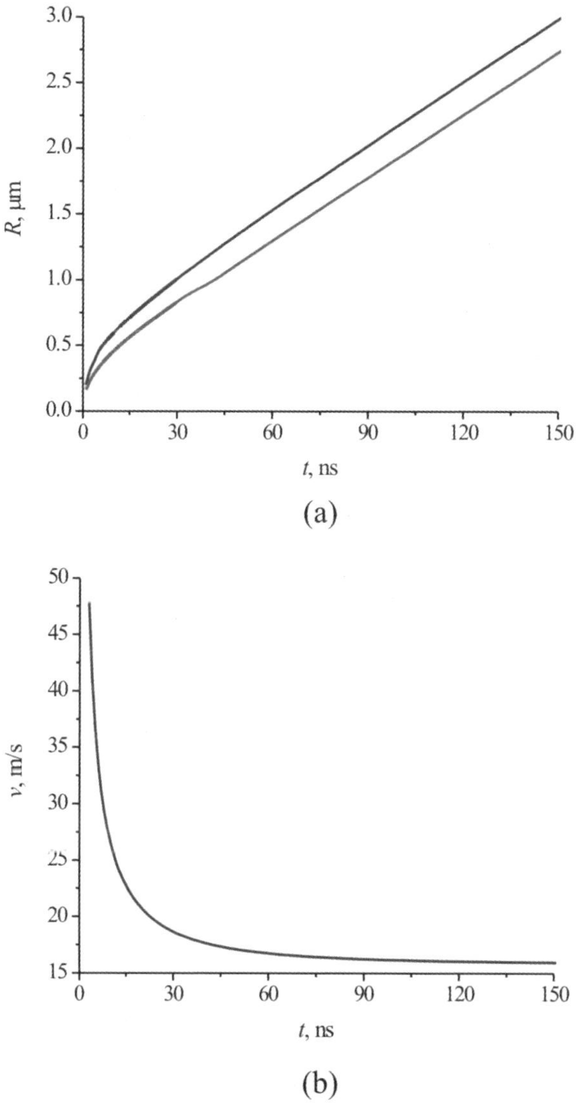

Fig. 6.10. (a) Bubble expansion around the 30 nm gold particle in blood irradiated by a laser pulse with energy density $E = 0.33$ J/cm^2 and pulse duration $\tau_L = 8$ ns at constant bubble temperature (*upper curve*) and under adiabatic conditions (*lower curve*). (b) bubble expansion rate, $v(t) = dR/dt$.

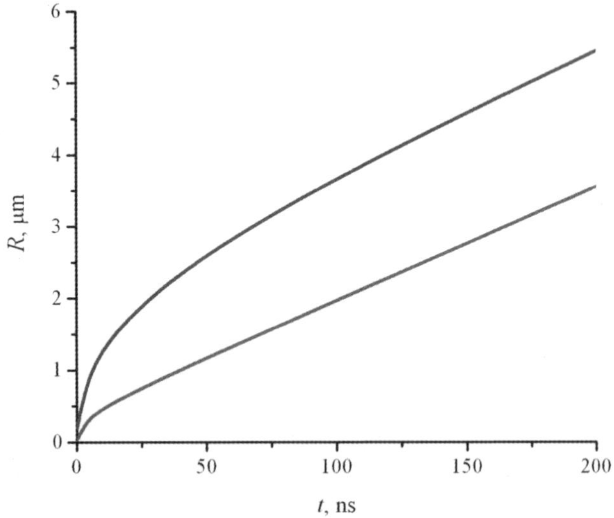

Fig. 6.11. Adiabatic expansion of the bubble produced by a single 30 nm particle (*lower curve*) and 190 nm continuous "cluster" (*upper curve*) irradiated by a laser pulse with energy density $E = 0.33$ J/cm^2 and pulse duration $\tau_L = 8$ ns.

It is important to compare the bubble expansion rate to the thermal conduction rate in the surrounding medium. Using the thermal properties of blood given in Table 6.1 and the size of the locally heated region $l \sim 1$ μm as a focal spot of the laser radiation, we find that the approximate speed for heat conduction is

$$v_{thermal} \approx \frac{\mu_0}{\rho C l} \approx 0.15 \text{ m/s}, \qquad (6.22)$$

which is more than two orders of magnitude less than the bubble expansion rate. Here, ρ, C and μ_0 are, respectively, the density, specific heat and thermal conductivity of the surrounding medium. So, during a laser pulse of $\tau_L = 8$ ns, thermal damage occurs over a negligible distance from the nanoparticle in comparison to the bubble damage. Thus, bubble expansion occurs on a time scale much shorter than heat loss, which justifies the use of an adiabatic treatment during bubble expansion. The thermal relaxation time is 1.75 μs for water and 1.56 μs for blood, which is much longer than the bubbles' lifetime of about 200 ns. This means that the cell damage by bubble expansion, it is sufficient to heat only the nanoparticles up to the critical threshold temperature for the bubble generation (423 K) without having to heat the surrounding cellular medium. Therefore, the relatively low optical flux of 0.33 J/cm^2 (τ_L = 8 ns) is enough for effective cell killing.

Similar bubble expansion dynamics is observed for the bubble produced by homogenous cluster illustrated on Fig. 6.11. Cluster's curve has both linear and nonlinear parts with notably higher expansion rate of more than 90 m/s in first 4 ns. During adiabatic expansion of the bubble produced by homogeneous cluster in blood the saturated expansion rate is 17 m/s and the final radius reaches ~ 5.5 μm for the bubble lifetime 200 ns. According to our calculations the heated 380 nm cluster is able to produce a one big bubble in the cell volume, whose damage area is comparable to the size of cancer cell (10–15 μm in diameter). Thus, the formation of just one large bubble around one large nanocluster aggregated in cell volume is good enough for cancer cell killing.

Thus, the bubble mode of the photothermolysis decreases the threshold optical flux up to 0.33 J/cm^2 (τ_L = 8 ns) for cancer cell killing and drastically increases the damage area due to rapid expanding microbubbles around the nanoparticles and cluster:

- The expansion of the bubble has a strong nonlinearity in the time range 0–30 ns due to active heat transfer from nanoparticles to the surrounding medium. During the first several nanoseconds, the bubble produced around a 30 nm particle rapidly expands with a high velocity of 40 m/s. Then, the expansion rate drops exponentially and reaches a constant saturation value of 16 m/s for $t > 30$ ns. Since there is no heating source in the system after 30 ns, the microbubble expands uniformly and linearly at a constant rate.
- The adiabatic expansion of the bubble produced around a single 30 nm particle reaches a final radius of 3.5 μm for the bubble lifetime 200 ns.

- The speed for a heat conduction in the cellular medium ($v_{thermal}$ = 0.15 m/s) is more than two orders of magnitude less than the bubble expansion rate ($v_{saturated}$ = 16 m/s), so that during a laser pulse of τ_L = *8* ns, thermal damage occurs over a negligible distance from the nanoparticle in comparison to the bubble damage.
- The heated 380 nm cluster is able to produce a single large bubble within the cell volume, whose damage area is comparable to the size of cancer cell (10–15 μm in diameter).

6.4. Laser-Induced Thermal Explosion Mode — "Nanobombs"

In this section, we discuss a new mechanism for selective laser killing of abnormal cells by laser thermal explosion of single nanoparticles — "*nanobombs*" — delivered to the cells. Thermal explosion of the nanoparticles is realized for ultrashort laser pulses when the heat is generated within the strongly-absorbing target more rapidly than the heat can diffuse away. Laser-induced rapid explosive evaporation of gold nanoparticles (GNs) enables the generation of stress transients, shock waves and high local pressure. A schematic picture of these complex physical effects is provided in Figs. 6.2 and 6.12. There are two main physical mechanisms that could lead to the laser-induced explosion of GNs — ***thermal explosion mode*** through electron–phonon excitation-relaxation and ***Coulomb explosion mode*** through multiphoton ionization.

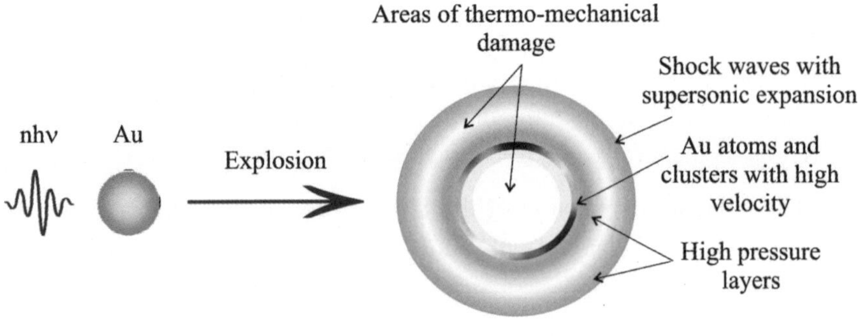

Fig. 6.12. Laser-induced thermal explosion of a gold nanoparticle.

6.4.1. *Thermal explosion mode*

Under the action of ultrashort laser pulses in the spectral range of the surface plasmon resonances (for a solid spherical GN, the maximum absorption is ~ 520 nm), GN atoms are excited to upper electronic states owing to the absorption of many photons. Through rapid (picosecond time scale) relaxation, GN atoms decay to their ground state with effective electron–phonon conversion of the absorbed photon energy into thermal energy. Depending on the GN temperature, T, the following scenarios, individually or in combination, can occur:

1. $T < T_{LV}$, where T_{LV} is the liquid vaporization temperature (water in many cases), ~ 150–350ºC [24]. This is thermal expansion of a single GN and surrounding thin liquid layer, which is accompanied by the generation of linear acoustic waves, known as the "classic" photoacoustic effect.
2. $T_{LV} \leq T < T_{GNM}$, where T_{GNM} is the GN melting point, ~ 1,063°C [25]. This is bubble formation with expansion and collapse, which is accompanied by the production of acoustic and shock waves [26–28].
3. $T_{GNM} \leq T < T_{GNB}$, where T_{GNB} is the GN boiling point of ~ 2,710°C [25], with GN melting.
4. $T \geq T_{GNB}$, which is GN boiling with the formation of gold vapor around liquid gold drops.

The photothermal process of scenario 3, and especially scenario 4, may lead to GN fragmentation into smaller parts — "nanobullets" — and to the thermal explosion of a GN into single atoms.

6.4.2. *Coulomb explosion mode*

For ultrashort laser pulses (e.g., femtosecond time scale) comparable to the time scale of electron–electron interactions and shorter than the electron–phonon interaction time, a non-photothermal mechanism of GN explosion can occur through multiphoton ionization when the absorbed photon energy is transferred directly to the electrons, leading to their ejection due to the Coulomb explosion mechanism [29,30]. High GN plasmon-resonance absorption may facilitate this effect due to thermionic electron emission.

6.4.3. *Time scale approximations*

The goal of the theoretical modeling is to estimate the threshold laser energy density, E_{expl}, required for realization of the thermal explosion mode of GNs, and to compare the calculated data with available experimental results. This mode is realized through the rapid overheating of a strongly-absorbing target during a short laser pulse when the influence of heat diffusion is minimal. Let us first estimate the time scale for thermal relaxation due to heat diffusion from the surface of the nanoparticle.

In the vicinity of GN $\leq 5R$, where R is the nanoparticle's radius, the thermal relaxation time for a spherical GN can be estimated as $\tau_T = R^2/6.75k$, where k is the thermal diffusivity [3]. For $R = 50$, 100 and 200 nm, estimates of τ_T (for water, $k = 1.44 \times 10^{-3}$ cm^2/s) are approximately 2.6, 10 and 41 ns, respectively. For a laser pulse duration $t_P \leq \tau_T$, heat is generated within the GN more rapidly than can be diffused away, so that we can neglect heat losses from the surface of the GN due to heat diffusion into the surrounding medium. This condition for smaller GNs is valid with picosecond and femtosecond laser pulses.

6.4.4. *Threshold intensity for laser-induced thermal explosion*

Consider a nanoparticle selectively delivered to a targeted site (e.g., cancer cell) to be irradiated by a laser of intensity I. The power of the absorbed electromagnetic field is $P = \sigma_{abs} I$, where σ_{abs} is the GN's cross-section of absorption. If σ_{abs} is large enough, thermal explosion of the GN may occur at certain values of the threshold laser intensity I_{expl}, which is less than the threshold intensity for optical plasma formation in the surrounding medium [13–15]. Under the thermal explosion of GNs, we understand the specific case for which the total energy absorbed by the GN during the time of its inertial retention in vapor state, $\tau_{expl} = R/u_s$, exceeds the energy required for the GN's complete evaporation,

$\rho qV = N_{Au} q_1 V$. Here, u_s is the sound velocity in Au vapor at the critical temperature $T_{cr} \sim T_{GNB}$; ρ is the volume density of GN; N_{Au} is the number of Au atoms per unit volume; q and q_1 are the particle's specific heat of evaporation per unit mass and per particle, respectively; and V is the GN's volume. We have a laser pulse duration of $\tau_L > \tau_{expl} = R/u_s$ (~ 1 ps for $R \sim 10$ nm), where τ_{expl} is an explosive evaporation time.

To compute I_{expl}, we can use the model described in the "Nanoscale Materials in Strong Ultrashort Laser Fields" chapter of this book. According to that model, the threshold intensity of the laser radiation for thermal explosion of the spherical GN can be expressed as

$$I_{\exp l}(R) \approx \frac{4\rho q u_s}{3K_{abs}(R)}, \tag{6.23}$$

where K_{abs} is the nanoparticle absorption efficiency. Thus, the solid nanoparticle in a relatively strong laser field with intensity $I \geq I_{\exp l}$ during the short time $\tau_{\exp l} \sim R/u_s$ transforms into a gas (vapor) sphere of radius ~ R, which has high temperature $T \sim T_{GNB}$ and high pressure $P_{vap} >> P_\infty$, where P_∞, is the ambient pressure.

We assume here that thermal explosion of a GN is accompanied by a generation of shock waves that expand with a supersonic velocity $u_s = \frac{R}{\tau_{\exp l}} = \frac{I_{\exp l}}{4\Delta H} \approx 10^5$ cm/s. Here, $\Delta H = \Delta H_{300-vap} + \Delta H_{vap}$, where $\Delta H_{300-vap}$ is the enthalpy change per unit volume for heating the GN from the ambient temperature to the vaporization temperature, and ΔH_{vap} is the vaporization enthalpy per unit volume. The shock waves could be waves of high acoustic or/and water vapor pressure, which spread out over long distances around an epicenter of explosion and produce irreparable mechanical cell damage [31–37]. It should be noted that the pressure produced by vaporization of a GN itself (e.g., ~10^{-2} atm [38]) is less than water vapor pressure around the hot GNs. Indeed, the number of atoms vaporized per unit of time and per unit of particle surface is given by [39]

$$\frac{N_{vap}}{dt} = u_s N_{\mathrm{Au}} = \frac{I_{\exp l}}{4\Delta H} N_{\mathrm{Au}} \ . \tag{6.24}$$

For example, the atomic density of bulk gold is 5.9×10^{22} cm^{-3}, while each GN contains approximately 10^9 atoms, leading to an average atom density of 10^{17} cm^{-3}. Higher gold vapor pressure can only be reached on the front end of the explosion, where explosion products are localized in a thin layer.

6.4.5. *Cell damage effects*

The therapeutic effect of laser-induced explosion of GNs for cancer treatments can be reached due to one or several phenomena, such as protein inactivation (e.g., through its denaturation or coagulation) around hot GNs [31], bubble formation [3], generation of acoustic and shock waves [32–37], and interaction with GN fragments and atoms. An important damage-related factor is not only the temperature but the vapor pressure produced by both water and gold vapors, accompanied by the cavitations and shock waves, with local pressure up to gigapascals [32–37]. The damage of specific cellular structures (e.g., plasmatic membranes, nuclei, cytoskeletons and organelles) in the laser explosion mode depends on GN parameters (composition, size and shape), their number and location, laser parameters (wavelength, fluence, pulse duration and number of pulses) and properties of the surrounding media (e.g., amount of water).

The time required to destroy an abnormal cell of size d by supersonic expansion of shock waves can be roughly estimated as $\tau_{sw} \approx d/u_s$. In the case of 15 μm diameter cancer cells [3,13–15], this time is $\tau_{sw} \approx 15$ ns, which is comparable to the lifetime of sound/pressure shock waves in an aqueous medium. We can estimate the lower level of the threshold energy density of a laser pulse required for thermal explosion of the GN using $E_{expl}(R) \approx I_{expl}(R) \times \tau_{expl}$, where the threshold intensity, I_{expl}, depends on the absorption efficiency of the nanoparticle. For the particular lasers used in the experiments [3,13–15] ($\lambda = 532$ nm, $\tau_L = 8$–10 ns), the threshold energy density for thermal explosion of a solid gold nanosphere of size $R = 35$ nm (absorption efficiency of $K_{abs} = 4.02$) is $E_{expl} = 38.5$ mJ/cm^2. This density strongly depends on the types of nanoparticles (e.g., gold solid nanospheres, nanoshells and nanorods). For example, gold nanorods have a near-IR resonance absorption efficiency of approximately 14 for an effective radius of 11.43 nm [4]. Due to the higher plasmon-resonance absorption efficiency of nanorods, the threshold energy density can be reduced by using gold nanorods up to E_{expl} = 25 mJ/cm^2. It is important to note that the estimated threshold energy densities for the thermal explosion of GNs at the given wavelength λ = 532 nm are 2.5 to 4 times less than the laser fluence of 100 mJ/cm^2 established as the safety standard for medical lasers [40].

The estimated values for E_{expl} are in good agreement with some available experimental results. Indeed, for spherical GNs with average size 45 nm irradiated with second-harmonic Nd:YAG laser pulses (532 nm for 7 ns), the laser fluence threshold for changing the GN shape and its fragmentation associated with GN melting and boiling phenomena are 16 and 30 mJ/cm^2, respectively [25]. The more intense GN fragmentation to small fragments of mainly 5–10 nm in size has been observed for laser fluences in the range 30–140 mJ/cm^2 and higher [25]. For a 30 ps pulse, 25 nm GN fragmentation has been observed at 23 mJ/cm^2, with a slight effect on changing the GN shape even at 2–5 mJ/cm^2 [41–42].

Thus, thermal explosion around GNs may give significant contribution to cell damage alone or together with conventional water vapor bubble formation in nanophotothermolysis, which is characterized by a higher laser fluence range of 50–500 mJ/cm^2 [1–3,24–25]. As previously mentioned, conventional bubble formation usually starts at the end of nanosecond laser pulse, while thermal explosion time scale is from a ps to a few ns [13–15]. A schematic time scale for shock wave and conventional bubble formation in thermal explosion mode is shown in Fig. 6.13. Thus, shock waves appear earlier than bubbles induced by the explosive vaporization of GNs in a liquid environment.

The laser-induced explosion effect can explain the experimental results [2,10] with notable cancer damage with 1.4 nm GNs inside viruses on a cell membrane with picosecond laser pulses (30 ps, 50–100 mJ/cm^2) when the probability of classic water bubble formation is very low. We believe that the explosion mode may be essential in selective nanophotothermolysis of DNA [10], and this mode is definitel becoming dominant in the absence of a sufficient amount of water around GN. This mode can be combined with a bubble-overlapping mode [3] as described above, where the explosion of few closely-located GNs in a GNC can

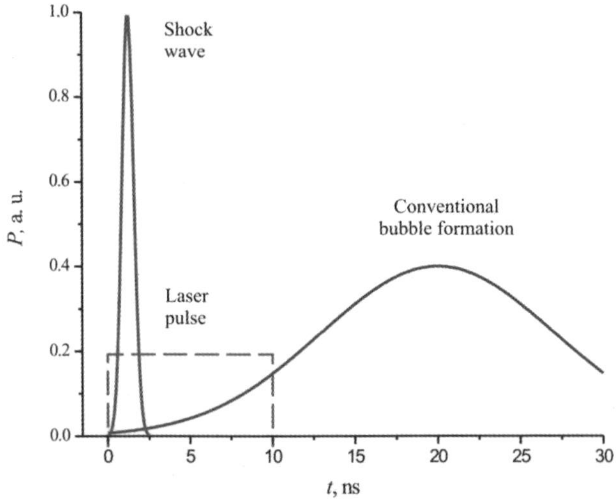

Fig. 6.13. Schematic time scale for shock wave and conventional bubble formation using the thermal explosion mode.

produce one large bubble with enhanced killing efficiency. The explosion of GNs on smaller particles or single gold atoms during a laser pulse may provide the condition for the interaction of the pulse with its atoms leading to their ionization and plasma formation. However, these effects should appear at relatively high laser fluences that are not safe for normal cells [27].

Thus, we have considered a new mechanism for selective laser killing of abnormal cells by laser thermal explosion of single nanoparticles (nanobombs) delivered to the cells. Thermal explosion is realized when heat is generated within a strongly-absorbing target more rapidly than the heat can diffuse away. On the basis of simple energy balance, it is shown that the threshold energy density of a single laser pulse required for the thermal explosion of a solid gold nanosphere is approximately 40 mJ/cm^2. The nanoparticle's explosion threshold energy density can be reduced further (up to 11 mJ/cm^2) by using large nanorods (and probably nanoshells) and many other advanced GNs whose optical plasmon resonance lies in the near-IR region, where the biological tissue transmissivity is the highest. Additionally, the effective therapeutic effect for cancer cell killing is achieved due to nonlinear phenomena that accompany the thermal explosion of the nanoparticles, such as the generation of GN explosion products with high kinetic energy (nanobullets) as well as of strong shock waves with supersonic expansion in the cell volume or producing optical plasma. It is important that most of these phenomena can explain some published experimental results whose interpretation was performed without taking into account this effect.

References

1. C. M. Pitsillides, E. K. Joe, X. Wei, R. R. Anderson and C. P. Lin, Selective cell targeting with light-absorbing microparticles and nanoparticles, *Biophys. J.* **84**, 4023–4032 (2003).
2. V. P. Zharov, V. Galitovsky and M. Viegas, Photothermal guidance of selective photothrmo;ysis with nanoparticles, *SPIE Proc.* **5319**, 291–300 (2004).
3. V. P. Zharov, R. R. Letfullin and E. Galitovskaya, Microbubbles-overlapping mode for laser killing of cancer cells with absorbing nanoparticle clusters, *J. Phys. D: Appl. Phys.* **38**, 2571–2581 (2005).
4. P. K. Jain, K. S. Lee, I. H. El-Sayed and M. A. El-Sayed, Calculated absorption and Scattering properties of gold nanoparticles of different size, shape, and composition: Applications in biological imaging and biomedicine, *J. Phys. Chem. B* **11**, 7238–7248 (2006).
5. G. Paciotti and L. Myer, Colloidal gold: a novel nanoparticle vector for tumor directed drug delivery, *Drug Delivery* **11**, 169–183 (2004).
6. I. Brigger, C. Dubernet and P. Couvreur, Nanoparticles in cancer therapy and diagnosis, *Adv. Drug Delivery Rev.* **54**, 631–651 (2002).
7. J. F. Hainfeld and J. M. Robinson, New frontier in gold labeling: symposium overview, *J. Histochem. Cytochem.* **48**, 459–475 (2000).
8. A. Karabutov, E. Savateeva and A. Oraevsky, Optoacoustic supercontrast for early cancer detection. *SPIE Proc.* **4256**, 179–187 (2001).
9. D. Leszczynski, C. M. Pitsillides, R. K. Pastila, R. R. Anderson and C. P. Lin, Laser-beam-triggered microcavitation: a novel method for selective cell destruction, *Radiat. Res.* **156**, 399–407 (2001).
10. V. P. Zharov, J-W. Kim, M. Everts and D. T. Curiel, Self-assembling nanoclusters in living systems: Application for integrated photothermal nanodiagnostics and therapy, *Nanomedicine* **1**, 326–345 (2005).
11. R. R. Letfullin, C. E. W. Rice and T. F. George, Bone tissue heating and ablation by short and ultrashort laser pulses, *SPIE Proc.* **7548F-173** (2010), in press.
12. V. P. Zharov, R. R. Letfullin, E. Galitovskay, Laser-induced synergistic effects around absorbing nanoclusters in live cells, *SPIE Proc.* **5695**, 43–50 (2005).

13. R. R. Letfullin, C. Joenathan, T .F. George and V. P. Zharov, Laser-induced explosion of gold nanoparticles: potential role for nanophotothermolysis of cancer, *Nanomedicine* **1**, 473–480 (2006).
14. R. R. Letfullin, V. P. Zharov, C. Joenathan and T. F. George, Laser induced thermal explosion mode for selective nano-photothermolysis of cancer cells, *SPIE Proc.* **6436**, 64360I-1-5 (2007).
15. R. R. Letfullin, V. P. Zharov, C. Joenathan and T. F. George, Nano-photothermolysis of cancer cells, *SPIE Newsroom (Society of Photo-Optical Instrumentation Engineers)* DOI: 10.1117/2.1200701.0634–1–2 (2007).
16. R. R. Letfullin, T. F. George, G. C. Duree and B. M. Bollinger, Ultrashort laser pulse heating of nanoparticles: comparison of theoretical approaches, *Advances in Optical Technologies* **2008**, ID 251718-1-8 (2008).
17. R. R. Letfullin, C. B. Iversen and T. F. George, Multipulse mode of heating nanoparticles by nanosecond, picosecond and femtosecond pulses, *SPIE Proc.* **7576–73** (2010), submitted.
18. R. R. Letfullin, C. E. W. Rice and T. F. George, Bone tissue heating and ablation by short and ultrashort laser pulses, *SPIE Proc.* **7548F–173** (2010), in press.
19. H. Goldenberg and C. J. Tranter, Heat flow in an infinite medium heated by a sphere", *British Journal of Applied Physics* **3**, 296–298 (1952).
20. M. C. Daniel and D. Astruc, Gold nanoparticles: Assembly, supramolecular chemistry, quantum-size-related properties and application toward biology, catalysis and nanotechnology, *Chem. Rev.* **104**, 293–346 (2004).
21. C. P. Lin, M. W. Kelly, S. A. Sibayan, M. A. Latina and R. R. Anderson, Selective cell killing by microparticle absorption of pulsed laser radiation, *IEEE J. Sel. Topics Quant. Electron.* **5**, 963–968 (1999).
22. C. E. Brennen, *Cavitation and Bubble Dynamics* (New York: Oxford University Press, 1995).
23. K. Aslan, J. R. Lakowicz and C. D. Geddes, Nanogold-plasmon-resonanse-based glucose sensing, *Analytical Biochemistry* **330**, 145–155 (2004).
24. J. Neumann and R. Brinkmann, Boiling nucleation on melanosomes and microbeads transiently heated by nanosecond and microsecond laser pulses, *J. Biomed. Opt.* **10**, 024001–024012 (2005).
25. Takami, H. Kurita and S. Koda, Laser-indiced size reduction of noble particle, *J. Phys. Chem. B* **103**, 1226–1232 (1999).
26. R. R. Anderson and J. A. Parrish, Selective photothermolysis: Precise microsurgery by selective absorption of pulsed radiation, *Science* **220**, 524–527 (1983).
27. V. Venugopalan, A. Gyerra III, K. Nahen and A. Vogel, Role of laser-induced plasma formation in pulsed cellular microsurgery and micromanipulation, *Phys. Rev. Lett.* **88**, 078103–1–4 (2002).
28. A. Vogel and V. Venugolapan, Mechanism of pulsed laser ablation of biological tissue, *Chem. Rev.* **103**, 577–644 (2003).

29. K. Yamada, Y. Tokumoto, T. Nagata and F. Mafune, Mechanism of laser-induced size-reduction of gold nanoparticles as studied by nanosecond transient absorption spectroscopy, *J. Phys. Chem. B* **110**, 11751–11756 (2006).
30. S. Inasawa, M. Sugiyama, S. Noda and Y. Yamaguchi, Spectroscopic study of laser-induced phase transition of gold nanoparticles on nanosecond time scale and longer, *J. Phys. Chem. B* **110**, 3114–3119 (2006).
31. G. Huttmann, B. Radt, J. Serbin and R. Birngruber, Inactivation of proteins by irradation of gold nanopartciles with nano- and picosecond laser pusles, *SPIE Proc.* **5142**, 88–95 (2003).
32. P. Steinbach, F. Hofstadter, H. Nicolai, W. Rossier and W. Wieland, *In vitro* investigations on cellular damage induced by high energy shock waves, *Ultrasound Med. Biol.* **18**, 691–699 (1992).
33. A. G. Doukas, D. J. McAuliffe and T. J. Flotte, Biological effects of laser-induced shock waves: Structural and functional cell damage in vitro. *Ultrasound Med. Biol.* **19**, 137–146 (1993).
34. T. Douki, S. Lee, K. Dorey, T. J. Flotte, T. F. Deutsch and A. G. Doukas, Stress-wave-induced injury to retinal pigment epithelium cells in vitro, *Lasers Surg. Med.*, **19**, 249–259 (1996).
35. T. Kodama, H. Uenohara and K. Takayama, Innovative technology for tissue disruption by explosive-induced shock waves, *Ultrasound Med. Biol.* **24**, 1459–1466 (1998).
36. A. Sonden, B. Svensson, N. Roman, H. Ostmark, B. Brismar, J. Palmblad and B. T. Kjellstrom, Laser-induced shock wave endothelial cell injury, *Lasers Surg. Med.*, **26**, 364–375 (2000).
37. E. Faraggi, B. S. Gerstman and J. Sun, Biophysical effects of pulsed lasers in the retina and other tissues containing strongly absorbing particles: Shockwave and explosive bubble generation, *J. Biomed. Opt.* **10**, 064029 (2005).
38. S. Inasawa, M. Sugiyama and Y. Yamaguchi, Bimodal size distribution of gold nanoparticles under picosecond laser pulses, *J. Phys. Chem. B* **109**, 9404–9410 (2005).
39. L. Boufendi, A. Bouchoule, B. Dubreuil, E. Stoffels, W. W. Stoffels and M .L. deGiorgi, Study of initial dust formation in an Ar-SiH_4 discharge by laser induced particle explosive evaporation, *J. Appl. Phys.* **76**, 148–153 (1994).
40. *2000 American National Standard for safe use of Lasers* **ANSIZ136.1**.
41. Z. Peng, T. Walther and K. Kleinermanns, Photofragmentation of phase-transferred gold nanoparticles by intense pulsed laser light, *J Phys Chem B* **109**, 15735–15740 (2005).
42. Z. Peng, T. Walther and K. Kleinermanns, Influence of intense pulsed laser irradiation on optical and morphological properties of gold nanoparticle aggregates produced by surface acid-base reactions, *Langmuir* **21**,4249–4253 (2005).

Chapter 7

New Direct Inhibitors and Their Computed Effect on the Dynamics of Thrombin Formation in Blood Coagulation

Liliana Braescu

West University of Timisoara

lilianabraescu@balint1.math.uvt.ro

Marius Leretter

Victor Babes University of Medicine and Pharmacy

mariusleretter@yahoo.com

Thomas F. George

University of Missouri–St. Louis

tfgeorge@umsl.edu

Thromboembolitic diseases, such as myocardial infarction, stroke and deep vein thrombosis, are a leading cause of death throughout the world. The past two decades have seen a great deal of progress in the development of antithrombotic agents, spurred by the severity of the problem and the medical need for life-saving treatments [1]. Three treatment approaches are the development of (1) antiplatelet agents, (2) compounds that aid in the lysis of blood clots, and (3) agents that affect the activity and generation of thrombin. Thrombin is a serine protease enzyme that is responsible for many aspects in the blood coagulation cascade. Many agents have been developed to control the generation of thrombin, including heparin, low molecular weight heparins and natural and semisynthetic thrombin inhibitors such as hirudin and hirulog. These agents all have the disadvantage that they must be administered as either intravenous infusions in a hospital or as subcutaneous injections several times a day. Warfarin is an oral agent utilized as an anticoagulant because of its effects on the vitamin-K-dependent coagulation pathway. Some disadvantages of warfarin include a slow onset of action, the need for dietary restrictions, and the possibility of drug-drug interactions. The need for better agents that can be used orally with infrequent patient monitoring still exists and has stimulated a great deal of interest in the pursuit of small-molecule inhibitors of thrombin. Direct thrombin inhibitors have been developed and tested in clinical trials for a variety of these thrombotic disorders. However, the numerous biochemical reactions that take place leading to the formation and lysis of clots, and the exact influence of hemodynamic factors in these reactions, are incompletely understood [2]. Mathematical modeling provides the opportunity to integrate and quantify reaction details which, in turn, aids in the design of the more expensive empirical experiments.

In this chapter, some of the direct thrombin inhibitors (DTIs) like argatroban, hirudin and melagatran are taken under study, and their effect on the thrombin formation is investigated. Toward this aim, a recent mathematical model [3] developed on the base of the biochemical reactions of both intrinsic and extrinsic pathways of the blood coagulation system is considered, in which new reactions corresponding to DTIs are introduced.

7.1. Introduction

Blood coagulation is a basic physiological defense mechanism that occurs in all vertebrates to prevent blood loss following vascular injury. The coagulation is composed of a set of pro- and anticoagulant systems that maintain the balance of blood fluidity. Defects in this balance can result in either thrombosis or bleeding tendencies. Qualitative or quantitative alterations in this haemostatic balance can have devastating effects, producing hemorrhagic diseases or thrombosis diseases. An understanding of the working mechanism of coagulation is based on the well-known scheme of the reactions cascade. This has two pathways — intrinsic and extrinsic — representing a series of reactions of coagulation factors (indicated by Roman numerals according to the standard system of designation — see Fig. 7.1) in which a stable form of a protein is activated to become an enzyme which then catalyzes the next reaction in the cascade.

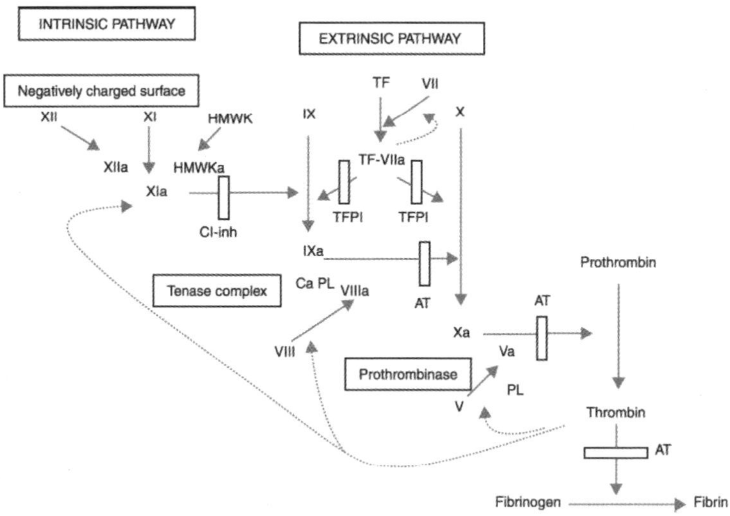

Fig. 7.1. Coagulation cascade showing both intrinsic and extrinsic activation, inhibitors and feedback activation (dashed lines) [4].

Both the extrinsic and intrinsic pathways serve the common purpose of proteolytic cleavage of factor X into the active form Xa, which in the presence of the factor Va, phospholipids and calcium ions forms the prothrombinase complex leading to a burst of thrombin generation (factor IIa):

(i) The intrinsic pathway [3–7] is activated when blood comes into contact with sub-endothelial connective tissues or with negatively-charged surfaces that are exposed as a result of tissue damage. Quantitatively, it is the more important of the two pathways, but is slower to cleave fibrin (factor Ia) than the extrinsic pathway. The Hageman factor (factor XII), factor XI and high-molecular-weight-kininogen (HMWK) are involved in this pathway of activation. The first step is the binding of factor XII to a sub-endothelial surface exposed by an injury. HMWK interacts with the exposed surface in close proximity to the bound factor XII, which becomes activated (XIIa). The free HMWK also binds to factor XI, and this complex adsorbs to the same surface as XIIa, resulting in the activation of factor XI (specifically, factor XIa). Next, XIa activates factor IX. Calcium ions (Ca^{++}) facilitate the binding of IXa, X and VIII to a negatively-charged phospholipids (PL) bilayer surface, such as the membrane of a procoagulant platelet. Once the factor VIII in this complex is activated by thrombin (factor IIa), activation of factor X ensues.

(ii) The extrinsic coagulation pathway [2–3,8–10] is initiated when cryptic tissue factor TF is exposed to circulating blood and binds plasma factor VIIa. The factor VIIa≡TF complex activates factors IX and factor X. Factor Xa, that is initially produced, generates picomolar amounts of thrombin, which activates platelets and factors V, VIII and XI. Additionally, factor Xa cleaves factor IX and generates factor IXa, which is also the intermediate product following the proteolytic cleavage of factor IX by either the VIIa≡TF complex or factor XIa. Formation of the enzymatic complex composed by factors IXa and VIIIa, negatively-charged phospholipids and Ca^{++} (intrinsic tenase) leads to further activation of factor X.

Thrombin (factor IIa) and Xa play a major role in the positive feedback mechanisms by catalyzing the production of almost all the intermediates required for their production. Thrombin activates platelets that then release ADP (adenosine diphosphate that circulates in the blood), leading in turn to the activation of other platelets. Thrombin activates factor XI — a zymogen that is linked to the intrinsic coagulation pathway, which in turn activates IX. Thrombin also plays a role in the inhibition of coagulation by catalyzing the formation of active protein C (APC) in plasma through the thrombin–thrombomodulin complex.

Consequently, thrombin is the essential enzyme product of the blood coagulation cascade leading to formation of fibrin (factor Ia) monomers that polymerize to form fibrin strands. These strands form a matrix that binds the platelet aggregates, red blood cells and white blood cells, which entrap the plasma and constitute a clot. The fibrin strands themselves, while providing structural integrity to the clot, actually constitute less than 1% by volume of the clot. Negative regulatory networks also regulate the thrombus formation, including both anticoagulant (inhibitors of coagulation: antithrombin III (AT), tissue factor pathway inhibitor (TFPI) or active protein C (APC)) and destructive (fibrinolytic) mechanisms.

Mathematical models able to capture many of the salient features of the phenomenon of clot formation have been developed; the disadvantage of these models is that they involve separately the intrinsic [4–7] and extrinsic [2,8–10] pathways. Because these pathways meet at a common point — factor X — and the production of X by an intrinsic pathway is 100 times more than what is produced in an extrinsic pathway alone, a more realistic mathematical model was reported in Ref. [3], based on the scheme of biochemical reactions, in which both intrinsic and extrinsic pathways were modeled. Numerical results displayed a great similarity in the relative amounts of thrombin produced under experimental conditions with similar peak values, and eliminated the major nonconformity reported by the early models in which the duration of the initiation phase is noticeably larger than those observed in the empirical experiments [9].

In order to compute the effect of DTI on the dynamic of the thrombin formation, a mathematical model containing DTI should be built. For this purposc, thc mathcmatical model reported in Ref. [3] is considered and modified according to the new reactions appearing due to the presence of DTI.

7.2. Mathcmatical and Computational Formulation

7.2.1. *Mathematical model without direct thrombin inhibitors*

The biochemical reactions of a coagulation system (see Fig. 7.2), where a stable form of a protein is activated to become an enzyme which then catalyzes the next reaction in the cascade, can be expressed by the following simplified scheme:

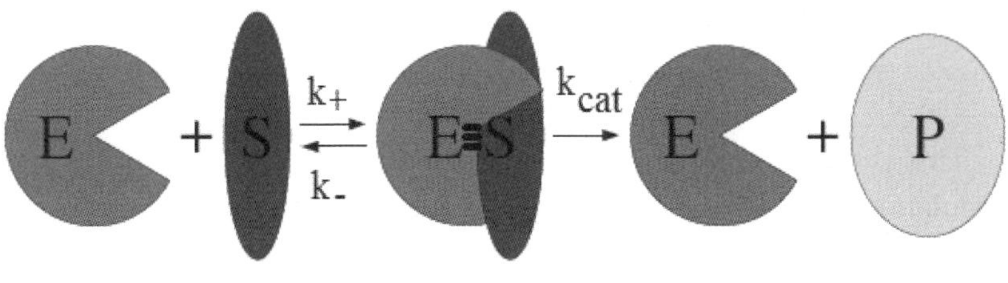

Fig. 7.2. Schematic enzyme reaction mechanism.

Here E and S the represent enzyme and substrate, respectively; they combine into an instable complex $E \equiv S$ which then breaks down into a product P and the original enzyme E. The first reaction is reversible in which the quantity k_+ is the rate constant of second order (expressed in $M^{-1}s^{-1}$), and k_- is the rate constant of first order (expressed in s^{-1}). The second reaction is irreversible in which the rate constant k_{cat}, called the catalyzed constant of the enzyme (expressed in s^{-1}), measures the rate at which the product is created.

In order to quantify the rate of change of chemical concentrations, the law of mass action is involved. Let $[S]$, $[E]$, $[E \equiv S]$ and $[P]$ denote the concentrations of S, E, $E \equiv S$, and P, expressed in M. Taking into account that we have reactions of the first order (k_- is expressed in s^{-1}), second order (k_+ is expressed in $M^{-1}s^{-1}$), and of the Michaelis–Menten type (k_{cat}), the reaction rates (expressed in s^{-1}) corresponding to the above biochemical reactions are the following:

$$v_{\mathrm{I}} = k_- \cdot [E \equiv S],\ v_{\mathrm{II}} = k_+ \cdot [E] \cdot [S] \text{ and } v_{M\text{-}M} = \frac{k_{cat} \cdot [E] \cdot [S]}{k_m + [S]},$$

where $k_m = \dfrac{k_-}{k_+}$.

Thus, the total reaction rates of change of the variables $[S]$, $[E]$, $[E \equiv S]$, $[P]$ are given by the differential equations

$$\begin{cases} \dfrac{d[E]}{dt} = -k_+ \cdot [E] \cdot [S] + k_- \cdot [E \equiv S] + \dfrac{k_{cat} \cdot [E] \cdot [S]}{k_m + [S]} \\ \dfrac{d[S]}{dt} = -k_+ \cdot [E] \cdot [S] + k_- \cdot [E \equiv S] \\ \dfrac{d[E \equiv S]}{dt} = +k_+ \cdot [E] \cdot [S] - k_- \cdot [E \equiv S] - \dfrac{k_{cat} \cdot [E] \cdot [S]}{k_m + [S]} \\ \dfrac{d[P]}{dt} = +\dfrac{k_{cat} \cdot [E] \cdot [S]}{k_m + [S]} \end{cases} \quad . \tag{7.1}$$

This is the central idea of the mathematical model. According to the coagulation cascade and with the biochemical reactions for intrinsic and extrinsic pathways, a dynamical system in which intrinsic and extrinsic pathways are modeled simultaneously is constructed.

The intrinsic pathway is described through 23 biochemical reactions [6]:

$$IX \xrightarrow[k_{i9}]{XIa} IXa\,;\ IXa + AT \xrightarrow{h_{i9}} (IXa \equiv AT)\,;\ VIII \xrightarrow[k_{i8}]{IIa} VIIIa\,;$$

$$VIII \xrightarrow{h_{i8}} VIIIai\,;\ VIII \xrightarrow[k_{ia}]{APC} VIIIi\,;\ V \xrightarrow[k_{i5}]{IIa} Va\,;\ V \xrightarrow{h_{i5}} Vai\,;$$

$$V \xrightarrow[k_{ia}]{APC} Vai\,;\ VIIIa + IXa \underset{h_{i8,9}}{\overset{k_{i8,9}}{\rightleftarrows}} (VIIIa \equiv IXa)\,;$$

$$(VIIIa \equiv IXa) \xrightarrow[k_{ia}]{APC} VIIIai + IXa\,;\ Va + Xa \underset{h_{i5,10}}{\overset{k_{i5,10}}{\rightleftarrows}} (Va \equiv Xa)\,;$$

$$(Va \equiv Xa) \xrightarrow[k_{ia}]{APC} Vai + Xa\,;\ X \xrightarrow[k_{i10}]{IXa} Xa\,;\ X \xrightarrow[\bar{k}_{i10}]{IXa \equiv VIIIa} Xa\,;$$

$$Xa + AT \xrightarrow{h_{i10}} (Xa \equiv AT)\,;\ II \xrightarrow[k_{i2}]{Xa} IIa\,;\ II \xrightarrow[\bar{k}_{i2}]{Xa \equiv Va} IIa\,;$$

$$IIa + AT \xrightarrow{h_{i2}} (IIa \equiv AT)\,;\ PC \xrightarrow[k_{iAPC}]{IIa} APC\,;$$

$$APC + AT \xrightarrow{h_{iAPC}} (APC \equiv AT)\,;\ I \xrightarrow[k_{i1}]{IIa} Ia\,;\ XI \xrightarrow[k_{i11}]{IIa} Xia\,;$$

$$XIa \xrightarrow{h_{i11}} XIai\,,$$

with the rate constants

$k_{i9} = 0.33\ s^{-1}$; $h_{i9} = 0.33\times10^{-2}\ s^{-1}$; $k_{i8} = 0.16\times10^{-6}\ s^{-1}$; $h_{i8} = 0.516\times10^{-2}\ s^{-1}$;
$k_{ia} = 2\times10^{7}\ M^{-1}s^{-1}$; $k_{i5} = 0.28\times10^{-2}\ s^{-1}$; $h_{i5} = 0.516\times10^{-2}\ s^{-1}$;
$k_{i8,9} = 1.667\times10^{9}\ M^{-1}s^{-1}$; $h_{i8,9} = 1.667\times s^{-1}$; $k_{i5,10} = 1.667\times10^{9}\ M^{-1}s^{-1}$;
$h_{i5,10} = 1.667\ s^{-1}$; $k_{i10} = 5\times10^{-5}\ s^{-1}$; $\bar{k}_{i10} = 8.33\ s^{-1}$; $h_{i10} = 0.017\ s^{-1}$;
$k_{i2} = 0.038\ s^{-1}$; $\bar{k}_{i2} = 33.33\ s^{-1}$; $k_{i2m} = 58\times10^{-9}\ M$; $\bar{k}_{i2m} = 210\times10^{-9}\ M$;
$h_{i2} = 0.022\ s^{-1}$; $k_{iAPC} = 2.33\times10^{-5}\ s^{-1}$; $h_{iAPC} = 1.66\times10^{-3}\ s^{-1}$; $k_{i1} = 0.047\ s^{-1}$;
$k_{i11} = 1.3\times10^{-4}\ s^{-1}$; $h_{i11} = 3.33\times10^{-3}\ s^{-1}$.

The extrinsic pathway is described through 27 biochemical reactions [9]:

$$TF{+}VII \underset{k_2}{\overset{k_1}{\rightleftarrows}} (TF \equiv VII)\,;\ TF{+}VIIa \underset{k_4}{\overset{k_3}{\rightleftarrows}} (TF \equiv VIIa)\,;$$

$$(TF \equiv VIIa) + VII \xrightarrow{k_5} (TF \equiv VIIa) + VIIa\,;\ Xa + VII \xrightarrow{k_6} Xa + VIIa\,;$$

$$IIa + VII \xrightarrow{k_7} IIa + VIIa\,;$$

$$(TF \equiv VIIa) + X \underset{k_9}{\overset{k_8}{\rightleftarrows}} (TF \equiv VIIa \equiv X) \xrightarrow{k_{10}} (TF \equiv VIIa \equiv Xa)\,;$$

$$(TF \equiv VIIa) + Xa \underset{k_{12}}{\overset{k_{11}}{\rightleftarrows}} (TF \equiv VIIa \equiv Xa)\,;$$

$$(TF \equiv VIIa) + IX \underset{k_{14}}{\overset{k_{13}}{\rightleftarrows}} (TF \equiv VIIa \equiv IX) \xrightarrow{k_{15}} (TF \equiv VIIa) + IXa\,;$$

$$Xa + II \xrightarrow{k_{16}} Xa + IIa\,;\ IIa + VIII \xrightarrow{k_{17}} IIa + VIIIa\,;$$

$$\mathrm{VIIIa + IXa \underset{k_{19}}{\overset{k_{18}}{\rightleftharpoons}} (IXa \equiv VIIIa)};$$
$$\mathrm{(IXa \equiv VIIIa) + X \underset{k_{21}}{\overset{k_{20}}{\rightleftharpoons}} (IXa \equiv VIIIa \equiv X) \xrightarrow{k_{22}} (IXa \equiv VIIIa) + Xa};$$
$$\mathrm{VIIIa \underset{k_{24}}{\overset{k_{23}}{\rightleftharpoons}} VIIIa_{1,L} + VIIIa_2};$$
$$\mathrm{(IXa \equiv VIIIa \equiv X) \xrightarrow{k_{25}} VIIIa_{1,L} + VIIIa_2 + X + IXa};$$
$$\mathrm{(IXa \equiv VIIIa) \xrightarrow{k_{25}} VIIIa_{1,L} + VIIIa_2 + IXa};$$
$$\mathrm{IIa + V \xrightarrow{k_{26}} IIa + Va};\ \mathrm{Xa + Va \underset{k_{28}}{\overset{k_{27}}{\rightleftharpoons}} (Xa \equiv Va)};$$
$$\mathrm{(Xa \equiv Va) + II \underset{k_{30}}{\overset{k_{29}}{\rightleftharpoons}} (Xa \equiv Va \equiv II) \xrightarrow{k_{31}} (Xa \equiv Va) + mIIa};$$
$$\mathrm{mIIa + (Xa \equiv Va) \xrightarrow{k_{32}} IIa + (Xa \equiv Va)};\ \mathrm{Xa + TFPI \underset{k_{34}}{\overset{k_{33}}{\rightleftharpoons}} (Xa \equiv TFPI)};$$
$$\mathrm{(TF \equiv VIIa \equiv Xa) + TFPI \underset{k_{36}}{\overset{k_{35}}{\rightleftharpoons}} (TF \equiv VIIa \equiv Xa \equiv TFPI)};$$
$$\mathrm{(TF \equiv VIIa) + (Xa \equiv TFPI) \xrightarrow{k_{37}} (TF \equiv VIIa \equiv Xa \equiv TFPI)};$$
$$\mathrm{Xa + AT \xrightarrow{k_{38}} (Xa \equiv AT)};\ \mathrm{mIIa{+}AT \xrightarrow{k_{39}} (mIIa \equiv AT)};$$
$$\mathrm{Xa + AT \xrightarrow{k_{40}} (IXa \equiv AT)};\ \mathrm{IIa + AT \xrightarrow{k_{41}} (IIa \equiv AT)};$$
$$\mathrm{(TF \equiv VIIa) + AT \xrightarrow{k_{42}} (TF \equiv VIIa \equiv AT)},$$

with the rate constants

$k_1 = 3.2\times10^6$ M^{-1}s^{-1}; $k_2 = 3.1\times10^{-3}$ s^{-1}; $k_3 = 2.3\times10^7$ M^{-1}s^{-1};

$k_4 = 3.1\times10^7$ s^{-1}; $k_5 = 4.4\times10^5$ s^{-1}; $k_6 = 1.3\times10^7$ M^{-1}s^{-1};

$k_7 = 2.3\times10^4$ M^{-1}s^{-1}; $k_8 = 2.5\times10^7$ M^{-1}s^{-1}; $k_9 = 1.05$ s^{-1}; $k_{10} = 6$ s^{-1};

$k_{11} = 2.2\times10^7$ M^{-1}s^{-1}; $k_{12} = 19$ s^{-1}; $k_{13} = 1\times10^7$ M^{-1}s^{-1}; $k_{14} = 2.4$ s^{-1};

$k_{15} = 1.8$ s^{-1}; $k_{16} = 7.5\times10^3$ M^{-1}s^{-1}; $k_{17} = 2\times10^7$ M^{-1}s^{-1}; $k_{18} = 1\times10^7$ M^{-1}s^{-1};

$k_{19} = 5\times10^{-3}$ s^{-1}; $k_{20} = 1\times10^8$ M^{-1}s^{-1}; $k_{21} = 1\times10^{-3}$ s^{-1}; $k_{22} = 8.2$ s^{-1};

$k_{23} = 6\times10^{-3}$ s^{-1}; $k_{24} = 2.2\times10^4$ M^{-1}s^{-1}; $k_{25} = 1\times10^{-3}$ s^{-1};

$k_{26} = 2\cdot10^7$ M^{-1}s^{-1}; $k_{27} = 4\cdot10^8$ M^{-1}s^{-1}; $k_{28} = 0.2$ s^{-1}; $k_{29} = 1\times10^8$ M^{-1}s^{-1};

$k_{30} = 103$ s^{-1}; $k_{31} = 63.5$ s^{-1}; $k_{32} = 1.5\times10^7$ M^{-1}s^{-1}; $k_{33} = 9\times10^5$ M^{-1}s^{-1};

$k_{34} = 3.6\times10^{-4}$ s^{-1}; $k_{35} = 3.2\times10^8$ M^{-1}s^{-1}; $k_{36} = 1.1\times10^{-4}$ s^{-1};

$k_{37} = 5\times10^7$ M^{-1}s^{-1}; $k_{38} = 1.5\times10^3$ M^{-1}s^{-1}; $k_{39} = 7.1\times10^3$ M^{-1}s^{-1};

$k_{40} = 4.9\times10^2$ M^{-1}s^{-1}; $k_{41} = 7.1\times10^3$ M^{-1}s^{-1}; $k_{42} = 2.3\times10^2$ M^{-1}s^{-1}.

Considering the above biochemical reactions for both intrinsic and extrinsic pathways and the law of mass action as it was presented for the system (7.1), according to Ref. [3], the dynamics of the coagulation factors' concentrations [*Ia*], [*APC*], [*TF*], [*VII*], [*TF*≡*VII*], [*VIIa*], [*TF*≡*VIIa*], [*X*], [*TF*≡*VIIa*≡*X*], [*TF*≡*VIIa*≡*Xa*], [*IX*], [*TF*≡*VIIa*≡*IX*], [*VIII*], [*IXa*≡*VIIIa*≡*X*], [$VIIIa_{1.L}$], [$VIIIa_2$], [*V*], [*Xa*≡*Va*≡*II*], [*mIIa*], [*TFPI*], [*Xa*≡*TFPI*], [*TF*≡*VIIa*≡*Xa*≡*TFPI*], [*AT*], [*Xa*≡*AT*], [*mIIa*≡*AT*], [*IXa*≡*AT*], [*IIa*≡*AT*], [*TF*≡*VIIa*≡*AT*], [*IXa*], [*Xa*], [*IIa*], [*II*], [*VIIIa*], [*Va*], [*Xa*≡*Va*] and [*IXa*≡*VIIIa*] are described by the following nonlinear system of the 36 differential equations:

$$\frac{d[Ia]}{dt} = k_{i1} \cdot [IIa] \tag{7.2}$$

$$\frac{d[APC]}{dt} = k_{iAPC} \cdot [IIa] - h_{iAPC} \cdot [APC] \tag{7.3}$$

$$\frac{d[TF]}{dt} = -k_1 \cdot [TF] \cdot [VII] + k_2 \cdot [TF \equiv VII] + k_4 \cdot [TF \equiv VIIa] - k_3 \cdot [VIIa] \cdot [TF] \tag{7.4}$$

$$\frac{d[VII]}{dt} = -k_1 \cdot [TF] \cdot [VII] + k_2 \cdot [TF \equiv VII] - k_5 \cdot [TF \equiv VIIa] \cdot [VII] - k_6 \cdot [Xa] \cdot [VII] - k_7 \cdot [IIa] \cdot [VII] \tag{7.5}$$

$$\frac{d[TF \equiv VII]}{dt} = k_1 \cdot [TF] \cdot [VII] - k_2 \cdot [TF \equiv VII] \tag{7.6}$$

$$\frac{d[VIIa]}{dt} = k_4 \cdot [TF \equiv VIIa] - k_3 \cdot [TF] \cdot [VIIa] + k_5 \cdot [TF \equiv VIIa] \cdot [VII] + k_6 \cdot [Xa] \cdot [VII] + k_7 \cdot [IIa] \cdot [VII] \tag{7.7}$$

$$\begin{aligned} \frac{d[TF \equiv VIIa]}{dt} = {} & -k_4 \cdot [TF \equiv VIIa] + k_3 \cdot [TF] \cdot [VIIa] - k_8 \cdot [TF \equiv VIIa] \cdot [X] + \\ & + k_9 \cdot [TF \equiv VIIa \equiv X] - k_{11} \cdot [TF \equiv VIIa] \cdot [Xa] + k_{12} \cdot [TF \equiv VIIa \equiv Xa] - \\ & - k_{13} \cdot [TF \equiv VIIa] \cdot [IX] + k_{14} \cdot [TF \equiv VIIa \equiv IX] + k_{15} \cdot [TF \equiv VIIa \equiv IX] - \\ & - k_{37} \cdot [TF \equiv VIIa] \cdot [Xa \equiv TFPI] - k_{42} \cdot [TF \equiv VIIa] \cdot [AT] \end{aligned} \tag{7.8}$$

$$\begin{aligned} \frac{d[X]}{dt} = {} & -k_8 \cdot [TF \equiv VIIa] \cdot [X] + k_9 \cdot [TF \equiv VIIa \equiv X] - k_{20} \cdot [IXa \equiv VIIIa] \cdot [X] + \\ & + k_{21} \cdot [IXa \equiv VIIIa \equiv X] + k_{25} \cdot [IXa \equiv VIIIa \equiv X] \end{aligned} \tag{7.9}$$

$$\frac{d[TF \equiv VIIa \equiv X]}{dt} = k_8 \cdot [TF \equiv VIIa] \cdot [X] - k_9 \cdot [TF \equiv VIIa \equiv X] - k_{10} \cdot [TF \equiv VIIa \equiv X] \tag{7.10}$$

$$\begin{aligned} \frac{d[TF \equiv VIIa \equiv Xa]}{dt} = {} & k_{10} \cdot [TF \equiv VIIa \equiv X] - k_{12} \cdot [TF \equiv VIIa \equiv Xa] + \\ & + k_{11} \cdot [TF \equiv VIIa] \cdot [Xa] - k_{35} \cdot [TF \equiv VIIa \equiv Xa] \cdot [TFPI] + \\ & + k_{36} \cdot [TF \equiv VIIa \equiv Xa \equiv TFPI] \end{aligned} \tag{7.11}$$

$$\frac{d[IX]}{dt} = k_{14} \cdot [TF \equiv VIIa \equiv IX] - k_{13} \cdot [TF \equiv VIIa] \cdot [IX] \tag{7.12}$$

$$\frac{d[TF \equiv VIIa \equiv IX]}{dt} = -k_{14} \cdot [TF \equiv VIIa \equiv IX] + k_{13} \cdot [TF \equiv VIIa] \cdot [IX] - \\ - k_{15} \cdot [TF \equiv VIIa \equiv IX] \tag{7.13}$$

$$\frac{d[VIII]}{dt} = -k_{17} \cdot [IIa] \cdot [VIII] \tag{7.14}$$

$$\frac{d[IXa \equiv VIIIa \equiv X]}{dt} = k_{20} \cdot [IXa \equiv VIIIa] \cdot [X] - k_{25} \cdot [IXa \equiv VIIIa \equiv X] - \\ - k_{22} \cdot [IXa \equiv VIIIa \equiv X] - k_{21} \cdot [IXa \equiv VIIIa \equiv X] \tag{7.15}$$

$$\frac{d[VIIIa_{1.L}]}{dt} = k_{23} \cdot [VIIIa] - k_{24} \cdot [VIIIa_{1.L}] \cdot [VIIIa_2] + k_{25} \cdot [IXa \equiv VIIIa \equiv X] + \\ + k_{25} \cdot [IXa \equiv VIIIa] \tag{7.16}$$

$$\frac{d[VIIIa_2]}{dt} = k_{23} \cdot [VIIIa] - k_{24} \cdot [VIIIa_2] \cdot [VIIIa_{1.L}] + k_{25} \cdot [IXa \equiv VIIIa \equiv X] + \\ + k_{25} \cdot [IXa \equiv VIIIa] \tag{7.17}$$

$$\frac{d[V]}{dt} = -k_{26} \cdot [IIa] \cdot [V] \tag{7.18}$$

$$\frac{d[Xa \equiv Va \equiv II]}{dt} = k_{29} \cdot [II] \cdot [Xa \equiv Va] - k_{30} \cdot [Xa \equiv Va \equiv II] - k_{31} \cdot [Xa \equiv Va \equiv II] \tag{7.19}$$

$$\frac{d[mIIa]}{dt} = k_{31} \cdot [Xa \equiv Va \equiv II] - k_{32} \cdot [mIIa] \cdot [Xa \equiv Va] - k_{39} \cdot [mIIa] \cdot [AT] \tag{7.20}$$

$$\frac{d[TFPI]}{dt} = -k_{33} \cdot [Xa] \cdot [TFPI] + k_{34} \cdot [Xa \equiv TFPI] - k_{35} \cdot [TF \equiv VIIa \equiv Xa] \cdot [TFPI] + \\ + k_{36} \cdot [TF \equiv VIIa \equiv Xa \equiv TFPI] \tag{7.21}$$

$$\frac{d[Xa \equiv TFPI]}{dt} = k_{33} \cdot [Xa] \cdot [TFPI] - k_{34} \cdot [Xa \equiv TFPI] - k_{37} \cdot [TF \equiv VIIa] \cdot [Xa \equiv TFPI] \tag{7.22}$$

$$\frac{d[TF \equiv VIIa \equiv Xa \equiv TFPI]}{dt} = k_{35} \cdot [TF \equiv VIIa \equiv Xa] \cdot [TFPI] - k_{36} \cdot [TF \equiv VIIa \equiv Xa \equiv TFPI] + \\ + k_{37} \cdot [TF \equiv VIIa] \cdot [Xa \equiv TFPI] \tag{7.23}$$

$$\frac{d[AT]}{dt} = -k_{38} \cdot [Xa] \cdot [AT] - k_{39} \cdot [mIIa] \cdot [AT] - k_{40} \cdot [AT] \cdot [IXa] - k_{41} \cdot [IIa] \cdot [AT] - \\ - k_{42} \cdot [TF \equiv VIIa] \cdot [AT] \tag{7.24}$$

$$\frac{d[Xa \equiv AT]}{dt} = k_{38} \cdot [Xa] \cdot [AT] \tag{7.25}$$

$$\frac{d[mIIa \equiv AT]}{dt} = k_{39} \cdot [mIIa] \cdot [AT] \tag{7.26}$$

$$\frac{d[IXa \equiv AT]}{dt} = k_{40} \cdot [IXa] \cdot [AT] \tag{7.27}$$

$$\frac{d[IIa \equiv AT]}{dt} = k_{41} \cdot [IIa] \cdot [AT] \tag{7.28}$$

$$\frac{d[TF \equiv VIIa \equiv AT]}{dt} = k_{42} \cdot [TF \equiv VIIa] \cdot [AT] \tag{7.29}$$

$$\frac{d[IXa]}{dt} = k_{15} \cdot [TF \equiv VIIa \equiv IX] - k_{18} \cdot [IXa] \cdot [VIIIa] + k_{19} \cdot [IXa \equiv VIIIa] + k_{25} \cdot [IXa \equiv VIIIa \equiv X] - -k_{40} \cdot [IXa] \cdot [AT] + k_{i9} \cdot [XIa] - h_{i9} \cdot [IXa] \tag{7.30}$$

$$\frac{d[Xa]}{dt} = -k_{11} \cdot [TF \equiv VIIa] \cdot [Xa] + k_{12} \cdot [TF \equiv VIIa \equiv Xa] + k_{22} \cdot [IXa \equiv VIIIa \equiv X] - -k_{27} \cdot [Xa] \cdot [Va] + k_{28} \cdot [Xa \equiv Va] - k_{33} \cdot [Xa] \cdot [TFPI] + k_{34} \cdot [Xa \equiv TFPI] - -k_{38} \cdot [Xa] \cdot [AT] + k_{i10} \cdot [IXa] + \bar{k}_{i10} \cdot [IXa \equiv VIIIa] - h_{i10} \cdot [Xa] \tag{7.31}$$

$$\frac{d[IIa]}{dt} = k_{32} \cdot [mIIa] \cdot [Xa \equiv Va] - k_{41} \cdot [IIa] \cdot [AT] + k_{16} \cdot [Xa] \cdot [II] + +k_{i2} \cdot \frac{[Xa] \cdot [II]}{[II] + k_{i2m}} + \bar{k}_{i2} \cdot \frac{[Xa \equiv Va] \cdot [II]}{[II] + \bar{k}_{i2m}} - h_{i2} \cdot [IIa] \tag{7.32}$$

$$\frac{d[II]}{dt} = -k_{16} \cdot [Xa] \cdot [II] - k_{29} \cdot [Xa \equiv Va] \cdot [II] + k_{30} \cdot [Xa \equiv Va \equiv II] - k_{i2} \cdot \frac{[Xa] \cdot [II]}{[II] + k_{i2m}} - -\bar{k}_{i2} \cdot \frac{[Xa \equiv Va] \cdot [II]}{[II] + \bar{k}_{i2m}} \tag{7.33}$$

$$\frac{d[VIIIa]}{dt} = k_{17} \cdot [IIa] \cdot [VIII] - k_{18} \cdot [VIIIa] \cdot [IXa] + k_{19} \cdot [IXa \equiv VIIIa] - k_{23} \cdot [VIIIa] + +k_{24} \cdot [VIIIa_{1.L}] \cdot [VIIIa_2] + k_{i8} \cdot [IIa] - h_{i8} \cdot [VIIIa] - k_{ia} \cdot [APC] \cdot ([VIIIa] + [IXa \equiv VIIIa]) \tag{7.34}$$

$$\frac{d[Va]}{dt} = k_{26} \cdot [IIa] \cdot [V] + k_{28} \cdot [Xa \equiv Va] - k_{27} \cdot [Xa] \cdot [Va] + k_{i5} \cdot [IIa] - h_{i5} \cdot [Va] - -k_{ia} \cdot [APC] \cdot ([Va] + [Xa \equiv Va]) \tag{7.35}$$

$$\frac{d[Xa \equiv Va]}{dt} = k_{27} \cdot [Xa] \cdot [Va] - k_{28} \cdot [Xa \equiv Va] - k_{29} \cdot [II] \cdot [Xa \equiv Va] + k_{30} \cdot [Xa \equiv Va \equiv II] + +k_{31} \cdot [Xa \equiv Va \equiv II] + k_{i5,10} \cdot [Va] \cdot [Xa] - h_{i5,10} \cdot [Xa \equiv Va] - k_{ia} \cdot [Xa \equiv Va] \cdot [APC] \tag{7.36}$$

$$\frac{d[IXa \equiv VIIIa]}{dt} = k_{18} \cdot [VIIIa] \cdot [IXa] - k_{19} \cdot [IXa \equiv VIIIa] - k_{20} \cdot [IXa \equiv VIIIa] \cdot [X] + +k_{21} \cdot [IXa \equiv VIIIa \equiv X] + k_{22} \cdot [IXa \equiv VIIIa \equiv X] - k_{25} \cdot [IXa \equiv VIIIa] + +k_{i8,9} \cdot [VIIIa] \cdot [IXa] - h_{i8,9} \cdot [IXa \equiv VIIIa] - k_{ia} \cdot [IXa \equiv VIIIa] \cdot [APC]. \tag{7.37}$$

Concentrations of those 36 coagulation factors were computed solving numerically the nonlinear system of differential equations (7.2)–(7.37) through MathCAD 13 Enterprise Edition software [3], for the following initial concentrations: $[Ia](0)=0$, $[APC](0)=0$, $[TF](0)=2.5 \cdot 10^{-11}$,

$[VII](0)=1\cdot 10^{-8}$, $[TF\equiv VII](0)=0$, $[VIIa](0)=1\cdot 10^{-10}$, $[TF\equiv VIIa](0)=0$, $[X](0)=1.7\cdot 10^{-7}$, $[TF\equiv VIIa\equiv X](0)=0$, $[TF\equiv VIIa\equiv Xa](0)=0$, $[IX](0)=9\cdot 10^{-8}$, $[TF\equiv VIIa\equiv IX](0)=0$, $[VIII](0)=7\cdot 10^{-10}$, $[IXa\equiv VIIIa\equiv X](0)=0$, $[VIIIa_{1.L}](0)=0$, $[VIIIa_2](0)=0$, $[V](0)=2\cdot 10^{-8}$, $[Xa\equiv Va\equiv II](0)=0$, $[mIIa](0)=0$, $[TFPI](0)=2.5\cdot 10^{-9}$, $[Xa\equiv TFPI](0)=0$, $[TF\equiv VIIa\equiv Xa\equiv TFPI](0)=0$, $[AT](0)=3.4\cdot 10^{-6}$, $[Xa\equiv AT](0)=0$, $[mIIa\equiv AT](0)=0$, $[IXa\equiv AT](0)=0$, $[IIa\equiv AT](0)=0$, $[TF\equiv VIIa\equiv AT](0)=0$, $[IXa](0)=0$, $[Xa](0)=0$, $[IIa](0)=0$, $[II](0)=1.4\cdot 10^{-6}$, $[VIIIa](0)=0$, $[Va](0)=0$, $[Xa\equiv Va](0)=0$, $[IXa\equiv VIIIa](0)=0$.

The computed concentration of thrombin $[IIa]$ shows that the dynamics of the thrombin peaks after a transition period (i.e., the initiation phase in the thrombin formation) and then decreases slowly (see Fig. 7.3). The peak value and the duration of the transition period are in agreement with the reported empirical data [9].

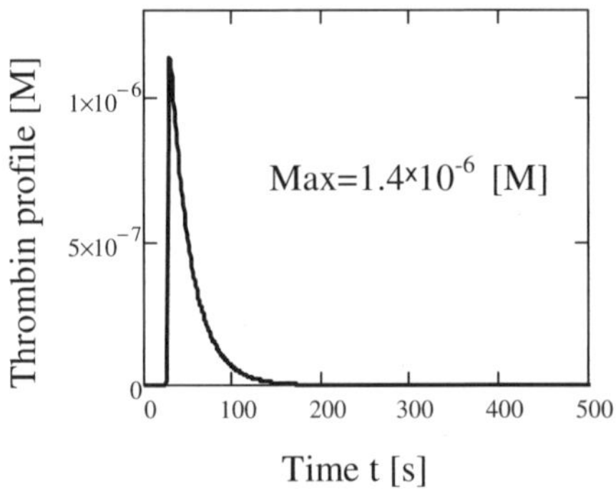

Fig. 7.3. Computed thrombin profile without DTI.

It should be emphasized that in previous reactions and hence differential equations, anticoagulants drugs (DTI) were not involved, but the inhibitors factors of the thrombin which are present within the human body itself were considered (AT, TFPI, APC) because under normal conditions when a wound occurs, a blood clot is desirable.

7.2.2. *Mathematical model in which direct thrombin inhibitors are included*

Thrombin plays such a pivotal role in haemostasis, and hence its level must be tightly controlled. The effect of anticoagulant drugs is to inhibit thrombin formation, and they are given to the patients suffering from certain clotting disorders. These people have a tendency to develop clots very quickly, and hence the aim of these drugs is to inhibit the production of thrombin and reduce the clots.

In the following, the effect of some drugs (argatroban, hirudin and melagatran) is computed, as well as the effect of the increasing dosage of each considered DTI. It is considered that drugs have the same type of reaction schemes, i.e., they have an effect only on factors IIa (thrombin) and mIIa (meizothrombin):

$$\mathrm{DTI+IIa} \underset{k_{44}}{\overset{k_{43}}{\rightleftarrows}} (\mathrm{DTI} \equiv \mathrm{IIa})$$

$$\mathrm{DTI+mIIa} \underset{k_{46}}{\overset{k_{45}}{\rightleftarrows}} (\mathrm{DTI} \equiv \mathrm{mIIa}) \cdot$$

The difference consists in the rate at which each drug reacts. More precisely: $k_{43} = 3.3\times10^{7}\,\mathrm{M^{-1}s^{-1}}$, $k_{44} = 3.3\times10^{-1}\,\mathrm{s^{-1}}$, $k_{45} = 3.3\times10^{7}\,\mathrm{M^{-1}s^{-1}}$, $k_{46} = 3.3\times10^{-1}\,\mathrm{s^{-1}}$ for argatroban; $k_{43} = 2.9\times10^{8}\,\mathrm{M^{-1}s^{-1}}$, $k_{44} = 1.7\times10^{-5}\,\mathrm{s^{-1}}$, $k_{45} = 2.9\times10^{8}\,\mathrm{M^{-1}s^{-1}}$, $k_{46} = 1.7\times10^{-5}\,\mathrm{s^{-1}}$ for hirudin; $k_{43} =1.2\times10^{7}\,\mathrm{M^{-1}s^{-1}}$, $k_{44} = 3.6\times10^{-2}\,\mathrm{s^{-1}}$, $k_{45} = 1.2\times10^{7}\,\mathrm{M^{-1}s^{-1}}$, $k_{46} = 3.6\times10^{-2}\,\mathrm{s^{-1}}$ for melagatran.

According to these two reactions, differential equations from the considered mathematical model, corresponding to the factors' concentrations $[mIIa]$ and $[IIa]$, are changed as follows:

$$\begin{aligned}\frac{d[mIIa]}{dt} &= k_{31}\cdot[Xa \equiv Va \equiv II] - k_{32}\cdot[mIIa]\cdot[Xa \equiv Va] - k_{39}\cdot[mIIa]\cdot[AT] \\ &\quad - k_{45}\cdot[mIIa]\cdot[DTI] + k_{46}\cdot[DTI \equiv mIIa]\end{aligned} \tag{7.20a}$$

$$\begin{aligned}\frac{d[IIa]}{dt} &= k_{32}\cdot[mIIa]\cdot[Xa \equiv Va] - k_{41}\cdot[IIa]\cdot[AT] + k_{16}\cdot[Xa]\cdot[II] + \\ &\quad + k_{i2}\cdot\frac{[Xa]\cdot[II]}{[II]+k_{i2m}} + \bar{k}_{i2}\cdot\frac{[Xa \equiv Va]\cdot[II]}{[II]+\bar{k}_{i2m}} - h_{i2}\cdot[IIa] - k_{43}\cdot[IIa]\cdot[DTI] + k_{44}\cdot[DTI \equiv IIa].\end{aligned} \tag{7.32a}$$

Also, the above reactions result in the new differential equations corresponding to the new factors DTI, $(\mathrm{DTI} \equiv \mathrm{mIIa})$ and $(\mathrm{DTI} \equiv \mathrm{IIa})$:

$$\frac{d[DTI]}{dt} = -k_{43}\cdot[IIa]\cdot[DTI] + k_{44}\cdot[DTI \equiv IIa] - k_{45}\cdot[DTI]\cdot[mIIa] + k_{46}\cdot[DTI \equiv mIIa] \tag{7.38}$$

$$\frac{d[DTI \equiv mIIa]}{dt} = k_{45}\cdot[mIIa]\cdot[DTI] - k_{46}\cdot[DTI \equiv mIIa] \tag{7.39}$$

$$\frac{d[DTI \equiv IIa]}{dt} = k_{43} \cdot [DTI] \cdot [IIa] - k_{44} \cdot [DTI \equiv IIa] \cdot \qquad (7.40)$$

Solving numerically the system of differential equations (7.2)–(7.40), in which equations (7.20) and (7.32) are replaced with (7.20a) and (7.32a), the coagulation factors are found in the case when anticoagulant drugs are used. The initial concentration of DTI is varied over $[DTI](0) = 0$; 1×10^{-9}; 1×10^{-8}; 1×10^{-7}; 2×10^{-7} M, and the initial concentrations for $(\mathrm{DTI} \equiv \mathrm{mIIa})$ and $(\mathrm{DTI} \equiv \mathrm{IIa})$ are equal to zero.

The dynamics of the thrombin formation under the effect of drugs and the evolution of the drug concentration for the considered DTI with different increasing dosages are shown in Figs. 7.4–6.

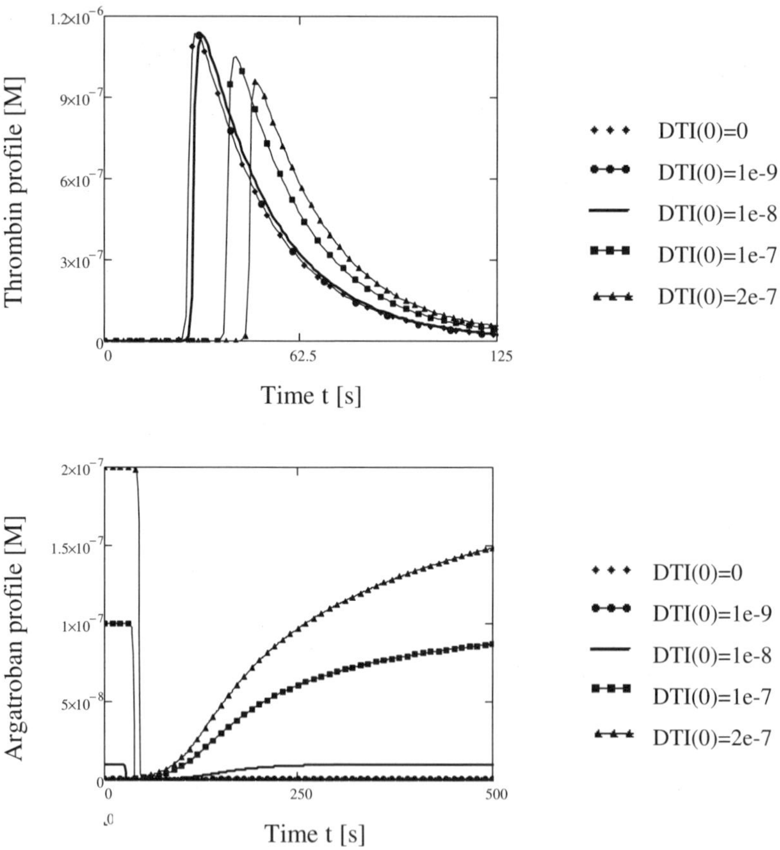

Fig. 7.4. Computed thrombin profiles at different argatroban concentrations and the corresponding drug profile.

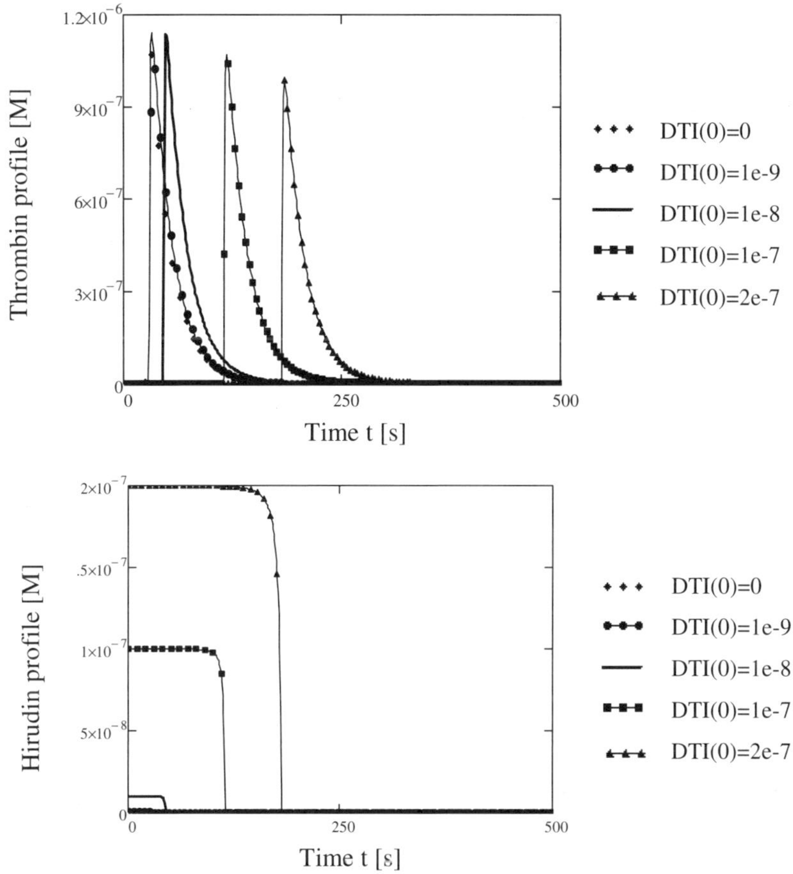

Fig. 7.5. Computed thrombin profiles at different hirudin concentrations and the corresponding drug profile.

Computations show that all drugs inhibit the thrombin production, i.e. the initiation phase of thrombin formation is delayed. More precisely, the mean value of time for the thrombin generation to peak becomes larger, and the peak thrombin values decrease when the DTI concentrations increase. Using similar drug concentrations for argatroban, hirudin and melagatran, it can be concluded that argatroban and melagatran lead to almost the same behavior of the thrombin dynamics, while hirudin has a very strong effect, proving that this drug dosage should be chosen carefully. This effect can be observed from the drug profile as well. The drug curves for argatroban and melagatran decrease slowly at the

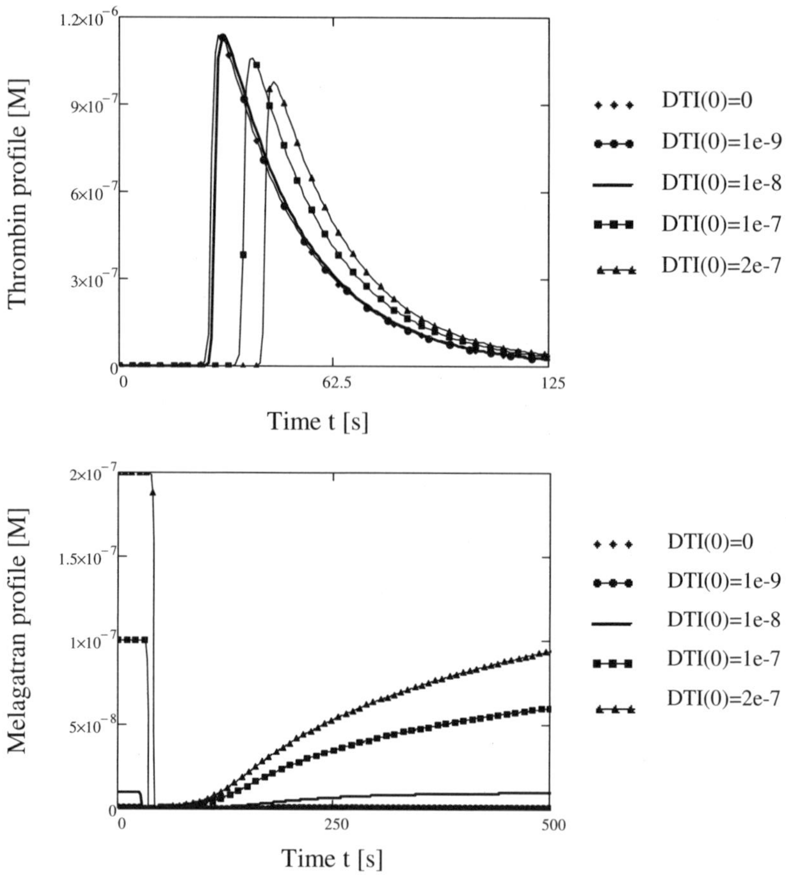

Fig. 7.6. Computed thrombin profiles at different melagatran concentrations and the corresponding drug profile.

beginning, indicating the absorption of the drug, but after some time, the levels start increasing due to a reverse of the reaction of the drug with thrombin. In the case of hirudin, after the decreasing of the drug curve, the reaction of the drug with thrombin is almost zero. However, there are several pharmacologic characteristics that differ between these drugs. Differences in the effects of melagatran and hirudin for the healthy male subjects were reported in Ref. [11], and between argatroban, hirudin and other DTIs in [12–17]. It is established that argatroban has a rapid onset of action and is rapidly dissociated from thrombin. In comparison with hirudin, argatroban has a shorter half-life, which seems to enable easy clinical administration of the drug ('turn-on/turn-off') and is reflected in a wider safety net (lesser bleeding) than that of hirudin. Also, hirudin is a protein and can generate antibodies that increase its anticoagulant activity by prolonging the half-life or antibodies that decrease its anticoagulant effect.

Acknowledgment

The authors are grateful to the Romanian National University Research Council NURC (PNCD II — NatComp project 11–028/14.09.2007) for support of this work.

References

1. J. P. Vacca, New advances in the discovery of thrombin and factor Xa inhibitors, *Current Opinion in Chemical Biology* **4**, 394–400 (2000).
2. M. Anand, K. Rajagopal and K. R. Rajagopal, A model for the formation and lysis of blood clots, *Pathophysiology of Haemostasis and Thrombosis* **34**, 109–120 (2005).
3. L. Braescu, T. F. George, C. Orbulescu and M. Leretter, Dynamics of thrombin formation using a mathematical model including both intrinsic and extrinsic pathways of blood coagulation, *Dynamics of Continuous, Discrete and Impulsive Systems (Series A)* **16** (S1), 75–82 (2009).
4. J. Jesty and Y. Nemerson, The pathways of blood coagulation, in *Williams Hematology*, 5th edn. (McGraw-Hill Health Professions Division, New York, 1995), pp. 1227–1239.
5. Z. Guo, K. M. Bussard, K. Chatterjee, R. Miller, E. A. Vogler and C. A. Siedlecki, Mathematical modeling of material-induced blood plasma coagulation, *Biomaterials* **27**, 796–806 (2006).
6. V. I. Zarnitsina, A. V. Pokhilko and F. I. Ataullakhanov, A mathematical model for spatio-temporal dynamics of intrinsic pathway of blood coagulation. I. The model description, *Thrombosis Research* **84**, 225–236 (1996).
7. V. I. Zarnitsina, A. V. Pokhilko and F. I. Ataullakhanov, A mathematical model for spatio-temporal dynamics of intrinsic pathway of blood coagulation. II. Results, *Thrombosis Research* **84**, 333–344 (1996).
8. G. T. Gerotziafas, T. Chakroun, F. Depasse, M. M. Samama and I. Elalamy, Effect of tissue factors and fondaparinux, an anti-thrombin dependent synthetic inhibitor of factor Xa, on thrombin generation in human blood, *Haema* **8**, 43–52 (2005).

9. M. F. Hockin, K. C. Jones, S. J. Everse and K.G. Mann, A model for the stoichiometric regulation of blood coagulation, *J. Biol. Chem.* **277**, 18322–18333 (2002).
10. K. C. Jones and K. G. Mann, A model for tissue factor pathway to thrombin; II. A mathematical simulation, *J. Biol. Chem.* **269**, 23367–23373 (1994).
11. T. C. Sarich, M. Wolzt, U. G. Eriksson, C. Mattsson, A. Schmidt, S. Elg, M. Andersson, M. Wollbratt, G. Fager and D. Gustafsson, Effects of ximelagatran, an oral direct thrombin inhibitor, r-hirudin and enoxaparin on thrombin generation and platelet activation in healthy male subjects, *J. Am College Cardiol.* **41**, 557–564 (2003).
12. B. E. Lewis and J. M. Walenga, Argatroban in HIT type II and acute coronary syndrome, *Pathophysiology of Haemostasis and Thrombosis* **32**, 46–55 (2002).
13. S. L. Bostrom, G. F. H. Hansson, T. C. Sarich and M. Wolzt, The inhibitory effect of melagatran, the active form of the oral direct thrombin inhibitor ximelagatran, compared with enoxaparin and r-hirudin on ex vivo thrombin generation in human plasma, *Thrombosis Research* **113**, 85–91 (2004).
14. K. A. Bauer, New anticoagulants, *Hematology* **1**, 450–456 (2006).
15. J. Hirsh, M. O'Donnell and J. I. Weitz, New anticoagulants, *Blood* **105**, 453–463 (2005).
16. J. I. Weitz and J. Hirsh, New anticoagulants drugs, *Chest* **119**, 95S-107S (2001).
17. M. M. Samama, G. T. Gerotziafas, I. Elalamy, M. H. Horellou and J. Conard, Biochemistry and clinical pharmacology, of new anticoagulant agents, *Pathophysiology of Haemostasis and Thrombosis* **32**, 218–224 (2002).

Chapter 8

Laser Ablation of Biological Tissue by Short and Ultrashort Pulses

Renat R. Letfullin

Rose-Hulman Institute of Technology
letfullin@rose-hulman.edu

Thomas F. George

University of Missouri–St. Louis
tfgeorge@umsl.edu

In this chapter, time-dependent thermal simulations are performed for short and ultrashort pulse laser ablation of biological tissue in single-pulse and multipulse (set of ultrashort pulses) modes of laser heating. A comparative analysis for both radiation modes is discussed for laser heating of different types of biological tissues: cancerous and healthy cell organelles in soft tissue and solid hard tissue, such as those found in bone and teeth, on the nanosecond, picosecond and femtosecond time scales. It is shown that ultrashort laser pulses with high-energy densities can ablate the biological tissue without heating tissues bordering the ablation creator. This reaction is particularly desirable as heat accumulation and thermal damage are the main factors affecting tissue regrowth rates, and thus patient recovery times.

8.1. Introduction

Soft tissue laser surgery is now common practice; many specific laser treatments and procedures have been developed. The use of laser scalpels in surgery has shown a reduced risk of infection as well as improved patient recovery time. This has been attributed to the laser's ability to seal small blood vessels and nerve endings at the cutting site, thus reducing pain and blood loss [1]. LASIK surgery utilizes excimer lasers to vaporize small layers of corneal stroma and thus reshape the lens of the eye [2]. Many applications have been found for laser resurfacing — from the treatment of sun damage and stretch-marks to the removal of scars and tattoos [3]. Lasers used for soft-tissue medical applications are chosen mainly for the strong absorption of their radiation by water, the primary constituent of soft-tissues. Similarly, development of hard-tissue applications for Er:YAG lasers is stressed because the 2.94 μm radiation is more strongly absorbed by bone than the radiation of other solid-state mid-infrared lasers [4]. Although not as strongly absorbed in hard-tissues, Holmium:YAG lasers at a wavelength of 2.12 μm have also been investigated because they can be transmitted through quartz optical fiber [5].

The success of soft-tissue laser surgery has promoted research for laser applications in hard-tissue. However, ablating hard tissues such as those found in bone and teeth impose different complications that must be overcome. One complication that is much more prominent in hard-tissue laser surgery is thermal necrosis — cell death caused by extreme temperatures [6]. In soft tissue, damaged cells are reabsorbed, and the materials are reused by the body relatively quickly. In hard tissues, however, dead cells leave behind a mineral lattice which slows the regrowth process in the damaged region. Additionally, excessive temperatures can also lead to dehydration or thermal expansion of the material, leading to the formation of microcracks, and possible weakening the bone [7]. Thus a balance must be found between laser energies required for effective ablation and energies that cause excessive thermal damage to surrounding tissue. The use of short and ultrashort laser pulses seeks to solve this problem by using pulses of sufficient energy to vaporize the target material in a timeframe that minimizes heat diffusion to surrounding tissues.

8.2. Laser Ablation of Hard Tissue

The difficulties in modeling laser heating and ablation of bone tissue arise from its complex structure which combines mineral lattice with an organic component. The inorganic mineral portion, which composes about 70% of total bone material, is formed mostly from calcium phosphate in a crystalline structure known as hydroxyapatite (HAP) whose formula is ($Ca_5(PO_4)_3(OH)$) [8]. In bones, HAP is formed over a collagen matrix creating a repeating unit macrostructure much larger than the individual HAP crystals. This unique structure and composition can lead to differences in the optical or thermal properties between bone tissues and mineral HAP.

Previous studies have modeled the laser pulse heating of bone tissue on the microsecond timescale by treating the incoming laser energy as a point heat source [9,10]. The heating of the surrounding tissue would then be modeled using 1D or 3D heat diffusion models for heat conduction in solid media [11]. The laser-matter interaction and subsequent thermal diffusion for microsecond timescales do not scale well to simulations of bone tissue heating by shorter pulses in the pico- and femtosecond range. In order to model the temperature of the area irradiated by short and ultrashort laser pulses, this chapter uses a different model which has been shown to provide an adequate description of laser heating of spherical volumes in the femtosecond, picosecond and nanosecond regimes [12].

8.2.1. *Laser heating model*

Consider a laser pulse given by its pulse shape $f(t)$ and initial intensity I_0, focused to a small spot size of radius r_0 on the solid tissue. To describe the heating and cooling kinetics of this tissue by short and ultrashort laser pulses, we will use the one-temperature model (OTM) [12]. This model describes the heat flow into and out of a small spherical volume (SV) which has a radius equal to the radius of the spot size of laser radiation on a target. In the experiment conducted by Barton *et al.* [9], the light is transmitted through a quartz fiber to the bone sample. The terminal end of the fiber core acts as an aperture to create a near-field diffraction pattern on the specimen's surface. The central peak of this pattern has an intensity profile resembling that of a focused beam with a spot size approximately equal to the diameter of the fiber's core, as shown in Fig. 8.1. It is assumed that the heat is evenly distributed across the SV. The rate at which heat energy is generated in SV depends on the optical

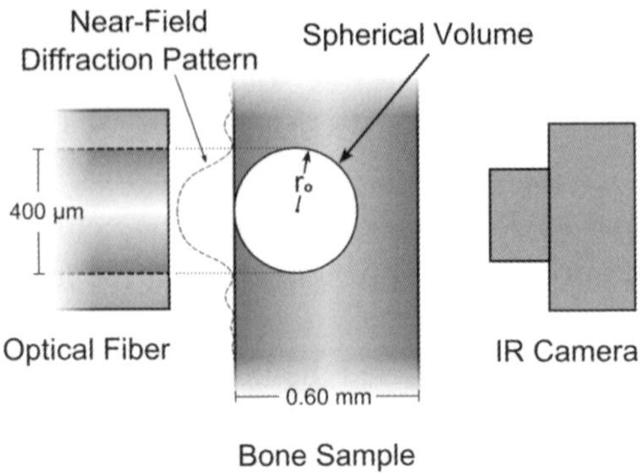

Fig. 8.1. Geometry for the OTM applied to the experimental setup of Barton *et al.* [9].

and thermal properties of the target material, namely, the absorption efficiency K_{abs}, specific heat $C_s(T_s)$ and density ρ_s. It is then assumed that heat is lost by diffusion across the surface area of the SV, with the rate of heat loss determined by the thermal conductivity μ_∞ of the surrounding medium, as well as the difference between the temperature T_∞ of the surrounding medium and the temperature T_S of the SV. Using the above assumptions, the rate equation takes the following form [12]:

$$\frac{d}{dt}T_s = \frac{3K_{abs}(r_0,\lambda)I_0 f(t)}{4r_0\rho_s C_s(T_s)} - \frac{\mu_\infty T_s}{(\alpha+1)r_0^2 C_s(T_s)\rho_s}\left[\left(\frac{T_s}{T_\infty}\right)^{\alpha+1} - 1\right]. \qquad (8.1)$$

Here, the first term on the right-hand side represents heat generation due to laser energy absorption, and the second represents thermal energy dissipated due to heat diffusion from the SV to the surrounding medium. The power exponent *s*, a constant, depends on thermal properties of the surrounding medium.

8.2.2. *Ablation model*

To model the ablation of material, we consider an Arrhenius-type model, which has been used successfully to model the ablation depth in metal and silicon targets due to ultrashort laser pulses [13,14]. This empirical model describes the evaporation kinetics as a temperature-stimulated material failure according to the equation [14]

$$V = V_0 e^{-\frac{\rho\Omega}{CT}}, \tag{8.2}$$

where ρ is the density, Ω the heat of vaporization, and C the heat capacity of the target media. V is the velocity of the evaporation front, with the constant V_0 depending on the choice of material. When the laser pulse is very short, the temperature of the material lattice can be determined from the average cooling time of the electrons [14], giving

$$T \approx \frac{F_a \alpha}{C} e^{-\alpha z}, \tag{8.3}$$

where F_a is the absorbed laser fluence, and α the material absorption coefficient. Substituting this into the above Arrhenius-type model and solving for the depth, z, gives the simple expression

$$L = \alpha^{-1} \ln\left(\frac{F_a}{T_{th}}\right), \tag{8.4}$$

where $F_{t\ h} = \rho\Omega/\alpha$ is the threshold fluence, and L represents the final depth of the crater created by one full pulse.
To model the time dependence of the process, we assume that the particle density of the undamaged tissue is proportional to the original tissue density, and that the rate of change is temperature dependent [9]:

$$\frac{dN}{dt} = N_0 A e^{-\frac{\Delta E}{RT_s}}. \tag{8.5}$$

Here, N_0 is the particle density of the material, R is the ideal gas constant, ΔE is the change in energy required to transition from the non-ablated to the ablated state, A is an experimentally determined constant, and T_S is obtained by solving the differential equation (8.1).

8.2.3. *Bone tissue heating results*

First, simulations are performed to test the temperature model against the experimental data obtained for the 270 μs pulse width timescale by Barton *et al.* [9]. In this experiment, pulses of 2.12 μm wavelength are transmitted through a quartz fiber with a core diameter of 400 μm, and the temperature is recorded with an infrared camera from the back side of the specimen as seen in Fig. 8.1. The thermal conductivity, specific heat and density of bone values used in our simulations based on the OTM are listed in Table 8.1. The laser pulse is modeled by a Gaussian curve with FWHM equal to the pulse duration of 270 μs.

The absorption efficiency of bone tissue for laser radiation of wavelength λ and given radius r_0 of the SV can be estimated by using Mie diffraction theory at the single-scattering approximation. The Mie formalism requires the use of two dimensionless input parameters, $\rho = 2\pi r_0/\lambda$ and $\delta = \rho \boldsymbol{m}$. Here ρ is the Mie parameter; $\boldsymbol{m} = m_0/m_1$ is the relative value of the refractive index of the bone; and $m_0 = n_0\text{-}ik_0$ and $m_1 = n_1\text{-}ik_1$ are, respectively, the complex refractive indices of the bone and the surrounding medium. As stated previously, the primary mineral constituent of bone is hydroxyapatite, $Ca_5(PO_4)_3(OH)$. The hydroxyl group is commonly replaced by carbonate, chloride, or fluoride. For the bulk of bone tissues, carbonate replaces the OH^- ion [18], so that bones are modeled as calcium carbonate ($CaCO_3$). Fluorinated dental treatments

Table 8.1. Input data used in the OTM hearing simulations.

Parameters	Values	Reference
pulse shape	$f_{nano}(t) = \sum_{n=0}^{N-1} e^{-(at-nb)^2}$ $a = 2.775\times10^7$ Hz (for 60 ns pulse) b = pulse separation term N = number of pulses $f_{pico}(t) = \sum_{n=0}^{N-1} e^{-(at-nb)^2}$ $a = 2.775\times10^{10}$ Hz (for 60 ps pulse width) $f_{femto}(t) = e^{-(at-b)^2} + ce^{-(at-d)^2}$ $a = 2.775\times10^{13}$ Hz (for 60 ps pulse width) $b = 3$, $c = 0.2$, $d = 1$	[12]
single pulse nergy	0.15 J	
K_{abs}	0.3	
r_0	200 μm	
ρ_0	0.00185 kg/cm^3	[15]
c_{bone}	1.33 kJ/kg K	[16]
μ_∞	0.00560 W/cm K	[17]
T_∞	300 K	

replace the hydroxyl group with a fluorine ion, so that tooth enamel properties are modeled as calcium fluoride (CaF_2) for these purposes.

Computer calculations of the absorption efficiency, K_{abs}, for spheres of calcium carbonate and calcium fluoride with radius $r_0 = 200$ μm have been conducted. The calculations show that both constituent materials are effective absorbers of the laser radiation over a wide range of the spectrum. The absorption efficiency of 2.12 μm wavelength radiation for both materials is $K_{abs} \approx 0.8$. This value for K_{abs} also has verified by fitting OTM temperature profiles to the experimental data.

With the above absorption efficiency, the OTM has been used to describe the temperature kinetics caused by short and ultrashort laser pulse heating with peak widths in the nanosecond, picosecond, and femtosecond timescales. The kinetics due to a multiple-pulse mode of heating also have been modeled.

The results of the simulation using a single pulse mode are shown in Fig. 8.2. As seen in Figs. 2(a–c), the heating kinetics of the bone SV heated by a single pulse are very fast. The maximum temperature is achieved at the end of the pulse's lifetime. As shown in Fig. 8.2(d), the cooling kinetics are relatively slow; temperatures fall to the half-maximum temperature approximately 0.03 s after the pulse. This cooling time is six orders of magnitude greater than the heating time for even the nanosecond pulse. Substantial cooling is realized approximately 0.1 to 0.2 s after the pulse. This slow rate of heat diffusion allows for the ablation of bone tissue without heating the surrounding tissue. Given a pulse of sufficiently high-energy density, the SV could be heated beyond its vaporization temperature (T = 1923 K), and due to the slow rate of heat diffusion, very little heat could be accumulated by the surrounding medium before the vaporized materials are expelled. This result has been confirmed by experiments, where Strassl *et al.* show that the complex microstructure of dentine is exposed after ablation without applying any etchant, showing no signs of melting or re-solidification of the surrounding tissues [19]. Girard *et al.* show that protein denaturation does not occur in living cortical bone adjacent to the removed tissue for femtosecond pulses [15], which

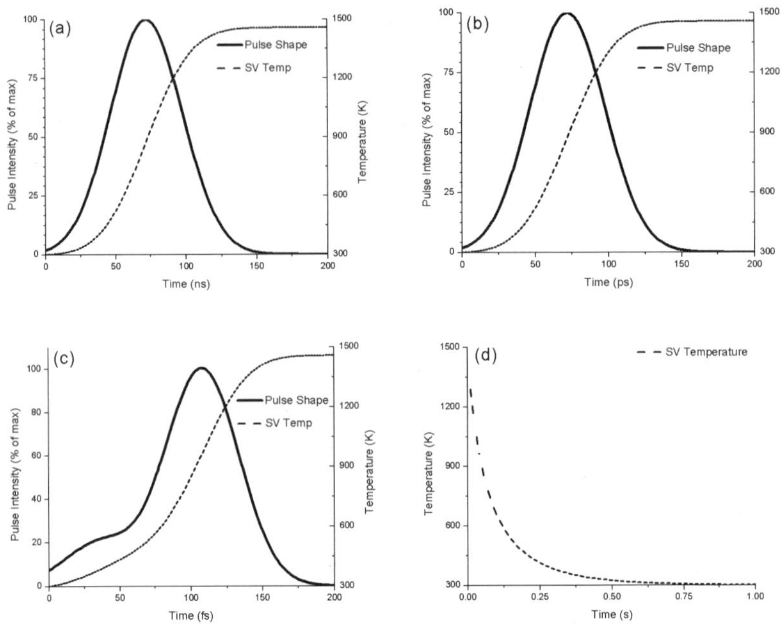

Fig. 8.2. Single pulse heating of bone tissue. The pulse shape is shown by the solid line and the heating kinetics by the dashed line for bone tissue using a single 0.3 J pulse with the pulsewidths (a) 60 ns, (b) 60 ps and (c) 60 fs, where (d) shows the cooling kinetics of the same 60 ns pulse.

This slow rate of cooling also allows for accumulation of heat by a multipulse-heating mode, as seen in Fig. 8.3, where five 0.03 J pulses reach a maximum temperature (T_{max} = 1460 K) approximately equal to that obtained by a single 0.15 J pulse. This method could be used to attain the higher temperatures required for tissue ablation using many pulses with low individual energies. Because the timescale for cooling is many orders of magnitude longer, it is possible to use tens, thousands or even more small energy pulses to reproduce virtually any single-pulse energy. This gives a large amount of control over the individual pulse energies, which is especially important in medical and dental laser applications where regulations may limit the maximum pulse energies for different procedures.

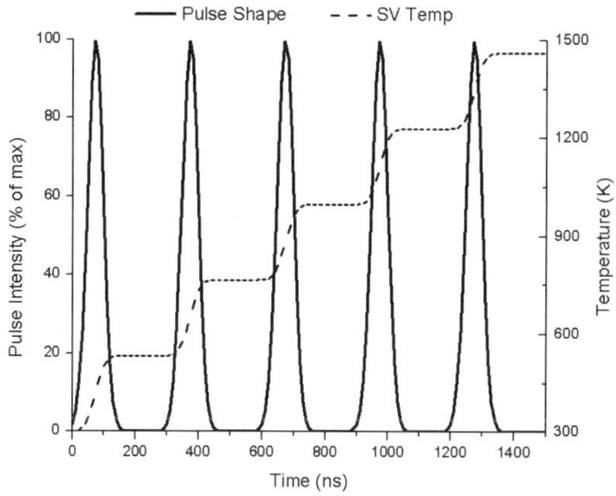

Fig. 8.3. Multipulse mode of heating of bone tissue. A representative multipulse function is drawn with a solid line and the heating kinetics with a dashed line, showing five 30 mJ pulses of 60 ns pulsewidth repeating every 300 ns.

8.2.4. *Bone tissue ablation results*

To simulate the ablation rate of bone tissue, the thermal and optical properties of HAP are again considered. Using experimental data for ultrashort ArF (193 nm) and KrF (248 nm) laser pulses focused on HAP targets [20], we can solve for the coefficients α and F_{th}, as shown in Fig. 8.4. Dividing F_{th} by the absorption coefficient and then dividing by the density of HAP, ρ = 0.003150 kg cm^3 [21], yields the heat of vaporization. Solving for the KrF data gives Ω = 317 J/g, and the ArF data gives 559 J/g. While these values differ by almost a factor of two, they end up close to the value of 456 J/g which is calculated using the molar density [22] and molar heat of vaporization[23] from the literature. The differences are minor, however, because approaching these temperatures, HAP begins to break down, releasing CaO, H_2O, and P_2O_5 vapors, rather than evaporating as a homogenous HAP vapor [24]. In fact, the heat of vaporization given by Samsonov [23] is actually the weighted sum of the aforementioned gases.

Bone tissues are composed of 70% inorganic material, mostly HAP, and 30% organic material [8]. To account for the change in composition, we multiply F_a by an accumulation factor k. The ablation rates and fluencies obtained from the experimental work performed by Strassl *et al.* [19] on human and bovine compacta using Yb:glass laser pulses, at

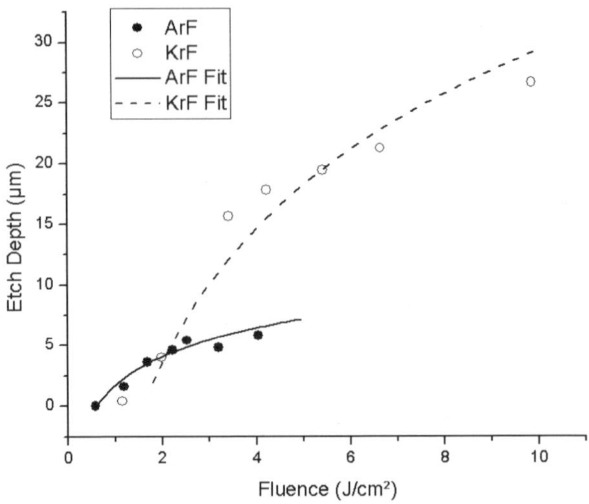

Fig. 8.4. Ablation data and fit curves for ArF (193 nm) and KrF (248 nm) laser pulses.

a wavelength of λ = 1040 nm, and that of Girard *et al.* [9] on bovine cortical bone using Ho:YAG laser pulses, at a wavelength λ = 2150 nm, were compared to the rates for pure HAP. Fitting a curve to the data yields the coefficients α and F_{th}/k (see Figs. 8.5(a) and 8.5(b)). Using the relationship $F_{th} = \rho\Omega/\alpha$ and dividing by the density of bone, ρ = 0.00185 kg/cm^3 [15], yields Ω/k = 3910 J/g for the Ho:YAG case and Ω/k = 3780 J/g for the Yb:glass case. Using the accepted value of Ω = 456 J/g, for pure HAP, gives the accumulation factors of k = 0.117 (Ho:YAG) and k = 0.121 (Yb:glass).

The differences in the fluencies and material absorption coefficients can be attributed the different laser wavelengths used and the differences in the experimental procedures. The optical properties could vary considerably from 1040 nm to 2150 nm in bone tissue. Unlike Girard *et al.* [9], the procedure of Strassl *et al.* [19] did not use single pulses, but rather many small pulses at a repetition rate of 1 kHz, giving the average depth per pulse in the data. It has been shown in Rosenfeld *et al.* [25] that multipulse interactions can have a dramatic effect on lowering the threshold fluence for ablation.

For the time progression of the ablation of particles due to a single pulse, we use the OTM to simulate the temperature kinetics for a 0.3 J pulse with a FWHM pulse width of 60 ns. This temperature function $T(t)$

is then used in Eq. (2.5), along with the parameters found in Table 8.2, for a simulation of the number density change for unablated particles at the ablation site, as plotted in Fig. 8.6. The constant *A* has been experimentally determined to be on the order of magnitude 10^{23}, as this results in the majority of particles being ablated on the 0–500 ns timescale as observed bt Serra *et al.* [26].

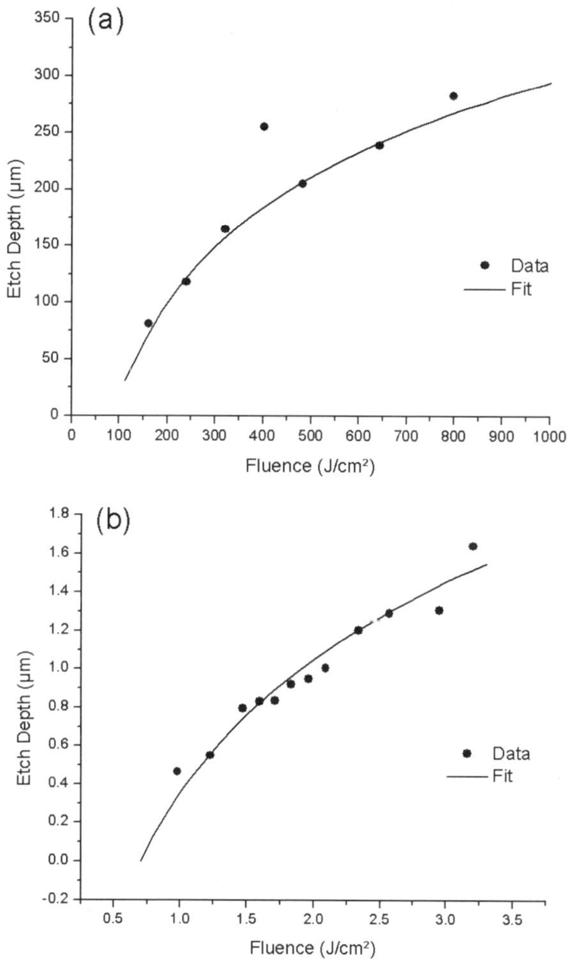

Fig. 8.5. Ablation data and fit curves for (a) Ho:YAG laser pulses (2150 nm) on bovine cortical bone and (b) Yb:glass laser pulses (1040 nm) on bovine compacta.

Table 8.2. Input data used in the time-dependent ablation simulations.

Parameters	Values	Reference
ΔE	458 kJ/mol	[23]
R	8.31 J/K mol	
N_0	$1.9 \cdot 10^{27}/m^3$	[23]
A	10^{23}	

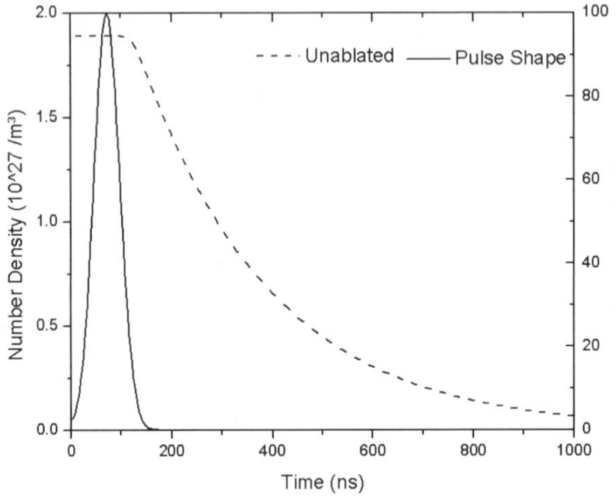

Fig. 8.6. Pulse shape and ablation dynamics through time.

Thus, the simulations using the OTM to model the heating of bone tissue with laser pulses in the nanosecond, picosecond and femtosecond regimes show the following:

- The absorption efficiency for bone tissues can be well approximated using only the optical properties of its mineral constituents, i.e., hydroxyapatite, rather than its organic components, i.e., proteins/water, when absorbing short and ultrashort high-energy density pulses.
- The cooling kinetics are much slower for short and ultrashort pulses than the heating process, even for nanosecond pulses, by several orders of magnitude.

- Slow cooling rates allow the use of multipulse modes of heating without causing collateral damage to surrounding tissues.
- Multipulse modes of heating can reach the same high temperatures as single pulse heating at much lower individual pulse energies.

Simulations using the Arrhenius model for the ablation of bone tissue using ultrashort laser pulses show the following:

- The Arrhenius model can readily describe hydroxyapatite ablation, and the addition of a simple coefficient can then modify the equation to model hard bone tissue.
- The logarithmic nature of the ablation rate means that gains in the ablation rate diminish for higher and higher fluencies.
- For ultrashort pulses we see that 80% of the particles are ablated within the first 500 ns.

Therefore, the use of multipulse techniques for both laser heating and laser ablation of bone tissue is recommended because the multipulse mode of heating allows greater flexibility in the choice of individual pulse energies without sacrificing the desired maximum temperature, and the threshold fluence required for ablation is reduced in multipulse ablation modes, maximizing etch depths with minimal thermal accumulation.

8.3. Laser Ablation of Soft Tissue

In this section, we perform time-dependent simulations of soft tissue-laser interaction to determine the potential for cancer treatment via organelle specific ablation. Simulating both single- and multipulse modes of heating, we calculated the time-temperature profiles for a variety of cancerous and healthy cell nuclei and mitochondria.

Technological improvements over the past several decades have led to rapid development of new equipment and techniques for all areas of medicine. Scientists of today have the opportunity to develop more effective and efficient techniques for cancer diagnosis and treatment with these new technologies and the accompanying theories. This is still a difficult feat due to the complexity of the human body, which lends itself to a wide variety of "abnormalities" that can cause cancer. Each form of cancer is in its own way distinct, so having a common treatment for all forms of cancer requires methods that also harm healthy cells, such as radiation or chemotherapy.

Biologically speaking, cancer is a result of abnormal cells in the body that rapidly and uncontrollably replicate. Normally cells undergo several checkpoints before division; these checkpoints are used to determine whether or not the cell should replicate its DNA and divide, whether it should undergo apoptosis due to faulty DNA, or if it should continue normal function. Inhibitory mechanisms will stop healthy cells from division when further growth would cause nutrient deprivation or stress the surrounding biological system in any other way. When the inhibitory mechanisms for cell replication malfunction, a cell can undergo rapid and uncontrolled replication regardless of the effects on the system. Such cell propagation can result in the starvation of the surrounding healthy cells as well as physical expansion into other biological systems, disturbing system function on multiple levels.

Given the effects of cancer on cell organelles, here we investigate if it is possible to selectively heat cancer organelles with lasers as a method of cancer treatment without using the nanoparticles. By utilizing the physical changes experienced by mitochondria and the nucleus in abnormal cells, we can determine how laser irradiation would affect the cancerous cells in comparison to healthy cells. The critical temperature for organelle heating is 425K [27] as that results in protein denaturing [28] and alters membrane permeability [29].

Our procedure consists of two primary steps: (1) determining the absorption characteristics of the organelle and (2) numerically solving the heat transfer equation to determine the temperature of the organellee over time. To determine the absorption characteristics using generalized Lorenz-Mie diffraction theory, we must assume the organelles are spherical and determine complex index of refraction data for the organelle and media. This data paired with the organelle size allows us to determine the absorption efficiency of the spherical organelle at a given wavelength.

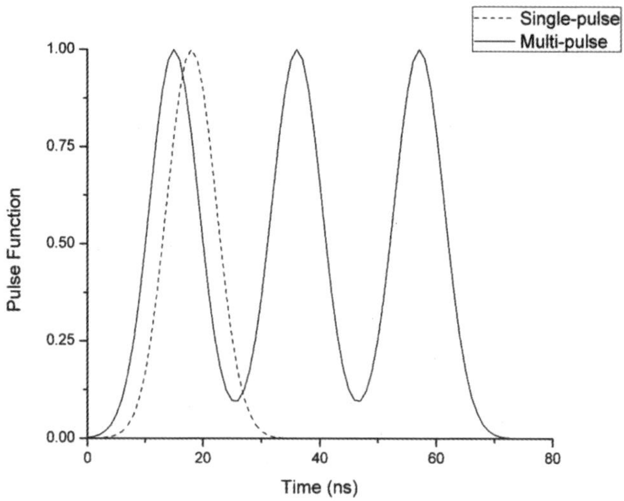

Fig. 8.7. Laser pulse functions, where both functions use a = 1.66511E8 to simulate a 12 ns pulse duration with 54.2 MHz repetition frequency. Single-pulse parameter: b=3. Multipulse parameters: b_1 = 2.5, b_2 = 6, b_3 = 9.5.

After determining the absorption efficiency, we use a one-temperature model [12] discussed in the previous section to calculate the organelle temperature as a function of time. We use Gaussian beam profiles in the form of $f(t) = e^{-(a*t-b)^2}$ for the laser pulse time shape, where the coefficients are adjusted to create a pulse profile of a standard pulse duration, as shown in Fig. 8.7. To simulate multiple pulses, we use the same function as in the single-pulse simulation, but with the addition of several values of b to shift each consecutive pulse: $f(t) = e^{-(a*t-b_1)^2} + e^{-(a*t-b_2)^2} + e^{-(a*t-b_3)^2}$.

Due to the complexity of the human biological system, it is not possible to make sweeping claims in terms of intercellular changes at the level of organelles. Furthermore, the morphometric analysis of organelles can produce a wide range of results for the same cancer depending on the developmental stage [30,31], which makes calculations tentative and case specific. For our analysis, we look at several specific cases from which we extrapolate the result to other forms of cancer.

8.3.1. *Hepatocellular carcinoma*

Hepatocellular carcinoma (HCC) is a cancer located in the liver that is normally a secondary effect of hepatitis or cirrhosis. It is historically difficult to treat due to the late recognition of tumors as well as difficulty removing the entire tumor upon discovery; HCC is not effectively treated by either chemotherapy or radiation therapy and stands to be an area where a new cancer treatment would be invaluable [32].

Torimura *et al.* [33] have used electron microscopy to observe 31 cases of hepatocellular carcinoma in an attempt to determine the usefulness of organelle morphometry via electron microscopy in cancer diagnosis, determining that as a standalone method it is ineffective. They have determined the mitochondrial and nuclear area for 4 cases of well differentiated HCC, 17 cases of moderately differentiated HCC, and 10 cases of poorly differentiated HCC. Using this area, we determine the effective radius of the mitochondria and nucleus, which is the term we use in our calculations for absorption efficiency and when solving the OTM (Eq. (2.4)). Using these values as shown in Table 8.3, we calculate the absorption efficiency (Table 8.4, Fig. 8.8) and time-temperature profiles using organelle specific data (Table 8.5) for each of the three levels of differentiation for both mitochondria and nucleus. Extra parameter values are as follows: the thermal conductivity for mitochondria and nucleus are 0.3W/mK [34], specific heat for mitochondria and nucleus are 3000 J/kg-K [35], and density for mitochondria and nucleus are 1100 kg/m^3 [36] and 1400 kg/m^3, respectively [36].

Looking at the single-pulse mode of heating (Fig. 8.9), the healthy nuclei are heated to the highest temperature of all the organelles. As the cell passes from well to poorly differentiated, the maximum nucleus temperature decreases. Alternatively, the healthy mitochondria reaches a maximum temperature slightly below the well differentiated and moderately differentiated cell mitochondria, while the poorly differentiated cell mitochondria reach the highest mitochondrial temperature.

Table 8.3. Experimental data for healthy and tumor cells.

Ref.	Cancer	Organelle	Healthy diameter (μm)	Well known diamet er (μm)	Moderatel y known diameter (μm)	Poorly known diameter (μm)
[33]	Hepatocellular carcinoma	Nucleus	7.5	8.5	8.8	10.3
[33]	Hepatocellular carcinoma	Mitochon dria	1.1	0.9	0.9	0.66
[37]	Colorectal adenocarcinoma	Nucleus	5.2	7.2	–	–
Ref.	**Cancer**	**Organelle**	**Benign diameter (μm)**	***In situ* diameter (μm)**	**Invasive diameter (μm)**	
[31]	Ductal breast carcinoma	Nucleus	7	11.6	8.33	
[38]	Intraductal breast cancer	Nucleus	8.58	–	8.97	
[39]	Breast cancer	Nucleus	5.6	–	5.8	

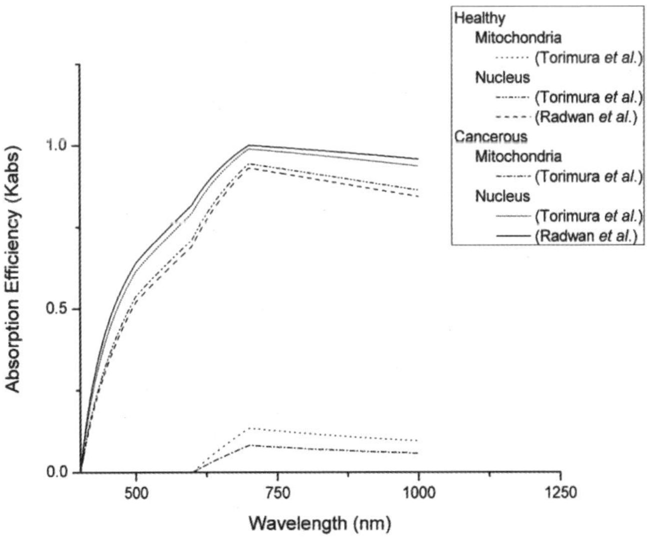

Fig. 8.8. Absorption efficiency (K_{abs}) profiles for mitochondria and nuclei. Organelle diameters are as follows: 1.1 nm (dot), 7 nm (dash), 7.5 nm (dash-dot-dot), 0.66 nm (dash-dot), 10.3 nm (short-dot), 11.6 nm (solid).

Table 8.4. Computational input data for the OTM calculations.

Ref	Cancer	Organelle	Organelle State	Diameter (μm)	Absorption efficiency (at $\lambda = 700$ nm)[a]
[33]	Hepatocellular carcinoma	Nucleus	Healthy	7.5	0.944183362
			Well differentiated	8.5	0.964865712
			Moderately differentiated	8.8	0.969903343
			Poorly differentiated	10.3	0.989209877
[33]	Hepatocellular carcinoma	Mitochon dria	Healthy	1.1	0.134234082
			Well differentiated	0.9	0.110949860
			Moderately differentiated	0.9	0.110949860
			Poorly differentiated	0.66	0.082130836
[31]	Ductal breast carcinoma	Nucleus	Benign	7.0	0.930908259
			In situ	11.6	1.000194442
			Invasive	8.33	0.961677234
[38]	Intraductal breast cancer	Nucleus	Benign	8.58	0.966255375
			Invasive	8.97	0.972557278
[39]	Breast cancer	Nucleus	Benign	5.6	0.880181004
			Invasive	5.8	0.889019082

Table 8.5. Organelle and cytoplasm properties.

Organelle	Diameter, d (μm)	Specific heat, C (J/kg-K)	Thermal conductivity, μ (W/m-K)	Density, p (kg/m^3)
Nucleus	5 [41], 5–10 [42], 7.5–10 [43], 5–10 [35]	3000 [34]	0.3 [34]	1050 [34], 1400 [35]
Mitochondria	0.85–1.15 [41], 1–3.57 [42],0.87–3.57 [43], 1–2 [35]	3000 [34]	0.3 [34]	1050 [34], 1100 [35]
Ribosome	0.025 [28], 0.025 [43], 0.02 [35]	3000 [34]	0.3 [34]	1050 [34], 1600 [35]
Microtubules	0.025 [41], 0.028 [42], 0.02 [43], 1–2 [35]	3000 [34]	0.3 [34]	1050 [34], 1100 [35]
Cytoplasm	–	4180 [34]	0.59 [34]	

[a] 700 nm was chosen because it is the wavelength with the greatest absorption efficiency difference between cancerous and healthy organelles at a wavelength in the range 400–1000 nm.

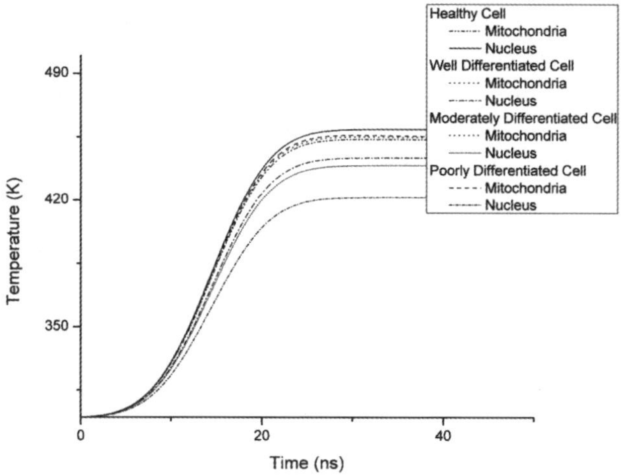

Fig. 8.9. Time-temperature profiles for single-pulse laser irradiation of the nucleus and mitochondria based on data from Torimura *et al.* [33]. Laser parameters: 12 ns duration, 0.5 J/cm^2 at $\lambda = 700$ nm.

This result indicates that trying to target hepatocellular carcinoma cells with single-pulse laser irradiation may not be an effective treatment due to the probable damage of healthy cell nuclei. Our multipulse calculations produce a similar result (Fig. 8.10), though in the multipulse mode of heating, the mitochondria reaches a higher temperature than the nuclei. The mitochondria reach a higher temperature than the nuclei as a result of the different densities used for the nucleus and the mitochondria; these densities are used in the OTM and affect the heating and cooling kinetics of the organelle. Keeping that in mind, due to the semipermeable membranes for both organelles, determining a density value for the organelle is difficult; the density is a dynamic quantity.

Anson *et al.* [34] used 1050 kg/m^3 for the density of all organelles, which would bring the maximum nucleus and mitochondria temperatures comparable when considering multi-pulse heating mode. As well as the variable nature of organelle density due to standard system function, mitochondrial cristae have been shown to be affected in tumor cells, decreasing in number or almost disappearing completely [36]. This alteration of internal structure of the mitochondria could contribute to permeability and density changes experienced by tumor mitochondria.

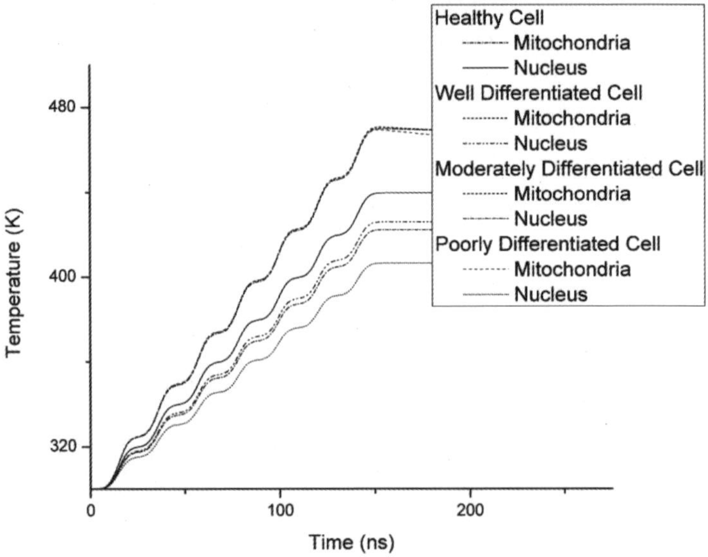

Fig. 8.10. Time-temperature profiles for multipulse laser irradiation of the nucleus and mitochondria based on data from Torimura *et al.* [33]. Laser parameters: 12 ns duration, 0.1 J/cm^2 at $\lambda = 700$ nm; $b_1 = 2.5$, $b_2 = 6$, $b_3 = 9.5$, $b_4 = 13$, $b_5 = 16.5$, $b_6 = 20$, $b_7 = 23.5$.

8.3.2. *Ductal breast carcinoma*

Ductal breast carcinoma is a prevalent form of breast cancer that has the therapeutic advantage of being near the surface, allowing it to be surgically removed or treated with radiation therapy. To test the results of nuclear heating in hepatocellular carcinoma against another form of cancer, we are using data acquired from the work of Radwan *et al.* [31] on the morphometric analysis of benign, *in situ* and invasive ductal breast carcinoma. *In situ* ductal breast carcinoma has a high prevalence of becoming invasive, so early treatment via laser irradiation would allow a safe prohibitive therapy that could decrease cancer incidence rates for those that are diagnosed early with *in situ* ductal carcinoma. For their analysis, Radwan *et al.* measured the area and perimeter, along with other characteristics, of the varying cell nuclei. For the radius used in our calculations, we determine an effective radius based on the area as well as for the perimeter values given for each of the three divisions (benign, *in situ* and invasive), and then average the two values. Additionally,

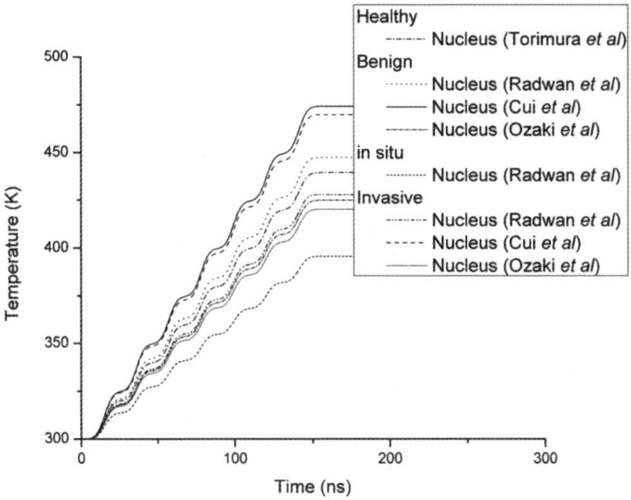

Fig. 8.11. Time-temperature profiles for multipulse laser irradiation of the nucleus based on data from Tables 8.1 and 8.2. Laser parameters: 12 ns duration, 0.1 J/cm^2 at λ=700 nm; $b_1 = 2.5$, $b_2 = 6$, $b_3 = 9.5$, $b_4 = 13$, $b_5 = 16.5$, $b_6 = 20$, $b_7 = 23.5$.

we compared the heating for these cancerous nuclei to two breast cancer studies that attempt to determine breast cancer progression from benign to invasive by looking at nuclear characteristics. Data for our calculations can be found in Tables 3 and 4, while Fig. 8.11 shows time-temperature profiles for various situations. Once again, the calculations show progressively higher temperatures the more benign the tumor becomes. This is due to the increased size of the nucleus that follows from cancer development. As Table 8.4 shows, the larger the organelle, the larger the absorption efficiency; in the cases analyzed, the increase in absorption efficiency is not enough to offset the increase in size (resulting in decreased overall temperature of the organelle).

8.3.3. *Nuclear heating, mitochondrial heating, and applications to therapy*

Due to biological variability, the availability of organelle characteristics are minimal, even for specific cases. While it is hard to be concrete, several generalities can be made when considering tumor cells versus healthy cells:

- The nucleus increases in size and the cell has an increased nucleus-to-cytoplasm ratio [31,33,37–39].
- Mitochondrial activity fluctuates while mitochondrial cristae decrease, and the total number of mitochondria decreases [40].

In terms of mitochondrial size, there does not appear to be a steady answer. Mintz *et al.* [40] find tumor mitochondria to lack respiratory control, which in part controls mitochondrial size, and find that the cancerous mitochondria swells as a result. This is at odds with Miller and Goldfeder [44], and furthermore, Torimura *et al.* [33] show a decreasing size of tumor mitochondria. These differences could all result from the numerous stages, grades and origins of cancer, but they elevate the difficulty of determining organelle ablation as a method of cancer treatment. Before more concrete heating analysis could be made, investigations into cancer organelle morphology would need to be done under specific conditions, with the intent and purpose being to provide data for calculations such as ours.

8.3.4. *Ribosomal heating*

Other organelles, such as the ribosome, have considerably less data published on the physical effects of cancer. The proper calculations needed to make authoritative statements about ribosome heating, as well as lysosome and cytoskeletal, would require more data, as mentioned in the last subsection. Because we cannot find specific data, we take an average size for a healthy ribosome as listed in various sources (Table 8.5) and add a quarter of the radius in one case (for organelle expansion) and subtract a quarter of the radius in another case (for organelle contraction) to see if either provides beneficial results in terms of selective killing. Figures 8.12 and 8.13 show ribosome of various diameters and the relative heating profiles they exhibit. For the single-pulse mode of heating, the moderately large ribosome (D = 2.9 nm) is heated the most of all the organelles. Alternatively, in the multipulse mode of heating, the largest ribosome (D = 4.6 nm) exhibits the highest level of heating. This change shows a similar dynamic between size (absorption efficiency) and level of heating. Due to the insignificant level

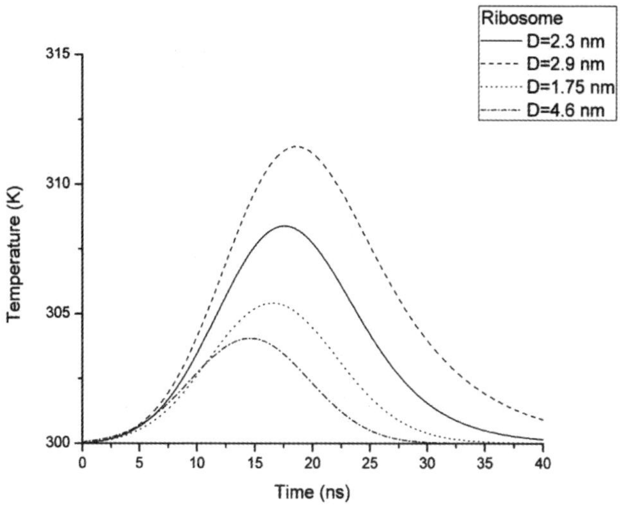

Fig. 8.12. Ribosomal time-temperature profiles at various diameter values. Laser parameters: 12 ns duration, 0.1 J/cm^2 at $\lambda = 700$ nm.

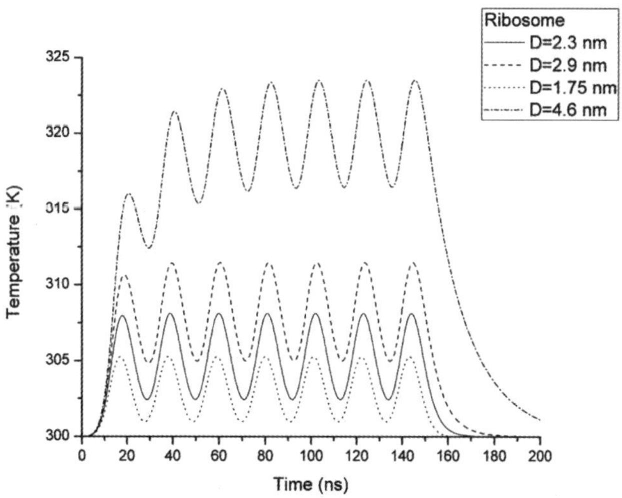

Fig. 8.13. Ribosomal time-temperature profiles at various diameter values. Laser parameters: 12 ns duration, 0.1 J/cm^2 at $\lambda = 700$ nm; $b_1 = 2.5$, $b_2 = 6$, $b_3 = 9.5$, $b_4 = 13$, $b_5 = 16.5$, $b_6 = 20$, $b_7 = 23.5$.

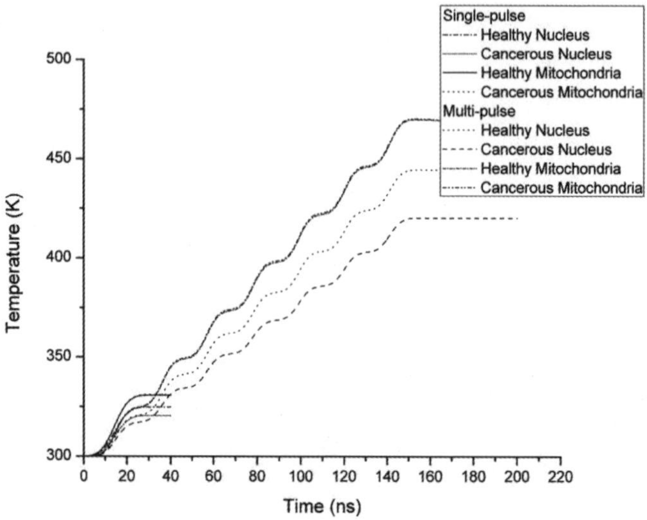

Fig. 8.14. Both single- and multipulse laser heating for 'average' organelles. Average values are: healthy nucleus (7.17 μm), cancerous nucleus (8.97 μm), healthy mitochondria (1.1 μm) and cancerous mitochondria (0.82 μm). The single-pulse laser parameters are the same as for Fig. 8.2, with a 0.1 J/cm^2 energy density. The multipulse laser parameters are: 12 ns duration, 0.1 J/cm^2 at λ = 700 nm; $b_1 = 2.5$, $b_2 = 6$, $b_3 = 9.5$, $b_4 = 13$, $b_5 = 16.5$, $b_6 = 20$, $b_7 = 23.5$.

of absorption, the energy density required to kill cancerous ribosome in either scenario (D = 2.9 nm or 4.6 nm) would also kill many healthy organelles. These results indicate that with a given increase in organelle size, the single- or multipulse mode of heating could be effective at killing the organelles. In the case of the nucleus, at a certain increase in size (after progression of cancer reaches a certain point), it may be possible to use single-pulse laser irradiation as a method of killing cancer cell nuclei. The very specific requirements make such a case unlikely and difficult to put into practical treatment applications.

Using an average value for both organelles in healthy and cancerous forms, we can determine the relative single- and multipulse heating characteristics (Fig. 8.14) that supports the individual cases. Due to the heating and cooling kinetics of mitochondria, we find that for single-pulse laser irradiation, tumor mitochondria will reach a higher maximum temperature than healthy mitochondria by only by a 1–2 K. On the other hand, cancerous nuclei will be as much as 40 K below the peak temperature of the healthy nuclei. This result seems counterintuitive due to the increased size (and thus increased absorption efficiency) of the cancerous nucleus as shown in Fig. 8.3. We believe the size parameter has a greater effect in the temperature kinetics because of the squared term in the OTM, likely as a function of the surface area of the organelle. Due to this result, the single-pulse mode of heating organelles would create more harm to healthy cells than to cancerous cells.

In terms of multipulse heating, a similar result occurs. As Fig. 8.10 shows and Fig. 8.14 reinforces, all types of mitochondria will heat to a temperature well above the nucleus. Additionally, the healthy nucleus once again reaches a higher maximum temperature than any grade of tumor nucleus tested. Due to the indecipherable difference between the maximum temperature of the healthy and cancerous mitochondria, the multipulse mode of heating also proves ineffective at selectively destroying cancer cell organelles.

References

1. D. H. Sliney and S. L. Trokel, *Medical Lasers and Their Safe Use* (Springer-Verlag, New York, 1992).
2. R. F. Steinert, T. F. Deutsch, F. Hillenkamp, E. J. Dehm and C. Adler, Excimer laser ablation of the cornea and lens, *Br. J. Ophthalmol.* **75**, 258–269 (1991).
3. D. Ratner, Y, Tse, N. Marchell, M. P. Goldman, R. E. Fitzpatrick and D. J. Fader, Cutaneous laser resurfacing, *J. Am. Acad. Dermatol.* **41**, 365–89 (1999).
4. J. T. Walsh and T. F. Deutsch, Er:YAG laser ablation of tissue: measurement of ablation rates, *Lasers Surg. Med.* **9**, 327–337 (1989).
5. E. Stein, T. Sedlacek, R. L. Fabian and N. S. Nishioka, Acute and chronic effects of bone ablation with a pulsed holmium laser, *Laser Surg. Med.* **10**, 384–388 (1990).
6. J. S. Nelson, A. Orenstein, L. L. Liaw, R. B. Zavar, S. Gianchandani and M. W. Berns, Ultraviolet 308-nm excimer laser ablation of bone: An acute and chronic study, *Appl. Opt.* **28**, 2350–2357 (1989).
7. D. Zhang, S. Mao, C. Lu, E. Romberg and D. Arola, Dehydration and the dynamic dimensional changes within dentin and enamel, *Dental Mater.* **25**, 937–945 (2009).
8. B. Girard, D. Yu, M. R. Armstrong, B. C. Wilson, C. M. L. Clokie and R. J. D. Miller, Effects of femtosecond laser irradiation on osseous tissues, *Lasers Surg. Med.* **39**, 273–285 (2007).

9. T. G. Barton, H. J. Foth, M. Christ and K. Hormann, Interaction of holmium laser radiation and cortical bone: ablation and thermal damage in a turbid medium, *Appl. Opt.* **36**, 32–43 (1997).
10. H. S. Carslaw and J. C. Jaeger, *Conduction of Heat in Solids* (Oxford University Press, Oxford, 1959), pp. 255–259.
11. A. L. McKenzie, Physics of thermal processes in laser-tissue interaction, *Phys. Med. Biol.* **35**, 1175–1209 (1990).
12. R. R. Letfullin, T. F. George, G. C. Duree and B. M. Bollinger, Ultrashort laser pulse heating of nanoparticles: comparison of theoretical approaches, *Advances in Optical Technologies* **2008**, ID 251718–1-8 (2008).
13. B. N. Chichkov, C. Momma, S. Nolte, F. V. Alvensleben and A. Tunnermann, Femtosecond, picosecond and nanosecond laser ablation of solids, *Appl. Phys. A* **63**, 109–115 (1996).
14. C. Momma, S. Nolte, B. N. Chichkov, F. V. Alvensleben and A. Tunnermann, Precise laser ablation with ultrashort pulses, *Appl. Surf. Sci.* **109–110**, 15–19 (1997).
15. P. L. Blanton and N. L. Biggs, Density of fresh and embalmed human compact and cancellous bone, *Am. J. Phys. Anthropol.* **29**, 39–44 (1968).
16. H. Fukushima, Y. Hashimoto, S. Yoshiya, M. Kurosaka, M. Matsuda, S. Kawamura and T. Iwatsubo, Conduction analysis of cement interface temperature in total knee arthroplasty, *Kobe J. Med. Sci.* **48**, 63–72 (2002).
17. S. R. H. Davidson and D. F. James, Measurement of thermal conductivity of bovine cortical bone, *Med. Eng. Phys.* **22**, 741–747 (2000).
18. A. Milev, G. S. K. Kannangara and B. Ben-Nissan,, Morphological stability of hydroxyapatite precursor, *Mater. Lett.* **57**, 1960–1965 (2003).
19. M. Strassl, V. Wieger, D. Brodoceanu, F. Beer, A. Moritz and E. Wintner, Ultrashort pulse laser ablation of biological hard tissue and biocompatibles, *J. Laser Micro/Nanoeng.* **3**, 30–40 (2008).
20. K. Nakata, M. Umehara and T. Tsumura, Excimer laser ablation of sintered hydroxyapatite, *Sur. & Coat. Technol.* **201**, 4943–4947 (2007).
21. O. R. Musaev, V. Dusevich, D. M. Wieliczka, J. M. Wrobel and M. B. Kruger, Nanoparticle fabrication of hydroxyapatite by laser ablation in water, *J. Appl. Phys.* **104**, 084316–1-5 (2008).
22. O. Guillot-Noel, R. G.-S. Roman and J. Perriѐe, Growth of apatite films by laser ablation: Reduction of the droplet areal density, *J. Appl. Phys.* **80**, 1803–1808 (1996).
23. G. J. Samsonov, *Physicochemical Properties of Oxides* (Metallurgiya, Moscow, 1978).
24. S. Dyshlovenko, B. Pateyron, L. Pawlowskia and D. Murano, Numerical simulation of hydroxyapatite powder behaviour in plasma jet, *Surf. Coat. Technol.* **179**, 110–117 (2004).

25. A. Rosenfeld, M. Lorenz, R. Stoian and D. Ashkenasi, Ultrashort-laser-pulse damage threshold of transparent materials and the role of incubation, *Appl. Phys. A* **69**, S373-S376 (1999).
26. P. Serra and J. L. Morenza, Imaging and spectral analysis of hydroxyapatite laser ablation plumes, *Appl. Surf. Sci.* 127–129, 662–667 (1998).
27. V. P. Zharov, R. R. Letfullin and E. N. Galitovskaya, Microbubbles-overlapping mode for laser killing of cancer cells with absorbing nanoparticle clusters, *J. Phys. D: Appl. Phys.* **38**, 2571–2581 (2005).
28. J. R. Lepock, H. E. Frey and K. P. Ritchie, Protein denaturation in intact hepatocytes and isolated cellular organelles during heat shock, *J. Cell Biol.* **122**,1267–1276 (1993).
29. C. Yao, R. Rahmanzadeh, E. Endl, Z. Zhang, J. Gerdes and G. Hüttmann, Elevation of plasma membrane permeability by laser irradiation of selectively bound nanoparticles, *J. Biomed. Opt.* **10**, 064012 (2005).
30. J. A. Chapman, N. A. Miller, H. L. A. Lickley, J. Quian, W. A. Christens-Barry, Y. Fu, Y, Yuan and D. E. Axelrod, Ductal carcinoma *in situ* of the breast (DCIS) with heterogeneity of nuclear grade: prognostic effects of quantitative nuclear assessment, *BMC Cancer* **7**,174–184 (2007).
31. M. M. Radwan, K. A. Amer, N. M. Mokhtar, M. A. Kandil, A. M. El-Barbary and H. A. Aiad, Nuclear morphometry in ductal breast carcinoma with correlation to cell proliferative activity and prognosis, *J. Egyptian Nat. Cancer Institute* **15**,169–182 (2003).
32. Z. Y. Tang, Y. Q. Yu, X. D. Zhou, Z. C. Ma, R. Yang, J. Z. Lu, Z. Y. Lin and B. H. Yang, Surgery of small hepatocellular carcinoma: analysis of 144 cases, *Cancer* **15**, 536–41 (1989).
33. T. Torimura, T. Ueno, S. Inuzuka, Y. Kimura, P. Ko, M. Kin, T. Minetoma, T. Majima, M. Sata, H. Abe and K. Tanikawa, Ultrastructural observation on hepatocellular carcinoma: correlation of tumor grade and degree of atypia of cell organelles by morphometry, *Med. Electron Microscopy* **26**, 19–28 (1993).
34. L. W. Anson and R. C. Chivers, Ultrasonic propagation in mammalian cell suspensions based on a shell model, *Physics in Medicine and Biology* **34**,1153–1167 (1989).
35. *Carolina Tips* (November 1, 1973).
36. A. Goldfeder and J. N. Selig, Radiosensitivity and biological properties of tumors: XIV. Correlation between mitochondrial structure, oxidative metabolism and radiosensitivity, *Radiation Research* **37**,499–524 (1969).
37. G. T. Deans, P. W. Hamilton, P C. H. Watt, M. Heatly, K. Williamson, C. C. Patterson, B. J. Rowlands, G. Parks and R. Spence, Morphometric analysis of colorectal cancer, *Diseases of the Colon and Rectum* **36**,450–456 (1993).

38. D. Ozaki and Y. Kondo, Comparative morphometric studies of benign and malignant intraductal proliferative lesions of the breast by computerized image analysis, *Human Pathology* **26**,1109–1113 (1995).
39. Y. Cui, E. A. Koop, P. J. van Diest, R. A. Kandel and T. E. Rohan, Nuclear morphometric features in benign breast tissue and risk of subsequent breast cancer, *Breast Cancer Research and Treatment* **104**,103–107 (2007).
40. H. A. Mintz, D. H. Yawn, B. Safer, E. Bresnick, A. G. Liebelt, Z. R. Blailock, E. R. Rabin and A. Schwartz, Morphological and biochemical studies of isolated mitochondria from fetal, neonatal, and adult liver and from neoplastic tissues, *J. Cell Biol.* **34**, 513–523 (1967).
41. N. Campbell and J. Reece, *The Cell: A Tour of the Cell Biology*, 8th edn. (Benjamin Cummings, San Francisco, 2007).
42. *BSCS Biology. Student Edition: A Molecular Approach*, BSCS Blue Version, 9th edn. (Glencoe/McGraw-Hill, Columbus, Ohio, 2006).
43. D. E. Sadava, *Cell Biology: Organelle Structure and Function* (Jones and Bartlett Publishers, Boston, 1993).
44. L. A. Miller and A.Goldfeder, Radiosensitivity and biological properties of two tumor types indigenous to the same host. IX. Swelling properties of mitochondira isolated from mouse tumor and liver tissue, *Proc. Soc. Exp. Biol. Med.* **119**, 759–763 (1965).

Chapter 9

Incorporating Protein Flexibility in Molecular Docking by Molecular Dynamics: Applications to Protein Kinase and Phosphatase Systems

Zunnan Huang and Chung F. Wong*

University of Missouri–Saint Louis

wongch@umsl.edu

This article reviews our application of molecular dynamics simulation to including protein flexibility in molecular docking, particularly in studying protein kinase and phosphatase systems. It demonstrates that incorporating protein flexibility could improve the identification of correct docking pose, help study the different protein conformations induced by the binding of diverse ligands, and provide insights into molecular docking pathways in atomic detail.

9.1. Introduction

Molecular docking has played an increasingly important role in understanding protein-ligand recognition and in computer-aided drug design. Earlier methods focused on docking rigid ligands to rigid receptors. Although these methods still play an important role in studying protein-ligand interactions and in virtual screening of drug candidates because they are cheaper to use, realistic docking experiments require adequate description of molecular flexibility for both the protein and the ligand — because the lock-and-key model fails in many cases and the induced-fit model is more realistic. With the rapid advance of high-performance computing technology, it has become increasingly feasible to treat both the ligands and their receptors as flexible molecules during

---

* Corresponding author.

docking. Recent progresses have also taken advantage of new simulation methods and protocols, suitable timesaving approximations, and refined and less expensive scoring and energy models. In this article, we focus on reviewing our application of simulated annealing molecular dynamics using implicit-solvent models to perform molecular docking on protein kinase and phosphatase systems.

Unlike many docking potentials solely designed for docking rigid or flexible ligands to rigid proteins, molecular dynamics (MD)-based methods take advantage of force fields already designed for simulating protein motion. However, to make it practical to use MD-based methods for docking, it is necessary to go beyond running MD simulations at room or physiological temperatures to improve configurational sampling. Recently, we have explored two variants of the simulated annealing method to enhance configurational sampling. These variants are characterized by running multiple short simulated annealing cycles rather than single long cycles and by randomizing velocities periodically to encourage a system to sample a larger configurational space.

In addition to using enhanced sampling molecular dynamics, we have also used implicit-solvent models — coordinates of the solvent not explicitly included — to reduce the number of particles to decrease simulation time. Researchers have now developed numerous implicit-solvent models, including simple distance-dependent dielectric models, generalized Born models, and Poisson–Boltzmann models. This review also provides a brief overview of these models.

Energy model is another important component determining the success of a docking simulation. Although, in principle, the total energy of a system should be used to distinguish correct docking poses from incorrect ones, it is noisy from molecular simulations when the protein is also flexible. Because the energy of the protein is much larger than that of the ligand and that of the interaction energy between the protein and the ligand, a slight non-optimal structure of the protein in a docking complex can reject a docking pose that is actually quite good. One common practice to reduce noise is to ignore the energy of the protein completely. This review summarizes some of our findings in using this approximation: Empirically, we found that this approximation worked better for docking small organic ligands to protein kinases and phosphatases but not for docking larger flexible peptides to these proteins. For the latter applications, including the energy of the protein could improve the identification of good docking pose, although sufficiently long simulations were essential to reduce sampling noise.

Another trick to perform practical flexible protein-flexible ligand docking is to apply appropriate restraints to the protein to known experimental data. This can reduce simulation costs as well as reducing artifacts produced by approximate energy and solvation models. While applying restraints would limit the range of ligands that can be docked successfully, allowing some protein flexibility is already a valuable step forward to dock a larger range of ligands properly to a protein target than a rigid-protein model can.

Lastly, although docking studies have focused on identifying good docking poses, we have also extended our analysis to studying protein-ligand docking pathways, and to providing preliminary estimate of activation barrier for protein-ligand docking.

9.2. Refined Simulated Annealing Protocols

Simulated annealing facilitates the search of the global energy minimum by heating a system to a high temperature and then gradually cooling the system down to 0 K. The success of simulated annealing in finding the global energy minimum depends on the temperature to which the system is heated up and on the subsequent cooling rate. Although heating a system to a high temperature and cooling the system slowly afterwards could increase the chance of finding the global minimum, practical considerations limit how slow a system could be cooled down because the slower the cooling, the longer the simulation time. The alternative that we have tested recently is to run many short simulated annealing cycles and reassign velocities at the beginning of each simulated annealing cycle to encourage a system to sample large configurational space. Here, we present two variations of this simulated annealing protocol that we have tested to show that they are more efficient than conventional simulated annealing (using one single long cooling cycle) [1–4] and the more recent replica-exchange (RE) method [5–8]. In the RE method, multiple simulations are performed simultaneously at different temperatures and the snapshots are allowed to exchange between temperature windows according to the Metropolis criteria as in Monte Carlo simulation. Because structures sampled at higher temperature windows could occasionally enter the lower temperature window at which results are sought, the lower temperature window could sample a larger configurational space than a single simulation running at the lower temperature could.

9.2.1. *NVE and NVT simulated annealing cycling protocols*

The two variants differed by running the simulation at two different thermodynamics ensembles between two time points that the system was disturbed by velocity reassignment. One was the micro-canonical ensemble or NVE ensemble (constant number of particles: N, constant volume: V, and constant energy: E). The other was the canonical ensemble or NVT ensemble (constant number of particles: N, constant volume: V, and constant temperature: T).

In the NVE variant, each simulated annealing cycling run [9] typically included several independent trajectories — six in most of our previous simulations. We started these trajectories from the same initial structure but with different initial atomic velocities and therefore these trajectories had different initial energies and temperatures. The simulation protocol can be summarized as follows [9]:

(1). Assigned initial velocities to an initial structure at six different temperatures, e.g., 10 K, 50 K, 100 K, 300 K, 600 K, or 1000 K, to initiate six different trajectories.
(2). Allowed the system to evolve for 1 ps in an NVE simulation.
(3). At the end of each 1-ps simulation, reassigned atomic velocities to reduce the temperature of the system by ~1/4 or to heat up the system to a chosen target temperature T_{target}. We usually chose T_{target} to lie between 1000 K and 2000 K, which we found to be high enough for many of our systems to go over large energy barriers easily.

Whether heating or cooling was performed after each 1 ps of NVE simulation was determined by whether the temperature T was below a pre-chosen low temperature, such as 10 K: if above, cooling was performed; if below, heating was done. By choosing a low threshold temperature below 10 K, one could take a system to within ~1–2 kcal/mol of local energy minima. Practically, cooling occurred more frequently while heating happened only occasionally when the system already explored a local energy minimum well.

In the NVT version, each simulated annealing cycling simulation was performed as follows [10–12]:

(1). Assigned initial velocities at a high temperature $T_{heating}$ such as 500 K.
(2). Allowed the system to evolve for 1 ps in the NVT ensemble. The Nosé-Hoover chain method [13] was used to maintain constant temperature.
(3). At the end of each 1-ps simulation, the temperature was reduced by half.
(4). Loop back to 1) unless the temperature of the simulation was lower than 5 K.

The above loop was continued until a prescribed simulation length was reached. Typically, each simulated annealing cycle approximately covered the following temperatures: 500 K, 250 K, 125 K, 62.5 K, 31.2 K, 15.6 K, 7.8 K, and 3.9 K when $T_{heating}$ was 500 K. The end of each simulated-annealing cycle identified a structure near a local or the global energy minimum. By performing many such short simulated annealing cycles in a molecular dynamics simulation, one could identify many energy minima. In addition, we usually ran multiple trajectories using different placements of the ligand in the protein, or/and by assigning different atomic velocities to the same starting structure using different random number-generating seeds.

9.2.2. *Comparing the performance of simulated annealing cycling, conventional simulated annealing, and replica-exchange (RE) simulation*

Using the docking of flexible balanol to rigid protein kinase A (PKA) as a test system, Huang *et al.* [9] compared the performance of the NVE simulated annealing cycling protocol with that of conventional SA, and that of RE simulation. They performed ten runs using each method. Each run of the NVE simulated annealing cycling simulation was composed of six trajectories and each trajectory lasted 2 ns. Each conventional SA run also included six trajectories and each trajectory was performed by cooling the system from 1000 K to 0 K at a constant rate of 0.5 K/ps. The RE simulations were performed with two different temperature ranges: 270–650 K, and 10–1000 K. For each temperature range, 24 exponentially distributed temperature windows were used. Each trajectory in the RE simulations also lasted 2 ns. The RE simulations with only six replicas were not used for comparison because of their low acceptance ratio and inefficient structural sampling. All MD simulations were conducted using CHARMM param19 force field [14] with a distance-dependent dielectric model: $\varepsilon(r) = 4r$ where r was the distance between two atoms (other computational details were reported in a previous publication [9]).

Table 9.1 summarizes the performance of the three types of simulations. In the table, $R_{contact}$ denotes the contact distance between the hydroxyl oxygen of the hydroxybenzamide moiety of the ligand and the peptide carbonyl oxygen of Glu 121. We used this distance to monitor the formation of the hydrogen bond observed between the ligand and the linker region connecting the N- and C-terminal lobes of the kinase domain in the crystal structure. RMSD represents the root-mean-square deviation between a docking structure and the crystal structure for the ligand. These two measures helped to monitor how well balanol was docked to the protein. Each set of RE simulations contained 10 runs with

Table 9.1. Comparison of the performance of simulated annealing cycling (SAC), regular simulated annealing (SA) and replica-exchange (RE) simulations.

Criteria / Simulation	RMSD ≤ 6Å $R_{contact}$ ≤ 10Å	RMSD ≤ 3Å $R_{contact}$ ≤ 5Å	RMSD ≤ 2Å $R_{contact}$ ≤ 3Å
10/60* SAC	10/33^{+}	9/11	1/1
10/60 SA	10/39	7/9	0/0
10/240* RE (270 K–650 K)	10/34	0/0	0/0
10/240 RE (10 K–1000 K)	9/22	0/0	0/0

(This table was modified from one published in *Proteins: Structure, Function, Bioinformatics*, **71**, 440–454 (2008).)

Notes. In the above table, each SAC and SA run included six trajectories and each RE simulation consisted of 24 replicas. *10/60 denotes 10 runs and 60 trajectories because each run contained six trajectories for the SAC and the SA simulation. *10/240 denotes 10 runs of RE simulation each containing 24 temperature windows, thus a total of 240 trajectories. $^{+}$10/33 indicates that 10 runs and 33 trajectories resulted in docking poses satisfying the criteria defined at the top of the column.

24 replicas in each run, making a total simulation time of 480 ns. On the other hand, each set of NVE simulated annealing cycling and conventional SA simulations consisted of 10 runs each including six trajectories, making a total simulation time of only 120 ns. Nevertheless, it is evident from Table 9.1 that the RE simulations were less successful than the other simulations in docking. None of the RE simulations was able to obtain docking pose within 3 Å of the crystal structure and with an $R_{contact}$ < 5 Å. In contrast, all other simulations found better docking poses. Among these simulations, NVE simulated annealing cycling was more efficient than the conventional SA method, by having more runs and more trajectories identifying docking poses within 3 Å of the crystal structure and with an $R_{contact}$ < 5 Å.

Table 9.2 compares the best docking structures obtained from the simulations. The NVE simulated annealing simulation identified the best structure to within 0.80 Å and found docking poses to within 3.0 Å of the experimental structure for any run. In contrast, 3 of 10 conventional SA runs failed to find docking poses within 3 Å of the experimental structure. On the other hand, the better one of the RE simulation, with temperature windows between 270 and 650 K, found structures within 4 Å but greater than 3 Å of experiment in only three of the ten runs. No runs from the other one of the RE simulation, with temperature windows between 10 and 1000 K, sampled any structure within 4 Ǻ of the crystal structure. These data show that the NVE simulated annealing cycling protocol performed better than conventional SA, which in turn worked better than the replica-exchange method.

Table 9.2. The lowest RMSD from experimental structure for the SAC, SA, and RE simulations.

$RMSD_{Lowest}$ in Å	1	2	3	4	5	6	7	8	9	10
SAC	**2.60**	**2.36**	**2.94**	**2.93**	**2.50**	**2.86**	**<u>0.80</u>**	3.01	**2.59**	**3.00**
SA	**2.42**	3.91	**2.67**	3.05	**2.53**	**2.53**	**2.90**	**2.59**	3.69	**2.53**
RE(270K-650K)	4.09	5.46	4.48	4.51	4.02	3.93	5.14	3.61	3.63	4.04
RE(10K-1000K)	5.61	5.90	4.37	5.18	4.78	4.71	4.27	4.99	4.21	6.26

(This table was modified from one published in Proteins: Structure, Function, Bioinformatics **71,** 440–454 (2008).)

Notes. In the above table, **<u>bold</u>** indicates docking poses with RMSD < 2 Å while **bold** indicates docking poses with RMSD < 3Å. 1 through 10 labels one of the ten runs.

9.2.3. *Applications of simulated annealing cycling simulations to docking small-molecule inhibitors and flexible peptides to protein kinases and phosphatases*

We have now applied simulated annealing cycling simulations to docking both small organic molecules and flexible peptides to several protein kinases and phosphatases and found that they were successful in identifying good docking poses [9–12]. We typically ran between 20 to 40 trajectories for docking each ligand to each protein. As the trajectories were completely independent, our modified scripts distributed these trajectories to multiple processors simultaneously so that we could obtain results within a couple of days for each protein-ligand docking simulation using modern computer clusters. Besides using efficient configurational sampling protocols and high-performance computing systems, we have also taken advantage of implicit-solvent models, identified combination of energy/solvation models that could discriminate correct poses from incorrect ones, and experimented with different restraints imposed on the protein to reduce simulation time and to overcome deficiency of approximate energy and solvation models. We now elaborate each in further detail.

9.3. Implicit Solvent Models

Because it is expensive to include many solvent molecules explicitly in simulating biomolecular systems, implicit-solvent models are often used to reduce simulation time. Gilson and Zhou had recently written a good review on these models [15]. Here, we focus on discussing two that we have tested extensively in docking ligands to protein kinases and phosphatases, and found to work well without incurring significant computational costs.

9.3.1. *Distance-dependent dielectric (DDD) models*

These models modify Coulomb's law by multiplying the distance between two atoms by a dielectric coefficient that varies with the distance between the two atoms. The $\varepsilon(r) = r$ model, where r is the distance between two atoms in a molecular system, was the first distance-dependent dielectric model employed in molecular dynamics simulation of proteins. Later, dielectric function that increased more rapidly with distance has become popular; $\varepsilon(r) = 4r$ model is one of them. This and other distance-dependent dielectric models are now widely used in programs such as CHARMM [14], DOCK [16]; FITTED [17] and others [18–22]. For protein-ligand docking, Ferrara *et al.* [23] showed that the $\varepsilon(r) = 4r$ model with the CHARMM force field distinguished near-native structures from decoys better than the scoring functions in AutoDock, Dock, and SYBYL(PMF, Gold, ChemScore) and no worse than the more expensive Generalized Born (GB) model in a test set of 189 protein-ligand complexes. In addition, we have found this model to work well in docking small organic ligands [9–11] and flexible peptides [12] to protein kinases and phosphatases.

9.3.2. *Generalized Born (GB) models*

The GB model was first proposed by Still and co-workers [24]. One popular form of their model calculates a solute-solvent electrostatic polarization term G_{pol} by

$$G_{pol} = -\frac{1}{2}\left(\frac{1}{\varepsilon_p} - \frac{1}{\varepsilon_w}\right)\sum_{ij}\frac{q_i q_j}{f_{gb}}$$

$$f_{gb} = \left[r_{ij}^2 + \alpha_i \alpha_j \exp\left(-\frac{r_{ij}^2}{4\alpha_i \alpha_j}\right)\right]^{\frac{1}{2}}$$

where ε_p represents the low dielectric constant inside a solute such as a protein, ε_w is the dielectric constant of water, r_{ij} is the separation distance between atom i and j, q_i and q_j are the charges of the atoms, and α_I and α_j are their corresponding effective Born radii. The effective Born radius describes the average distance of an atomic charge from the continuum dielectric boundary. In the equation, f_{gb} is a function dependent on the distance r_{ij} and the effective Born radii α_i and α_j. It becomes r_{ij} at large separating distances and the square root of the product of α_i and α_j as r_{ij} approaches zero [24].

There are various ways of calculating the effective Born radii α_i and α_j [25–34], among them is the Generalized Born using Molecular Volume (GBMV) [31,32] model which we used in our docking simulations. Generalized Born models are also usually supplemented by terms for describing hydrophobic and nonpolar effects. Earlier models used terms dependent only on surface area [24].

Because Generalized Born models consume much more computational time than simple distance-dependent dielectric models do, they are still not commonly used in molecular docking, especially in larger scale virtual screening. On the other hand, it has been used to refine and rescore docking poses. For example, Lee and Sun recently reported [35] that further optimizing and rescoring docking poses obtained from the DOCK program could improve the identification of correct docking pose in a test set of 79 protein-ligand complexes. In addition, Huang and Wong [12] showed in peptide-protein docking that rescoring poses obtained from the DDD model with the GBMV model could improve the identification of docking structures. Huang and Wong also found that rescoring structures obtained from simulations using the GBMV model with the $\varepsilon(r) = 4r$ model could find better docking poses, as did rescoring structures obtained from simulations using the DDD model with the GBMV model. If no re-scoring was done, they found that using the $\varepsilon(r) = 4r$ model by itself performed better than using the GBMV model by itself.

9.4. Application of Experimental Restraints to Protein Structures

We have explored several ways of applying restraints. All are similar in imposing restraints only to the α carbons but not on the side chains. They differ by the number of reference structures used or whether the same restraint is applied to all parts of the protein.

9.4.1. *Protein restraint based on one reference structure*

Here, we restrained the conformational sampling of a protein structure to be near one single experimental structure. We achieved this by applying a restraint potential to the α carbons according to [9,10,12]:

$$V_{potential} = k_{restraint} RMSD^2$$

where *RMSD* was the Root-Mean-Square-Deviation of a dynamics snapshot from the reference structure, and $k_{restraint}$ was a suitable force constant to control the extent that the protein could move. $k_{restraint}$ was usually chosen to be 1000 kcal/(mol·Å^2).

Even with this relatively restrictive restraint, a protein could still experience relatively large conformational change — as much as ~7–8 Å from the reference structure for simulations that we have performed so far [9,10].

Although not completely general, this semi-flexible receptor model allows a wider range of ligands to dock properly than a completely rigid receptor model can. In docking balanol to protein kinase A from the protein surface, Huang *et al.* found balanol to move much more easily inside the protein with this model than with the rigid receptor model [9]. In addition, it was feasible to dock balanol using reference protein structures selected from crystal structures with other ligands bound; this failed more often with the rigid-receptor model. [9]

Although this semi-flexible protein model performed much better than the rigid-protein model, restraining to one single reference structure was too restrictive for some applications when the reference structure deviated significantly from the structure that a ligand prefers to bind. As an example, balanol failed to dock into the binding pocket of protein kinase A from the surface if the protein was restrained to a tightly closed conformation found in PDB entry 1FMO [9]. On the other hand, if the open *apo* structure from PDB entry 1J3H was used as the reference structure, the glycine-rich loop of protein kinase A had difficulty closing up sufficiently upon ligand binding [11]. This limitation led us to explore less restrictive restraints.

9.4.2. *Protein restraint based on two reference structures*

By using two extreme structures observed experimentally to construct restraints, one could guide a docking simulation to sample conformations near any of these two structures as well as those in between (schematically shown in Fig. 9.1). In the figure, the two smaller dark-gray circles represented the conformational space sampled by the one-reference structural model using either one of the reference structures and the larger oval circle depicted that sampled by the two-reference-structure model. We have constructed restraining potential using two reference structures according to [11]:

$$V_{potential} = k_{restraint}[(\frac{RMSD_1 + RMSD_2}{2}) - OFFSET]^2$$
$$OFFSET = \frac{RMSD_{1\text{-}2}}{2}$$

where $RMSD_1$ and $RMSD_2$ were the RMSDs of a dynamics snapshot from reference structures 1 and 2 respectively, and $RMSD_{1-2}$ was the RMSD between the two reference structures. Again, we applied restraints only to the α carbons. The extent of deviation from the reference structures was controlled by the force constant $k_{restraint}$, which we chose to be 1000 kcal/(mol·Å^2) in our previous simulations.

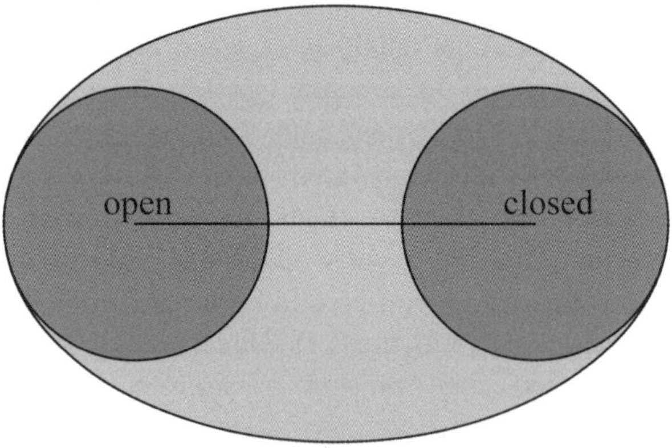

Fig. 9.1. Conformational space sampled by the one-reference-structure restrained models (dark circles) and the two-reference-structure restrained model (larger ellipsoid).
(This figure was published previously in the Supplementary Material of *J. Comput. Chem* **30**, 631-644 (2009).)

We [11] tested this model by docking four diverse ligands to protein kinase A. These docking studies provided a good test case because the conformations of the protein with these ligands bound were quite different. During the docking simulations, the protein structures from PDB codes 1J3H [36] and 1FMO [37] was used as reference structures because they represented the most open unliganded form and a tightly closed conformation respectively. This two-reference-structure model allowed ligands to dock successfully to the protein no matter which protein structure was used as the starting protein conformation. [11] Ligands that failed to dock properly with the one-reference-structure model, such as docking balanol to the crystal structure in PDB entry 1FMO, were able to dock correctly with the two-reference-structure model. In addition, we [11] found that the proper conformation of the protein was induced upon ligand binding. In particular, the glycine-rich loop was formed properly when different ligands were bound.

9.4.3. *Protein restraint based on one reference structure but with more than one force constants*

The restraint model using two reference structures could allow a wider range of ligands to dock properly than using only one reference structure. On the other hand, realizing that the largest conformational change is usually associated with flexible loops, such as the glycine-rich loop of PKA [38] and the WPD-loop [39,40] of protein tyrosine phosphatase (PTP), it could be effective to use only one single reference structure but applying less restrictive restraints to the loops than to the α carbons.

We have tried using two restraining potentials having different force constants $k_{restraint}$'s in different parts of a protein. A smaller force constant of 100 kcal/(mol·Å^2) was applied to the flexible glycine-rich loop of PKA. The larger force constant $k_{restraint}$ = 1000 kcal/(mol·Å^2) was applied to the rest of the protein.

Although the restraint model using one reference structure described above could not dock balanol properly into the protein structure in PDB entry 1FMO, which accommodates another ligand, a modified model using less restrictive restraint on the glycine-rich loop was successful (unpublished data) and worked better than another model without any restraint applied to the glycine-rich loop [9].

9.4.4. *Other restraints used in flexible-protein docking*

When suitable experimental data are available, one could introduce additional restraints to facilitate docking. For example, to prevent a ligand from moving too far away from the protein surface, Huang *et al.* [9,11] applied a quartic potential [5 kcal/(mol·$Å^4$)] × $(rcm\text{-}rcenter)^4$ to the ligand, where *rcm* was the center of mass of the ligand and *rcenter* was a point in the binding pocket, when *rcm-rcenter* was larger than 15 Å.

In docking peptides to protein-ATP complexes, one can restrain ATP to known experimental structures as its structure is not expected to change significantly upon peptide binding [12]. Applying such a restraint can further reduce simulation time.

9.5. Scoring Functions and Energy Models

Scoring function is a critical component in docking simulations. Good scoring functions need to be able to distinguish correct docking poses from incorrect ones. Many scoring functions have already been introduced but it appears that no one single scoring function works well for all systems [23,41–44]. Scoring functions can be roughly grouped into three classes: force field methods such as CHARMM [14], AMBER [45], AutoDock [46], G-score in GOLD [47] and D-score in DOCK [16]; empirical scoring functions such as F-score [48], ChemScore [49], LUDI [50], PLP [51], and X-score [52]; knowledge-based potentials such as PMF [53], DrugScore [54] and SMoG [55].

For our applications in incorporating protein flexibility, force fields methods are particularly useful because they have been developed for simulating the dynamics of biomolecular systems and are thus already suitable for directly simulating the conformational change of a protein during protein-ligand docking. In contrast, empirical scoring functions and knowledge-based potentials usually do not include protein energy and are thus less suitable for describing the conformational change of the receptors. (Many [48–53] even ignore the intra-molecular energy of the ligand.) Protein energy is not needed in rigid-protein docking because the energy of the protein remains the same no matter where the ligand binds.

This is no longer true in flexible-protein docking. In predicting which conformation of a protein a ligand prefers to bind, including the conformational energy of the protein can be important. For example, although a ligand interacts favorably with one conformation, it may still not prefer to bind to this conformation if the energy costs for reaching the conformation are high. On the other hand, incorporating protein energy directly in scoring introduces additional challenges. Because the magnitude of the energy of the protein is typically larger than that of the sum of the energy of the ligand and the protein-ligand energy, insufficient sampling of protein conformations could introduce significant noise, making it difficult to use the total energy of a protein-ligand system to identify good docking poses. Nevertheless, it is important to carry out actual simulations to test whether this is really the case, or approximations such as neglecting the energy of the protein completely can still provide a reasonable model for identifying good docking poses from flexible ligand-flexible protein docking. As our interests have focused on studying protein kinases and phosphatases, we have focused on testing several scoring models for these systems. Even before one can make general conclusions for many proteins, identifying models that work specifically for particular classes of proteins can still be fruitful.

In the past few years, we have tested three energy models for identifying good docking poses:

(1) Energy Model I: total energy of the whole protein-ligand system.
(2) Energy Model II: total energy minus the protein energy (including molecules that are tightly bound to the protein such as ATP). In other words, only the intra-molecular energy of the ligand and protein-ligand interaction energy are included.
(3) Energy Model III: Energy of whole system minus the energy of protein and the ligand, i.e., only protein-ligand interaction energy is included.

Note that these models have only been used in the final stage in distinguishing good docking poses from bad ones. The total energy has always been used during flexible ligand-flexible protein docking.

Variations of the above three models include running simulation with one solvation energy model but rescoring the resulting structures with another solvation model. After some experimentations with several implicit-solvent models, we have focused on testing the distance dielectric model $\varepsilon(r) = 4r$ and the GBMV model.

Huang and Wong [9–12] found that the best energy models could be different for small-molecule ligands and for peptide ligands. For small organic ligands [9–11], Energy Model II (excluding the energy of the protein) usually worked well in identifying good docking poses, whereas Energy Model I (total energy of the whole system) did not work as well. On the other hand, for peptide ligands [12], Energy Model I performed significantly better. However, we noticed that sufficient sampling of protein conformations was essential in using Energy Model I to separate signals from noise.

In addition, our unpublished data indicated that Energy Model III was close to Energy Model II in performance, suggesting that including ligand energy was less crucial. Nevertheless, our unpublished data found that Energy Model III worked slightly less well than Energy Model II. As the computational costs were similar for these two energy models, we prefer to use Energy Model II.

The above discussions focused on using one solvation energy model. We have also rescored structures obtained from simulations performed with one solvation energy model ($\varepsilon(r) = 4(r)$ or GBMV) with another solvation energy model. Re-scoring generally improved the identification of good docking poses for both small-molecule ligands [10] and larger peptide ligands [12].

9.6. Docking Pathways

Although more work has been done on identifying where in a target protein a ligand might bind than on elucidating how the ligand gets there, new methods such as steered [56] and biased [57] molecular dynamics have been developed to facilitate the exploration of docking pathways. These approaches apply biases to steer a ligand to enter or release from a protein, and can reveal how the conformational change of the protein couples with ligand entry and release.

One application on protein kinase was performed by Ozkirimli and Post [58]. They carried out biased molecular dynamics simulation to study the conversion of the Src kinase from the inactive to the active form. By analyzing the structural changes that occurred, they identified an electrostatic switch that might play a role in Src kinase activation. In another study, Lu *et al.* [59] performed force-pulling MD simulation to study the release of ADP from protein kinase A and found that structural changes distant from the product release channel could also couple with ADP movement. Previously, Wong *et al.* [60] found in the linear response limit that structural correlation between two regions could also be viewed as structural response of one region when a small force was applied to the other. Coupling this finding with the force-pulling results suggested that molecules binding to and interacting with surface regions far away from the binding pocket could also influence ADP release.

Although the steered and biased molecular dynamics simulations are quite useful in revealing docking pathways, they are expensive to do especially if one wants to repeat simulations with different starting conditions or with different biasing forces. We have therefore explored using results from our simulated annealing cycling simulations to map out docking pathways, as such simulations sample configurational space extensively. We achieved this by first clustering structures from such simulations and then suitably connecting them to form pathways. The averaged energies of structures within clusters along a pathway then provided a crude estimate of the energy profile along the pathway. In one version, the clustering was done in two stages to identify reaction paths:

(1) We first clustered the structures based on one reaction coordinate, R_{path}, that measured the distance of the system from the stable protein-ligand complex determined experimentally or from docking simulations.
(2) We then performed a second round of clustering for structures that fell within a small bin centering around a given value of R_{path}. If multiple clusters were found, multiple pathways were possible at this value of R_{path}. To connect these clusters to form continuous pathways, we performed similarity analysis between representative structures (chosen to be the centroid structure of each cluster, for example) in adjacent bins to determine how they should be connected.

The similarity measure was simply the RMSD between centroid structures in adjacent bins. By analyzing the representative structures of the clusters along each pathway, one could gain atomistic insights into the structural change associated with a docking process along each pathway.

One advantage of the simulated annealing cycling technique is that it can reduce the possibility of a simulated system being trapped into local energy minima, because the simulation protocol periodically heats the system to a high temperature to help the system to move away from local energy wells. On the other hand, using an implicit-solvent model in the simulation and simply using the averaged energy of cluster structures to estimate energy profile might make the result less quantitative. Nevertheless, the pathways obtained could provide useful starting points for performing explicit-solvent simulations to estimate potential of mean force using the umbrella sampling technique [61–63], for example.

Our first practical test of this idea involved mapping docking pathways between para-nitrocatechol sulfate (pNCS) and the *Yersinia* protein tyrosine phosphatase YopH. This bacterial protein tyrosine phosphatase is responsible for causing human diseases ranging from gastrointestinal syndromes to bubonic plague [39,64,65]. pNCS is a specific inhibitor of YopH that displays >10x selectivity towards YopH over a panel of mammalian protein tyrosine phosphatases [66]. A second test involved mapping out the docking pathways between a hexapeptide substrate with the sequence GDYMNM and the activated form of the catalytic domain of the insulin receptor protein kinase (IRK) [67]. During the simulated annealing cycling simulations, suitable experimental restraints were applied as described earlier in docking studies. In addition, we used the cheaper distance-dependent dielectric model $\varepsilon(r) = 4r$, instead of the more expensive GBMV model, because it worked well in our docking studies.

9.6.1. *Docking pathways between pNCS and YopH*

In this application, we used the distance between the center-of-mass of pNCS and the center of the binding site of YopH as R_{path} [10]. Figure 9.2(a) gives the preliminary estimate of the energy profile along R_{path}. From Fig. 9.2(a), one can see that the energy barrier for ligand entry was only ~4 kcal/mol whereas the energy barrier for ligand release was significantly larger at ~23 kcal/mol.

To examine whether there could be multiple pathways of entry and exit, we performed the clustering and similarity analysis described above [10]. Figure 9.2(b) demonstrates multiple pathways converging into a single one leading to the co-crystal structure. Although not shown in Fig. 9.2(b), our previous analysis also found that the ligand could arrive at the docking pocket with the wrong docking structure and needed to move away from the pocket a bit before finding a pathway that could lead it to the correct docking pose [10].

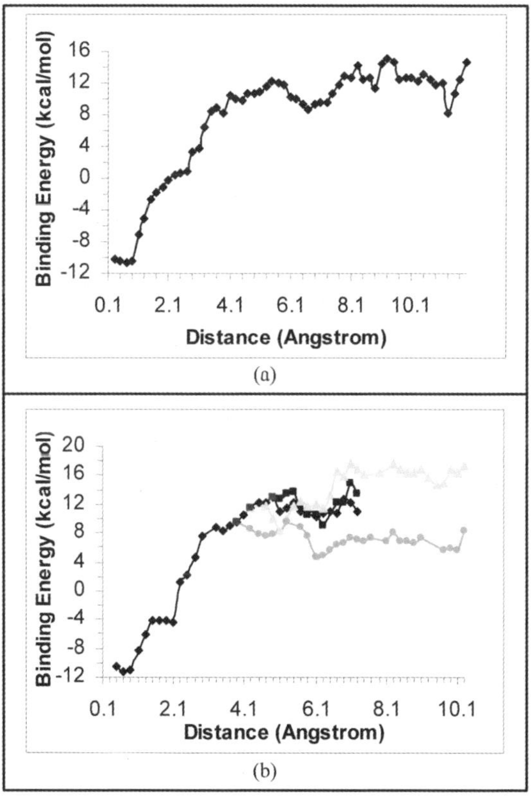

Fig. 9.2. (a) Energy profile along R_{path}. (b) Major docking pathways leading to the correctly docking pose between pNCS and YopH. Each line represents a docking pathway. (This figure was modified from *Biophys. J.*, **93**, 4141–4150 (2007).)

Figure 9.3 illustrates the structural changes associated with the docking pathways for R_{path} between 0.7 Å to 6.3 Å. Each structure represented one closest to the centroid of a cluster. The coupled movement of the side chains of nine residues (Phe-229, Ile-232, Asp-356, Gln-357, Arg-404, Ala-405, Val-407, Arg-409, and Gln-446) is also demonstrated in the figure. These residues interacted with the ligand at one point or another. The large movement of some residues such as Ile-223 and Arg-404 is evident. These large movements might have functional significance. For instance, Arg-404 maintained interactions with the ligand from the surface to the binding pocket; it might help the ligand to move between the binding pocket and the protein surface. Likewise, our previous analysis revealed that the sulfate group of pNCS might also play a role in directing the ligand into the binding pocket from the protein surface besides providing favorable interactions between the ligand and the protein in the bound form of the protein-ligand complex [10].

9.6.2. *Docking pathways between a hexapeptide and the insulin receptor tyrosine kinase*

Here, we performed the NVT simulated annealing cycling simulation described in Sec. 9.2.1 to sample configurations for constructing pathways. The hexapeptide substrate had the sequence GDYMNM. We acetylated the N terminus of the hexapeptide and blocked its C terminus so that it was in the *N*-methylamide form. We ran three sets of independent trajectories starting from the same structure and temperature but with different initial atomic velocities assigned with different random number generating seeds. In one set, containing ten independent trajectories, we used the crystal structure as the starting structure. In the second set, containing twenty trajectories, we used an extended form of the hexapeptide located near the binding site. In the third set, containing twenty trajectories, we placed the hexapeptide in the extended form at the surface of the protein. Each trajectory lasted 2 ns. The total aggregate simulation time of these 50 trajectories was thus 100 ns. These trajectories generated many structures at local energy minima from which we could construct docking pathways by clustering and similarity analysis described above. The CHARMM param27 force

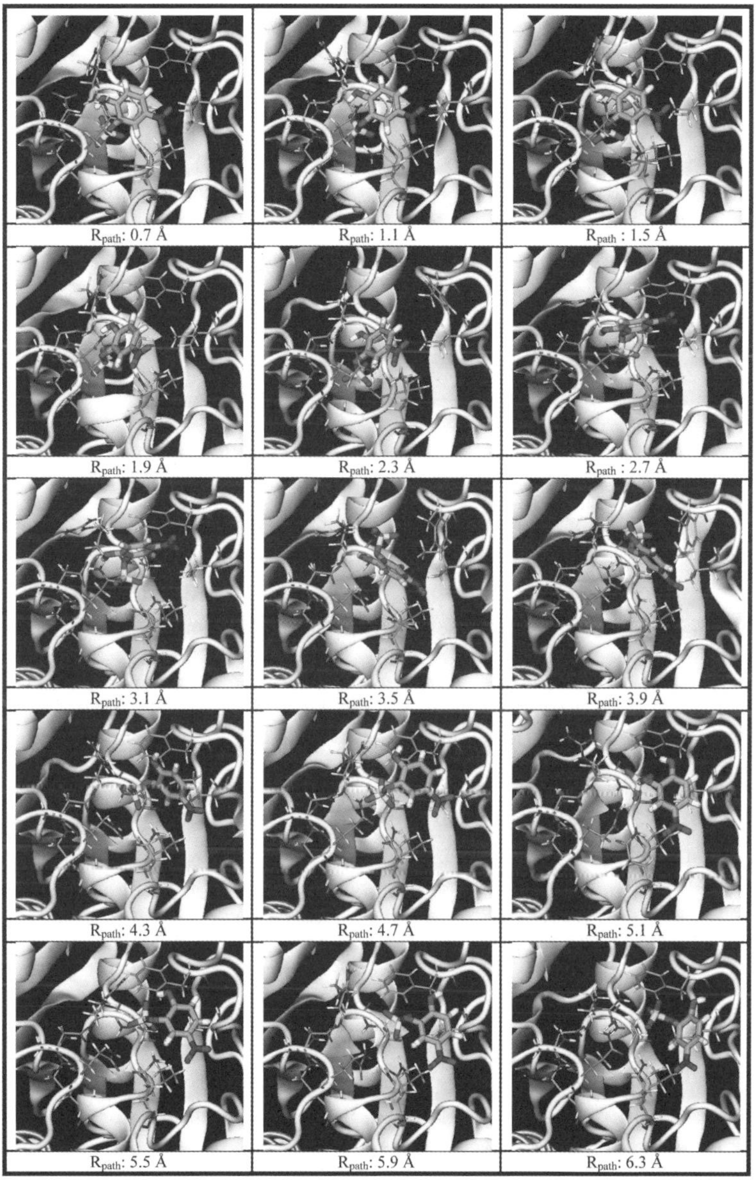

Fig. 9.3. Structural change associated with the docking pathways in Fig. 9.2(b). pNCS and nine residues (Phe-229, Ile-232, Asp-356, Gln-357, Arg-404, Ala-405, Val-407, Arg-409, and Gln-446) in or near the binding pocket are shown. (This figure was modified from *Biophys. J.*, **93**, 4141–4150, (2007).

field [68,69] was used. We also used the $\varepsilon(r) = 4r$ model in the simulation. In addition, we used a nonbonded cutoff distance of 14 Å, a switching function for the electrostatic interactions that began at 10 Å and ended at 12 Å, and a shifting function for the Lennard-Jones potential.

Figure 9.4(a) shows that the $\varepsilon(r) = 4r$ model with Energy model I (total energy of the system) was capable of identifying the correct docking pose to within 3.29 Å of the crystal structure. We then performed pathway analysis as described above for the docking between pNCS and YopH except that we defined R_{path} differently:

$$R_{path} = \sqrt{\sum_{i} \left(\frac{\left(R_i - R_{i\_ref}\right)^2}{N} \right)}$$

where R_{i_ref} was the distance between a heavy atom on the peptide and its closest protein heavy atom in the crystal structure, R_i was the corresponding distance for any structure sampled during the simulation.

Figure 9.4(b) shows the energy profile along R_{path}. (The resolution of the bin was 0.4 Å.) It shows that the energy barrier for peptide entry was ~13 kcal/mol whereas the energy barrier for ligand release was larger at ~24 kcal/mol.

Figure 9.5 illustrates the structural change associated with one major docking pathway, with R_{path} between 1.0 Å and 9.8 Å. Each structure represents a structure closest to the centroid of a cluster. The structure at the smallest values of R_{path} (1.0 Å) resembles the crystal structure, only 1.32 Å away. The movement of the side chains of thirteen residues (Lys-1085, Asp-1132, Arg-1136, Gly-1169, Leu-1170, Leu-1171, Pro-1172, Val-1173, Met-1176, Leu-1181, Asn-1215, Glu-1216, and Leu-1219) in or near the binding site is also shown. In the crystal structure, or the docking structure similar to it (Fig. 9.5(a)), the γ-phosphorus of ATP was ~5 Å from the hydroxyl oxygen of the tyrosine of the peptide, Tyr(P), that undergoes phosphorylation. In addition, the methionines at the P+1 and P+3 positions fitted into two adjacent hydrophobic pockets. The pocket accommodating the methionine at the P+1 position comprised

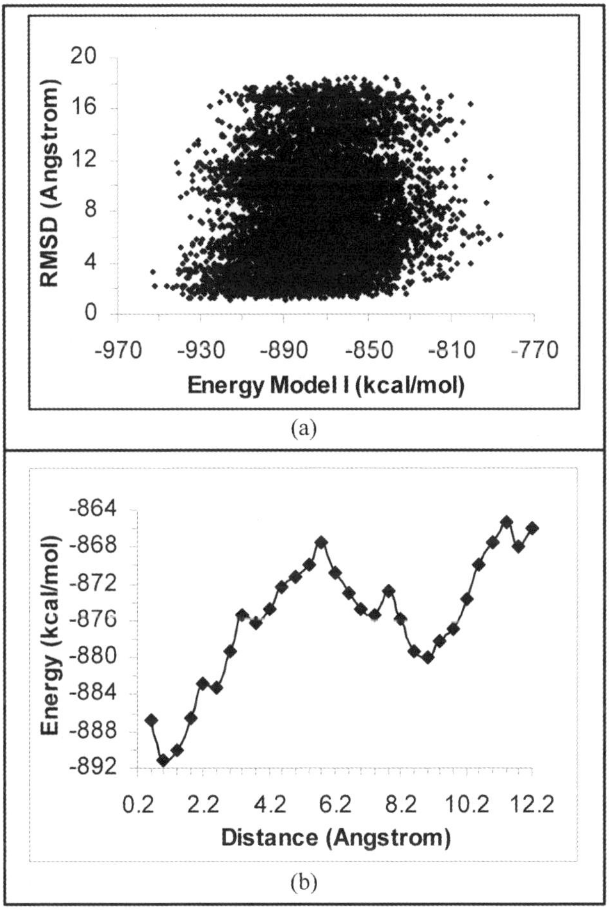

Fig. 9.4. (a) Root-mean-square deviation from experimental structure (heavy atoms only) versus energy (using Energy Model I) for the docking of the hexapeptide GDYMNM to the catalytic domain of the insulin receptor tyrosine kinase. The plot included only structures near local energy minima obtained below 5 K. (b) Energy profile along R_{path}.

Val-1173, Leu-1219 and the aliphatic side chains of Asn-1215 and Glu-1216. On the other hand, the pocket binding the methionine at the P+3 position was composed of the side chains of residues Leu-1171, Val-1173, Met-1176, Leu-1181 and Leu-1219. As the peptide was releasing from the binding pocket, it moved toward ATP. The long hydrophobic side chain of the methionine at the P+1 position moved out first,

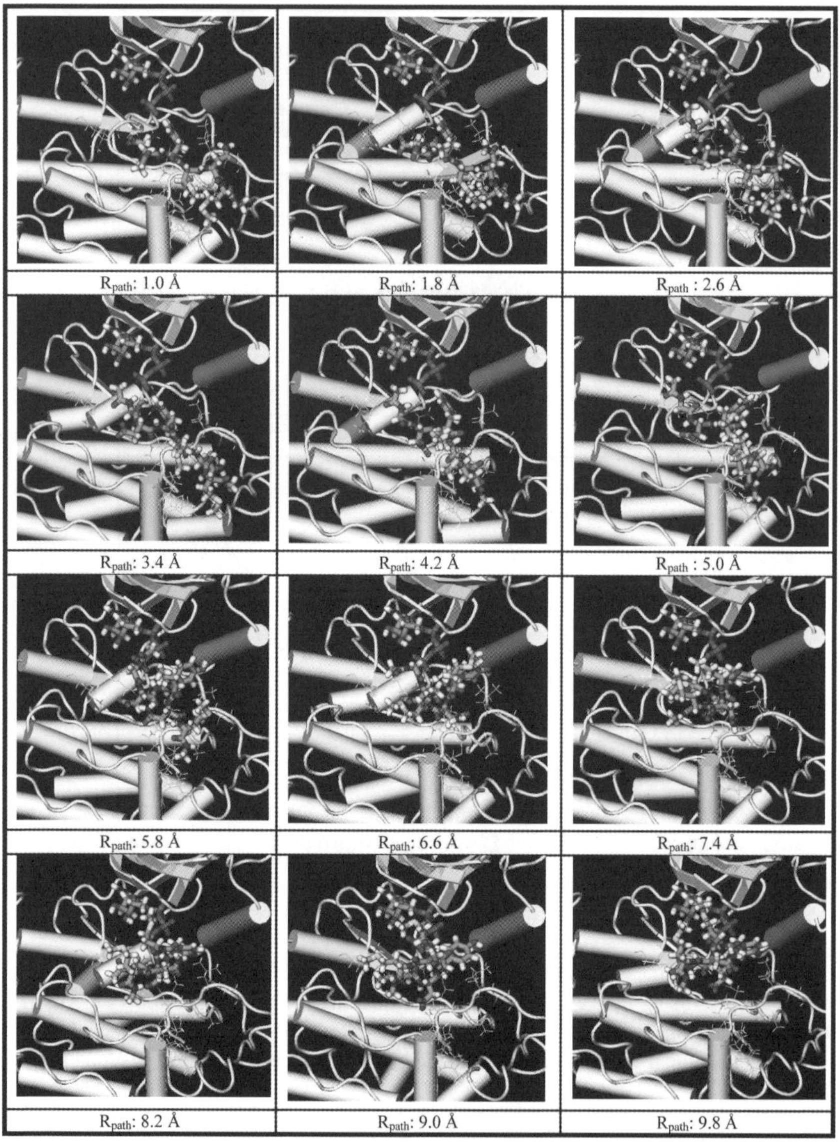

Fig. 9.5. Structural change along one major docking pathway for the docking of a hexapeptide to the insulin receptor tyrosine kinase. The hexapeptide and thirteen residues (Lys-1085, Asp-1132, Arg-1136, Gly-1169, Leu-1170, Leu-1171, Pro-1172, Val-1173, Met-1176, Leu-1181, Asn-1215, Glu-1216, and Leu-1219) in the binding pocket are shown.

followed by the other methionine. The peptide moved in a nearly extended form and reached R_{path} = 4.2 Å when the distance between the γ-phosphorus of ATP and the hydroxyl oxygen of Tyr(P) decreased to 3.76 Å. This structure might mimic one poised for reaction. When R_{path} = 5.0 Å, this same distance increased to >7 Å and the peptide began to take a more compact form. The peptide first formed a turn at the junction of Tyr(P) and the methionine at position P+1, with hydrogen bonds formed between the aspartate at position P-1 and the asparagine at position P+2. Afterwards, the compact structure further extended to including the methionine at position P+3, followed by the glycine at position P-2. Subsequently, at R_{path} = 9 Å, the compact structure adjusted further to bind to a big pocket formed by the activation loop (residues 1149–1170), the catalytic loop (residues 1130–1137), the nucleotide-binding loop (residues 1003–1008), and the αC helix. The peptide then began to release from the protein after R_{path} = 9.8 Å. Coupled with the peptide release, large conformational change of the protein was evident.

9.7. Concluding Remarks

Although it is still challenging to dock flexible ligands to flexible proteins, our recent explorations with molecular dynamics-based methods have found encouraging results. By using a simulated annealing cycling strategy to improve configurational sampling, by employing simple but rapid implicit-solvent models to reduce the number of particles, and by imposing suitable restrains to keep the conformational sampling of the protein near known experimental structures to reduce computational time and to avoid artifacts due to approximate force fields, we could obtain good protein-ligand structures in docking small organic compounds and short peptides to protein kinases and phosphatases. Mutual induced fit effects between a protein and a ligand have also been captured. In addition, as our docking simulations sampled configurations well beyond the lowest-energy docking poses, we could use them to construct docking pathways between a ligand and a protein, by performing suitable clustering and similarity analysis on the resulting structures. These docking pathways have provided a detailed atomistic view on how a ligand might enter or exit a protein, and how ligand movement could couple with protein motion. Furthermore, crude initial estimates of energy profile associated with ligand loading and unloading were provided. The pathways obtained could also serve as a starting point for more elaborate but expensive simulations to refine the pathways and the associated activation barriers further. Although it is not prudent to generalize to all proteins, our applications of modified molecular dynamics-based techniques to protein kinase and phosphatase systems have found that our simulation protocols and approximations were useful, so that docking poses similar to experiment could be identified and useful atomistic insights into docking pathways could be obtained. Such simulations should continue to improve with the continuing development of high-performance computing technology and improvement of simulation methods and protocols.

Acknowledgments

This research was supported by a Research Award from the University of Missouri-Saint Louis, a Research Board Award from the University of Missouri System, and the National Institutes of Health. We also thank the University of Missouri Bioinformatics Consortium and the University of Missouri-Saint Louis Information Technology Services for providing computational resources.

Reference

1. S. Kirkpatrick, C. D. Gelatt Jr. And M. P. Vecchi, Optimization by simulated annealing, *Science* **220**, 671–680 (1983).
2. A. Basu and L. N. Frazer, Rapid determination of the critical temperature in simulated annealing inversion, *Science* **249**, 1409–1412 (1990).
3. G. A. Huber and J. A. Mccammon, Weighted-ensemble simulated annealing: faster optimization on hierarchical energy surfaces, *Phys. Rev. E* **55**, 4822–4825 (1997).
4. A. Roitberg and R. Elber, Modeling side chains in peptides and proteins: application of the locally enhanced sampling and the simulated annealing methods to find minimum energy conformations, *J. Chem. Phys.* **95**, 9277–9287 (1991).
5. A. Mitsutake, Y. Sugita and Y. Okamoto, Generalized-ensemble algorithms for molecular simulations of biopolymers, *Biopolymers* **60**, 96–123 (2001).

6. A. Mitsutake and Y. Okamoto, Helix-coil transitions of amino-acid homo-oligomers in aqueous solution studied by multicanonical simulations, *J. Chem. Phys.* **112**, 10638–10647 (2000).
7. Y. Sugita, A. Kitao and Y. Okamoto, Multidimensional replica-exchange method for free-energy calculations, *J. Chem. Phys.* **113**, 6042–6051 (2000).
8. Y. Sugita and Y. Okamoto, Replica-exchange molecular dynamics method for protein folding, *Chem. Phys. Let.* **314**, 141–151 (1999).
9. Z. Huang, C. F. Wong and R. A. Wheeler, Flexible protein-flexible ligand docking with disrupted velocity simulated annealing, *Proteins: Struct. Funct. Bioinform.* **71**, 440–454 (2008).
10. Z. Huang and C. F. Wong, A mining-minima approach to exploring the docking pathways of p-nitrocatechol sulfate to YopH, *Biophys. J.* **93**, 4141–4150 (2007).
11. Z. Huang and C. F. Wong, Conformational selection of protein kinase a revealed by flexible-ligand flexible-protein docking, *J. Comput. Chem.* **30**, 631–644 (2009).
12. Z. Huang and C. F. Wong, Docking flexible peptide to flexible protein by molecular dynamics using two implicit-solvent models: an evaluation in protein kinase and phosphatase systems, *J. Phys. Chem. B.* **113**, 14343–14354 (2009).
13. G. J. Martyna, M. L. Klein and M. E. Tuckerman, Nosé-Hoover chains: the canonical ensemble via continuous dynamics, *J. Chem. Phys.* **97**, 2635–2643 (1992).
14. B. R. Brooks, R. E. Bruccoleri, B. D. Olafson, D. J. States, S. Swaminathan and M. Karplus, CHARMM: A program for macromolecular energy, minimization, and dynamics calculations, *J. Comput. Chem.* **4**, 187–217 (1983).
15. M. K. Gilson and H.-X. Zhou, Calculation of protein-ligand binding affinities, *Ann. Rev. Biophys. Biomol. Struct.* **36**, 21–42 (2007).
16. T. J. A. Ewing, S. Makino, A. G. Skillman and I. D. Kuntz, DOCK 4.0: Search strategies for automated molecular docking of flexible molecule databases, *J. Comput. Aided Mol. Des.* **15**, 411–428 (2001).
17. C. R. Corbeil, P. Englebienne and N. Moitessier, Docking ligands into flexible and solvated macromolecules. 1. Development and validation of FITTED 1.0, *J. Chem. Inf. Model.* **47**, 435–449 (2007).
18. S. Grosdidier, C. Pons, A. Solernou and J. Fernández-Recio, Prediction and scoring of docking poses with pydock, *Proteins: Struct. Funct. Bioinform.* **69**, 852–858 (2007).
19. M. Bueno and C. J. Camacho, Acidic groups docked to well defined wetted pockets at the core of the binding interface: a tale of scoring and missing protein interactions in CAPRI, *Proteins: Struct. Funct. Bioinform.* **69**, 786–792 (2007).
20. M. Glick, D. D. Robinson, G. H. Grant and W. G. Richards, Identification of ligand binding sites on proteins using a multi-scale approach, *J. Am. Chem. Soc.* **124**, 2337–2344 (2002).
21. T. Wang and R. C. Wade, Implicit solvent models for flexible protein-protein docking by molecular dynamics simulations, *Proteins: Struct. Funct. Genet.* **50**, 158–169 (2003).

22. N. Moitessier, E. Therrien and S. Hanessian, A method for induced-fit docking, scoring, and ranking of flexible ligands. Application to peptidic and pseudopeptidic β-secretase (BACE 1) inhibitors, *J. Med. Chem.* **49**, 5885–5894 (2006).
23. P. Ferrara, H. Gohlke, D. J. Price, G. Klebe and C. L. Brooks III, Assessing scoring functions for protein-ligand interactions, *J. Med. Chem.* **47**, 3032–3047 (2004).
24. W. C. Still, A. Tempczyk, R. C. Hawley and T. Hendrickson, Semianalytical treatment of solvation for molecular mechanics and dynamics, *J. Am. Chem. Soc.* **112**, 6127–6129 (1990).
25. M. Schaefer and M. Karplus, A comprehensive analytical treatment of continuum electrostatics, *J. Phys. Chem.* **100**, 1578–1599 (1996).
26. D. Qiu, P. S. Shenkin, F. P. Hollinger and W. C. Still, The GB/SA continuum model for solvation. A fast analytical method for the calculation of approximate Born radii, *J. Phys. Chem. A* **101**, 3005–3014 (1997).
27. B. N. Dominy and C. L. Brooks III, Development of a Generalized Born model parametrization for proteins and nucleic acids, *J. Phys. Chem. B* **103**, 3765–3773 (1999).
28. V. Tsui and D. A. Case, Theory and applications of the Generalized Born solvation model in macromolecular simulations, *Biopolymers* **56**, 275–291 (2001).
29. A. Onufriev, D. Bashford and D. A. Case, Modification of the Generalized Born model suitable for macromolecules, *J. Phys. Chem. B* **104**, 3712–3720 (2000).
30. B. Jayaram, D. Sprous and D. L. Beveridge, Solvation free energy of biomacromolecules: parameters for a modified Generalized Born model consistent with the AMBER force field, *J. Phys. Chem. B* **102**, 9571–9576 (1998).
31. M. S. Lee, M. Feig, F. R. Salsbury Jr. and C. L. Brooks III, New analytic approximation to the standard molecular volume definition and its application to Generalized Born calculations, *J. Comput. Chem.* **24**, 1348–1356 (2003).
32. M. S. Lee, F. R. Salsbury Jr. and C. L. Brooks III, Novel Generalized Born methods, *J. Chem. Phys.* **116**, 10606–10614 (2002).
33. J. Chocholoušová and M. Feig, Balancing an accurate representation of the molecular surface in Genelized Born formalisms with integrator stability in molecular dynamics simulations, *J. Comput. Chem.* **27**, 719–729 (2006).
34. J. Mongan, C. Simmerling, J. A. McCammon, D. A. Case and A. Onufriev, Generalized Born model with a simple, robust molecular volume correction, *J. Chem. Theory Comput.* **3**, 156–169 (2007).
35. M. R. Lee and Y. Sun, Improving docking accuracy through molecular mechanics Generalized Born optimization and scoring, *J. Chem. Theory Comput.* **3**, 1106–1119 (2007).
36. P. Akamine, Madhusudan, J. Wu, N.-H. Xuong, L. F. Ten Eyck and S. S. Taylor, Dynamic features of cAMP-Depdentdent protein kinase revealed by apoenzyme crystal structure, *J. Mol. Biol.* **327**, 159–171 (2003).

37. N. Narayana, S. Cox, S. Shaltiel, S. S. Taylor and N.-H. Xuong, Crystal structure of a polyhistidine-tagged recombinant catalytic subunit of cAMP-dependent protein kinase complexed with the peptide inhibitor PKI(5–24) and adenosine, *Biochemistry* **36**, 4438–4448 (1997).
38. D. A. Johnson, P. Akamine, E. Radzio-Andzelm, Madhusudan and S. S. Taylor, Dynamics of cAMP-dependent protein kinase, *Chem. Rev.* **101**, 2243–2270 (2001).
39. J. A. Stuckey, H. L. Schubert, E. B. Fauman, Z.-Y. Zhang, J. E. Dixon and M. A. Saper, Crystal structure of *Yersinia* protein tryrosine phosphatase at 2.5 Å and the complex with tungstate, *Nature* **370**, 571–575 (1994).
40. X. Hu and C. E. Stebbins, Dynamics of the WPD loop of the *Yersinia* protein tyrosine phosphatase, *Biophys. J.* **91**, 948–956 (2006).
41. C. Bissantz, G. Folkers and D. Rognan, Protein-based virtual screening of chemical databases. 1. Evaluation of different docking/scoring combinations, *J. Med. Chem.* **43**, 4759–4767 (2000).
42. G. E. Terp, B. N. Johansen, I. T. Christensen and F. S. Jørgensen, A new concept for multidimensional selection of ligand conformations (MultiSelect) and multidimensional scoring (MultiScore) of protein-ligand binding, *J. Med. Chem.* **14**, 2333–2343 (2001).
43. M. Stahl and M. Rarey, Detailed analysis of scoring functions for virtual screening, *J. Med. Chem.* **44**, 1035–1042 (2001).
44. R. Wang, Y. Lu and S. Wang, Comparative evaluation of 11 scoring functions for molecular docking, *J. Med. Chem.* **46**, 2287–2303 (2003).
45. D. A. Case, T. A. Darden, T. E. Cheatham III, C. L. Simmerling, J. Wang, R. E. Duke, R. Luo, K. M. Merz, B. Wang, D. A. Pearlman, M. Crowley, S. Brozell, V. Tsui, H. Gohlke, J. Mongan, V. Hornak, G. Cui, P. Beroza, C. Schafmeister, J. W. Caldwell, W. S. Ross and P. A. Kollman, *AMBER 8*, University of California, San Franciso, (2004).
46. G. M. Morris, D. S. Goodsell, R. S. Halliday, R. Huey, W. E. Hart, R. K. Belew and A. J. Olson, Automated docking using a Lamarckian genetic algorithm and an empirical binding free energy function, *J. Comput. Chem.* **19**, 1639–1662 (1998).
47. G. Jones, P. Willett, R. C. Glen, A. R. Leach and R. Taylor, Development and validation of a genetic algorithm for flexible docking, *J. Mol. Biol.* **267**, 727–748 (1997).
48. M. Rarey, B. Kramer, T. Lengauer and G. Klebe, A fast flexible docking method using an incremental construction algorithm, *J. Mol. Biol.* **261**, 470–489 (1996).
49. M. D. Eldridge, C. W. Murray, T. R. Auton, G. V. Paolini and R. P. Mee, Empirical scoring functions: I. The development of a fast empirical scoring function to estimate the binding affinity of ligands in receptor complexes, *J. Comput. Aided Mol. Des.* **11**, 425–445 (1997).

50. H.-J. Böhm, The development of a simple empirical scoring function to estimate the binding constant for a protein-ligand complex of known three-dimensional structure, *J. Comput. Aided Mol. Des.* **8**, 243–256 (1994).
51. D. K. Gehlhaar, G. M. Verkhivker, P. A. Rejto, C. J. Sherman, D. R. Fogel, L. J. Fogel and S. T. Freer, Molecular recognition of the inhibitor AG-1343 by HIV-1 protease: conformationally flexible docking by evolutionary programming, *Chem. Biol.* **2**, 317–324 (1995).
52. R. Wang, L. Lai and S. Wang, Further development and validation of empirical scoring functions for structure-based binding affinity prediction, *J. Comput. Aid. Mol. Des.* **16**, 11–26 (2002).
53. I. Muegge and Y. C. Martin, A general and fast scoring function for protein-ligand interactions: a simplified potential approach, *J. Med. Chem.* **42**, 791–804 (1999).
54. H. Gohlkea, M. Hendlicha and G. Klebe, Knowledge-based scoring function to predict protein-ligand interactions, *J. Mol. Biol.* **295**, 337–356 (2000).
55. A. V. Ishchenko and E. I. Shakhnovich, SMall Molecule Growth 2001 (SMoG2001): an improved knowledge-based scoring function for protein-ligand interactions, *J. Med. Chem.* **45**, 2770–2780 (2002).
56. B. Isralewitz, S. Izrailev and K. Schulten, Binding pathway of retinal to bacterio-opsin: a prediction by molecular dynamics simulations, *Biophys. J.* **73**, 2972–2979 (1997).
57. E. Paci and M. Karplus, Forced unfolding of fibronectin type 3 modules: an analysis by biased molecular dynamics simulations, *J. Mol. Biol.* **288**, 441–459 (1999).
58. E. Ozkirimli and C. B. Post, Src kinase activation: a switched electrostatic network, *Prot. Sci.* **15**, 1051–1062 (2006).
59. B. Lu, C. F. Wong and J. A. McCammon, Release of ADP from the catalytic subunit of protein kinase A: a molecular dynamics simulation study, *Prot. Sci.* **14**, 159–168 (2005).
60. C. F. Wong, C. Zheng, J. Shen, J. A. McCammon and P. G. Wolynes, Cytochrome c: a molecular proving ground for computer simulations, *J. Phys. Chem.* **97**, 3100–3110 (1993).
61. G. M. Torrie and J. P. Valleau, Nonphysical sampling distributions in Monte Carlo free-energy estimation: umbrella sampling, *J. Comp. Phys.* **23**, 187–199 (1997).
62. S. H. Northrup, M. R. Pear, C.-Y. Lee, J. A. McCammon and M. Karplus, Dynamical theory of activated processes in globular proteins, *Proc. Natl. Acad. Sci.* **79**, 4035–4039 (1982).
63. J. A. Mccammon and S. C. Harvey, *Dynamics of Proteins and Nuclei Acids* (Cambridge University Press, Cambridge, 1987).
64. K. L. Guan and J. E. Dixon, Protein tyrosine phosphatase activity of an essntial virulence determinant in *Yersinia*, *Science* **249**, 553–556 (1990).
65. T. Bulter, *Textbook of Medicine* (W. B. Saunders, Philadelphia, PA, 1985).

66. J.-P. Sun, L. Wu, A. A. Fedorov, S. C. Almo and Z.-Y. Zhang, Crystal structure of the *Yersinia* protein-tyrosine phosphatase YopH complexed with a specific small molecule inhibitor, *J. Biol. Chem.* **278**, 33392–33399 (2003).
67. S. R. Hubbard, Crystal structure of the activated insulin receptor tyrosne kinase in complex with peptide substrate and ATP analog, *EMBO J.* **16**, 5572–5581 (1997).
68. A. D. MacKerell Jr., M. Feig and C. L. Brooks III, Extending the treatment of backbone energetics in protein force fields: limitations of gas-phase quantum mechanics in reproducing protein conformational distributions in molecular dynamics simulations, *J. Comput. Chem.* **25**, 1400–1415 (2004).
69. A. D. MacKerell Jr., D. Bashford, M. Bellott, R. L. Dunbrack Jr., J. D. Evanseck, M. J. Field, S. Fischer, J. Gao, H. Guo, S. Ha, D. Joseph-McCarthy, L. Kuchnir, K. Kuczera, F. T. K. Lau, C. Mattos, S. Michnick, T. Ngo, D. T. Nguyen, B. Prodhom, W. E. Reiher III, B. Roux, M. Schlenkrich, J. C. Smith, R. Stote, J. Straub, M. Watanabe, J. Wiorkiewicz-Kuczera, D. Yin and M. Karplus, All-atom empirical potential for molecular modeling and dynamics studies of proteins, *J. Phys. Chem. B* **102**, 3586–3616 (1998).

Chapter 10

Spin Valves in Conjugated Polymeric Light-Emitting Diodes

Sheng Li* and Guo-Ping Tong†

Zhejiang Normal University

**shenglee@zjnu.edu.cn*

†tgp6463@zjnu.cn

Thomas F. George

University of Missouri–St. Louis

tfgeorge@umsl.edu

An electric field-induced spin accumulation phenomenon is presented for electroluminescent conjugated polymers as light-emitting diodes (LEDs). When an electric field is applied along a polymer chain and exceeds a critical value, it quenches the luminescence and dissociates the singlet exciton into two carriers with opposite spin signs. Simultaneously, the field drives these two opposite spin carriers to move in opposite directions, leading to spin accumulation at the two ends of the organic material LED, which can be detected through Kerr rotation microscopy. Two optically-controlled spin transfer effects are proposed for π-conjugated polymers. When such a polymeric molecule undergoes two-photon excitation, the charge of a spin carrier can be reversed, and simultaneously an applied external electric field drives the charge-reversed spin carrier to move in the opposite direction. As for a spinless carrier, the photoexcitation dissociates it into two spin carriers, forming entanglement. The coupling between the newly-produced spin carriers and a ferromagnet will change the magnetoresistance. By combining an electric field, magnetic field and photoexcitation, two generic optically-controlled ultrafast response organic spin valves are designed.

10.1. Introduction

During the past two decades, giant magnetoresistance and tunneling magnetoresistance were discovered in metallic spin valves [1,2], launching a new research field known as "spintronics" [3]. Different from an inorganic semiconductor, the flexibility of an organic semiconductor induces its carrier to be a composite particle characterized by lattice distortion [4,5]. Furthermore, by virtue of the extremely weak spin-orbit interaction and weak hyperfine interaction in π-conjugated organic semiconductors (OSEs), the electron spin diffusion length is especially long [6], which makes it possible to realize spin-coherent transport in OSEs. For spin injection in OSEs, Xie *et al.* [7] proposed a simple physical picture to illustrate its mechanism: at a manganite/polymer interface, when the Fermi level of the manganite lies below the bipolaron level of the polymer, the completely polarized spin can be injected into the polymer [7]. The planar junction LSMO/T6/LSMO (LSMO stands for La0.7Sr0.3MnO3 and T6 for sexa-thiophene) has been nanostructured for spin injection [8]. Following this, a comprehensive theory was developed, providing a detailed explanation of the observed magnetoresistance and I-V characteristics in this structure [9]. So far, several new spin-dependent devices based on OSEs have been fabricated as spin valves and organic tunable magnetoresistances [10,11].

The proposed spin Hall effect paved a new way to utilize an external field to control spin currents in inorganic semiconductors and to have spin accumulation [12–14]. A new question arises as to whether we can apply an external field to realize the spin accumulation effect in OSEs. In order to answer this question, we briefly review previous research on polymer light-emitting diodes (LEDs). In the past decade, electroluminescence was first observed in poly(paraphenylene vinylene) (PPV) and its derivatives [15]. This effect arises from a bound electron–hole pair in the conjugated polymer, known as a singlet exciton [15]. When an electric field is applied to a polymer LED and exceeds a critical value, it dissociates the singlet exciton and quenches the luminescence [16–18], leading to new carriers (charged polarons) [19]. For this specific electric-field-induced effect, if we focus our attention on the spin properties, we have a method to control the spin current and to form spin accumulation in polymer LEDs. Here, we propose a method to control the spin current and to form spin accumulation in polymer LEDs, and to demonstrate electric-field-induced spin accumulation in a polymer LED.

Since 2004, scientists have conducted extensive research on organic magnetoresistance [19–21]. There are two approaches to realize this phenomenon: one is to use the spin effect of an exciton [14,15], and the other is to capitalize on the spin properties of the charged carriers as bipolarons or polarons [22]. In OSEs, the introduction of spin-orbit coupling largely suppresses the production of triplet excitons, yielding more singlet excitons [23], which is called the electrophosphorescent effect. Following this principle, the mixture of the strong spin-orbit coupling molecule fac-tris(2-phenylpyridinato) iridium [Ir(ppy)3] and polymer poly(*N*-vinyl carbazole)(PVK) changes the ratio between singlet and triplet excitons. Due to the different spin polarizations of the singlet and triplet excitons, the magnetoresistance effect can be tuned by a modification of the singlet to triplet exciton ratio [24]. Besides the effect caused by excitons, recent experiments have confirmed that the mobility of holes and electrons, i.e., positive and negative polarons, can lead to positive and negative magnetoresistance OSEs [25]. Here, we show that organic spintronics can be controlled by changing the transport of carriers in organic materials. The main shortcoming of the above method is the slow response of the charge injection into the conjugated polymers. In order to resolve this, we consider on another typical characteristic of the conjugated polymer — the self-trapping effect.

As is well known, once a conducting polymeric molecule is doped, a charged carrier, such as a positive polaron or bipolaron, can be formed along the polymer chain. Once these carriers undergo photoexcitation, the self-trapping effect can induce a change in their properties [26]. It is appropriate to apply photoexcitation to control the properties of plastic spintronics. Furthermore, the ultrafast process of photoexcitation will highly improve the response speed of the plastic spintronics. Here, we suggest two methods of controlling the spin transport, based on photoinduced carrier fission [27], to lead to an optically-controlled spin valve effect [28,29].

10.2. Model

10.2.1. *Extended Su–Schrieffer–Heeger–Hubbard Model*

Following the development of the extended Su–Schreiffer–Heeger–Hubbard Hamiltonian [4], the prominent self-trapping effect and strong electron–phonon coupling of conjugated polymers can now be quantitatively described. A Brazovskii–Kirova symmetry-breaking term can be added to this model to describe the confinement effect of a non-degenerate polymer:

$$H=-\sum_{l,s}\{t_0+\alpha(u_{l+1}-u_l)+(-1)^l t_e\}\times[c_{l+1,s}^{+}c_{l,s}+H.c.]$$
$$+\frac{K}{2}\sum_{l}(u_{l+1}-u_l)^2+H'+H_E, \tag{10.1}$$

$$H_{e-e}=U\sum_{l,s}(a_{l,s}^{+}a_{l,s}-\frac{1}{2})(a_{l,-s}^{+}a_{l,-s}-\frac{1}{2})$$
$$+V\sum_{l,s,s'}(a_{l,s}^{+}a_{l.s}-\frac{1}{2})(a_{l+1,s'}^{+}a_{l+1,s'}-\frac{1}{2}) \tag{10.2}$$

$$H_E=\sum_{l,s}Ee(l-\frac{N+1}{2})an_{l,s}. \tag{10.3}$$

The parameters used in the above Hamiltonian are conventional values as determined by previous research on conducting polymers: t_0 is a hopping constant; α is an electron–lattice coupling constant; $c_{l,s}^{+}(c_{l,s})$ denotes the electron creation (annihilation) operator at site l with spin s; a is a lattice constant; u_l is the displacement of atom l with mass M; K is an elastic constant; and t_e is the Brazovskii–Kirova term reflecting the confinement effect in a polymer with a non-degenerate ground state, which ensures that composite particles in the polymer, such as bipolarons or polarons, are stable [21]. H' is the electron–electron interaction, which can be treated by the Hartree–Fock approximation since the polymer is not a strongly-correlated system [22]. H_E is the interaction of the electrons with the electric field $\vec{E}$ directed along the polymer chain.

10.2.2. *Electron interaction*

In a conjugated polymeric molecule, since the bandwidth $W > 10$ eV of the polymer is much larger than the strength $U \sim 3$–5 eV of the electron–electron interaction, the polymer is not a strongly-correlated system. Thus, the electron–electron interaction term in Eq. (10.4) can be treated by the Hartree–Fock approximation:

$$\begin{aligned} H_{e-e} &= U\sum_{l,s}(a_{l,s}^{+}a_{l,s}-\frac{1}{2})(a_{l,-s}^{+}a_{l,-s}-\frac{1}{2}) \\ &+V\sum_{l,s,s'}(a_{l,s}^{+}a_{l.s}-\frac{1}{2})(a_{l+1,s'}^{+}a_{l+1,s'}-\frac{1}{2}) \end{aligned} . \qquad (10.4)$$

According to the Wick theorem, the Fermi operators α, β, γ and δ can be expanded as follows:

$$\begin{aligned} \alpha^{+}\beta^{+|}\gamma\sigma &= \langle\alpha^{+}\sigma\rangle\beta^{+}\gamma+\langle\beta^{+}\gamma\rangle\alpha^{+}\sigma-\langle\alpha^{+}\gamma\rangle\beta^{+}\sigma \\ &-\langle\beta^{+}\sigma\rangle\alpha^{+}\gamma-\langle\alpha^{+}\sigma\rangle\langle\beta^{+}\gamma\rangle+\langle\alpha^{+}\gamma\rangle\langle\beta^{+}\sigma\rangle \end{aligned} . \qquad (10.5)$$

The last two terms in Eq. (10.5) are constant, which cannot change the energy structure, and can be omitted. But, the Hubbard term i $U\sum_{l,s,s'}(a_{l,s}^{+}a_{l.s}-\frac{1}{2})(a_{l,-s}^{+}a_{l,-s}-\frac{1}{2})$ as part of the electron–electron interaction becomes:

$$\begin{aligned} &(a_{l,s}^{+}a_{l,s}-\frac{1}{2})(a_{l,-s}^{+}a_{l,-s}-\frac{1}{2}) \\ &= a_{l,s}^{+}a_{l,s}a_{l,-s}^{+}a_{l,-s}-\frac{1}{2}a_{l,s}^{+}a_{l,s}-\frac{1}{2}a_{l,-s}^{+}a_{l,-s}+\frac{1}{4} \end{aligned} . \qquad (10.6)$$

On the basis of Eq. (10.4), $a_{l,s}^{+}a_{l,s}a_{l,-s}^{+}a_{l,-s}$ can be expanded as:

$$\begin{aligned} a_{l,s}^{+}a_{l,s}a_{l,-s}^{+}a_{l,-s} &= \langle a_{l,s}^{+}a_{l,s}\rangle a_{l,-s}^{+}a_{l,-s}+\langle a_{l,-s}^{+}a_{l,-s}\rangle a_{l,s}^{+}a_{l,s} \\ &-\langle a_{l,s}^{+}a_{l,-s}\rangle a_{l,-s}^{+}a_{l,s}-\langle a_{l,-s}^{+}a_{l,s}\rangle a_{l,s}^{+}a_{l,-s}-\frac{1}{2}a_{l,s}^{+}a_{l,s}-\frac{1}{2}a_{l,-s}^{+}a_{l,-s} \end{aligned} . \qquad (10.7)$$

Due to the Pauli repulsion, $\langle a_{l,s}^{+}a_{l,-s}\rangle a_{l,-s}^{+}a_{l,s}$ and $\langle a_{l,-s}^{+}a_{l,s}\rangle a_{l,s}^{+}a_{l,-s}$ are 0. Thus, $U\sum_{l,s}(a_{l,s}^{+}a_{l,s}-\frac{1}{2})(a_{l,-s}^{+}a_{l,-s}-\frac{1}{2})$ is transformed to the following expression:

$$
\begin{aligned}
&U\sum_{l,s}(a_{l,s}^{+}a_{l,s}-\frac{1}{2})(a_{l,-s}^{+}a_{l,-s}-\frac{1}{2})= \\
&U\sum_{l,s}\left(\left\langle a_{l,s}^{+}a_{l,s}\right\rangle-\frac{1}{2}\right)a_{l,-s}^{+}a_{l,-s}+\left(\left\langle a_{l,-s}^{+}a_{l,-s}\right\rangle-\frac{1}{2}\right)a_{l,s}^{+}a_{l,s}
\end{aligned}
\quad . \tag{10.8}
$$

According to the same process, the extended term of the electron – electron interaction $V(a_{l,s}^{+}a_{l.s}-\frac{1}{2})(a_{l+1,s'}^{+}a_{l+1,s'}-\frac{1}{2})$ also can be expanded, so that

$$
\begin{aligned}
&(a_{l,s}^{+}a_{l.s}-\frac{1}{2})(a_{l+1,s'}^{+}a_{l+1,s'}-\frac{1}{2})=\left\langle a_{l,s}^{+}a_{l,s}\right\rangle a_{l+1,s'}^{+}a_{l+1,s'} \\
&+\left\langle a_{l+1,s'}^{+}a_{l+1,s'}\right\rangle a_{l,s}^{+}a_{l,s}-\left\langle a_{l,s}^{+}a_{l+1,s'}\right\rangle a_{l+1,s'}^{+}a_{l,s}-\left\langle a_{l+1,s'}^{+}a_{l,s}\right\rangle a_{l,s}^{+}a_{l+1,s'} \\
&-\frac{1}{2}a_{l,s}^{+}a_{l,s}-\frac{1}{2}a_{l+1,s'}^{+}a_{l+1,s'}
\end{aligned}
\quad , \tag{10.9}
$$

where all of constant terms have been omitted. When $s'=s$, Eq. (10.9) becomes

$$
\begin{aligned}
&\left(\left\langle a_{l,s}^{+}a_{l,s}\right\rangle-\frac{1}{2}\right)a_{l+1,s}^{+}a_{l+1,s}+\left(\left\langle a_{l+1,s}^{+}a_{l+1,s}\right\rangle-\frac{1}{2}\right)a_{l,s}^{+}a_{l,s} \\
&-\left\langle a_{l,s}^{+}a_{l+1,s}\right\rangle a_{l+1,s}^{+}a_{l,s}-\left\langle a_{l+1,s}^{+}a_{l,s}\right\rangle a_{l,s}^{+}a_{l+1,s}
\end{aligned}
\quad . \tag{10.10}
$$

If the first term is changed to $\left(\left\langle a_{l-1,s}^{+}a_{l-1,s}\right\rangle-\frac{1}{2}\right)a_{l,s}^{+}a_{l,s}$, Eq. (10.9) becomes

$$
\begin{aligned}
&\left(\left\langle a_{l-1,s}^{+}a_{l-1,s}\right\rangle+\left\langle a_{l+1,s}^{+}a_{l+1,s}\right\rangle-1\right)a_{l,s}^{+}a_{l,s} \\
&-\left\langle a_{l,s}^{+}a_{l+1,s}\right\rangle a_{l+1,s}^{+}a_{l,s}-\left\langle a_{l+1,s}^{+}a_{l,s}\right\rangle a_{l,s}^{+}a_{l+1,s}
\end{aligned}
\quad . \tag{10.11}
$$

If $s'=-s$, Eq. (10.9) changes to

$$\left(\left\langle a_{l,s}^{+}a_{l,s}\right\rangle-\frac{1}{2}\right)a_{l+1,-s}^{+}a_{l+1,-s}+\left(\left\langle a_{l+1,-s}^{+}a_{l+1,-s}\right\rangle-\frac{1}{2}\right)a_{l,s}^{+}a_{l,s} \quad . \tag{10.12}$$
$$-\left\langle a_{l,s}^{+}a_{l+1,-s}\right\rangle a_{l+1,-s}^{+}a_{l,s}-\left\langle a_{l+1,-s}^{+}a_{l,s}\right\rangle a_{l,s}^{+}a_{l+1,-s}$$

Due to different spins, $\left\langle a_{l,s}^{+}a_{l+1,-s}\right\rangle$ and $\left\langle a_{l+1,-s}^{+}a_{l,s}\right\rangle$ are 0, and the first term of Eq. (10.12) can be written as

$$\left(\left\langle a_{l-1,-s}^{+}a_{l-1,-s}\right\rangle+\left\langle a_{l+1,-s}^{+}a_{l+1,-s}\right\rangle-1\right)a_{l,s}^{+}a_{l,s} \; . \tag{10.13}$$

If $\{ Z_{\mu,l} \}$ is the electronic wavefunction in the μ energy spectrum, the electron - electron interaction under the Hartree–Fock approximation can be expressed as:

$$H_{e-e}=$$
$$\sum_{l,s}\left\{U\left(\sum_{\mu}^{occ}\left|Z_{l,\mu}^{-s}\right|^{2}-\frac{1}{2}\right)+V\left[\sum_{s'}\left(\sum_{\mu}^{occ}\left|Z_{l-1,\mu}^{s'}\right|^{2}+\sum_{\mu}^{occ}\left|Z_{l+1,\mu}^{s'}\right|^{2}-2\right)\right]\right\}a_{l,s}^{+}a_{l,s}$$
$$-\sum_{l,s}\left(V\sum_{\mu}^{occ}Z_{l,\mu}^{s}Z_{l+1,\mu}^{s}\right)\left(a_{l+1,s}^{+}a_{l,s}+H.C\right). \tag{10.14}$$

Inserting this into the original model of Eqs. (10.1) and (10.2) gives

$$\begin{aligned}\varepsilon_{\mu}^{s}Z_{l,\mu}^{s}=&\left[U\left(\rho_{l}^{-s}-\frac{1}{2}\right)+V\left(\sum_{s'}\rho_{l-1}^{s'}+\sum_{s'}\rho_{l+1}^{s'}-2\right)+Ee\left(l-\frac{N+1}{2}\right)a\right]Z_{l,\mu}^{s}\\&-\left[VW_{l-1}^{s}+t_{0}+(-1)^{l-1}\alpha(\varphi_{l-1}+\varphi_{l})+(-1)^{l-1}t_{e}\right]Z_{l-1,\mu}^{s}\\&-\left[VW_{l+1}^{s}+t_{0}+(-1)^{l}\alpha(\varphi_{l}+\varphi_{l+1})+(-1)^{l}t_{e}\right]Z_{l+1}^{s}.\end{aligned} \tag{10.15}$$

Based on Eq. (10.15), the electronic wave function can be obtained.

10.3. Electric Field Induced Spin Accumulation

Electroluminescence, that is, the generation of light by electrical excitation, was first reported in 1990 for the conjugated polymer poly(*p*-phenylene vinylene) (PPV) [15], whose structure is a single semiconductor layer between metallic electrodes. When a polymer LED is biased, an exciton should be formed with triplet and singlet configurations in the ratio 3:1. For the singlet exciton, its localized lattice configuration is shown in Fig. 10.1(a) for a chain whose length is 200 sites, which produces two localized electronic states in the gap, Φ_u and Φ_d. These two states are occupied collectively by two electrons with opposite spin-polarization directions, indicating a spinless carrier, as illustrated in Fig. 10.1(b). These localized electron–hole pairs surrounding the lattice distortion form a sea of neutral carrier excitons. Different from the singlet exciton, the same spin polarizations in two localized states cause a triplet exciton to be a neutral spin carrier. Additionally, due to the Pauli exclusion principle, the luminescence is mostly generated through the decay of a singlet exciton.

When an electric field is applied along the polymer chain, at first an increase in the field strength does not destroy the localized lattice structure. However, once E exceeds 4.5 MV/cm, the high electric field dissociates the singlet exciton. Just within 150 fs, the singlet exciton is completely separated into two parts, as shown in Fig. 10.2, where one is a positively-charged polaron and the other is negatively charged. The triplet exciton, however, is less susceptible to the external electric field

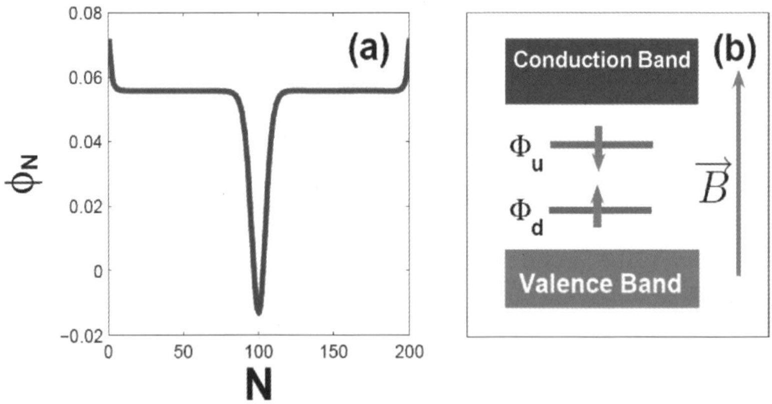

Fig. 10.1. (a) Lattice configuration of a singlet exciton contributing to electroluminescence in a polymer LED. The vertical axis is the lattice configuration, and N refers to the sites. (b) Electronic spectrum of the singlet exciton

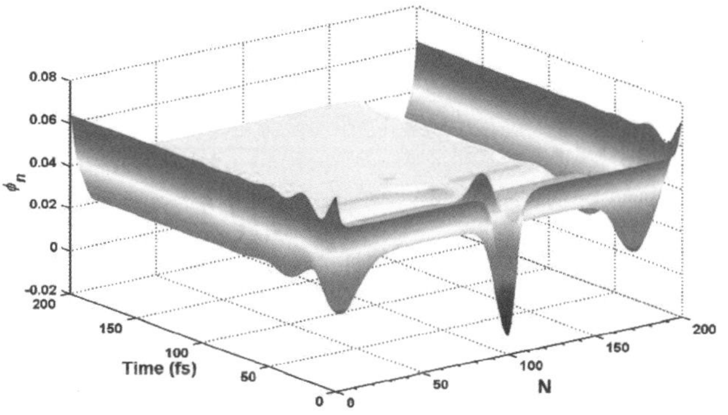

Fig. 10.2. Time-dependent lattice configuration of a conjugated polymer after an external electric field dissociates the original singlet exciton.

than the singlet. The electric field cannot dissociate the triplet exciton until E exceeds 24 MV/cm. In these results, the relaxation time and critical value of the electric field agree precisely with the experimental observations [16–19].

Once the introduced electric field approaches and exceeds the critical threshold E =4.5 MV/cm, the applied electric field above the critical value splits the spinless carrier (a singlet exciton) in the polymer LED into two spin carriers, where one is a spin-up polaron and the other is a spin-down polaron. For the different electron populations of the positive and negative polarons, as shown in Fig. 10.3, the original degeneracy of the positive and negative polarons without electron interaction is eliminated, thus disentangling the spin coupling between these two polarons. We recognize that the positively-charged polaron carries the spin-up sign, and the negative polaron takes the spin-down sign. Afterward, the electric field drives the polaron with the spin-up sign in the direction of the electric field, and simultaneously, the polaron with the spin-down sign moves in the opposite direction. After lattice and charge relaxation, the opposite spin polarization carriers are accumulated at the two ends of the polymer LED. The time-dependent evolution of the spin distribution is shown in Fig. 10.4. Compared with the singlet exciton, the electron interaction makes the triplet exciton less susceptible to the electric field, which is still not destroyed until E approaches 24 MV/cm.

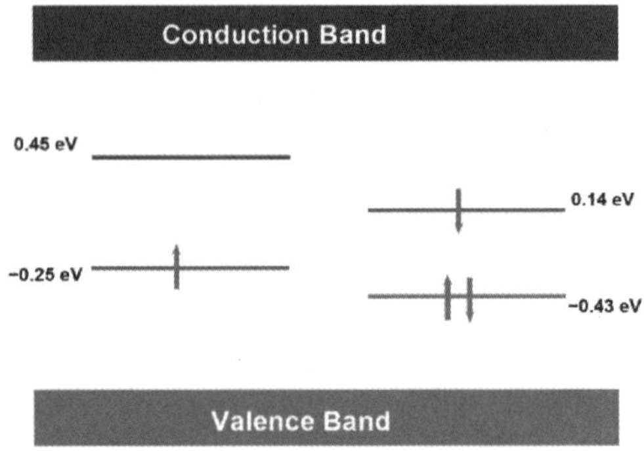

Fig. 10.3. Electronic spectrum after an external electric field induces singlet exciton dissociation.

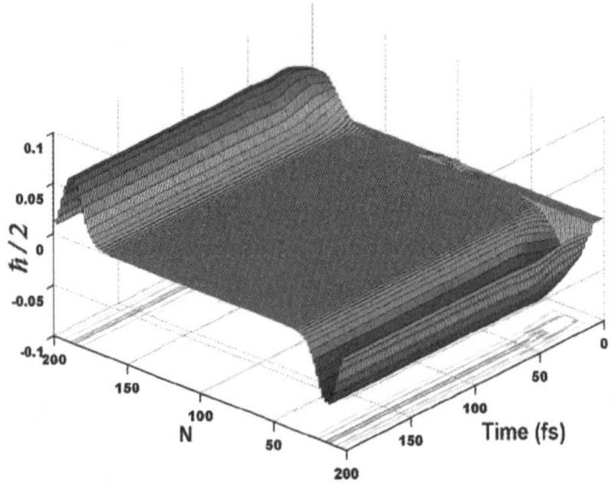

Fig. 10.4. Time-dependent spin distribution after an external electric field induces singlet exciton dissociation. The direction of the electric field goes in the direction from lower to higher *N* (atomic site) along the polymer chain.

On the basis of our findings and the experiments on field-induced dissociation of excitons in a polymer LED [16–19], this spin accumulation at the two ends of the LED can be observed through the Kerr rotation effect. We can design a generic spintronic scheme, as illustrated in Fig. 10.5. We place a sample in a Kerr microscope with the channel oriented perpendicularly to the external applied in-plane magnetic field, which is a general method to determine the relative spin polarizations of the states Φ_u and Φ_d in the gap in Fig. 10.1(b) [8,10]. Due to the extremely-weak spin-orbit interaction and weak hyperfine interaction in π-conjugated organic semiconductors [6], the spin coherence length is as high

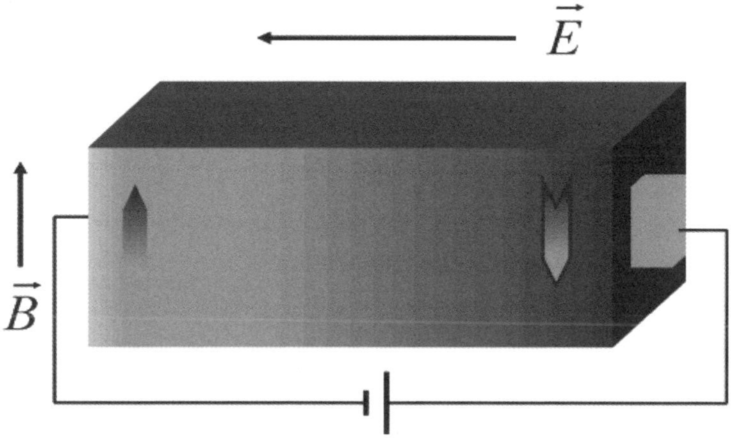

Fig. 10.5. Proposed device designed to detect spin accumulation in polymer LEDs.

as 10^2–10^3 nm, even at room temperature [8], which is much larger than the distance of 10 nm between polymer chains. This means that when a spin hops from one polymer chain to the other, the spin of the polaron still maintains the same direction. This also makes it easier to have spin accumulation at the ends of the polymer LED than in an inorganic semiconductor [14].

10.4. Optically-Controlled Spin Valves in Conjugated Polymers

When a conjugated polymeric molecule is in its ground state, the homogeneous dimerization of the lattice configuration leads to valence and conduction bands, reflecting its semiconducting property. After one electron at the top of valence band (the highest-occupied molecular orbital known as HOMO) is removed, the original homogeneous dimerization of the lattice configuration is no longer stable, yielding a localized configuration of the bond structure as shown in Fig. 10.6(a). The strong electron–lattice interaction then produces two localized states at the center of the gap between the valence and conduction bands, Φ_u and Φ_d, where Φ_d is occupied by only one electron, and the other is empty, as shown in Fig. 10.6(b), indicating a spin carrier, namely, a

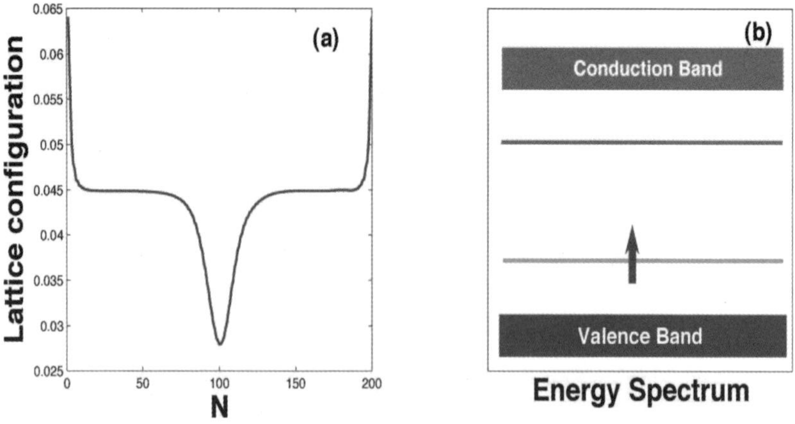

Fig. 10.6. (a) Lattice configuration of a positive polaron in a polymer light-emitting diode. The vertical axis is the lattice configuration in the unit Å, and N refers to the sites. (b) Electronic spectrum of the positive polaron.

positively-charged polaron. Afterwards, if one more electron is removed from Φ_d of the positive polaron, the distortion of the localized lattice is more serious, producing a new carrier-bipolaron which is a spinless carrier with two positive charges, and both Φ_u and Φ_d are empty, all of which are illustrated in Fig. 10.7.

Here, without the external electric field, when the positive polaron undergoes two-photon excitation, two electrons in the HOMO are excited to Φ_u through STIRAP (stimulated Raman adiabatic passage) technology or an external laser pump. The induced strong lattice oscillation completely changes the original charge distribution and lattice configuration, leading to carrier fission. The time-dependent charge distribution during its relaxation process following the two-photon excitation of a positive polaron is depicted in Fig. 10.8. From this figure, the original carrier is separated into two parts, where the one at the left is a positively-charged carrier, and the one at the right is negatively charged. In order to conserve total charge, it is found that for the resultant carriers, the positive carrier is a bipolaron with two charges, and the other is a negative polaron lying on the right of the polymer chain. This effect is the result of symmetry breaking [20].

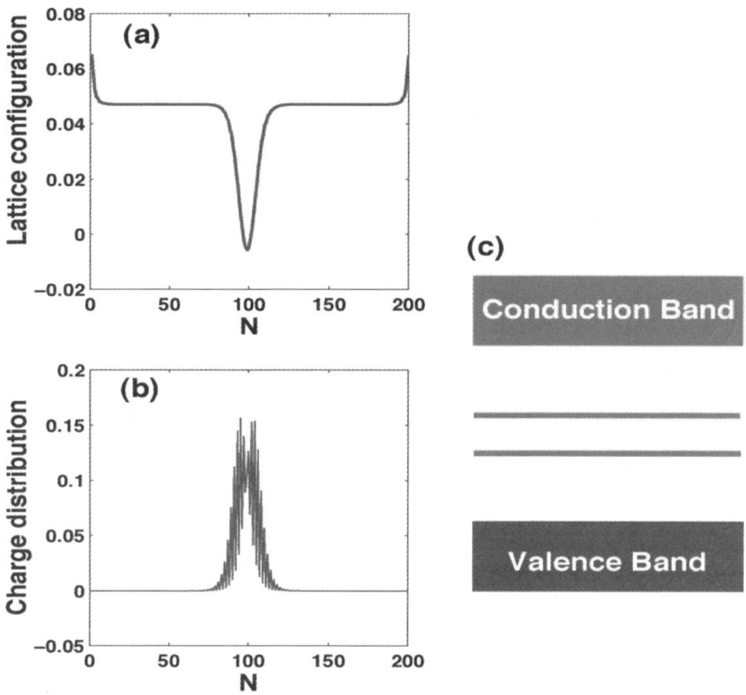

Fig. 10.7. (a) Lattice configuration of a positive bipolaron in a polymer light-emitting diode. The vertical axis is the lattice configuration in the unit Å, and N refers to the sites. (b) Charge distribution of a positive bipolaron in a polymer light-emitting diode. The vertical axis is the charge distribution in the unit +|e|, and N refers to the sites. (c) Electronic spectrum of the positive bipolaron.

As mentioned above, when the Fermi level of manganite lies below the bipolaron level at the interface of the manganite/conducting polymer, the completely-polarized spin can be injected into the polymer to form spin carriers, i.e., polarons. If the original polaron is positively charged, the spin carrier is a positive polaron with $\frac{\hbar}{2}$ spin. After two-photon excitation, dynamical relaxation splits the original carrier into two parts: one is a spinless carrier with $+2|e|$ charge (positive bipolaron), and the other is a spin carrier with $-|e|$ charge (negative polaron), which also conserves the total spin $\frac{\hbar}{2}$.

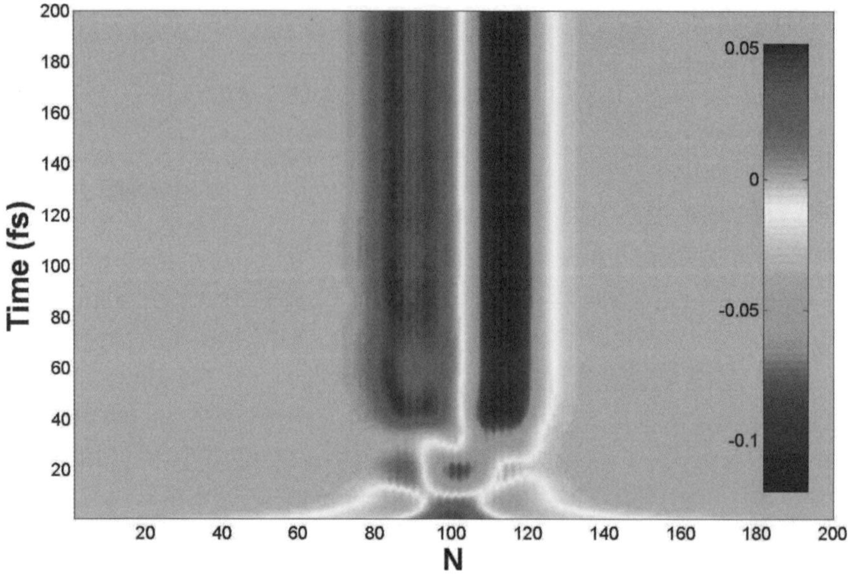

Fig. 10.8. Time-dependent charge distribution of a conjugated polymer after the original positive polaron undergoes two-photon excitation.

We now look at the spin property of the conjugated polymers. Through photoexcitation and dynamic relaxation, the charge of the spin carrier in a conducting polymer can be reversed from $+|e|$ to $-|e|$. When an electric field is applied along with the polymer chain with $E = 5.0 \times 10^{-4}$ V/cm, the spin of the positively-charged polaron is driven to move in the same direction as the field. However, at 300 fs when the spin carrier undergoes two-photon excitation, the relaxation splits this spin carrier into two parts and reverses the charge sign of the spin-positive polaron to a negative one, but still keeps the same spin sign. The electric field drives the spin in the opposite direction along the polymer chain. The whole process is shown in Fig. 10.9. Combining lattice relaxation and the spin transport process, the spin carrier-positive polaron can be separated into two parts within 100 fs, shown in Fig. 10.8. At the same time, the spin-transfer direction is also reversed, which apparently is an ultrafast process.

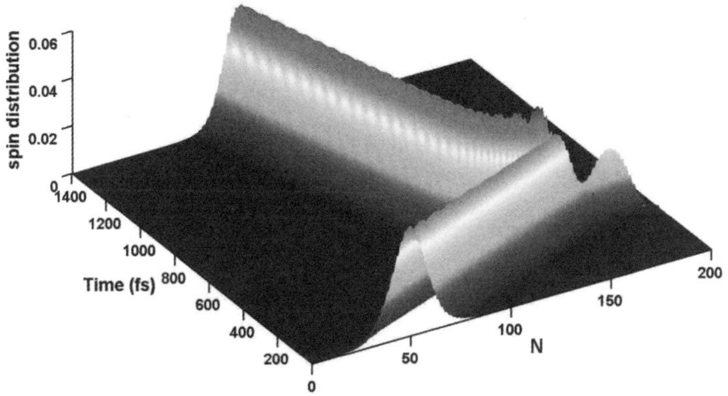

Fig. 10.9. Motion of a spin polaron under an external field $E = 5.0 \times 10^4$ V/cm.

Based on this optically-controlled backward spin-transfer effect in a conjugated polymer, a generic polymer spin valve can be designed as illustrated in Fig. 10.10. The valve consists of two different conjugated polymeric materials, A and B. Here, we apply an external magnetic field to determine the spin polarizations. Meanwhile, the bias voltage between A and B always makes the voltage in A higher than in B. Without photoexcitation, the positive spin polaron remains in B due to the applied bias voltage, as shown in Fig. 10.10(a). Once the positive polaron undergoes photoexcitation, the charge of the spin polaron is reversed, namely, the excitation transforms the positive spin into a negative-spin polaron. Then, the bias voltage in the junction drives the spin to transfer backward, easily injecting the spin of a negative polaron from B into A, as shown in Fig. 10.10(b).

Let us now turn our focus to the other spinless charged carrier-bipolaron. Once the positive bipolaron undergoes photoexcitation, namely one electron from the HOMO is excited to the state Φ_d, the original lattice configuration is no longer stable. It is found that the original spinless carrier is split into two spin carriers, where one is a polaron with positive spin sign, and the other corresponds to a negative spin sign, as shown in Fig. 10.11(a). In this case, although spin can be generated through external excitation, the total spin is still conserved. Most importantly, the resultant carriers caused by the external excitation are still carriers in the ground state, which maintains the generated spin

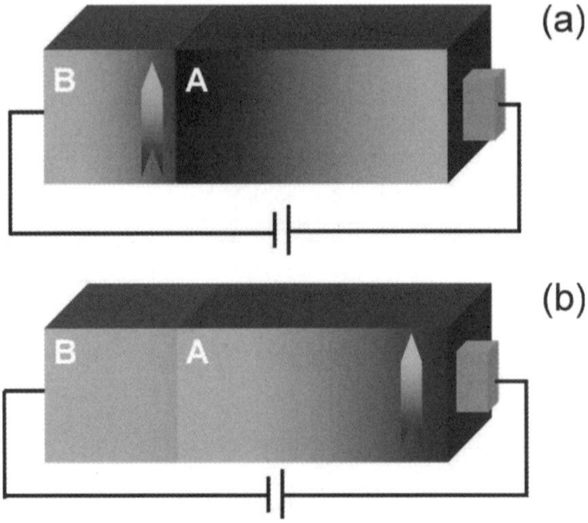

Fig. 10.10. Organic optically-controlled spin valve. Before photoexcitation, the positive spin polaron remains in the polymeric material B as shown in (a), and after photoexcitation, the spin is injected into material A in (b).

for a long time, thus prolonging the spin coherent time. Moreover, the charge distribution of the positive bipolaron is completely separated into two parts after 100 fs relaxation, as shown in Fig. 10.11(a). Combining the properties of spin and charge, the resultant carriers correspond to two positively-charged polarons.

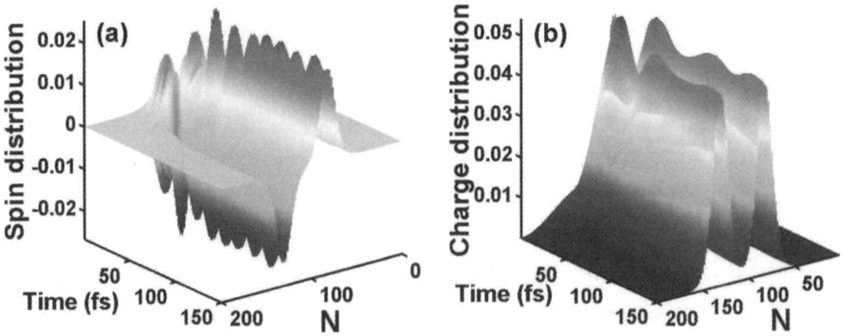

Fig. 10.11. (a) Time-dependent spin distribution of a conjugated polymer after the original positive bipolaron undergoes photoexcitation. (b) Time-dependent charge distribution of a conjugated polymer after the original positive bipolaron undergoes photoexcitation. The unit of charge distribution is +|e|, and *N* refers to the sites.

Actually, the result of the photoexcitation is an entanglement state. If the polaron with spin up is written as $P(\uparrow)$ and $P(\downarrow)$ stands for the one with spin down, this state becomes $P(\uparrow)P(\downarrow)+P(\downarrow)P(\uparrow)$. Once a given polaron's spin sign is fixed by an external magnetic field, the other polaron's spin sign is the opposite. Due to the same charge of the resultant polarons, these carriers with opposite spins will keep far away from each other, as shown in Fig. 10.11(b). Because recent experiments have shown that the magnetoresistance in organic semiconductors is dominated by charged polarons while the spinless bipolaron does not contribute to this effect [25], the photoexcitation not only induces bipolarons to produce polarons, but also provides an appropriate method to control the magnetoresistance, which is an ultrafast response spin valve effect.

The above photoinduced spin generation provides a possible approach to realize an ultrafast response organic spin valve mentioned in the Introduction. A conventional organic magnetoresistance device is a sandwich A/B/A structure as illustrated in Fig. 10.12(a), where A is a pinned layer, B is a spacer, and C is a free layer. Generally, a pinned layer is made of ferromagnetic materials, whose spin direction is fixed. A spacer consists of over-doped conjugated polymeric materials. Because of the over-dopping, the carriers in spacer are spinless bipolarons. The free layer is also composed of ferromagnetic materials whose spin direction, however, can be changed by the external magnetic field B. Before photoexcitation, the device is shown as Fig. 10.12(a). After the polymeric materials spacer undergoes the photoexcitation, the spinless bipolaron in the spacer will be split into two polarons with opposite spins. Following this, the ferromagnetic materials in the pinned and free layers couples with the newly-produced spin polarons. The coupling induces a change of magnetoresistance and finally changes the current in this circuit, as shown in Fig. 10.12(b). The whole process finishes within 100 fs, which should be an ultrafast response valve effect.

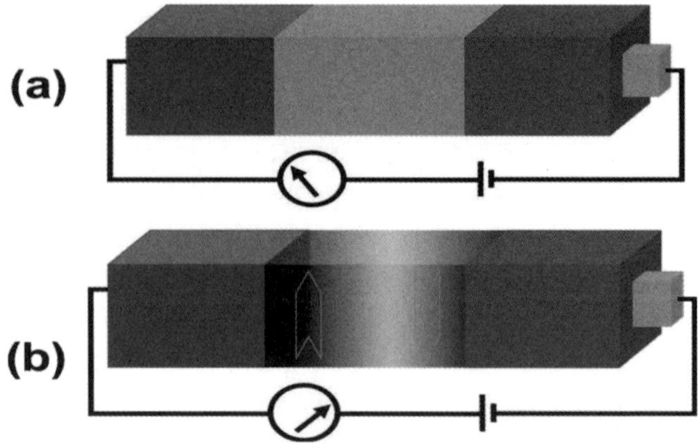

Fig. 10.12. Organic optically-controlled spin valve. Before photoexcitation, there is no spin in the spacer B as shown in (a), and after photoexcitation, the spin is generated in the spacer as shown in (b).

10.4. Conclusions

The results presented in this chapter lead to the following main conclusions:

i. We predict an electric-field-induced spin accumulation effect in a polymer LED. When an electric field is applied along the electroluminescent polymer chain and exceeds the critical value 4.5 MV/cm, the luminescence is quenched, and the singlet exciton contributing luminescence is dissociated into two polarons with opposite spin signs and charges. After relaxation, the electric field drives these two opposite spin carriers moving along opposite directions, finally with spin accumulation at the two ends of the LED. Furthermore, if we keep the electric field along the polymer LED with sufficient bias, the spin continues to accumulate at the ends of the LED. This effect can be detected through Kerr rotation microscopy.

ii. A photoexcitation-induced backward spin-transfer effect in a π-conjugated polymer is predicted: When a spin-polarized positive polaron absorbs two photons, the charge of the spin carrier can be reversed, and simultaneously, the applied external electric field drives the charge-reversed spin carrier to move in the opposite direction. Meanwhile, when a conducting conjugated polymeric molecule undergoes photoexcitation, the positively-charged spinless bipolaron dissociates into two positive polarons with opposite spin signs, forming an entanglement. The coupling between the newly-produced spin polaron and ferromagnet will change the magnetoresistance. By combining an electric field, magnetic field and photoexcitation, two generic optically-controlled ultrafast response organic spin valves are designed.

Acknowledgments

This work was supported by the National Science Foundation of China under Grants 20804039 and 21074118, the Zhejiang Provincial Natural Science Foundation of China under Grant Y4080300, and the Zhejiang Provincial Qianjiang Talent Project of China under Grant 2010R10019.

References

1. M. N. Baibich, J. M. Broto, A. Fert, Van Dau F. Nguyen and F. Petroff, Giant magnetoresistance of (001)Fe/(001)Cr magnetic superlattices, *Phys. Rev. Lett.* **61**, 2472–2475 (1988).
2. J. Moodera, L. Kinder, T. Wong and R. Meservey, Large magnetoresistance at room temperature in ferromagnetic thin film tunnel junctions, *Phys. Rev. Lett.* **74**, 3273–3276 (1995).
3. S. A. Wolf, D. D. Awschalom, R. A. Buhrman, J. M. Daughton, von S. Molner, M. L. Roukes, A. Y. Chtchelkanova and D. M. Treger, Spintronics: A spin-based electronics vision for the future, *Science* **294**, 1488–1491 (2001); J. M. Kikkawa and D. D. Awschalom, Lateral drag of spin coherence in gallium arsenide, *Nature* **397**, 139–141 (1999); Y. Ohno, D. K. Young, B. Beschoten, F. Matsukura, H.Ohno and D. D. Awschalom, Electrical spin injection in a ferromagnetic semiconductor heterostructure, *Nature* **402**, 790–792 (1999); I. Zutić, J. Fabian and S. D. Sarma, Spintronics: fundamentals and applications, *Rev. Mod. Phys.* **76**, 323–410 (2004).
4. A. J. Heeger, S. Kivelson, J. R. Schrieffer and W. P. Su, Solitons in conducting polymers, *Rev. Mod. Phys.* **60**, 781–850 (1988).
5. J. C. Scott, P. Pfluger, M. T. Krounbi and G. B. Street, Electron-spin-resonance studies of pyrrole polymers: Evidence for bipolarons, *Phys. Rev.* B **28**, 2140–2145

(1983); J. C. Scott, J. L. Brédas, K. Yakushi, P. Pfluger and G. B. Street, The evidence for bipolarons in pyrrole polymers, *Synth. Met.* **9**, 165–172 (1984).
6. V. I. Krinichnyi, 2-mm Waveband electron paramagnetic resonance spectroscopy of conducting polymers, *Synth. Met.* **108**, 173–222 (2000).
7. S. J. Xie, K. H. Ahn, D. L. Smith, A. R. Bishop and A. Saxenal, Ground-state properties of ferromagnetic metal/conjugated polymer interfaces, *Phys. Rev.* B **67**, 125202–125208 (2003).
8. V. Dediu, M. Murgia, F. C. Matacotta, C. Taliani and S. Barbanera, Room temperature spin polarized injection in organic semiconductor, *Solid State Commun.* **122**, 181–184 (2002).
9. Z. G. Yu, M. A. Berding and S. Krishnamurthy, Spin drift, spin precession, and magnetoresistance of noncollinear magnet-polymer-magnet structures, *Phys. Rev.* B **71**, 060408–060411 (2005).
10. Z. H. Xiong, Di Wu, Z. V. Vardeny and J. Shi, Giant magnetoresistance in organic spin-valves, *Nature* **427**, 821–824 (2004).
11. F. Luo, W. Song, Z. M. Wang and C. H. Yan, Room-temperature saturated ferroelectric polarization in $BiFeO_3$ ceramics synthesized by rapid liquid phase sintering, *Appl. Phys. Lett.* **84**, 1719–1721 (2004)
12. J. E. Hirsch, Spin hall effect, *Phys. Rev. Lett.* **83**, 1834–1837 (1999).
13. S. Murakami, N. Nagaosa and S. C. Zhang, Dissipationless quantum spin current at room temperature, *Science* **301**, 1348–1351 (2003).
14. Y. K. Kato, R. C. Myers, A. C. Gossard and D. D. Awschalom, Observation of the spin hall effect in semiconductors, *Science* **306**, 1910–1913 (2004).
15. J. H. Burroughes, D. D. C. Bradley, A. R. Brown, R. N. Marks, K. Mackay, R. H. Friend, P. L. Burns and A. B. Holmes, Light-emitting diodes based on conjugated polymers, *Nature* **347**, 539–541 (1990); R. H. Friend, R. W. Gymer, A. B. Holmes, J. H. Burroughes, R. N. Marks, C. Taliani, D. D. C. Bradley, D. A. Dos Santos, J. L. Bredas, M. Logdlund and W. R. Salaneck, Electroluminescence in conjugated polymers, *Nature* **397**, 121–128 (1999).
16. R. Kersting, U. Lemmer, M. Deussen, H. J. Bakker, R. F. Mahrt, H. Kurz, V. I. Arkhipov, H. Bässler and E. O. Göbel, Direct observation of ultrafast field-induced charge generation in ladder-type poly(para-phenylene), *Phys. Rev. Lett.* **73**, 1440–1443 (1994).
17. M. Duessen, M. Scheidler and H. Bassler, Electric field-induced photoluminescence quenching in thin-film light-emitting diodes based on poly(phenyl-p-phenylene vinylene), *Synth. Met.* **73**, 123–129 (1995).
18. W. Graupner, G. Cerullo, G. Lanzani, M. Nisoli, E. J. W. List, G. Leising and S. De Silvestri, Direct observation of ultrafast field-induced charge generation in ladder-type poly(para-phenylene), *Phys. Rev. Lett.* **81**, 3259–3262 (1998); W. Graupner, C. Zenz, E. J. W. List, G. Leising, G. Cerullo, S. De Silvestri and G. Lanzani, Electric field-assisted femtosecond pump-probe spectroscopy in organic light-emitting

diodes, *Synth. Met.* **101**, 277–280 (1999); V. Gulbinas, Y. Zaushitsyn, V. Sundström, D. Hertel, H. Bässler and A. Yartsev, Dynamics of the electric field-assisted charge carrier photogeneration in ladder-type poly(para-phenylene) at a low excitation intensity, *Phys. Rev. Lett.* **89**, 107401-1-4 (2002).

19. V. N. Prigodin, J. D. Bergeson, D. M. Lincoln and A. J. Epstein, Anomalous room temperature magnetoresistance in organic semiconductors, *Synth. Met.* **156**, 757–761 (2006).
20. P. Desai, P. Shakya, T. Kreouzis, W. P. Gillin, N. A. Morley and M. R. J. Gibbs, Magnetoresistance and efficiency measurements of Alq_3-based OLEDs, *Phys. Rev. B* **75**, 094423–094427 (2007).
21. Y. Wu, Z. Xu, B. Hu and J. Howe, Tuning magnetoresistance and magnetic-field-dependent electroluminescence through mixing a strong-spin-orbital-coupling molecule and a weak-spin-orbital-coupling polymer, *Phys. Rev. B* **75**, 035214–035219 (2007).
22. P. A. Bobbert, T. D. Nguyen, F. W. A. van Oost, B. Koopmans and M. Wohlgenannt, Bipolaron mechanism for organic magnetoresistance, *Phys. Rev. Lett.* **99**, 216801–216804 (2007).
23. J. Kido, H. Haromichi, K. Hongawa, K. Nagai and K. Okuyama, Bright red light-emitting organic electroluminescent devices having a europium complex as an emitter, *Appl. Phys. Lett.* **65**, 2124–2126 (1994); S. Hoshino and H. Suzuki, Electroluminescence from triplet excited states of benzophenone, *Appl. Phys. Lett.* **69**, 224–226 (1996); X. Zhang, R. Q. Zheng, Sun, T. Kobayashi and W. Li, Temperature-dependent electroluminescence from (Eu, Gd) coordination complexes, *Appl. Phys. Lett.* **71**, 2596–2598 (1997).
24. Y. Sheng, T. D. Nguyen, G. Veeraraghavan, O. Mermer and M. Wohlgenannt, Effect of spin-orbit coupling on magnetoresistance in organic semiconductors, *Phys. Rev. B* **75**, 035202–035207 (2007).
25. F. L. Bloom, W. Wagemans, M. Kemerink and B. Koopmans, Separating positive and negative magnetoresistance in oganic semiconductor devices, *Phys. Rev. Lett.* **99**, 257201–257204 (2007); F. L. Bloom, W. Wagemans and B. Koopmans, Temperature dependent sign change of the organic magnetoresistance effect, *J. Appl. Phys.* **103**, 07F320–07F323 (2008).
26. X. Sun, R. L. Fu, K. Yonemitsu and K. Nasu, Photoinduced polarization inversion in a polymeric molecule, *Phys. Rev. Lett.* **84**, 2830–2832 (2000).
27. S. Li, L. S. Chen, T. F. George and X. Sun, Photoinduced carrier fission in polymers with a nondegenerate ground state, *Phys. Rev. B* **70**, 075201-1-9 (2004).
28. S. Li, T. F. George and X. Sun, Electric field-induced spin accumulation in polymer light-emitting, *J. Phys. Chem. B (Letters)* **111**, 6097–6100 (2007).
29. S. Li, X. L. He, T. F. George, B. P. Xie and X. Sun, Optically-controlled spin valves in conjugated polymers, *J. Phys. Chem. B (Letters)* **113**, 400–404 (2009).

Chapter 11

Optical Properties of Wurtzite ZnO-based Quantum Well Structures with Piezoelectric and Spontaneous Polarizations

Seoung-Hwan Park

Catholic University of Daegu
shpark@cu.ac.kr

Doyeol Ahn

University of Seoul
dahn@uos.ac.kr

Sam Nyung Yi

Korea Maritime University
snyi@hhu.ac.kr

Tae Won Kang* and Seung Joo Lee†

Dongguk University
**twkang@dongguk.edu*
†leesj@dongguk.edu

Optical properties of ZnO/MgZnO quantum well (QW) structures, considering piezoelectric and spontaneous polarizations, are investigated by using the non-Markovian gain model with many-body effects. The spontaneous polarization constant for MgO was determined from a comparison with the experiment, which gives a value of about -0.070 C/m^2. The optical matrix element of the ZnO/MgZnO QW structure decreases with the inclusion of Mg. This is attributed to an increase in the spatial separation between the electron and the hole wave functions due to the large internal field. However, the ZnO/MgZnO QW structure with a relatively high Mg composition ($x = 0.3$) is found to have a larger optical gain than that with a relatively low Mg composition. This can be explained by the fact that the quasi-Fermi-level separation ΔE_{fc} in the conduction band increases with the inclusion of Mg. The increase in ΔE_{fc} is because the QW structure with a high Mg composition has a larger energy spacing in the conduction band. We also know that the exciton binding energy of ZnO/MgZnO QW structures is much larger than that of GaN/AlGaN QW structures. This can be explained by the fact that ZnO/MgZnO QW structures have a larger matrix element than the GaN/AlGaN QW structures and by the smaller dielectric constant.

11.1. Introduction

The wide band-gap wurtzite semiconductors have attracted much attention due to their potential applications for optoelectronic devices in the blue and the ultraviolet (UV) regions. So far, practical short-wavelength light-emitting diodes or laser diodes have been fabricated using GaN-related materials. Recently, on the other hand, ZnO and related oxides have been proposed as other wide band-gap semiconductors for short-wavelength optoelectronic applications because they have several advantages compared to GaN-related materials [1–5]. For example, the growth temperature of ZnO is usually around 500°C, which is much lower than typical growth temperature, 1000°C, of GaN [1]. Also, the ZnO system has a very large exciton binding energy (~60 meV), which permits excitonic recombination even at room temperature [2]. In principle, a lower pumping threshold can be expected if an exciton-related recombination, rather than an electron-hole plasma recombination, is used. Experimentally, several groups reported the successful growth of ZnO/MgZnO multiple quantum wells (QWs) and showed that these systems are promising for blue and ultraviolet optical applications [1,2,6,7]. In fact, room-temperature lasing in ZnO epilayers on sapphire (0001) substrates has been experimentally demonstrated [6].

In the case of GaN-based QW structures, the optical matrix element and the optical gain are largely reduced by the spatial separation between the conduction and the valence wave functions [8]. In the case of ZnO/MgZnO QW structures with a small Mg composition of about 0.1, on the other hand, the internal fields due to piezoelectric (PZ) and spontaneous (SP) polarizations for ZnO on ZnMgO are expected to be much smaller than those for GaN-related system because of the small difference between ZnO and ZnMgO [5,9]. However, a large internal field due to the spontaneous and the piezoelectric polarizations has been reported to exist in wurtzite ZnO/MgZnO QW structures with relatively high Mg compositions and thick well widths. On the theoretical side, these results suggest that an understanding of the roles of piezoelectric and spontaneous polarizations in wurtzite ZnO-based QW structures is very important in order to give guidelines on sample growth and a device design [9–12].

In this chapter, we review on optical properties of ZnO/ZnMgO QW structures numerically by using the non-Markovian gain model with many-body effects. These results are compared with those for GaN/AlGaN QW structures with the spontaneous and the piezoelectric polarizations grown on GaN substrates. We consider a QW structure with a ZnMgO buffer layer, which is used to as the substrate for the growth of QWs. Thus, a ZnO well is under a compressive strains and a ZnMgO barrier is lattice-matched to the substrate. The self-consistent (SC) band structures and wave functions for the QW structures are obtained by solving the Schrödinger equation for electrons and the 3×3 Hamiltonian for holes [13,14].

11.2. Theory

11.2.1. *6×6 effective-mass Hamiltonian H for the (0001)-oriented wurtzite crystal*

The Hamiltonian for the valence-band structure has been derived by using the $\mathbf{k}\cdot\mathbf{p}$ method. The 6×6 Hamiltonian for the (0001)-oriented wurtzite crystal can be written as [13]

$$H(\mathbf{k},\epsilon) = \begin{pmatrix} F & -K^* & -H^* & 0 & 0 & 0 \\ -K & G & H & 0 & 0 & \Delta \\ -H & H^* & \lambda & 0 & \Delta & 0 \\ 0 & 0 & 0 & F & -K & H \\ 0 & 0 & \Delta & -K^* & G & -H^* \\ 0 & \Delta & 0 & H^* & -H & \lambda \end{pmatrix} \begin{matrix} |U_1> \\ |U_2> \\ |U_3> \\ |U_4> \\ |U_5> \\ |U_6>, \end{matrix} \tag{11.1}$$

where

$$\begin{aligned}
F &= \Delta_1 + \Delta_2 + \lambda + \theta, \\
G &= \Delta_1 - \Delta_2 + \lambda + \theta, \\
\lambda &= \frac{\hbar^2}{2m_o}[A_1 {k_z}^2 + A_2({k_x}^2 + {k_y}^2)] + \lambda_\epsilon, \\
\theta &= \frac{\hbar^2}{2m_o}[A_3 {k_z}^2 + A_4({k_x}^2 + {k_y}^2)] + \theta_\epsilon, \\
K &= \frac{\hbar^2}{2m_o} A_5 (k_x + ik_y)^2 + D_5 \epsilon_+, \\
H &= \frac{\hbar^2}{2m_o} A_6 (k_x + ik_y)(k_z) + D_6 \epsilon_{z+},
\end{aligned}$$

$$\begin{aligned}
\lambda_\epsilon &= D_1(\epsilon_{zz}) + D_2(\epsilon_{xx} + \epsilon_{yy}),\\
\theta_\epsilon &= D_3(\epsilon_{zz}) + D_4(\epsilon_{xx} + \epsilon_{yy}),\\
\epsilon_+ &= \epsilon_{xx} - \epsilon_{yy} + 2i\epsilon_{xy},\\
\epsilon_{z+} &= \epsilon_{xz} + i\epsilon_{yz},\\
\Delta &= \sqrt{2}\Delta_3.
\end{aligned} \tag{11.2}$$

Here, the A_i's are the valence-band effective-mass parameters which are similar to the Luttinger parameters in a zinc-blende (ZB) crystal, the D_i's are the deformation potentials for wurtzite crystals, k_i is the wave vector, ϵ_{ij} is the strain tensor, Δ_1 is the crystal-field split energy, and Δ_2 and Δ_3 account for spin-orbit interactions. The bases for the Hamiltonian are defined as

$$\begin{aligned}
|U_1> &= -\frac{1}{\sqrt{2}}|(X+iY)\uparrow>,\\
|U_2> &= \frac{1}{\sqrt{2}}|(X-iY)\uparrow>,\\
|U_3> &= |Z\uparrow>,\\
|U_4> &= \frac{1}{\sqrt{2}}|(X-iY)\downarrow>,\\
|U_5> &= -\frac{1}{\sqrt{2}}|(X+iY)\downarrow>,\\
|U_6> &= |Z\downarrow>\,.
\end{aligned} \tag{11.3}$$

The block-diagonalized Hamiltonian for (0001)-oriented wurtzite crystal is given by

$$H = \begin{pmatrix} H^U & 0 \\ 0 & H^L \end{pmatrix}. \tag{11.4}$$

The upper three-by-three Hamiltonian is

$$H^U = \begin{pmatrix} F & K_t & -iH_t \\ K_t & G & \Delta - iH_t \\ iH_t & \Delta + iH_t & \lambda \end{pmatrix} \begin{matrix} |1> \\ |2> \\ |3> \end{matrix} \tag{11.5}$$

and the lower three-by-three Hamiltonian is

$$H^L = \begin{pmatrix} F & K_t & iH_t \\ K_t & G & \Delta + iH_t \\ -iH_t & \Delta - iH_t & \lambda \end{pmatrix} \begin{matrix} |4> \\ |5> \\ |6> \end{matrix} \tag{11.6}$$

where

$$
\begin{aligned}
F &= \Delta_1 + \Delta_2 + \lambda + \theta, \\
G &= \Delta_1 - \Delta_2 + \lambda + \theta, \\
\lambda &= \frac{\hbar^2}{2m_o}(A_1 k_z^2 + A_2 k_t^2) + \lambda_\epsilon, \\
\lambda_\epsilon &= D_1 \epsilon_{zz} + D_2(\epsilon_{xx} + \epsilon_{yy}), \\
\theta &= \frac{\hbar^2}{2m_o}(A_3 k_z^2 + A_4 k_t^2) + \theta_\epsilon, \\
\theta_\epsilon &= D_3 \epsilon_{zz} + D_4(\epsilon_{xx} + \epsilon_{yy}), \\
K_t &= \frac{\hbar^2}{2m_o} A_5 k_t^2, \\
H_t &= \frac{\hbar^2}{2m_o} A_6 k_z k_t, \\
\Delta &= \sqrt{2}\Delta_3,
\end{aligned}
\tag{11.7}
$$

where $k_t = \sqrt{k_x^2 + k_y^2}$ is the magnitude of the wavevector in the k_x-k_y plane.

The bases for the block-diagonalized Hamiltonians in Eqs. (11.5) and (11.6) are defined as

$$
\begin{aligned}
|1> &= \alpha^* \left| -\frac{1}{\sqrt{2}}(X + iY)\uparrow \right\rangle + \alpha \left| \frac{1}{\sqrt{2}}(X - iY)\downarrow \right\rangle, \\
|2> &= \beta \left| \frac{1}{\sqrt{2}}(X - iY)\uparrow \right\rangle + \beta^* \left| -\frac{1}{\sqrt{2}}(X + iY)\downarrow \right\rangle, \\
|3> &= \beta^* \left| Z\uparrow \right\rangle + \beta \left| Z\downarrow \right\rangle, \\
|4> &= \alpha^* \left| -\frac{1}{\sqrt{2}}(X + iY)\uparrow \right\rangle - \alpha \left| \frac{1}{\sqrt{2}}(X - iY)\downarrow \right\rangle, \\
|5> &= \beta \left| \frac{1}{\sqrt{2}}(X - iY)\uparrow \right\rangle - \beta^* \left| -\frac{1}{\sqrt{2}}(X + iY)\downarrow \right\rangle, \\
|6> &= -\beta^* \left| Z\uparrow \right\rangle + \beta \left| Z\downarrow \right\rangle,
\end{aligned}
\tag{11.8}
$$

where

$$
\begin{aligned}
\alpha &= \frac{1}{\sqrt{2}} exp\left[i\left(\frac{3\pi}{4} + \frac{3\phi}{2} \right) \right], \\
\beta &= \frac{1}{\sqrt{2}} exp\left[i\left(\frac{\pi}{4} + \frac{\phi}{2} \right) \right],
\end{aligned}
\tag{11.9}
$$

and $\phi = tan^{-1}(k_x/k_y)$ is the azimuthal angle in the k_x-k_y plane and covers a range between 0 and 2π.

11.2.2. *Electric fields in the well and barrier*

The electric fields in the well and barrier due to the piezoelectric and spontaneous polarizations along the (0001)-direction for (0001)-oriented wurtzite structures can be estimated from the periodic boundary condition for a superlattice structure as follows [15,16]:

$$\begin{aligned} F_z^w &= \frac{(P_{SP}^b + P_{PZ}^b - P_{SP}^w - P_{PZ}^w)}{\epsilon^w + \epsilon^b (L_w/L_b)}, \\ F_z^b &= -\frac{L_w}{L_b} F_z^w, \end{aligned} \tag{11.10}$$

where superscripts w and b represent the well and barrier, and L and ϵ are the layer thickness and the static dielectric constant, respectively. Although Eq. (11.10) applies to the superlattice structure, it can provide a good approximation for the MQW structure. The strain-induced piezoelectric polarization P_{PZ}^i $(i = w, b)$ is given by [17]

$$P_{PZ}^i = 2d_{31}^i \left(C_{11}^i + C_{12}^i - \frac{2{C_{13}^i}^2}{C_{33}^i} \right) {\epsilon^i}_{||}, \tag{11.11}$$

where d_{31}^i is the piezoelectric constant, $\epsilon_{||}^i = (a_s - a_e)/a_e$, a_s and a_e are the lattice constants of the substrate and the epilayer materials, respectively, and C_{11}, C_{12}, C_{13}, and C_{33} are the stiffness constants for the wurtzite structure.

11.2.3. *Self-consistent calculations with the screening effect*

The self-consistent band structures and wave functions are obtained by iteratively solving the Schrödinger equation for electrons, the block-diagonalized Hamiltonian for holes, and Poisson's equation [18,19]. The total potential profiles for the electrons and the holes are

$$\begin{aligned} V_c(z) &= V_{cw}(z) + |e|F_z z - |e|\phi(z), \\ V_v(z) &= V_{vw}(z) + |e|F_z z - |e|\phi(z), \end{aligned} \tag{11.12}$$

where $V_{cw}(z)$ and $V_{vw}(z)$ are the square-well potentials for the conduction band and the valence band, respectively, F_z is the piezoelectric field, and $\phi(z)$ is the screening potential induced by the charged carriers and satisfies the Poisson equation

$$\frac{d}{dz}\left(\epsilon(z)\frac{d}{dz}\right)\phi(z) = -|e|[p(z) - n(z)], \tag{11.13}$$

where $\epsilon(z)$ is the dielectric constant. The electron and the hole concentrations, $p(z)$ and $n(z)$, are related to the wave functions of the n-th conduction subband and the m-th valence subband by

$$n(z) = \frac{kTm_e}{\pi\hbar^2}\sum_n |f_n(z)|^2 \ln\left(1 + e^{[E_{fc}-E_{cn}(0)]/kT}\right) \tag{11.14}$$

and

$$p(z) = \sum_{\sigma=U,L}\sum_m \int dk_{||}\frac{k_{||}}{2\pi}\sum_\nu |g_{mk_{||}}^{\sigma(\nu)}(z)|^2\left(\frac{1}{1+e^{[E_{fv}-E_{vm}(k_{||})]/kT}}\right), \tag{11.15}$$

where m_e is the effective mass of electrons, $\hbar$ is Planck's constant divided by 2π, n and m are the quantized subband indices for the conduction and the valence bands, E_{fc} and E_{fv} are the quasi-Fermi levels of the electrons and the holes, respectively, $E_{cn}(0)$ is the quantized energy level of the electrons, $E_{vm}(k_{||})$ is the energy for the m-th subband in the valence band, σ denotes the upper (U) and the lower (L) blocks of the Hamiltonian, $k_{||}$ is the in-plane wave vector, ν refers to the new bases for the Hamiltonian, and $f_n(z)$ and $g_{mk_{||}}^{\sigma(\nu)}(z)$ are the envelope functions in the conduction and the valence bands, respectively. The potential $\phi(z)$ is obtained by integration [19]:

$$\phi(z) = -\int_{-L/2}^{z} E(z')dz', \tag{11.16}$$

where

$$E(z) = \int_{-L/2}^{z} \frac{1}{\epsilon(z)}\rho(z')dz'. \tag{11.17}$$

The procedures for the SC calculations consist of the following steps: (i) Start with the potential profiles $V_c(z)$ and $V_v(z)$ with $\phi^{(0)}(z) = 0$ in Eq. (11.12); (ii) Solve the Schrödinger equation (for electrons) and the block-diagonalized Hamiltonian (for holes) with the potential profiles $\phi^{(n-1)}(z)$ in step (i) to obtain band structures and wave functions; (iii) For a given carrier density, obtain the Fermi-energies by using the band structures and the charge distribution by using the wave functions; (iv) Solve Poisson's equation to find $\phi^{(n)}(z)$; (v) Check if $\phi^{(n)}(z)$ converges to $\phi^{(n-1)}(z)$. If not, set $\phi^{(n)}(z)= w\phi^{(n)}(z)+ (1-w)\phi^{(n-1)}(z)$, $n = n+1$; then, return to step (ii). If yes, the band structures and the wave functions obtained with $\phi^{(n-1)}(z)$ are solutions. An adjustable parameter w $(0 < w < 1)$ is typically set to be 0.5 at low carrier densities. With increasing carrier densities, a smaller value of w is needed for rapid convergence.

11.2.4. *Optical momentum matrix elements*

The optical momentum matrix elements are given by

$$|\hat{\mathbf{e}} \cdot \mathbf{M}^{\eta}|^2 = |\langle \Psi_l{}^{c\eta} | \hat{\mathbf{e}} \cdot \mathbf{p} | \Psi_m^v \rangle|^2, \tag{11.18}$$

where Ψ^c and Ψ^v are wave functions for the conduction and the valence bands, respectively, and $\eta = \uparrow$ and $\downarrow$ for both electron spins. The polarization-dependent interband momentum-matrix elements can be written as
TE-polarization ($\hat{\mathbf{e}} = \hat{\mathbf{x}}$) :

$$|\hat{\mathbf{e}} \cdot \mathbf{M}^{\uparrow}|^2 = \left| -\frac{1}{\sqrt{2}} P_x \langle g_m^{(1)} | \phi_l \rangle + \frac{1}{\sqrt{2}} P_x \langle g_m^{(2)} | \phi_l \rangle \right|^2 \tag{11.19}$$

$$|\hat{\mathbf{e}} \cdot \mathbf{M}^{\downarrow}|^2 = \left| \frac{1}{\sqrt{2}} P_x \langle g_m^{(4)} | \phi_l \rangle - \frac{1}{\sqrt{2}} P_x \langle g_m^{(5)} | \phi_l \rangle \right|^2 \tag{11.20}$$

TM-polarization ($\hat{\mathbf{e}} = \hat{\mathbf{z}}$):

$$|\hat{\mathbf{e}} \cdot \mathbf{M}^{\uparrow}|^2 = \left| P_z \langle g_m^{(3)} | \phi_l \rangle \right|^2 \tag{11.21}$$

$$|\hat{\mathbf{e}} \cdot \mathbf{M}^{\downarrow}|^2 = \left| P_z \langle g_m^{(6)} | \phi_l \rangle \right|^2, \tag{11.22}$$

where $g_m^{(\nu)}$ ($\nu = 1, 2, 3, 4, 5,$ and 6) is the wave function for the mth subband. Also,

$$\begin{aligned} P_x &= P_y = \langle S|p_x|X\rangle = \langle S|p_y|Y\rangle = \frac{m_o}{\hbar} P_2, P_z = \langle S|p_z|Z\rangle = \frac{m_o}{\hbar} P_1, \\ P_1^2 &= \frac{\hbar^2}{2m_o}\left(\frac{m_o}{m_e^z} - 1\right) \frac{(E_g + \Delta_1 + \Delta_2)(E_g + 2\Delta_2) - 2\Delta_3^2}{E_g + 2\Delta_2}, \\ P_2^2 &= \frac{\hbar^2}{2m_o}\left(\frac{m_o}{m_e^t} - 1\right) \frac{E_g[(E_g + \Delta_1 + \Delta_2)(E_g + 2\Delta_2) - 2\Delta_3^2]}{(E_g + \Delta_1 + \Delta_2)(E_g + \Delta_2) - \Delta_3^2}, \end{aligned} \tag{11.23}$$

where m_e^t and m_e^z are the electron effective masses perpendicular (t) and parallel (z) to the growth direction, respectively.

11.2.5. *Non-Markovian gain model with many-body effects*

The optical polarization dependent gain spectra are calculated by the non-Markovian gain model with many-body effects [20,21]. The optical gain with many-body effects including the effects of anisotropy on the valence band dispersion is given by

$$g(\omega) = \sqrt{\frac{\mu_o}{\epsilon}} \left(\frac{e^2}{m_o^2 \omega}\right) \int_0^{2\pi} d\phi \int_0^{\infty} dk_{||} \frac{2k_{||}}{(2\pi)^2 L_w} |M_{nm}(k_{||}, \phi)|^2 \\ \times [f_n^c(k_{||}, \phi) - f_m^v(k_{||}, \phi)] L(\omega, k_{||}, \phi), \quad (11.24)$$

where ω the angular frequency, μ_o the vacuum permeability, ϵ the dielectric constant, $k_{||}$ and ϕ the magnitude and angle of the in-plane wave vector in the QW plane, respectively, L_w the well thickness, $|M_{nm}|^2$ the momentum matrix element in the strained QW, f_n^c and f_m^v the Fermi functions for the conduction band states and the valence band states, and $\hbar$ is the Planck constant. The Gaussian line shape function $L(\omega, k_{||}, \phi)$ with many-body effects is given by

$$L(\omega, k_{||}, \phi) = \\ \frac{(1 - \mathrm{Re}Q(k_{||}, \hbar\omega))\mathrm{Re}\Xi(E_{lm}(k_{||}, \hbar\omega)) - \mathrm{Im}Q(k_{||}, \hbar\omega)\mathrm{Im}\Xi(E_{lm}(k_{||}, \hbar\omega))}{(1 - \mathrm{Re}Q(k_{||}, \hbar\omega))^2 + (\mathrm{Im}Q(k_{||}, \hbar\omega))^2}, \quad (11.25)$$

where [20,21]

$$\mathrm{Re}[L(E_{lm}(k_{\|}, \hbar\omega))] = \sqrt{\frac{\pi \tau_{in}(k_{\|}, \hbar\omega)\tau_c}{2\hbar^2}} \exp\left(-\frac{\tau_{in}(k_{\|}, \hbar\omega)\tau_c}{2\hbar^2} E_{lm}^2(k_{\|}, \hbar\omega)\right) \quad (11.26)$$

and

$$\mathrm{Im}[L(E_{lm}(k_{\|}, \hbar\omega))] = \frac{\tau_c}{\hbar} \int_0^{\infty} \exp\left(-\frac{\tau_c}{2\tau_{in}(k_{\|}, \hbar\omega)} t^2\right) \times \\ \sin\left(\frac{\tau_c E_{lm}(k_{\|}, \hbar\omega)}{\hbar} t\right) dt. \quad (11.27)$$

In the above, $E_{lm}(k_{||}, \hbar\omega) = E_l^c(k_{||}) - E_m^v(k_{||}) + E_g + \Delta E_{SX} + \Delta E_{CH} - \hbar\omega$ is the renormalized transition energy between electrons and holes, where E_g is the band-gap of the material, ΔE_{SX} and ΔE_{CH} are the screened exchange and Coulomb-hole contributions to the band-gap renormalization, and $Q(k_{||}, \hbar\omega)$ is the term related to the excitonic or Coulomb enhancement of the interband transition probability [22]. The correlation time τ_{co} is related to the non-Markovian enhancement of optical gain [20] and is assumed to be constant. The τ_{in} and the τ_{co} used in the calculation are 25 and 10 fs, respectively.

The bandgap renormalization is given as a summation of the Coulomb-hole self energy and the screened-exchange shift [22]. The ϕ dependence of the bandgap renormalization is very small and neglected for simplicity. The Coulomb-hole contribution to the bandgap renormalization is written as

$$\Delta E_{CH} = -2E_R a_o \lambda_s \ln\left(1+\sqrt{\frac{32\pi N L_w}{C\lambda_s^3 a_o}}\right), \tag{11.28}$$

where N is the carrier density, λ_s is the inverse screening length, and C is a constant usually taken between 1 and 4. The Rydberg constant E_R and the exciton Bohr radius a_o are given by

$$\Delta E_R(eV) = 13.6\frac{m_o/m_r}{(\epsilon/\epsilon_o)^2} \tag{11.29}$$

and

$$a_o(\text{Å}) = 0.53\frac{\epsilon/\epsilon_o}{m_o/m_r}, \tag{11.30}$$

where m_r is the reduced electron-hole mass defined by $1/m_r$=$1/m_e$+$1/m_h$. The exchange contribution to the bandgap renormalization is given by

$$\Delta E_{SX} = -\frac{2E_R a_o}{\lambda_s}\int_0^\infty dk_{||}k_{||}\frac{1+\frac{C\lambda_s a_o k_{||}^2}{32\pi N L_w}}{1+\frac{k_{||}}{\lambda_s}+\frac{C a_o k_{||}^3}{32\pi N L_w}}\left[f_n^c(k_{||})+1-f_m^v(k_{||})\right] \tag{11.31}$$

The factor $1/(1-Q(k_{||},\hbar\omega))$ represents the Coulomb enhancement in the Padé approximation. Here, the factor $Q(k_{||},\hbar\omega)$ is given by [21]

$$\begin{aligned} Q(k_{||},\hbar\omega) = {} & i\frac{a_o E_R}{\pi\lambda_s|M_{nm}(k_{||})|}\int_0^\infty dk'_{||}k'_{||}|M_{nm}(k'_{||})|(f_n^c(k'_{||})-f_m^v(k'_{||})) \\ & \times\,\Xi(E_{lm}(k'_{||},\hbar\omega))\Theta(k_{||},k'_{||}), \end{aligned} \tag{11.32}$$

where

$$\Theta(k_{||},k'_{||}) = \int_0^\infty d\theta\left(1+\frac{C\lambda_s a_o q_{||}^2}{32\pi N L_w}\right)\left(1+\frac{q_{||}}{\lambda_s}+\frac{C a_o q_{||}^3}{32\pi N L_w}\right)^{-1} \tag{11.33}$$

and

$$q_{||} = |\mathbf{k}_{||}-\mathbf{k}_{||}|.$$

11.2.6. *Exciton binding energies*

In the QW semiconductor, the free electron and hole states interact through the screened Coulomb interaction to form QW excitons. The exciton wave function is made up of a linear combination of direct products of the subband states for electrons and holes:

$$\Psi_{ex} = \sum_{n,m} \sum_{\mathbf{k}_\parallel} \sum_{\mathbf{k}'_\parallel} F_{nm}(\mathbf{k}_\parallel, \mathbf{k}'_\parallel) |\mathbf{k}_\parallel, n\rangle |\mathbf{k}'_\parallel, m\rangle, \tag{11.34}$$

where n and m are the subband indices for electrons and holes, respectively, and the summation runs over the in-plane wave vectors $\mathbf{k}_{||}(\mathbf{k'}_{||})$. $F_{nm}(\mathbf{k}_\parallel, \mathbf{k}'_\parallel)$ denotes the envelope function of the nm exciton, where is described by

$$F_{nm}(\mathbf{k}_\parallel, \mathbf{k}'_\parallel) = \delta(\mathbf{k}_\parallel + \mathbf{k}'_\parallel) G_{nm}(\mathbf{k}_\parallel), \tag{11.35}$$

where $G_{nm}(\mathbf{k}_\parallel)$, the exciton relative motion envelope function, satisfies

$$\begin{aligned} &[E_n^e(\mathbf{k}_\parallel) - E_m^h(\mathbf{k}_\parallel)] G_{nm}(\mathbf{k}_\parallel) \\ &+ \sum_{n',m'} \sum_{\mathbf{k}'_\parallel} \frac{\bar{V}_{n'm'}^{nm}(\mathbf{k}_\parallel, \mathbf{k}'_\parallel)}{\epsilon(|\mathbf{k}_\parallel - \mathbf{k}'_\parallel|)} G_{n'm'}(\mathbf{k}'_\parallel) = E G_{nm}(\mathbf{k}_\parallel). \end{aligned} \tag{11.36}$$

In this expression, $E_n^e(\mathbf{k}_\parallel)$ and $E_m^h(\mathbf{k}_\parallel)$ are the energies of the n-th conduction and the m-th valence subbands, respectively. $\bar{V}_{n'm'}^{nm}(\mathbf{k}_\parallel, \mathbf{k}'_\parallel)$ is the Coulomb potential between the nm and the $n'm'$ excitons and is given by [23]

$$\begin{aligned} &\bar{V}\,{}_{n'm'}^{nm}(\mathbf{k}_\parallel, \mathbf{k}'_\parallel) \\ &= \frac{e^2}{2A\sqrt{|\mathbf{k}_\parallel - \mathbf{k}'_\parallel|^2 + \lambda_s^2}} \int\int dz_e dz_h \psi_{n'}^*(z_e) \psi_n(z_e) \\ &\times \sum_\nu g_{m'}^{*\nu}(\mathbf{k'}_\parallel, z_h) g_m^\nu(\mathbf{k}_\parallel, z_h) e^{-\sqrt{|\mathbf{k}_\parallel - \mathbf{k}'_\parallel|^2 + \lambda_s^2}\,|z_e - z_h|}, \end{aligned} \tag{11.37}$$

where A is the interface area of the sample, ψ_n and g_m^ν are the envelope functions for electrons and holes, respectively, and ν denotes the bases for a block-diagonalized Hamiltonian for holes. The inverse screening length λ_s is given, for the carrier injection case where electrons and holes exist simultaneously, as follows [23]:

$$\begin{aligned} \lambda_s^2 = \frac{e^2}{\pi\hbar^2\epsilon} \sum_j \Big[&\sqrt{2m_e E_j^e} m_e f_e(E_j^e)/\pi\hbar \\ &+ \sqrt{2m_h E_j^h} m_h f_h(E_j^h)/\pi\hbar \Big], \end{aligned} \tag{11.38}$$

Table 11.1. Physical parameters for wurtzite ZnO and MgO.

Parameters	ZnO	MgO
Energy parameters		
E_g (eV)	3.35 [24,25]	5.289 [26]
Δ_1 (meV)	30.5 [27]	–
$\Delta_3 = \Delta_2$ (meV)	4.2 [27]	–
Conduction band effective masses [28]		
m_e/m_o	0.24	–
Valence band effective-mass parameters		
A_1	−3.78 [29]	–
A_2	−0.44 [29]	–
A_5	−3.13 [30]	–
Deformation potentials [31,32] (eV)		
a_c	−6.05	–
D_1	−3.90	–
D_2	−4.13	–
D_3	−1.15	–
D_4	1.22	–
D_5	1.53	–
D_6	2.83	–
Elastic stiffness constant [33,34] ($\times 10^{11}$ dyn/cm^2)		
C_{11}	20.97	22.2
C_{12}	12.11	9.0
C_{13}	10.51	5.8
C_{33}	21.09	10.9
C_{44}	4.247	10.5
Dielectric constant [28,35]		
ϵ	8.1	9.8
Piezoelectric constant [34,36]		
$d_{31}(\times 10^{-12}$m/V)	−5.0	−7.887
Spontaneous polarization constant		
P_{sp}(C/m^2)	−0.05 [37]	−0.068 [9]

where f_e and f_h are the Fermi distribution functions for the conduction- and the valence-band states, respectively, and the sum over j means all the subbands within the QW are considered in the calculation of λ_s.

The material parameters of ZnO and MgO used in the calculation are listed in Table 11.1. The parameters for $Zn_{1-x}Mg_xO$ are obtained from the linear combination between the parameters of ZnO and MgO. We used 3.2466+0.0163x + 0.1097x^2 (Å) as a lattice constant for $Mg_xZn_{1-x}O$, as shown in Fig. 11.1, which was estimated from the comparison with experimental data, and the spontaneous polarization constant for MgO was obtained from the comparison with the experiment [9]. However, many of

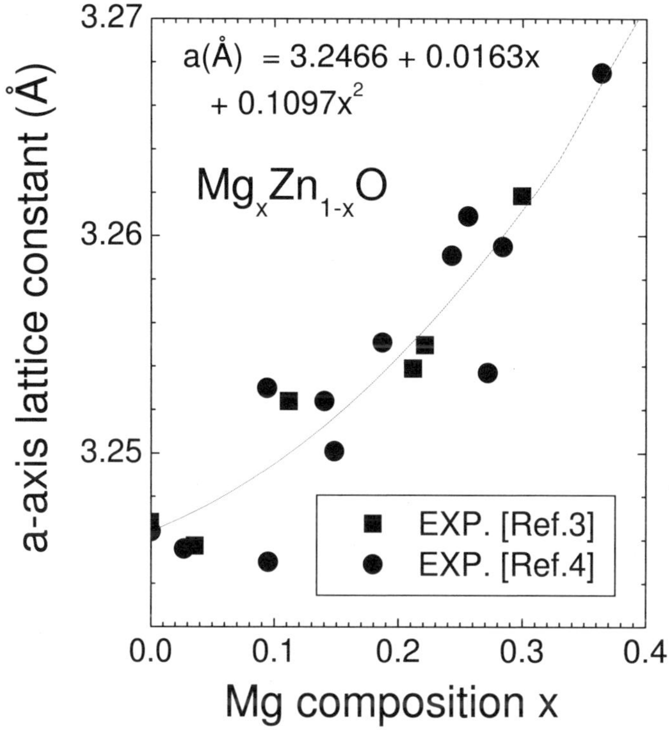

Fig. 11.1. (Mg composition dependence of the a-axis lattice constant of $Mg_xZn_{1-x}O$, which was taken from Ref. [3]. The solid line is the fitted result obtained by using $a = 3.2466+0.0163x + 0.1097x^2$ (©APL 87 2005).

the material parameters for MgO are not well known and we assumed such parameters to be equal to those of ZnO as a first approximation for cases in which published data were lacking because the Mg composition ratio in the MgZnO barrier is relatively small ($x = 0.2$). We used 65/35 as the ratio between the conduction and valence band offsets (ΔE_c /ΔE_v) in the ZnO/ZnMgO heterostructure.

11.3. Results and discussion

Figure 11.2 shows (a) energy shift due to the internal field caused by piezoelectric and spontaneous polarizations for several Mg compositions as a function of well width for ZnO/$Mg_xZn_{1-x}O$ QW structures and (b) comparison between the calculated exciton energy and the experimental data

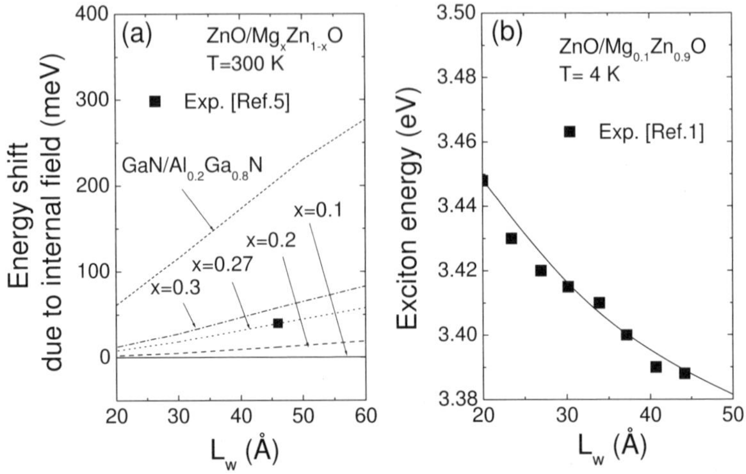

Fig. 11.2. (a) Energy shift due to the internal field caused by piezoelectric and spontaneous polarizations for several Mg compositions as a function of well width for $ZnO/Mg_xZn_{1-x}O$ QW structures and (b) comparison between the calculated exciton energy and the experimental data [1,5] at T= 4 K for $ZnO/Mg_{0.1}Zn_{0.9}O$ QW structures. Here, the barrier width is set to be 50 Å. For comparison, we plotted the result for $GaN/Al_{0.2}Ga_{0.8}N$ QW structure (©APL 87 2005).

given from Ref. [1] at T= 4 K for $ZnO/Mg_{0.1}Zn_{0.9}O$ QW structures. Here, the barrier width is set to be 50 Å. For comparison, we plotted the result for $GaN/Al_{0.2}Ga_{0.8}N$ QW structure. Recently, Makino *et al.* reported that the energy shift due to piezoelectric and spontaneous polarizations is about 40 meV for wurtzite $ZnO/Mg_{0.27}Zn_{0.73}O$ QWs with the well width of L_w =46.5 Å [5]. The spontaneous polarization constant for MgO estimated from a comparison with the experiment is about -0.070 C/m^2, which is larger than value (-0.050 C/m^2) for ZnO. The variation of the spontaneous polarization constant for MgO was also investigated as a function of the ratio of the conduction offset to valence offset to see how the best fitting parameter with the experiment changes. However, the variation of the spontaneous polarization constant was found to be very small for the ratio range from 0.4 to 0.9. In the case of ZnO/MgZnO QW structures with relatively low Mg composition ($x < 0.2$), the energy shift due to the internal field caused by piezoelectric and spontaneous polarizations is negligible in the investigated range of the well width. This can be explained by the fact that the sum of piezoelectric and spontaneous polarizations in the well is canceled by that in the barrier, as discussed below. How-

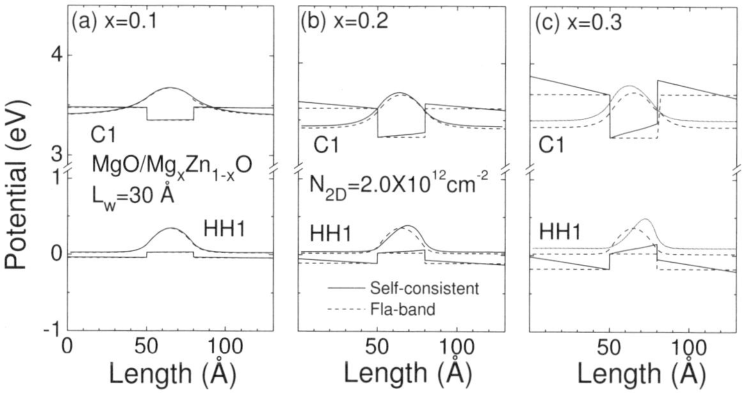

Fig. 11.3. (a) Energy shift due to the internal field caused by piezoelectric and spontaneous polarizations for several Mg compositions as a function of well width for ZnO/Mg$_x$Zn$_{1-x}$O QW structures and (b) comparison between the calculated exciton energy and the experimental data [1] at T= 4 K for ZnO/Mg$_{0.1}$Zn$_{0.9}$O QW structures. Here, the barrier width is set to be 50 Å. For comparison, we plotted the result for GaN/Al$_{0.2}$Ga$_{0.8}$N QW structure (©APL 87 2005).

ever, with increasing Mg composition, the large energy shift is observed for ZnO/MgZnO QW structures with relatively thick well width ($L_w > 40$ Å). We know that GaN/AlGaN QW structure shows significantly larger energy shift than ZnO/MgZnO QW structures due to the large internal field. Thus, we expect that important properties such as the optical gain can be enhanced by using ZnO/MgZnO QW system. The exciton energy in the QW is calculated by $E_{ex} = E_g + \Delta E_c + \Delta E_v - E_b$ where E_g is the band gap energy of ZnO, ΔE_c and ΔE_v are the subband energies of electrons and holes, respectively, E_b is the exciton binding energy in QWs. The calculated exciton energies at T= 4 K are in good agreement with experimental results. Here, the exciton binding energy of 65 meV was used in the calculation.

Figure 11.3 shows the potential profiles and the wave functions (C1 and HH1) at the zone center for the SC model with piezoelectric and spontaneous polarizations for 30-Å ZnO/Mg$_x$Zn$_{1-x}$O QW structures with (a) $x = 0.1$, (b) $x = 0.2$, and (c) = 0.3. Here, the barrier width is set to be 50 Å. The SC solutions are obtained at a sheet carrier density of $N_{2D} = 2 \times 10^{12}$ cm^{-2}, and the dashed line corresponds to the potential profiles for the flat-band model without the spontaneous and the piezoelectric polarizations. The MgZnO barrier is under a compressive strain and we assume that the layers have a zinc plane. For the Zn face, the spontaneous polarization is directed towards the substrate, and the alignment of the piezoelectric and the spontaneous polarizations is antiparallel in the case of compressive strain. If the barrier is under tensile strain, the alignment of the piezoelectric and spontaneous polarizations becomes parallel. In the case of ZnO/MgZnO QW structures with a relatively low Mg composition ($x = 0.1$), the potential profile change due to the internal field caused by the piezoelectric and the spontaneous polarizations is shown to be negligible. Thus, the result of the SC model is nearly the same as that of the flat-band model. On the other hand, in the case of ZnO/MgZnO QW structures with a relatively high Mg composition ($x = 0.3$), the potential profiles show a large internal field exists in the well due to the difference in the spontaneous polarization between the well and the barrier, although the piezoelectric field in the QW is zero. The estimated field in the well is -0.53 MV/cm for a (0001)-oriented ZnO/MgZnO QW with $L_w = 30$ Å. However, this value is relatively smaller than that (~ 1.55 MV/cm) of a (0001)-oriented GaN/AlGaN QW structure [8]. The QW structure shows a large spatial separation of the electron and the hole wave functions. Also, the ground-state transition energy, C1-HH1, is redshifted compared to the flat-band model due to the quantum-confined Stark effect. Thus, we expect important properties such as the effective mass and the optical moment matrix elements to be greatly affected by the Mg composition.

Figure 11.4 shows the interband transition energy between E_{C1} and E_{HH1} at $k_{||} = 0$ for several well widths as a function of Mg composition for ZnO/$Zn_{1-x}Mg_xO$ QW structures. For comparison, results for (0001)-oriented wurtzite GaN/$Al_xGa_{1-x}N$ are also plotted. Here, the barrier width is set to be 70 Å. The transition energy is shown to increase with Mg composition ratio. This is mainly attributed to the fact that the conduction band edge is shifted upward due to the increase in compressive strain with increasing Mg composition. The thin wells show a slightly faster change in transition energy with composition ratio than the thick wells. On the other hand, the transition energy of the GaN/AlGaN structure with a relatively large well width rapidly decreases with increasing Al composition due to the increase in internal field. In the case of a well width of L_w=30 Å, both lasers show nearly the same transition energies for composition ratios near $x = 0.16$. Here, we compare results of QW structures with x=0.20 and L_w=30 Å.

Figure 11.5 shows the valence-band structures for the SC model with piezoelectric and spontaneous polarizations for 30-Å ZnO/$Mg_xZn_{1-x}O$ QW structures with (a) $x = 0.1$, (b) $x = 0.2$, and (c) $x = 0.3$. The SC solutions

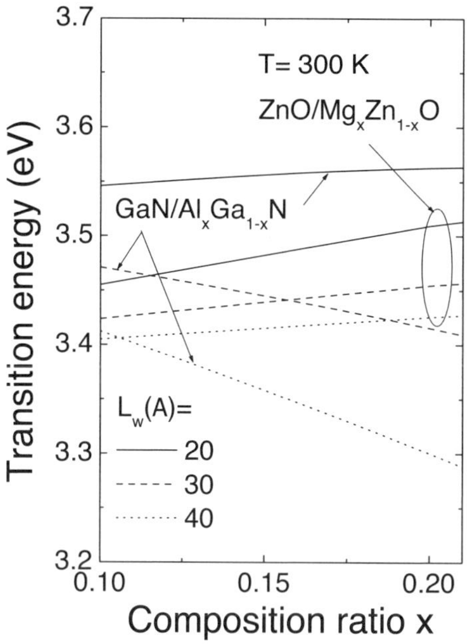

Fig. 11.4. Interband transition energy between E_{C1} and E_{HH1} at $k_{||} = 0$ for several well widths as a function of Mg composition ratio for $ZnO/Zn_{1-x}Mg_xO$ QW structures. For comparison, results for (0001)-oriented wurtzite $GaN/Al_xGa_{1-x}N$ are also plotted (©JJAP 44 2005).

are obtained at a sheet carrier density of $N_{2D} = 2 \times 10^{12}$ cm^{-2}. We know that the subband energies lie closer to each other than those in the more widely studied GaAs- and InP-based QW lasers. For example, the subband energy spacing (~10 meV) between the first two subbands is similar to that of ZnO-based QWs and is very small compared to that (~70 meV) of InP-based QWs. This is due to the heavy effective masses of MgO-based QWs. In the case of ZnO/MgZnO QW structures with a high Mg composition ($x = 0.3$), the valence-band structures show a larger energy spacing between the first two subbands (HH1 and LH1) and higher subbands (HH2 and LH2). In general, the increase in the subband energy spacing reduces the carrier population in the higher subbands. However, this effect will be compensated by the decrease in the optical matrix element due to the spatial separation of the wave functions. Also, the heavy-hole effective masses around the topmost valence band of ZnO-based QWs are much larger than those ($\sim 0.2m_0$) of InP- or GaAs-based QWs. Here, to estimate the magnitude of the hole effective mass, we considered a parabolic band fitted to the

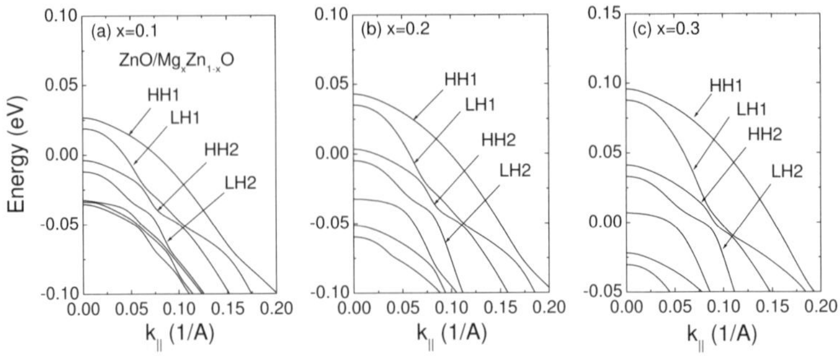

Fig. 11.5. Valence-band structures for the self-consistent (SC) model with piezoelectric and spontaneous polarizations for 30-Å $ZnO/Mg_xZn_{1-x}O$ QW structures with (a) $x = 0.1$, (b) $x = 0.2$, and (c) $x = 0.3$. The SC solutions are obtained at sheet carrier density of $N_{2D} = 2 \times 10^{12}$ cm^{-2}. (©JKPS 50 2007)

topmost valence subband of the exact band structure. The effective mass is determined so that, for a given carrier density and the quasi-Fermi level for holes, the carrier density and the quasi-Fermi level agree with those of the exact band structure. Hence, the effective mass of the fitted parabolic band reflects an average density of states. For example, the heavy-hole effective masses of ZnO/MgZnO QW structures with $x = 0.1$, 0.2, and 0.3 are $0.88m_0, 0.86m_0$, and $0.85m_0$, respectively. However, these values are smaller than those ($\sim 1.50m_0$) of GaN-based QW structures.

Figure 11.6 shows (a) the C1-HH1 transition wavelength as a function of the sheet carrier density and (b) the optical matrix elements as a function of $k_{||}$ for 30-Å $ZnO/Mg_xZn_{1-x}O$ QW structures with several Mg compositions. In the case of ZnO/MgZnO QW structures with relatively low Mg compositions ($x \leq 0.2$), the transition wavelength is nearly independent of the carrier density. This can be explained by the fact that the internal fields due to the piezoelectric and the spontaneous polarizations in the well are negligible, as discussed above. On the other hand, in the case of ZnO/MgZnO QW structures with a relatively high Mg composition ($x = 0.3$), the transition wavelength is gradually shifted to a short wavelength because the internal field is screened by charged carriers with increasing carrier density. The matrix element is nearly constant in the range of $k_{||} < 0.15$ while it rapidly decreases when the in-plane wave vector exceeds 0.15. Also, the ZnO/MgZnO QW structure with a composition of $x = 0.3$ is found to have a smaller optical matrix element than that with a low Mg composition ($x = 0.1$ or 0.2). For example, we find that the normalized optical matrix elements at the zone center are 0.22 and 0.17 for

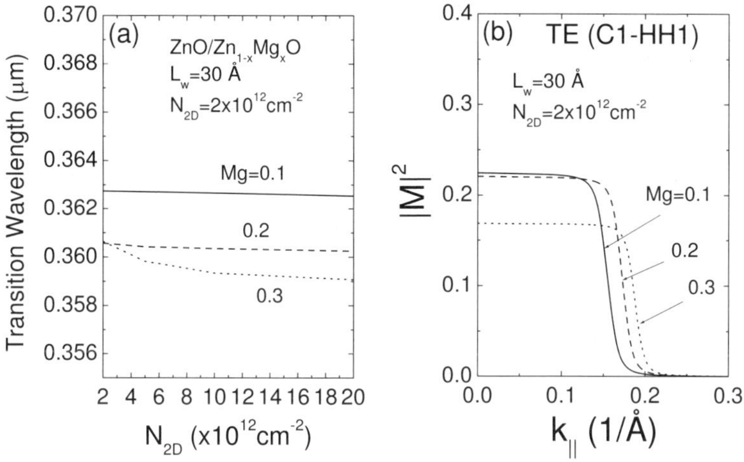

Fig. 11.6. (a) C1-HH1 transition wavelength as a function of the sheet carrier density and (b) optical matrix elements as a function of $k_{\|}$ for 30-Å ZnO/Mg$_x$Zn$_{1-x}$O QW structures with several Mg compositions (©JKPS 50 2007).

$x = 0.2$ and $x = 0.3$, respectively, This is attributed to the increase in the spatial separation between the electron and the hole wave functions due to the large internal field.

Figure 11.7 shows (a) the TE optical gain with many-body effects as a function of the wavelength for several Mg compositions and (b) the quasi-Fermi level separation, $\Delta E_{fc}+\Delta E_{fv}$, as a function of the Mg composition for 30-Å ZnO/Mg$_x$Zn$_{1-x}$O QW structures. Here, the optical gain and the quasi-Fermi level separation are calculated at a carrier density of 20×10^{12} cm^{-2}. The optical gain spectra have only one peak, which corresponds to the C1-HH1 transition. The peak wavelength is shifted to shorter wavelength with increasing Mg composition. The ZnO/MgZnO QW structure with a relatively high Mg composition ($x = 0.3$) is found to have a larger optical gain than that with a relatively low Mg composition. This can be explained by the fact that the quasi-Fermi-level separation is improved with the inclusion of Mg. Here, the quasi-Fermi-level separation ΔE_{fc} (ΔE_{fv}) is defined as the energy difference between the quasi-Fermi level and the ground-state energy in the conduction band (the valence band). We know that the improvement of the total quasi-Fermi-level separation ($\Delta E_{fc} + \Delta E_{fv}$) is mainly attributed to the increase in the quasi-Fermi-level separation ΔE_{fc} in the conduction band. In the case of the valence band, the quasi-Fermi-level separation ΔE_{fv} is nearly independent of the Mg compo-

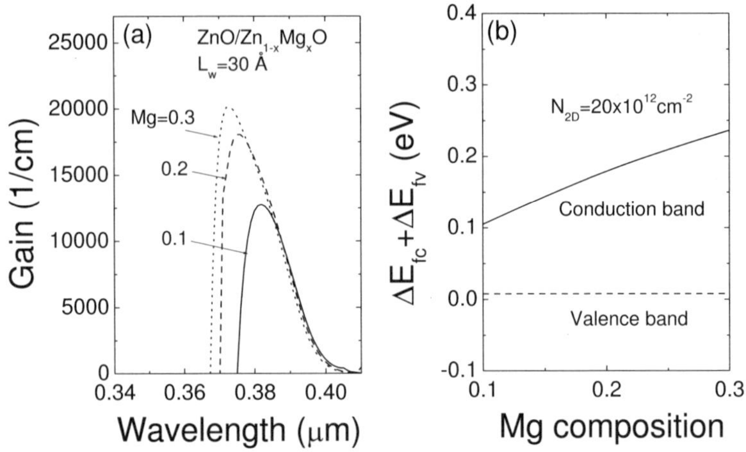

Fig. 11.7. (a) TE optical gain with many-body effects as a function of the wavelength for several Mg compositions and (b) quasi-Fermi-level separation $\Delta E_{fc}+\Delta E_{fv}$ as a function of the Mg composition for 30-Å ZnO/$Mg_xZn_{1-x}O$ QW structures (©JKPS 50 2007).

sition. The increase in ΔE_{fc} is due to the fact that the QW structure with a high Mg composition has a larger energy spacing in the conduction band.

Figure 11.8 shows exciton binding energies as functions of the sheet carrier density for ZnO/$Zn_{0.8}Mg_{0.2}O$ and GaN/$Al_{0.2}Ga_{0.8}N$ QW structures with $L_w = 30$ Å. For comparison, results (dashed line) for the flat-band model without the internal field due to piezoelectric and spontaneous polarizations are also plotted. With increasing sheet carrier density, both QW structures show an the exciton binding energy that is significantly reduced, suggesting that excitons are nearly bleached at typical densities (approximately 10^{13} cm^{-2}) for which lasing occurs. That is, excitons in ZnO-based or GaN-based QWs are not stable at high carrier densities. In particular, the many-body bleaching of the exciton occurs at densities for which the screening of the internal field is still not effective. The flat-band model of GaN/AlGaN QW structures shows that the exciton binding energy is much larger than that of the self-consistent model at low carrier densities. On the other hand, in the case of ZnO/MgZnO QW structures, the difference in the exciton binding energies between the flat-band and the self-consistent models is small compared to that of GaN/AlGaN QW structures. The difference in the exciton binding energy between the flat-band and self-consistent models gradually decreases with increasing carrier density. The exciton binding energy of ZnO/MgZnO QW structures is shown to be much larger than that of GaN/AlGaN QW structures.

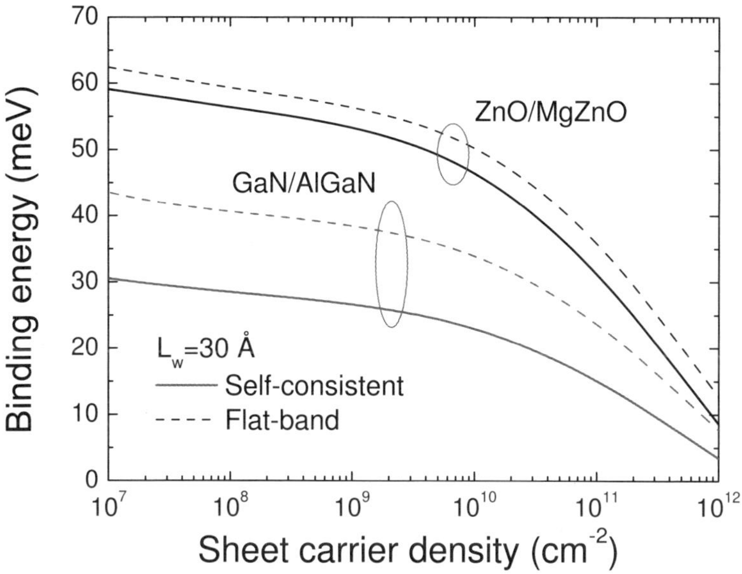

Fig. 11.8. Exciton binding energies as functions of the sheet carrier density for $ZnO/Zn_{0.8}Mg_{0.2}O$ and $GaN/Al_{0.2}Ga_{0.8}N$ QW structures with $L_w = 30$ Å. For comparison, results (dashed line) for the flat-band model without the internal field due to piezoelectric and spontaneous polarizations are also plotted (©JKPS 51 2007).

Figure 11.9 shows the exciton binding energy as a function of the well width for $ZnO/Zn_{0.8}Mg_{0.2}O$ and $GaN/Al_{0.2}Ga_{0.8}N$ QW structures with $L_w = 30$ Å. For comparison, results (dashed line) for the flat-band model without the internal field due to piezoelectric and spontaneous polarizations of both QW structures are also plotted. The self-consistent results are calculated at a sheet carrier density of 1×10^9 cm^{-2}. The flat-band model of the GaN/AlGaN QW structure shows that, with decreasing well width, the exciton binding energy increases due to the increasing confinement effect in the well. However, we know that a further decrease in well width results in a decrease in exciton binding energy. A similar tendency is also observed in the case of the ZnO/MgZnO QW structure. This decrease is due to the envelope function spreading into the barrier region and the three-dimensional character of the exciton being restored. The self-consistent model of both QW structures shows that, with the inclusion of the internal field, the exciton binding energy is substantially reduced compared to that of the flat-band value. The existence of the internal field causes a change

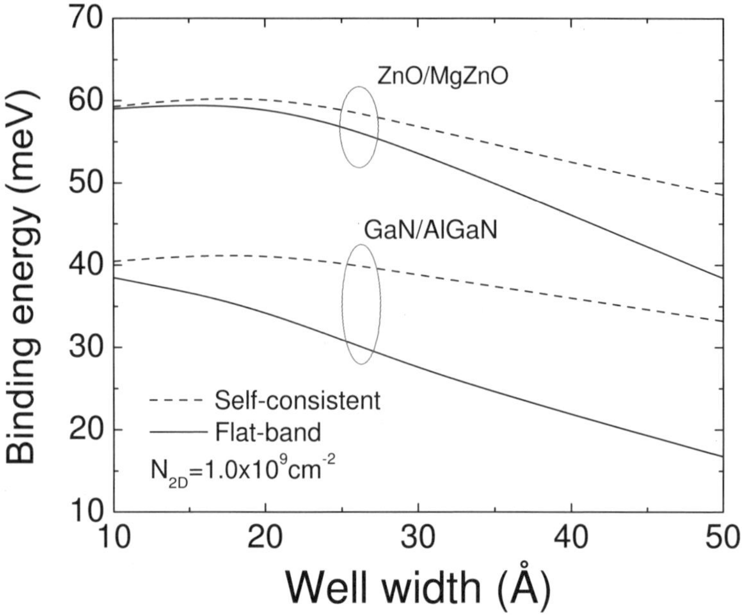

Fig. 11.9. Exciton binding energies as functions of the sheet carrier density for $ZnO/Zn_{0.8}Mg_{0.2}O$ and $GaN/Al_{0.2}Ga_{0.8}N$ QW structures with $L_w = 30$ Å. For comparison, results (dashed line) for the flat-band model without the internal field due to piezoelectric and spontaneous polarizations are also plotted (©JKPS 51 2007).

in the shape of the well width dependence of the exciton binding energy with respect to the flat-band model. That is, the exciton binding energy for the self-consistent model is very sensitive to well width while that for the flat-band model is relatively insensitive. For the very narrow QW, no significant variations due to the internal field were observed in the exciton binding energy because the reduction in wave function overlap due to the internal field was relatively less sensitive compared with that for a wider QW. However, in the case of the ZnO/MgZnO QW structure, we know that the decreasing effect of the exciton binding energy due to the internal field is largely reduced compared to the GaN/AlGaN QW structure. This is mainly due to the fact that the internal field effect in the ZnO well is largely reduced due to the cancellation of the sum of piezoelectric and spontaneous polarizations between the well and the barrier [9].

Figure 11.10 shows the momentum matrix elements $|M_{nm}|^2$ as functions of the well width of $ZnO/Zn_{0.8}Mg_{0.2}O$ and $GaN/Al_{0.2}Ga_{0.8}N$ QW structures with $L_w = 30$ Å. For comparison, results (dashed line) for the

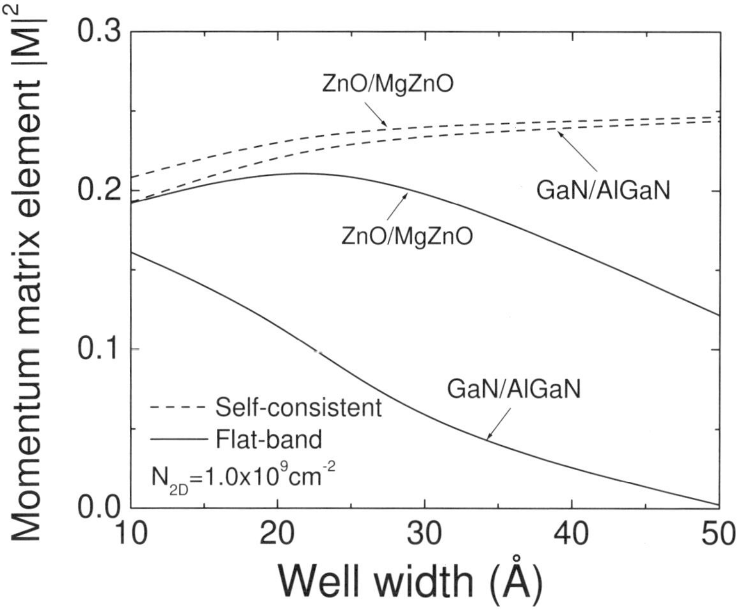

Fig. 11.10. Momentum matrix elements $|M_{nm}|^2$ as functions of the well width for $ZnO/Zn_{0.8}Mg_{0.2}O$ and $GaN/Al_{0.2}Ga_{0.8}N$ QW structures with $L_w = 30$ Å. For comparison, results (dashed line) for the flat-band model without the internal field due to piezoelectric and spontaneous polarizations of both QW structures are also plotted. The results for the self-consistent model (solid curve) are calculated at a sheet carrier density of 1×10^9 cm^{-2} (©JKPS 51 2007).

flat-band model without the internal field due to piezoelectric and spontaneous polarizations of both QW structures are also plotted. The results for the self-consistent model (solid curve) have been calculated at a sheet carrier density of 1×10^9 cm^{-2}. The matrix element for the flat-band model (dashed curve) is a slow function of the well width, though it shows a slight increase in the region of narrow wells. However, the matrix element for the self-consistent model involving the polarization effects largely decreases as the well width increases. This results from the smaller overlaps due to the larger spatial separation between the conduction and the valence wave functions with increasing well width. In the case of the ZnO-based QW structure, the decrease in the matrix element with the well width is observed to be significantly reduced because the internal field in the ZnO-based QW structure is relatively smaller than that in the GaN/AlGaN QW structure. Hence, the larger exciton binding energy observed in ZnO/MgZnO QW structure can be explained by the matrix element being larger than it is for the GaN/AlGaN QW structure. That is also partially attributed to the fact that the ZnO/MgZnO QW structure has a smaller dielectric constant ($8.1\epsilon_o$) than that ($10.0\epsilon_o$) of the GaN/AlGaN QW structure.

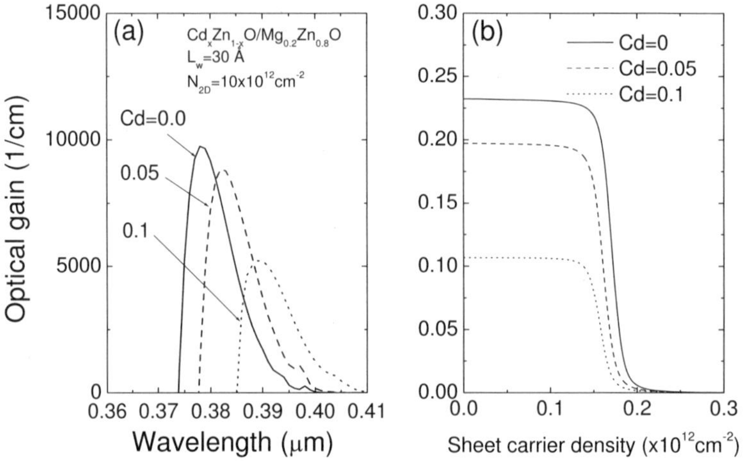

Fig. 11.11. (a) TE optical gain spectra as a function of the wavelength and (b) optical matrix elements as a function of the in-plane wave vector $k_{||}$ for several Cd compositions and (b) peak gain as a function of the sheet carrier density for the 30 Å $Cd_xZn_{1-x}O/Mg_{0.2}Zn_{0.8}O$ QW structures. The results are obtained at the sheet carrier density of $N_{2D} = 10 \times 10^{12}$ cm^{-2}.

Figure 11.11 shows (a) TE optical gain spectra as a function of the wavelength and (b) optical matrix elements as a function of the in-plane wave vector $k_{||}$ for several Cd compositions and (b) peak gain as a function of the sheet carrier density of 30 Å $Cd_xZn_{1-x}O/Mg_{0.2}Zn_{0.8}O$ QW structures. The results are obtained at the sheet carrier density of $N_{2D} = 10 \times 10^{12}$ cm^{-2}. For CdO, we used those given by Gopal and Spaldin [34] except for the band gap. We used 2.2 eV as the band gap energy for CdO [38]. The spontaneous polarization constant for CdO used in the calculation is -0.099 C/m^2 [34]. The optical gain spectra have only one peak which corresponds to C1-HH1 transition. The peak wavelength is shifted to longer wavelength with increasing Cd composition. The CdZnO/MgZnO QW structure with relatively high Cd composition is found to have smaller optical gain than that with relatively low Cd composition. The matrix element is nearly constant in a range of $k_{||} < 0.15$ while it rapidly decreases when the in-plane wave vector exceeds 0.15. Also, the matrix element of the CdZnO/MgZnO QW structures largely decreases with increasing Cd composition. This is attributed to the increase of the spatial separation between the electron and hole wave functions due to the large internal field. Thus, the decrease of the optical gain observed for a large Cd composition can be explained by the fact that the internal field of piezoelectric and spontaneous polarizations in the well increases with increasing Cd composition.

11.4. Summary

In summary, optical properties of ZnO/MgZnO QW structures, considering piezoelectric and spontaneous polarizations, are investigated by using the non-Markovian gain model with many-body effects. The optical matrix element of the ZnO/MgZnO QW structure decreases with the inclusion of Mg. This is attributed to an increase in the spatial separation between the electron and the hole wave functions due to the large internal field. However, the ZnO/MgZnO QW structure with a relatively high Mg composition ($x = 0.3$) is found to have a larger optical gain than that with a relatively low Mg composition. This can be explained by the fact that the quasi-Fermi-level separation ΔE_{fc} in the conduction band is improved with the inclusion of Mg. On the other hand, in the case of the valence band, the quasi-Fermi-level separation ΔE_{fv} is nearly independent of the Mg composition. The increase in ΔE_{fc} is because the QW structure with a high Mg composition has a larger energy spacing in the conduction band. With increasing sheet carrier density, both ZnO/MgZnO and GaN/AlGaN QW structures show a significantly reduced exciton binding energy. We also know that the exciton binding energy of ZnO/MgZnO QW structures is much larger than that of GaN/AlGaN QW structures. This can be explained by the fact that ZnO/MgZnO QW structures have a larger matrix element than the GaN/AlGaN QW structures and by the smaller dielectric constant. The CdZnO/MgZnO QW structure with relatively high Cd composition is found to have smaller optical gain than that with relatively low Cd composition.

Acknowledgments

This research was supported partially by Basic Science Research Program through the National Research Foundation of Korea(NRF) funded by the Ministry of Education, Science and Technology (2009-0071446) and the KOSEF grant funded by the Korean Government (MEST)(NO.R17-2008-024-01000-0).

References

1. B. P. Zhang, N. T. Binh, K. Wakatsuki, C. Y. Liu, Y. Segawa and N. Usami, Growth of ZnO/MgZnO quantum wells on sapphire substrates and observation of the two-dimensional confinement effect, *Appl. Phys. Lett.* **86**, 032105/1-032105/3 (2005).
2. T. Makino, C. H. Chia, Nguen T. Tuan, H. D. Sun, Y. Segawa, M. Kawasaki, A. Ohtomo, K. Tamura and H. Koinuma, Room-temperature luminescence of excitons in ZnO/(MgZn)O multiple quantum wells on lattice-matched substrates, *Appl. Phys. Lett.* **77**, 975–977 (2000).
3. T. Makino, Y. Segawa, M. Kawasaki, A. Ohtomo, R. Shiroki, K. Tamura, T. Yasuda and H. Koinuma, Band gap engineering based on $Mg_xZn_{1-x}O$ and $Cd_yZn_{1-y}O$ ternary alloy films, *Appl. Phys. Lett.* **78**, 1237–1239 (2001).
4. A. Ohtomo and A. Tsukazaki, Pulsed laser deposition of thin films and superlattices based on ZnO, *Semicond. Sci. Technol.* **20**, S1-S12 (2005).
5. T. Makino, K. Tamura, C. H. Chia and Y. Segawa, M. Kawasaki, A. Ohtomo and H. Koinuma, Radiative recombination of electron-hole pairs spatially separated due to quantum-confined Stark and Franz-Keldish effects in $ZnO/Mg_{0.27}Zn_{0.73}O$ quantum wells, *Appl. Phys. Lett.* **81**, 2355–2357 (2002).
6. P. Yu, Z. K. Tang, G. K. L. Wong, M. Kawasaki, A. Ohtomo, H. Koinuma and Y. Segawa, Ultraviolet spontaneous and stimulated emissions from ZnO microcrystallite thin films at room temperature, *Solid State Commun.* **103**, 459–463 (1997).
7. Th. Gruber, C. Kirchner, R. Kling, F. Reuss and A. Waag, ZnMgO epilayers and ZnO-ZnMgO quantum wells for optoelectronic applications in the blue and UVspectral region, *Appl. Phys. Lett.* **84**, 5359–5361 (2004).
8. S.-H. Park and S.-L. Chuang, Spontaneous polarization effects in wurtzite GaN/AlGaN quantum wells and comparison with experiment, *Appl. Phys. Lett.* **76**, 1981–1983 (2000).
9. S.-H. Park and D. Ahn, Spontaneous and piezoelectric polarization effects in wurtzite ZnO/MgZnO quantum lasers, *Appl. Phys. Lett.* **87**, 253509.1-253509/3 (2005).
10. S.-H. Park, K. J. Kim, S. Y. Yi, D. Ahn and S. J. Lee, Optical gain in wurtzite ZnO/ZnMgO quantum well lasers *Jpn. J. Appl. Phys.* **46**, L1403-L1406 (2005).
11. S.-H. Park, K. J. Kim, S. Y. Yi, D. Ahn and S. J. Lee, ZnO/ZnMgO quantum well lasers for optoelectronic applications in the blue and the UV spectral regions, *J. Korean Phys. Soc.* **47**, 448–453 (2005).
12. S. H. Park, Electronic and optical properties of ZnO/ZnMgO quantum well lasers with piezoelectric and spontaneous polarization, *J. Korean Phys. Soc.* **50**, 16–20 (2007).
13. S. L. Chuang and C. S. Chang, K·p method for strained wurtzite semiconductors, it Phys. Rev. B **54**, 2491–2504 (1996).
14. S. H. Park and S. L. Chuang, Crystal-orientation effects on the piezoelectric field and electronic properties of strained wurtzite semiconductors, *Phys. Rev. B* **59**, 4725–4737 (1999).

15. A. Bykhovski, B. Gelmont and S. Shur, Strain and charge distribution in GaN-AIN-GaN semiconductor-insulator-semiconductor structure for arbitrary growth orientation, *Appl. Phys. Lett.* **63** 2243–2245 (1993).
16. S. H. Park and S. L. Chuang, Comparison of zinc-blends and wurtzite GaN semiconductors with spontaneous polarization and piezoelectric field effects, *J. Appl. Phys.* **87**, 353–364 (2000).
17. G. Martin, A. Botchkarev, A. Rockett and H. Morkoç, Valence-band discontinuities of wurtzite GaN, AIN, and InN heterojunctions measured by x-ray photoemission spectroscopy, *Appl. Phys. Lett.* **68**, 2541–2543 (1996).
18. S. H. Park and S. L. Chuang, Piezoelectric effects on electrical and optical properties of wurtzite GaN/AIGaN quantum well lasers, *Appl. Phys. Lett.* **72**, 3103–3105 (1998).
19. S. L. Chuang, *Physics of Optoelectronic Devices* (Wiley, New York, 1995), Chap. 4.
20. D. Ahn, Theory of non-Markovian optical gain in quantum-well lasers, *Prog. Quantum Electron.* **21**, 249–287 (1997).
21. S. H. Park, S. L. Chuang and D. Ahn, Intraband relaxation time effects on non-Markovian gain with many-body effects and comparison with experiment, *Semicond. Sci. Technol.* **15**, 203–208 (2000).
22. W. W. Chow, S. W. Koch and M. Sargent III, *Semiconductor-Laser Physics* (Springer, Berlin, 1994), Chap. 4.
23. M. Asada, Intraband relaxation effect on optical spectra, in *Quantum Well Lasers*, edited by P. S. Zory, Jr. (Academic, San Diego, CA, 1993), pp. 97–130.
24. S. J. Pearton, C. R. Abernathy, M. E. Overberg, G. T. Thaler, D. P. Norton, Y. D. Park, F. Ren, J. Kim and L. A. Boatner, Wide band gap ferromagnetic semiconductors and oxides, *J. Appl. Phys.* **93**, 1–13 (2003).
25. S. Adachi, Ed., *Handbook on Physical Properties of Semiconductors*, Vol. 3 (Kluwer Academic, Boston, 2004), p. 65.
26. A. Ohtomo, M. Kawasaki, T. Koida, K. Masubuchi, H. Koinuma, Y. Sakurai, Y. Yoshida, T. Yasuda and Y. Segawa, $Mg_xZn_{1-x}O$ as a II-VI widegap semiconductor alloy, *Appl. Phys. Lett.* **72**, 2466 (1998).
27. P. Gil, Oscillator strengths of A, B, and excitons in ZnO films, *Phys. Rev. B* **64**, 201310-1/-3 (2001).
28. G. Coli and K. K. Bajaj, Excitonic transitions in ZnO/MgZnO quantum well, Appl. Phys. Lett. **78**, 2861–2863 (2001).
29. W. R. L. Lambrecht, A. V. Rodina, S. Limpijumnong, B. Segall and B. K. Meyer, Valence-band ordering and magneto-optic exciton fine structure in ZnO, *Phys. Rev. B* **65**, 075207-1/-12 (2002).
30. T. Dietl, H. Ohno and F. Matsukura, Hole-mediated ferromagnetism in tetrahedrally coordinated semiconductors, *Phys. Rev. B* **63**, 195205-1/-21 (2001).
31. W. Shan, W. Walukiewicz, J. W. Ager III, K. M. Yu, Y. Zhang, S. S. Mao, R. Kling, C. Kirchner and A. Waag, Pressure dependent photoluminescence study of ZnO nanowires, *Appl. Phys. Lett.* **86**, 153117-1/-3 (2005).
32. J. Wrzesinski and D. Fröhlich, Two-photon and three-photon spectroscopy of ZnO under uniaxial stress, *Phys. Rev. B* **56**, 13087–13093 (1997).

33. B. A. Auld, *Acoustic Fields and Waves in Solids* (Krieger, Malabar, Florida, 1990), p. 378.
34. P. Gopal and N. A. Spaldin, Polarization, piezoelectric constants and elastic constants of ZnO, MgO and CdO, pp 1–8 (available on-line, cond-mat/0507217).
35. M. P. Verma and S. K. Agrawal, Three-body-force shell model and the lattice dynamics of magnesium oxide, *Phys. Rev. B* **8**, 4880–4884 (1973).
36. T. Makino, T. Yasuda, Y. Segawa, A. Ohtomo, K. Tamura, M. Kawasaki and H. Koinuma, Strain effects on exciton resonance energies of ZnO epitaxial layers, *Appl. Phys. Lett.* **79**, 1282–1284 (2001).
37. A. D. Corso, M. Posternak, R. Resta and A. Baldereschi, *Ab initio* study of piezoelectricity and spontaneous polarization in ZnO, *Phys. Rev. B* **50**, 10715–10721 (1994).
38. T. Makino, Y. Segawa, M. Kawasaki and H Koinuma, Optical properties of excitons in ZnO-based quantum well heterostructures, *Semicond. Sci. Technol.* **20**, 578–591 (2005).

Chapter 12

Tailoring Electronic and Optical Properties of TiO_2: Nanostructuring, Doping and Molecular-Oxide Interactions

Letizia Chiodo*, Juan Maria García-Lastra†, Duncan John Mowbray and Angel Rubio‡,§

Universidad del País Vasco

§*Friz-Haber-Institut der Max-Planck-Gesellschaft*

**lchiodo78@gmail.com;*

letizia.chiodo@ehu.es

†*juanmaria.gracia@ehu.es*

‡*arubio@ehu.es*

Amilcare Iacomino

Università degli Studi Roma Tre

amilcaraiacomino@gmail.com

Titanium dioxide is one of the most widely investigated oxides. This is due to its broad range of applications, from catalysis to photocatalysis to photovoltaics. Despite this large interest, many of its bulk properties have been sparsely investigated using either experimental techniques or *ab initio* theory. Further, some of TiO_2's most important properties, such as its electronic band gap, the localized character of excitons, and the localized nature of states induced by oxygen vacancies, are still under debate. We present a unified description of the properties of rutile and anatase phases, obtained from *ab initio* state of the art methods, ranging from density functional theory (DFT) to many body perturbation theory (MBPT) derived techniques. In so doing, we show how advanced computational techniques can be used to quantitatively describe the structural, electronic, and optical properties of TiO_2 nanostructures, an area of fundamental importance in applied research. Indeed, we address one of the main challenges to TiO_2-photocatalysis, namely band gap narrowing, by showing how to combine nanostructural changes with doping. With this aim we compare TiO_2's electronic properties for 0D clusters, 1D nanorods, 2D layers, and 3D bulks using different approximations within DFT and MBPT calculations. While quantum confinement effects lead to a widening of the energy gap, it has been shown that substitutional doping with boron or nitrogen gives rise to (meta-)stable structures and the introduction of dopant and mid-gap states which effectively reduce the band gap. Finally, we report how *ab initio* methods can be applied to understand the important role of TiO_2 as electron-acceptor in dye-sensitized solar cells. This task is made more difficult by the hybrid organic-oxide structure of the involved systems.

12.1. Introduction

TiO_2 has been one of the most studied oxides over the past few years. This is due to the broad range of applications it offers in strategic fields of scientific, technological, environmental, and commercial relevance. In particular, TiO_2 surfaces and nanocrystals provide a rich variety of suitably tunable properties from structure to opto-electronics. Special attention has been paid to TiO_2's optical properties. This is because TiO_2 is regarded as one of the best candidate materials to efficiently produce hydrogen via photocatalysis [1,2]. TiO_2 nanostructures are also widely used in dye-sensitized solar cells, one of the most promising applications in the field of hybrid solar cells [1].

Since the first experimental formation of hydrogen by photocatalysis in the early 1980s [2], TiO_2 has been the catalyst of choice. Reasons for this include the position of TiO_2's conduction band above the energy of hydrogen formation, the relatively long lifetime of excited electrons which allows them to reach the surface from the bulk, TiO_2's high corrosion resistance compared to other metal oxides, and its relatively low cost [1,3,4]. However, the large optical band gap of bulk TiO_2 (≈ 3 eV) means that only high energy UV light may excite its electrons. This effectively blocks most of the photons which pierce the atmosphere, typically in the visible range, from participating in any bulk TiO_2 based photocatalytic reaction. On the other hand, the difference in energy between excited electrons and holes, i.e. the band gap, must be large enough ($\gtrsim 1.23$ eV) to dissociate water into hydrogen and oxygen. For these reasons it is of great interest to adjust the band gap ε_{gap} of TiO_2 into the range $1.23 \lesssim \varepsilon_{gap} \lesssim 2.5$ eV, while maintaining the useful properties mentioned above [5].

With this aim, much research has been done on the influence of nanostructure [6–11] and dopants [5,11–19] on TiO_2 photocatalytic activity. For low dimensional nanostructured materials, electrons and holes have to travel shorter distances to reach the surface, allowing for a shorter quasi-particle lifetime. However, due to quantum confinement effects, lower dimensional TiO_2 nanostructures tend to have *larger* band gaps [20]. On the other hand, although doping may introduce mid-gap states, recent experimental studies have shown that boron and nitrogen doping of bulk TiO_2 yields band gaps *smaller* than the threshold for water splitting [12,13]. This suggests that low dimensional structures with band gaps larger than about 3.0 eV may be a better starting point for doping.

Recently, several promising new candidate structures have been proposed [11]. These small ($R \lesssim 5$ Å) TiO_2 nanotubes, with a hexagonal ABC PtO_2 structure (HexABC), were found to be surprisingly stable, even in the boron and nitrogen doped forms. This stability may be attributed to their structural similarity to bulk rutile TiO_2, with the smallest nanotube having the same structure as a rutile nanorod.

A further difficulty for any photocatalytic system is controlling how electrons and holes travel through the system [21]. For this reason, methods for reliably producing both n-type and p-type TiO_2 semiconducting materials are highly desirable. So far, doped TiO_2 tends to yield only n-type semiconductors. However, it has recently been proposed that p-type TiO_2 semiconducting materials may be obtained by nitrogen doping surface sites of low dimensional materials [11]. In this chapter we will discuss in detail the effects of quantum confinement and doping on the optical properties of TiO_2.

TiO_2 nanostructures are also one of the main components of hybrid solar cells. In a typical Grätzel cell [1], TiO_2 nanoparticles with average diameters around 20 nm collect the photoelectron transferred from a dye molecule adsorbed on the surface [22,23]. Such processes are favoured by a proper energy level alignment between solid and organic materials, although the dynamic part of the process also plays an important role in the charge transfer. Clearly, TiO_2's characteristics of long quasi-particle lifetimes, high corrosion resistance, and relatively low cost, must be balanced with control of its energy level alignment with molecular states, and a fast electron injection at the interface.

Despite all the engineering efforts, the main scientific goal remains to optimize the efficiency of solar energy conversion into readily available electricity. Different research approaches have been devoted to benefit from quantum size properties emerging at the nanoscale [24,25], find an optimal donor–acceptor complex [24,25], mix nanoparticles and one-dimensional structures, such as nanotubes or nanowires [26,27], and control the geometry of the TiO_2 nano-assembly [28].

A clear theoretical understanding of TiO_2's optoelectronic properties is necessary to help unravel many fundamental questions concerning the experimental results. In particular, the properties of excitons, photo-injected electrons, and surface configuration in TiO_2 nanomaterials may play a critical role in determining their overall behaviour in solar cells. For TiO_2 at both the nanoscale and macroscale regime, it is necessary to first have a complete picture of the optical properties in order to clarify the contribution of excitons.

Despite the clear importance of its surfaces and nanostructures, investigations of TiO_2 bulk (see Fig. 12.1) electronic and optical properties have not provided, so far, a comprehensive description of the material. Important characteristics, such

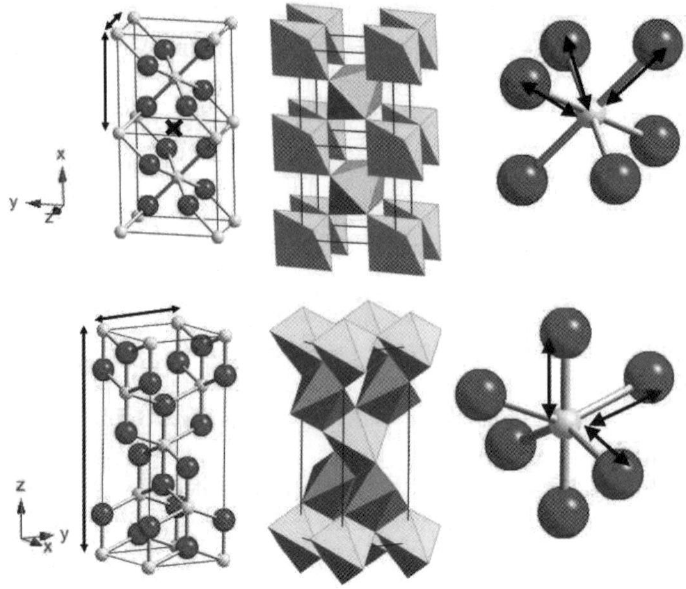

Fig. 12.1. (Color online). Unit cell (left), crystallographic structure (center) and TiO_6 octahedrons (right) of rutile (top) and anatase (bottom). Light blue small dots denote Ti atoms, red large dots denote O atoms. The lattice parameters in Å are denoted a, b, and c, while R_ax and R_eq are the distances in Å between a Ti^{4+} ion and its nearest and next-nearest neighbour O^{2-} ions, respectively. In the case of rutile the interstitial Ti impurity site is shown with a black cross (see left top).

as the electronic band gap, are still undetermined. Most of the experimental and computational work has been focused on synthesis and analysis of systems with reduced dimensionality.

The experimental synthesis and characterization of nanostructured materials is in general a costly and difficult task. However, predictions of a dye or nanostructure's properties from simulations can prove a great boon to experimentalists. Modern large-scale electronic structure calculations have become important tools by providing realistic descriptions and predictions of structure and electronic properties for systems of technological interest.

Although it will not be treated in this chapter, it is important to mention the problem of electron localization in reduced titania [29]. This will provide a glimpse of the complexity faced, from the theoretical point of view, when studying transition metal atoms. The localization of d electrons makes the accurate description of their exchange–correlation interaction a difficult task [30]. The electron localization in defective titania has been an open question from both experimental and theoretical points of view, and caused much controversy during the past few years [31–33]. Oxygen vacancies are quite common in TiO_2, and their presence and behaviour can significantly affect the properties of nanostructures. When an oxygen vacancy is created in TiO_2, i.e. when TiO_2 is reduced, the two electrons coming from the removed O^{2-} ion must be redistributed within the structure. One possibility is that these two extra electrons remain localized onto two Ti ions close to the O^{2-} vacancy. In this way a pair of Ti^{4+} ions become Ti^{3+} ions. Another option is for the two extra electrons to delocalize along the whole structure, i.e. they do not localize on any particular Ti ion. Finally, an intermediate situation, with one electron localized and the other spread, is also possible. Concerning the TiO_2 bulk, conventional density functional theory (DFT) calculations using either local density approximations (LDA) or generalized gradient approximations (GGA) for the exchange-correlation (xc)-functionals show a scenario with both electrons delocalized. On the other hand, hybrid functionals and Hartree-Fock calculations give rise to a situation with both electrons localized. For GGA+U calculations, the results are very sensitive to the value of the U parameter. For certain values of U both electrons remain localized, while for others there is an intermediate situation [34,35].

Experimentally, there are electron paramagnetic resonance (EPR) measurements suggesting that the extra electrons are mainly localized on interstitial Ti^{3+} ions [36]. These interstitial Ti^{3+} ions are impurities placed at the natural interstices of the rutile lattice (see Fig. 12.1) and, similarly to the Ti^{4+} ions of the pure lattice, they also form TiO_6 octahedrons. Recent STM and PES experiments have shown that the interstitial Ti^{3+} ions play a key role in the localization of the electrons when a bridge oxygen is removed from the TiO_2 (110) surface [37]. These experiments concluded the controversial discussion about the localization of electrons in the bridge oxygen defective TiO_2 (110) surface (see Refs. 31–33] for more details). However, the problem remains unresolved for the bulk case.

In summary, in this chapter we first analyze the full *ab initio* treatment of electronic and optical properties in Sec. 12.2 and Sec. 12.3, before applying it to the two most stable bulk phases, rutile and anatase in Sec. 12.4. These are also the phases most easily found when nanostructures are synthesized. We will focus on their optical properties and excitonic behaviour. We then explore the possibility of tuning the oxide band gap using quantum confinement effects and dimensionality, by analyzing atomic clusters, nanowires and nanotubes in Sec. 12.5. A further component whose effect has to be evaluated is that of doping, which may further tune the optical behaviour by introducing electronic states in the gap, as presented in Sec. 12.6. Combining the effects of quantum confinement and doping is hoped to produce a refined properties control. In Sec. 12.7, we report some details on modeling for dye-sensitized solar cells, before providing a summary and our concluding remarks.

12.2. Ground state properties through Density Functional Theory

DFT is a many-body approach, successfully used for many years to calculate the ground-state electronic properties of many-electron systems. However, DFT is by definition a ground state theory, and is not directly applicable to the study of excited states. To describe these types of physical phenomena it becomes necessary to include many-body effects not contained in DFT through Green's function theory. The use of many-body perturbation theory [38], with DFT calculations as a zero order approximation, is an approach widely used to obtain quasi-particle excitation energies and dielectric response in an increasing number of systems, from bulk materials to surfaces and nanostructures. We present a short, general discussion of the theoretical framework, referring the reader to the books and the reviews available in the literature for a complete description (see, for instance, Refs. 39] and [40]), before applying these methods to rutile and anatase TiO_2.

As originally introduced by Hohenberg and Kohn (HK) in 1964 [41], DFT is based on the theorem that the ground state energy of a system of N interacting electrons in an external potential $V_{ext}(\mathbf{r})$ is a unique functional of the ground state electronic density. The Kohn and Sham [42] formulation demonstrates how the the HK Theorem may be used in practice, by self-consistently solving a set of one-electron equations (KS equations),

$$\left[-\frac{1}{2}\nabla^2 + V_{eff}^{KS}[\rho(\mathbf{r})]\right]\phi_i(\mathbf{r}) = \varepsilon_i^{KS}\phi_i(\mathbf{r}), \quad \rho(\mathbf{r}) = \sum_i^N |\phi_i(\mathbf{r})|^2, \quad (12.1)$$

where $\rho(\mathbf{r})$ is the electronic charge density, $\phi_i(\mathbf{r})$ are the non-interacting KS wavefunctions, and $V_{eff}^{KS}[\rho(\mathbf{r})] = V_{ext}(\mathbf{r}) + V_H[\rho(\mathbf{r})] + V_{xc}[\rho(\mathbf{r})]$ is the effective one-electron potential. Here, V_H is the Hartree potential and V_{xc} is the exchange–correlation potential defined in terms of the xc-functional E_{xc} as $V_{xc} = \frac{\delta E_{xc}}{\delta \rho(r)}$, which contains all the many-body effects. V_{xc} is usually calculated in either LDA [43,44] or GGA [45] approximations. However, semi-empirical functionals are also available, called hybrids [46,47], which somewhat correct the deficiencies of LDA and GGA for describing exchange and correlation. This is accomplished by including an exact exchange contribution.

Other than the highest occupied molecular orbital (HOMO), KS DFT electronic levels do not correspond to the electronic energies of the many electron system. Indeed, the calculated KS band gaps of semiconductors and insulators severely underestimate the experimental ones. Experimentally, occupied states are accessible by direct photoemission, where an electron is extracted from the system ($N-1$ ground state), while unoccupied states are accessible by inverse photoemission, where one electron is added to the system ($N+1$ ground state). For isolated systems with a finite number of electrons, the electronic gap may be obtained from the DFT calculated ground-state energies with $N+1$, N, and $N-1$ electrons [48]. However, for periodic systems, adding or removing an electron is a non-trivial task. We therefore summarize in the following a rigorous method for describing excitations based on the Green's function approach. This method allows us to properly describe the electronic band structure. Further information and details about the Green's function approach may be found elsewhere [40,49–51].

12.3. Excited States Within Many Body Perturbation Theory

When a bare particle, such as an electron or hole, enters an interacting system, it perturbs the particles in its vicinity. In essence, the particle is "dressed" by a polarization cloud of the surrounding particles, becoming a so-called quasi-particle. Using this concept, it is possible to describe the system through a set of quasi-particle equations by introducing a non-local, time-dependent, non-Hermitian operator called the self-energy Σ. This operator takes into account the interaction of the particle with the system via

$$\left[-\frac{1}{2}\nabla^2 + V_{ext} + V_H\right]\Psi_i(\mathbf{r},\omega) + \int \Sigma(\mathbf{r},\mathbf{r}',\omega)\Psi_i(\mathbf{r}',\omega)d\mathbf{r}' = E_i(\omega)\Psi_i(\mathbf{r}), \tag{12.2}$$

where $\Psi_i(\mathbf{r})$ is the quasi-particle wavefunction.

Since the operator Σ is non-Hermitian, the energies $E_i(\omega)$ are in general complex, and the imaginary part of Σ is related to the lifetime of the excited particle [52]. The most often used approximation to calculate the self-energy is the so-called GW method. It may be derived as the first-step iterative solution of the Hedin integral equations (see Refs. [38], [53], and [54]), which link the Green's function G, the self-energy Σ, the screened Coulomb potential W, the polarization P and the vertex Γ. In practice Eq. (3.1) is not usually solved directly, since the KS wavefunctions are often very similar to the GW ones [53]. For this reason it is often sufficient to calculate the quasi-particle (QP) corrections within first-order perturbation theory (G_0W_0) [53,54]. Moreover, the energy dependence of the self-energy is accounted for by expanding Σ in a Taylor series, so that the QP energies $\varepsilon_i^{G_0W_0}$ are then given by

$$\varepsilon_i^{G_0W_0} = \varepsilon_i^{KS} + Z_i\langle\phi_i^{KS}|\Sigma(\varepsilon_i^{KS}) - V_{xc}^{KS}|\phi_i^{KS}\rangle, \tag{12.3}$$

where Z_i are the quasi-particle renomalization factors described in Refs. [40] and [49]. For a large number of materials, the G_0W_0 approximation of Σ works quite well at correcting the KS electronic gap from DFT.

Concerning optical properties, the physical quantity to be determined in order to obtain the optical spectra is the macroscopic dielectric function $\epsilon_M(\omega)$. This may be calculated at different levels of accuracy within a theoretical *ab initio* approach. A major component in the interpretation of the optical measurements of reduced dimensional systems are the local-field effects (LFE). These effects are especially important for inhomogeneous systems. Here, even long wavelength external perturbations produce microscopic fluctuations of the electric field, which must be taken into account. However, LFE are also important for bulk phases such as anatase and rutile TiO_2. They must be taken into account in the evaluation of optical absorption, and to calculate the screened interaction W used in GW. The effect becomes increasingly important when going to lower dimensional systems.

It is well known [55] that for inhomogeneous materials $\epsilon_M(\omega)$ is not simply the average of the corresponding microscopic quantity, but is related to the inverse of the microscopic dielectric matrix by

$$\epsilon_M(\omega) = \lim_{\mathbf{q}\to 0} \frac{1}{\epsilon^{-1}_{\mathbf{G}=\mathbf{0},\mathbf{G}'=\mathbf{0}}(\mathbf{q},\omega)}. \tag{12.4}$$

The microscopic dielectric function may be determined within the linear response theory [56], the independent-particle picture by the random phase approximation (RPA), and using eigenvalues and eigenvectors of a one-particle scheme such as DFT or GW. There is also a different formulation which includes LFE in the macroscopic dielectric function. This becomes useful when the electron–hole interaction is included in the polarization function. This formulation allows us to include, via the Bethe–Salpeter equation (BSE), excitonic and LFE on the same footing. In so doing, inverting of the microscopic dielectric matrix is avoided. The complete derivation may be found in Appendix B of Ref. [57].

So far, in RPA, we have treated the quasi-particles as non-interacting. To take into account the electron–hole interaction, a higher order vertex correction needs to be included in the polarization. In other words, the BSE, which describes the electron–hole pair dynamics, needs to be solved. As explained in Ref. [57], the BSE may be written as an eigenvalue problem involving the effective two-particle Hamiltonian

$$\begin{aligned} H^{(n_1,n_2),(n_3,n_4)}_{exc} &= (E_{n_2} - E_{n_1})\delta_{n_1,n_3}\delta_{n_2,n_4} - i(f_{n_2} - f_{n_1}) \\ \times \int d\mathbf{r}_1\, d\mathbf{r}'_1\, d\mathbf{r}_2\, d\mathbf{r}'_2\, &\phi_{n_1}(\mathbf{r}_1)\, \phi^*_{n_2}(\mathbf{r}'_1)\, \Xi(\mathbf{r}_1,\mathbf{r}'_1,\mathbf{r}_2,\mathbf{r}'_2)\, \phi^*_{n_3}(\mathbf{r}_2)\, \phi_{n_4}(\mathbf{r}'_2). \end{aligned}$$

The kernel Ξ contains two contributions: $\bar{v}$, which is the bare Coulomb interaction with the long range part corresponding to a vanishing wave vector not included and W, the attractive screened Coulomb electron–hole interaction. Using this formalism and considering only the resonant part of the excitonic Hamiltonian [57], the macroscopic dielectric function may be expressed as [57]

$$\epsilon_M(\omega) = 1 + \lim_{\mathbf{q}\to 0} v(\mathbf{q}) \sum_{\lambda} \frac{\left| \sum_{v,c;\mathbf{k}} \langle v, \mathbf{k}-\mathbf{q} | e^{-i\mathbf{q}\mathbf{r}} | c, \mathbf{k} \rangle A_{\lambda}^{(v,c;\mathbf{k})} \right|^2}{(E_{\lambda} - \omega)}. \tag{12.5}$$

In Eq. (3.4) the dielectric function, differently from the RPA approximation, is given by a mixing of single particle transitions weighted by the excitonic eigenstates A_λ. These are obtained by the diagonalization of the excitonic Hamiltonian. Moreover, the excitation energies in the denominator are changed from $\epsilon_c - \epsilon_v$ to E_λ. The electronic levels are mixed to produce optical transitions, which are no longer between pairs of independent particles. The excitonic calculation is in general, from the computational point of view, very demanding because the matrix to be diagonalized may be very large. The relevant parameters which determine its size are the number of $\mathbf{k}$-points in the Brillouin zone, and the number of valence and conduction bands, N_v and N_c respectively, which form the basis set of pairs of states.

Calculations performed for insulators and semiconductors show that the inclusion of the electron–hole Coulomb interaction yields a near-quantitative agreement with experiment. This is not only true below the electronic gaps, where bound excitons are generally formed, but also above the continuum edge. The same results apply to the titania-based materials investigated here, as shown in the following section.

12.4. The Bulk Phases of TiO_2: Role of Many Body Effects

Despite the importance of its surfaces and nanostructures, the most recent measurements of TiO_2's bulk (see Fig. 12.1) electronic and optical properties were performed in the 1960s, with a few exceptions. Here we aim to review the existing results obtained using a variety of experimental techniques and *ab initio* calculations, in order to elucidate the known properties of TiO_2. Previous data will be compared with a complete, consistent *ab initio* description, which includes many body effects when describing electronic and optical properties. Most experimental and theoretical data reported refers to the rutile phase, while anatase in general has been less studied. However, anatase has received more attention recently, due in part to its greater stability at the nanoscale compared to rutile.

As we will see, while a general agreement seems to exist concerning the optical absorption edge of these materials, values for basic electronic properties such as the band gap still have a large degree of uncertainty.

The electronic properties of valence states of rutile TiO_2 have been investigated experimentally by angle-resolved photoemission spectroscopy [58], along the two high symmetry directions (Δ and Σ) in the bulk Brillouin zone. The valence band of TiO_2 consists mainly of O $2p$ states partially hybridized with Ti $3d$ states. The metal $3d$ states constitute the conduction band, with a small amount of mixing with O $2p$ states. This photoemission data was compared to calculations performed with both pseudopotentials and linear muffin-tin orbital (LMTO) methods, which gave a direct gap of 2 eV in both cases. On the other hand, within the linear combination of atomic orbitals (LCAO) method, a gap of 3 eV was obtained.

From the symmetry of the TiO_6 octahedrons (see Fig. 12.1), d states are usually grouped into low energy t_{2g} and high energy e_g sub-bands. It is important to note that, from ultraviolet photoemission spectroscopy (UPS) data, it has been deduced that the electronic gap for rutile is at least 4 eV. This is the observed binding energy of the first states below the Fermi energy. This is in agreement with previous reported data from electron energy loss spectroscopy [59], and from other UPS [60–62] results. The electronic structure of rutile bulk has also been described using other experimental techniques, such as electrical resistivity [63], electroabsorption [64,65], photoconductivity and photoluminescence [66,67], X-ray absorption spectroscopy (XAS) [68–73], resonant Raman spectra [67,74] photoelectrochemical analysis [75,76] and UPS [77]. All of these experiments have provided many important details of its electronic properties, in particular concerning the hybridization between Ti $3d$ and O $2p$ states. However, the electronic band gap, corresponding to the difference between the valence band maximum (VBM) and the conduction band minimum (CBM), has not been obtained directly from any experimental data. Although the electronic band gap could be measured using combined photoemission and inverse photoemission experiments, such experiments do not appear in the literature.

The same discussion is valid for the anatase crystalline phase. Even though there are several XAS measurements concerning its electronic structure [73,78, 79], photoemission data is completely lacking for anatase. In the absence of more recent and refined experimental results for rutile, and due to the lack of results for anatase, we are left with an estimate of 4 eV for the electronic gap of rutile TiO_2. It is this value which we shall use as a reference in the following discussion.

We will now review the experimental results for optical properties of TiO_2. Such measurements are of great interest for the photocatalytic and photovoltaic applications of this material. From the optical absorption spectra of both phases [80], the room temperature optical band gap is found to be 3.0 eV for rutile, and 3.2 eV for anatase. The absorption edge has been investigated in detail for rutile by combining absorption, photoluminescence, and Raman scattering techniques [81]. These techniques provide a value for the edge of 3.031 eV associated to a $2p_{xy}$ exciton, while a lower energy $1s$ quadrupolar exciton has been identified below 3 eV. The first dipole allowed gap is at 4.2 eV [81] according to the combined results of these three techniques. Concerning anatase [82,83], the optical spectrum have been recently re-evaluated [83], confirming the 3.2 eV value for the edge. The fine details of anatase's spectrum have also been recently investigated [84]. Data on the Urbach tail has revealed that excitons in anatase are self-trapped in the octahedron of coordination of the titanium atom. This is in contrast to the rutile phase, where excitons are known to be free due to the different packing of rutile's octahedra. [84]

In general, measurements of optical properties can be significantly affected by the presence of defects, such as oxygen vacancies, and by phonons. Both defects and phonons will be present in any experimental sample of the material at finite temperatures. These observations have to be kept in mind when directly comparing experimental measurements with the theoretical results presented in the following. Moreover, there is a general trend in theoretical-computational studies to compare the theoretical electronic band gaps with the experimental optical gap values [29,85,85–88] derived from the above mentioned experiments.

It should be remembered that almost by definition, the optical gap is always smaller than the electronic band gap. This is because the two types of experiments (photoemission, and optical absorption) provide information on two different physical quantities. Reverse photoemission experiments involve a change in the total number of electrons in the material ($N \rightarrow N + 1$), while optical absorption experiments do not ($N \rightarrow N^*$). The latter involves the creation of an electron–hole pair in the material, with the hole stabilizing the excited electron. For this reason, comparison between experimental and theoretical data, and the resulting discussion, must take into account the proper quantities. The theoretical investigations presented in the literature of TiO_2's structural, electronic, and optical properties are at varying levels of theory and thus somewhat inconsistent. A comprehensive description of properties of both phases in the same theoretical and computational framework is still missing. Here we present, in a unified description and by treating with the same method for the two phases, the electronic and optical properties of the two most stable phases of bulk titania. The *ab intio* calculations performed yield results [89] in quite good agreement with the few available pieces of experimental data.

A combination of DFT [41] and many body perturbation theory (MBPT) methods is a reliable and well established toolkit to obtain a complete analysis of electronic and optical properties for a large class of materials and structures. In this DFT + GW + BSE framework, the properties of the two bulk phases of TiO_2 may be properly analyzed. Their structural and energetic properties have been calculated [89] using DFT, as a well established tool for the description of ground state properties. However, the DFT gap is, as expected, significantly smaller than the experimental gap, with the relative positions of the s, p, and d levels also affected by this description. To address this, standard G_0W_0 calculations may be applied to obtain the quasi-particle corrections to the energy levels, starting from DFT eigenvalues and eigenfunctions. Finally the electron–hole interaction is included, to properly describe the optical response of the system.

The description of ground state properties [86,90] performed in the framework of DFT are generally quite good, with the structural description of TiO_2 systems in reasonable agreement with experiments [89]. The lattice constants are within 2% of experiment, while bulk modulii are within 10% of the experimental results [86, 90], as is often found for DFT. However, DFT incorrectly predicts the anatase phase to be more stable than the rutile one, even for a small energy difference, independently of the xc-functional used [86].

Even if DFT is not an excited state method, the KS wavefunctions are often used to evaluate the band structure along the high symmetry directions (*cf.* Fig. 12.2 for rutile), the density of states, and the spatial behaviour of wavefunctions involved in the relevant bonds in the system [58,89,91,92]. The KS electronic gap, corresponding to the difference between the VBM and the CBM, is 1.93 eV and 2.16 eV for rutile (direct gap) and anatase (indirect gap), respectively. These are underestimations by almost 2 eV of the available experimental data [58]. However, the overall behaviour of the band dispersion of KS levels is reasonable, with valence bands mainly given by O $2p$ states, and Ti $3d$ states forming the conduction bands. The application of standard GW methods gives gaps of 3.59 and 3.97 eV for rutile and anatase [89], respectively. The value for rutile is again smaller than the one given by the UPS estimation [58], but still close to the experimental value of 4 eV.

There exist a number of theoretical works, with calculations performed at different levels of DFT or including MBPT descriptions, for the electronic gap of rutile and anatase TiO_2 [85,85,87,91–94]. Therefore in literature it is possible to find for the electronic gap a quite large range of possible values, attributed to the gap of titanium dioxide, which are often erroneously compared with the experimental optical gap.

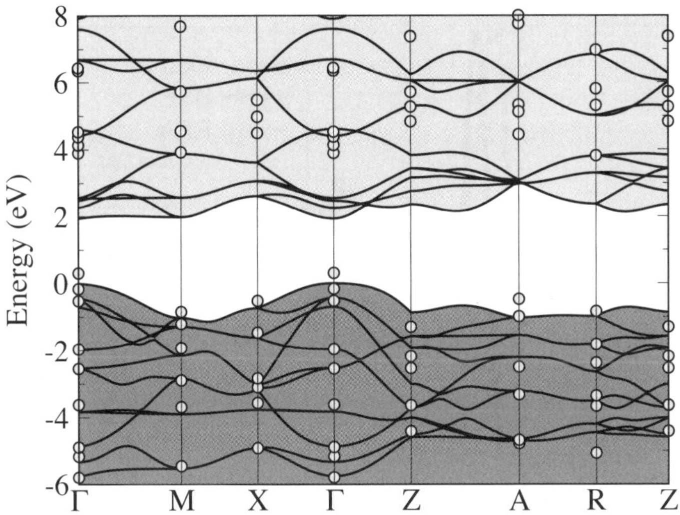

Fig. 12.2. (Color online). Electronic band structure of rutile bulk, along the high symmetry directions of the irreducible Brillouin zone, from GGA calculations (black lines ——). and including the G_0W_0 correction (yellow dots).

The DFT-GGA values calculated [89] are comparable to the ones obtained with a variety of different DFT approaches, with different functionals, by using plane waves or localized basis methods, and all-electron or pseudopotential approaches. Only the hybrid PBE0 [46] and B3LYP [47] xc-functionals give larger values.

From quasi-particle calculations, the electronic gap of anatase has been estimated to be 3.79 eV [94] by G_0W_0. However the more refined computational approach, because of its inclusion of a self-consistent evaluation of GW, yields a gap of 3.78 eV for rutile TiO_2. [85,85]

Moving to optical properties, and by applying the RPA method to both KS and QP energies, we obtain spectra (Fig. 12.3) that do not in overall behaviour agree with experiment. Differences are clear both in absorption edge determination, and in the overall shape of the spectra. The inclusion of quasi-particle corrections at the GW level yields a rigid shift of the absorption spectrum, moving the edge at higher energies, due to the opening of the gap. However, the shape of the absorption is quite unaffected, because the interaction is still treated using an independent quasi-particle approximation. A substantially better agreement may be obtained [89,95] by solving the BSE, which takes into account both many body interactions and excitonic effects. Indeed, it produces a good description of absorption spectra and excitons, as shown in Fig. 12.3

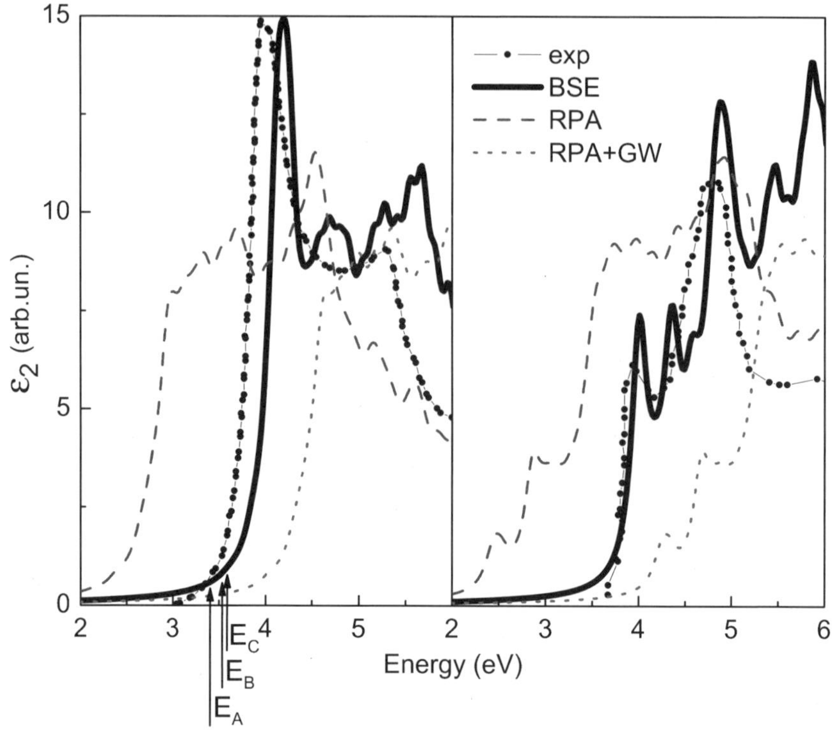

Fig. 12.3. (Color online). Imaginary part of the dielectric constant for rutile (left), and anatase (right), in-plane polarization, calculated by GGA RPA (blue dashed line – – –), using G_0W_0 on top of GGA (red dotted line · · · · · ·), and via the Bethe-Salpeter equation (BSE) (black line ——). The experimental spectrum (black dots — • —) from Refs. [80] and [92] is also shown for comparison.

The optical absorption spectra calculated for the two phases, with polarization along the x-direction of the unit cells, are provided in Fig. 12.3. The spectra given by independent-particle transitions present two characteristic features. First, the band edge is underestimated, due to the electronic gap underestimation in DFT. Second, the overall shape of the spectrum is, for both phases, and both orientations, different from the experiment, in the sense that the oscillator strengths are not correct. The inclusion of the quasi-particle description, which should improve the electronic gap description, does not improve the overall shape of the spectrum. The absorption edge is, however, shifted at higher energies, even higher than expected from experiments. The description of optical properties within the two interacting quasi-particle approximation (by solving the BSE) definitely improves the results. The absorption edge is now comparable to the experimental one, with the optical gap estimated from our calculations in good agreement with the available data. Further, the shape of the spectrum is now well described, with a redistribution of transitions at lower energies. The agreement is generally good for both phases [89].

The nature of the exciton is still under debate in TiO_2 materials [89]. The experimental binding energy is of 4 meV and some uncertainty exists for the exact determination of optical edge. Moreover, the exciton is localized in one of the two phases, and delocalized in the other one, at least based on experimental results. However, an explanation for this behaviour is so far missing. Refined measurements [67] give an exciton of $2p_{xy}$ character at 3.031 eV, and a $1s$ quadrupolar exciton at energy lower than 3 eV. From *ab initio* calculations with x polarization, two dark (optically inactive) or quasi dark excitons are located in rutile at 3.40 and 3.55 eV (denoted by E_A, E_B in Fig. 12.4). At the same time, the optically active exciton is located at 3.59 eV (E_C in Fig. 12.4). The spatial distribution of the first three excitons is plotted in Fig. 12.4. The transitions are from O $2p$ states to Ti $3d$ states of the triplet t_{2g}, as expected. While the first two optically forbidden transitions involve Ti atoms farther away from the excited O atom, the optical active transition involves states of the nearest neighbour Ti atoms.

In this section we have endeavoured to clarify though the application of a consistent description, the properties of the two main crystalline phases of titania. Particular attention has been taken to how the inclusion of a proper description of exchange and correlation effects can improve the description of both electronic and optical properties of TiO_2.

Now that bulk properties are known, from a theoretical point of view, at the level that the state-of-the-art *ab initio* techniques allow us to reach, we can attempt to describe how quantum confinement induced by reduced dimensionality, and doping both effect the properties of TiO_2. Our final aim is to demonstrate how we

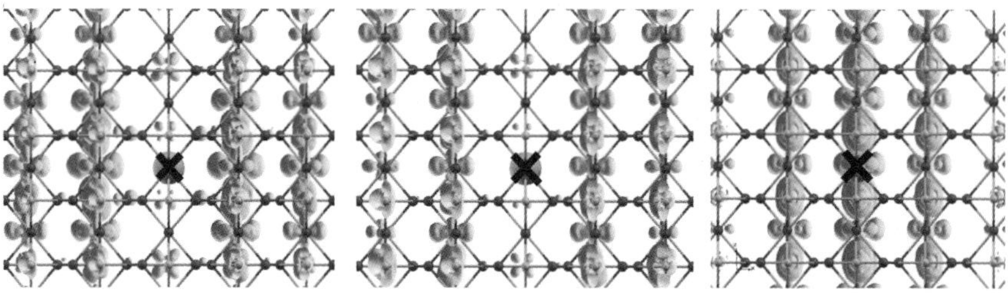

Fig. 12.4. (Color online). Spatial distribution (yellow isosurfaces in arbitrary units) of the partial dark, dark, and optically active first three excitons in rutile. The hole position, corresponding to an O atom, is denoted by the black cross on-top of the light green dot.

may tune the electronic and optical gap of nanostructures for photocatalytic and photovoltaic applications.

12.5. Using Nanostructure to Tune the Energy Gap

Having described in the previous section the electronic structure of the two bulk TiO_2 phases, we will now turn our attention to the influence of nanostructure on the energy gap. This is an area which has received considerable attention in recent years [6–11], in part due to the inherently high surface to volume ratio of nanostructures. It is hoped that this will allow materials with shorter quasi-particle lifetimes to function effectively for photocatalytic activities, since excitons are essentially formed at the material's surface. However, this advantage is partly countered by quantum confinement effects, which tend to increase the energy gap in nanostructures. These competing factors make the accurate theoretical determination of energy gaps in nanostructured materials a thing of great interest.

However, as shown in the previous section, standard DFT calculations tend to underestimate electronic band gaps for bulk TiO_2 by approximately 2 eV, due in part to self-interaction errors [96,97]. These errors arise from an incomplete cancellation of the electron's Coulomb potential in the exchange-correlation (xc)-functional.

This may be partially addressed by the use of hybrid functionals such as B3LYP [47], which generally seem to improve band gaps for bulk systems [32,87,98]. However, such calculations are computationally more expensive, due to the added dependence of the xc-functional on the electron's wavefunction. Moreover, B3LYP calculations for TiO_2 clusters largely underestimate the gap relative to the more reliable difference between standard DFT calculated ionization potential I_p and electron affinity E_a. For isolated systems such as clusters, the needed energetics of charged species are quantitatively described by standard DFT. Also, B3LYP and RPBE [99] calculations provide the same qualitative description of the *trends* in the energy gaps for TiO_2, as seen in Fig. 12.5. Another methodology is thus needed to describe the band gaps of periodic systems.

G_0W_0 is probably the most successful and generally applicable method for calculating quasi-particle gaps. For clusters it agrees well with $I_p - E_a$, and it has been shows to produce reliable results for bulk phases. On the other hand, G_0W_0 calculations describe an $N \to N+1$ transition where the number of charges is not conserved, rather than the electron–hole pair induced by photoabsorption, which is a neutral process. Indeed, a description in terms of electronic gap cannot be compared with, or provide direct information on the optical gap, which is the most investigated quantity, due to its critical importance for photocatalytic processes.

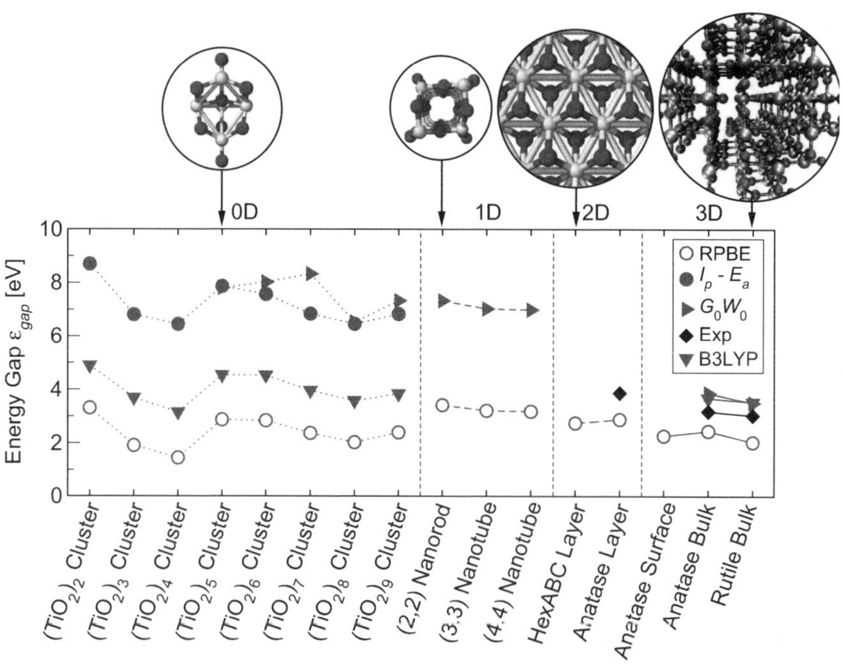

Fig. 12.5. (Color online). Energy gap ε_{gap} in eV versus TiO_2 structure for 0D $(TiO_2)_n$ clusters ($n \leq 9$), 1D TiO_2 (2,2) nanorods, (3,3) nanotubes, (4,4) nanotubes, 2D HexABC and anatase layers, and 3D anatase surface, anatase bulk, and rutile bulk phases. DFT calculations using the highest occupied and lowest unoccupied state gaps with the standard GGA RPBE xc-functional (○) and the hybrid B3LYP xc-functional (▼), are compared with DFT $I_p - E_a$ calculations (●), G_0W_0 quasi-particle calculations (▶), and experimental results (◆) [6,11,13,20,94]. Schematics of representative structures for each dimensionality are shown above and taken from Ref. [11].

Figure 12.5 shows that for both 3D and 2D systems, RPBE gaps underestimate the experimental optical gaps by approximately 1 eV. For 1D and 0D systems, there is a much larger difference of about 4 eV and 5 eV respectively, between the RPBE gaps and the $I_p - E_a$ and G_0W_0 results. This increasing disparity may be attributed to the greater quantum confinement and charge localization in the 1D and 0D systems, which yield greater self-interaction effects. The B3LYP gaps also tend to underestimate this effect, simply increasing the RPBE energy gaps for both 0D and 3D systems by about 1.4 eV.

On the other hand, the RPBE gaps reproduce qualitatively the structural dependence of the $I_p - E_a$, G_0W_0, and experimental results for a given dimensionality, up to a constant shift. This is true even for 3D bulk systems, where standard DFT does not correctly predict rutile to be the most stable structure [86].

To summarize, quantum confinement effects seem to increase the energy gap significantly for both 0D and 1D systems, while 2D and 3D systems may be much less affected. This suggests that 2D laminar structures are viable candidates for reducing the minimal quasi-particle, while leaving the band gap nearly unchanged. However, a more accurate description of the photoabsorption properties of these novel nanostructures, perhaps using the methodologies recently applied to bulk TiO_2 [89], still remains to be found.

12.6. Influence of Boron and Nitrogen Doping on TiO_2's Energy Gap

The doping of TiO_2 nanostructures has received much recent attention as a possible means for effectively tuning TiO_2's band gap into the visible range [5,11–19]. Recent experiments suggest substituting oxygen by boron or nitrogen in the bulk introduces mid-gap states, allowing lower energy excitations. However, to model such systems effectively requires large supercells, both to properly describe the experimental doping fractions of $\lesssim 10\%$, and to ensure dopant–dopant interactions are minimized.

Figure 12.6 shows the DFT calculated DOS and structures for the most stable boron doped and nitrogen doped TiO_2 (2,2) nanorods. The highest occupied state is also shown as isosurfaces of $\pm 0.05e/\text{Å}^3$ in the side views of the doped structures.

As with TiO_2 clusters, the influence of boron dopants on TiO_2 nanorods may be understood in terms of boron's weak electronegativity, especially when com-

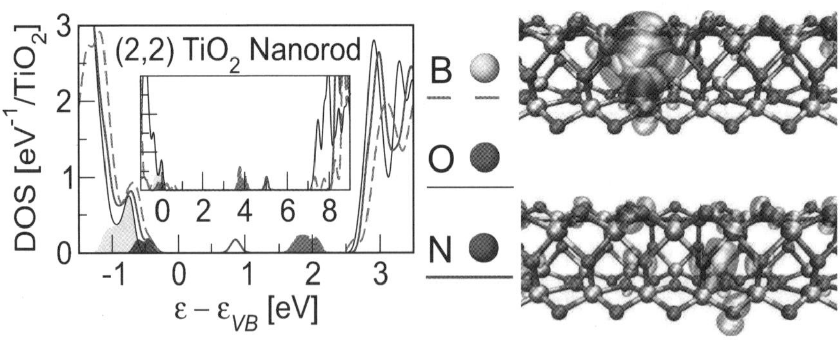

Fig. 12.6. (Color online). Total density of states (DOS) in eV^{-1} per TiO_2 functional unit vs. energy ε in eV for undoped (thin black line ——), boron doped (dashed magenta line – – –) and nitrogen doped (thick blue line ——)(2,2) TiO_2 nanorods from standard DFT GGA RPBE xc-functional and (inset) G_0W_0 quasi-particle calculations. The highest occupied states for boron and nitrogen doped (2,2) TiO_2 nanorods are depicted by isosurfaces of $\pm 0.05e/\text{Å}^3$ in the structure diagrams to the left.

pared with the strongly electronegative oxygen. Boron prefers to occupy oxygen sites which are 2-fold coordinated to neighbouring titanium atoms. However, as with the 0D clusters, boron's relatively electropositive character induces significant structural changes in the 1D structures, creating a stronger third bond to a neighbouring three-fold coordinated oxygen via an oxygen dislocation, as shown in Fig. 12.6. This yields three occupied mid-gap states localized on the boron dopant, which overlap both the valence band O $2p_\pi$ and conduction band Ti $3d_{xy}$ states, as shown in Fig. 12.6. Boron dopants thus yield donor states near the conduction band, which may be photocatalytically active in the visible region. However, the quantum confinement inherent in these 1D structures may stretch these gaps, as found for the G_0W_0 calculated DOS shown in the inset of Fig. 12.6.

On the other hand, nitrogen dopants prefer to occupy oxygen sites which are 3-fold coordinated to Ti, as was previously found for the rutile TiO_2 surface [17,18]. This yields one occupied state at the top of the valence band and one unoccupied mid-gap state in the same spin channel. Both states are localized on the nitrogen dopant but overlap the valence band O $2p_\pi$ states, as shown in Fig. 12.6. Nitrogen dopants thus act as acceptors, providing localized states well above the valence band, as is also found for the G_0W_0 calculated DOS shown in the inset of 12.6.

Although nitrogen dopants act as acceptors in TiO_2 1D structures, such large gaps between the valence band and the unoccupied mid-gap states would not yield p-type semiconductors. This may be attributed to the substantial quantum confinement in these 1D structures. However, for 2D and 3D systems, it is possible to produce both p-type and n-type classical semiconductors, as discussed in Ref. [11]. This has recently been shown experimentally in Ref. [19], where co-doping of anatase TiO_2 with nitrogen and chromium was found to improve the localization of the acceptor states, and reduce the effective optical gap. By replacing both Ti^{4+} and O^{2-} atoms with dopants in the same TiO_6 octahedral, it should be possible to "tune" the optical band gap to a much finer degree.

Whether calculated using RPBE, $I_p - E_a$ or G_0W_0, the energy gaps for both boron and nitrogen doped TiO_2 nanostructures are generally narrowed, as shown in Fig. 12.7(a) and (b). However, for nitrogen doped $(TiO_2)_n$ clusters where nitrogen acts as an acceptor ($n = 5, 6, 9$), the energy gap is actually *increased* when spin is conserved, compared to the undoped clusters in RPBE. This effect is not properly described by the $N \to N+1$ transitions of $I_p - E_a$, for which spin is not conserved for these nitrogen doped clusters. On the other hand, when nitrogen acts as a donor ($n = 7, 8$) the smallest gap between energy levels does conserve spin. The boron doped TiO_2 nanorods and nanotubes have perhaps the most promising energy gap results of the TiO_2 structures, as seen in Fig. 12.7(a). Boron dopants introduce in the nanorods localized occupied states near the conduction band edge in both RPBE (*cf.* Fig. 12.6) and G_0W_0 (*cf.* inset of Fig. 12.6) calculations. On the

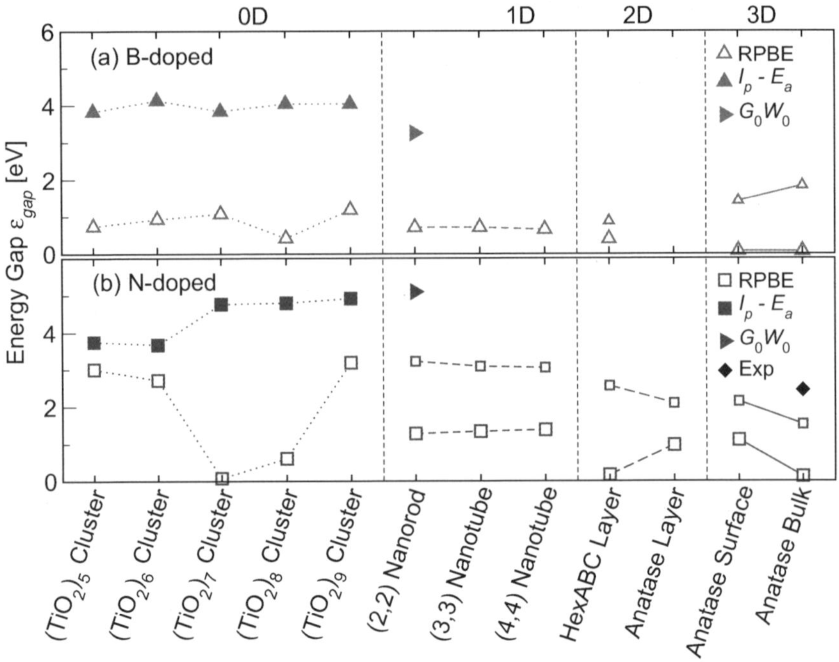

Fig. 12.7. (Color online). Influence of doping on the energy gap ε_{gap} in eV versus TiO_2 structure for 0D $(TiO_2)_n$ clusters ($n \leq 9$), 1D TiO_2 (2,2) nanorods, (3,3) nanotubes, (4,4) nanotubes, 2D HexABC and anatase layers, and 3D anatase surface and bulk. The energy gaps for (a) boron doped systems from DFT calculations using the highest occupied and lowest unoccupied state gaps with the standard GGA RPBE xc-functional (△) are compared with DFT $I_p - E_a$ calculations (▲), and G_0W_0 quasi-particle calculations (▶), and (b) nitrogen doped systems from DFT calculations using the highest occupied and lowest unoccupied state gaps with the standard GGA RPBE xc-functional (□) are compared with DFT $I_p - E_a$ calculations (■), and G_0W_0 quasi-particle calculations (▶), and experimental (◆) results [6,11,13,20,94]. Small open symbols denote transitions between highest fully occupied states and the conduction band.

other hand, nitrogen doping of nanorods introduces well defined mid-gap states, as shown in Fig. 12.6. However, to perform water dissociation, the energy of the excited electron must be above that for hydrogen evolution, with respect to the vacuum level. This is not the case for such a mid-gap state. This opens the possibility of a second excitation from the mid-gap state to the conduction band. However, the cross section for such an excitation may be rather low.

For boron doping of 2D and 3D structures, the highest occupied state donates its electron almost entirely to the conduction band, yielding an n-type semiconductor. Thus at very low temperatures, the RPBE band gap is very small. The same is true for n-type nitrogen doped bulk anatase. For these reasons the energy gap between the highest fully occupied state and the conduction band, which may be more relevant for photoabsorption, is also shown. These RPBE gaps are still generally smaller than those for their undoped TiO_2 counterparts, shown in Fig. 12.5.

In summary, for both boron and nitrogen doped clusters we find RPBE gaps differ from $I_p - E_a$ by about 3 eV, while for nitrogen doped anatase the RPBE gap differs from experiment by about 0.6 eV. Given the common shift of 1 eV for undoped 2D and 3D structures, this suggests that both boron and nitrogen doped 2D TiO_2 structures are promising candidates for photocatalysis. Further, the boron and nitrogen doped 1D nanotube results also warrant further experimental investigation.

12.7. Solar Cells from TiO_2 Nanostructures: Dye-Sensitized Solar Cells

TiO_2 is so far the most widely used solid material in the development of solar cell devices based on hybrid architectures[1,100]. In these devices, the dye, synthetic or organic, absorbs light, and electrons excited by the phonons are injected into the underlying oxide nanostructure. The hybrid system must therefore satisfy several requirements: (1) a proper absorption range for the dye, (2) a fast charge transfer in the oxide, (3) a slow back-transfer process, and (4) an easy collection and conduction of electrons in the oxide.

While absorption properties may be easily tuned at the chemical level by changing or adding functional groups, a critical point is to understand, and therefore control, the process at the interface. Indeed, simply having a good energy level alignment is not sufficient because the fast electron injection process is dynamical. Since the experimental characterization of complex systems (networks of nanostructures, with adsorbed dyes, and in solution) is quite complicated, the theoretical description of such systems can be of fundamental importance in unravelling the processes governing the behaviour of dye-sensitized solar cells (DSSC).

The most popular technique for studying these dynamical processes is time dependent DFT (TDDFT). For a detailed discussion of the success and possible limitations of this method when applied to hybrid systems, we refer the reader to Ref. [101]. TDDFT is a generalization of DFT which allows us to directly describe excited states. For example, it has previously been used successfully to calculate the optical absorption of large organic molecules (Fig. 12.8), such as Ru-dyes, or indolines. This same technique has also been applied recently to hybrid systems used for photovoltaic applications. We want here to highlight that the charge injection transfer has also been modelled for molecules adsorbed on TiO_2

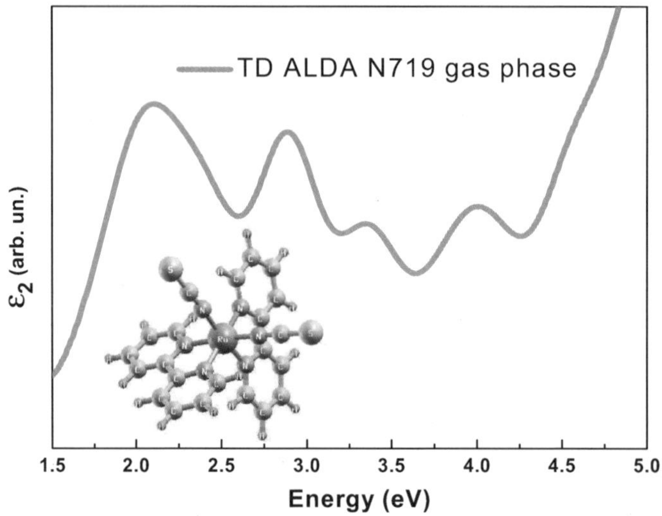

Fig. 12.8. (Color online). Absorption spectrum calculated by TD Adiabatic LDA of the Ru-dye N719, whose structure is shown in the inset.

clusters[102–111], giving estimations as fast as 8 fs for the injection time [107]. TDDFT is therefore a powerful tool to understand dynamical electronic effects in systems as large and complex as hybrid organic-oxide solar cells.

12.8. Conclusion

In this chapter we showed how an oxide of predominant importance in nanotechnological and environmental fields, such as photocatalysis and photovoltaics, can be successfully investigates by state-of-the-art *ab initio* techniques. The optical and excitonic properties of the bulk phases can be properly described by MBPT techniques. We also showed how the electronic properties of TiO_2 may be "tailored" using nanostructural changes in combination with boron and nitrogen doping. While boron doping tends to produce smaller band gap n-type semiconductors, nitrogen doping produces p-type or n-type semiconductors depending on whether or not nearby oxygen atoms occupy surface sites. This suggests that a p-type TiO_2 semiconductor may be produced using nitrogen doping in conjunction with surface confinement at the nanoscale. We also showed that it has been proved how, in the field of photovoltaics, TDDFT is a powerful tool to understand the mechanism of charge injection at organic-inorganic interfaces.

Acknowledgments

We acknowledge funding by "Grupos Consolidados UPV/EHU del Gobierno Vasco" (IT-319-07), the European Community through e-I3 ETSF project (Contract Number 211956). We acknowledge support by the Barcelona Supercomputing Center, "Red Espanola de Supercomputacion", SGIker ARINA (UPV/EHU) and Transnational Access Programme HPC-Europe++. J.M. G-L. acknowledges funding from the Spanish MEC through the 'Juan de la Cierva' program. L.C. acknowledges funding from UPV/EHU through the 'Ayudas de Especialización para Investigadores Doctores' program. We also thank H. Petek, S. Ossicini, M. Wanko, M. Piacenza, and M. Palummo for helpful discussions.

References

1. M. Grätzel, Photoelectrochemical cells, *Nature* **414**, 338–344 (2001).
2. A. Heller, Hydrogen-evolving solar cells, *Science* **223**, 1141–1148 (1984).
3. M. R. Hoffmann, S. T. Martin, W. Y. Choi and D. W. Bahnmann, Environmental applications of semiconductor photocatalysis, *Chem. Rev.* **95**, 69–96 (1995).
4. S. U. M. Khan, M. Al-Shahry and W. B. I. Jr., Efficient photochemical water splitting by a chemically modified n-TiO_2, *Science* **297**, 2243–2245 (2002).
5. Y. Gai, J. Li, S.-S. Li, J.-B. Xia and S.-H. Wei, Design of narrow-gap TiO_2: A passivated codoping approach for enhanced photoelectrochemical activity, *Phys. Rev. Lett.* **102**, 036402–1–4 (2009).
6. Z.-W. Qu and G.-J. Kroes, Theoretical study of the electronic structure and stability of titanium dioxide clusters $(TiO_2)_n$ with $n = 1$–9, *J. Phys. Chem. B* **110**, 8998–9007 (2006).
7. H.-J. Zhai and L.-S. Wang, Probing the electronic structure and band gap eveolition of titanium oxide clusters $(TiO_2)_n^-$ ($n = 1$–10) using photoelectron spectroscopy, *J. Am. Chem. Soc.* **129**, 3022–3026 (2007).
8. H. Imai, Y. Takei, K. Shimizu, M. Matsuda and H. Hirashima, Direct preparation of anatase TiO_2 nanotubes in porous alumina membranes, *J. Mater. Chem.* **9**, 2971–2972 (1999).
9. T. Kasuga, M. Hiramatsu, A. Hoson, T. Sekino and K. Niihara, Titania nanotubes prepared by chemical processing, *Adv. Mater.* **11**, 1307–1311 (1999).
10. G. Mogilevsky, Q. Chen, H. Kulkarni, A. Kleinhammes, W. M. Mullins and Y. Wu, Layered nanostructures of delaminated anatase: Nanosheets and nanotubes, *J. Phys. Chem. C* **112**, 3239–3246 (2008).
11. D. J. Mowbray, J. I. Martínez, J. M. García Lastra, K. S. Thygesen and K. W. Jacobsen, Stability and electronic properties of TiO_2 nanostructures with and without B and N doping, *J. Phys. Chem. C* **113**, 12301–12308 (2009).
12. S. In, A. Orlov, R. Berg, F. Carcía, S. Pedrosa-Jimenez, M. S. Tikhov, D. S. Wright and R. M. Lambert, Effective visible light-activated B-doped and B,N-codoped TiO_2 photocatalysts, *J. Am. Chem. Soc.* **129**, 13790–13791 (2007).

13. G. Liu, F. Li, D.-W. Wang, D.-M. Tang, C. Liu, X. Ma, G. Q. Lu and H.-M. Cheng, Electron field emission of a nitrogen-doped TiO_2 nanotube array, *Nanotechnology* **19**, 025606–025611 (2008).
14. R. Asahi, T. Morikawa, T. Ohwaki, K. Akoki and Y. Taga, Visible-light photocatalysis in nitrogen-doped titanium oxides, *Science* **298**, 269–271 (2001).
15. A. Ghicov, J. M. Macak, H. Tsuchiya, J. Kunze, V. Hæublein, S. Kleber, and P. Schmuki, TiO_2 nanotube layers: Dose effects during nitrogen doping by ion implantation, *Chem. Phys. Lett.* **419**, 426–429 (2006).
16. Y. Chen, S. Zhang, Y. Yu, H. Wu, S. Wang, B. Zhu, W. Huang and S. Wu, Synthesis, characterization, and photocatalytic activity of N-doped TiO_2 nanotubes, *J. Disp. Sci. Tech.* **29**, 245–249 (2008).
17. A. Nambu, J. Graciani, J. A. Rodriguez, Q. Wu, E. Fujita and J. F. Sanz, N doping of TiO_2(110): Photoemission and density-functional studies, *J. Chem. Phys.* **125**, 094706–1–8 (2006).
18. J. Graciani, L. J. Álvarez, J. A. Rodriguez and J. F. Sanz, N doping of rutile TiO_2(110) surface. a theoretical DFT study, *J. Phys. Chem. C* **112**, 2624–2631 (2008).
19. W. Zhu, X. Qiu, V. Iancu, X.-Q. Chen, H. Pan, W. Wang, N. M. Dimitrijevic, T. Rajh, H. M. M. III, M. P. Paranthaman, G. M. Stocks, H. H. Weitering, B. Gu, G. Eres and Z. Zhang, Band gap narrowing of titanium oxide semiconductors by noncompensated anion-cation codoping for enhanced visible-light photoactivity, *Phys. Rev. Lett.* **103**, 226401–1–4 (2009).
20. D. V. Bavykin, J. M. Friedrich and F. C. Walsh, Protonates titanates and TiO_2 nanostructured materials: Synthesis, properties, and applications, *Adv. Mater.* **18**, 2807–2824 (2006).
21. O. Khaselev and J. A. Turner, A monolithic photovoltaic-photoelectrochemical device for hydrogen production via water splitting, *Science* **280**, 425–427 (1998).
22. M. K. Nazeeruddin, A. Kay, I. Rodicio, R. Humphry-Baker, E. Müller, P. Liska, N. Vlachopoulos and M. Grätzel, Conversion of light to electricity by *cis*-X_2Bis(2,2'-bipyridyl-4,4'-dicarboxylate)ruthenium(ii) charge-transfer sensitizers ($X = Cl^-$, Br^-, I^-, CN^-, and SCN^-) on nanocrystalline TiO_2 electrodes, *J. Am. Chem. Soc.* **115**, 6382–6390 (1993).
23. P. Wang, S. M. Zakeeruddin, J. E. Moser, R. Humphry-Baker, P. Comte, V. Aranyos, A. Hagfeldt, M. K. Nazeeruddin and M. Grätzel, Stable new sensitizer with improved light harvesting for nanocrystalline dye-sensitized solar cells, *Adv. Mater.* **16**, 1806–1811 (2004).
24. X. Chen and S. S. Mao, Titanium dioxide nanomaterials: Synthesis, properties, modifications, and applications, *Chem. Rev.* **107**, 2891–2959 (2007).
25. D. F. Watson and G. J. Meyer, Electron injection at dye-sensitized semiconductor electrodes, *Ann. Rev. Phys. Chem.* **56**, 119–156 (2005).
26. J.-J. Wu, G.-R. Chen, C.-C. Lu, W.-T. Wu, and J.-S. Chen, Performance and electron transport properties of TiO_2 nanocomposite dye-sensitized solar cells, *Nanotechnology* **19**, 105702–105708 (2008).
27. S. H. Kang, S.-H. Choi, M.-S. Kang, J.-Y. Kim, H.-S. Kim, T. Hyeon and Y.-E. Sung, Nanorod-based dye-sensitized solar cells with improved charge collection efficiency, *Adv. Mater.* **20**, 54–58, (2008).

28. M. Adachi, Y. Murata, J. Takao, J. Jiu, M. Sakamoto and F. Wang, Highly efficient dye-sensitized solar cells with a titania thin-film electrode composed of a network structure of single-crystal-like TiO_2 nanowires made by the "oriented attachment" mechanism, *J. Am. Chem. Soc.* **126**, 14943–14949 (2004).
29. M. V. Ganduglia-Pirovano, A. Hofmann and J. Sauer, Oxygen vacancies in transition metal and rare earth oxides: Current state of understanding and remaining challenges, *Surf. Sci. Rep.* **62**, 219–270, (2007).
30. M. Gatti, F. Bruneval, V. Olevano and L. Reining, Understanding correlations in vanadium dioxide from first principles, *Phys. Rev. Lett.* **99**, 266402–1–4 (2007).
31. P. Krüger, S. Bourgeois, B. Domenichini, H. Magnan, D. Chandesris, P. L. Fèvre, A. M. Flank, J. Jupille, L. Floreano, A. Cossaro, A. Verdini, and A. Morgante, Defect states at the TiO_2(110) surface probed by resonant photoelectron diffraction, *Phys. Rev. Lett.* **100**, 055501–1–4 (2008).
32. C. Di Valentin, G. Pacchioni and A. Selloni, Electronic structure of defect states in hydroxylated and reduced rutile TiO_2(110) surfaces, *Phys. Rev. Lett.* **97**, 166803–1–4 (2006).
33. T. Minato, Y. Sainoo, Y. Kim, H. S. Kato, K. ichi Aika, M. Kawai, J. Zhao, H. Petek, T. Huang, W. He, B. Wang, Z. Wang, Y. Zhao, J. Yang and J. G. Hou, The electronic structure of oxygen atom vacancy and hydroxyl impurity defects on titanium dioxide (110) surface, *J. Chem. Phys.* **130**, 124502–1–11 (2009).
34. E. Finazzi, C. D. Valentin, G. Pacchioni and A. Selloni, Excess electron states in reduced bulk anatase TiO_2: Comparison of standard GGA, GGA + U, and hybrid DFT calculations, *J. Chem. Phys.* **129**, 154113–1–9 (2008).
35. G. Pacchioni, Modeling doped and defective oxides in catalysis with density functional theory methods: Room for improvements, *J. Chem. Phys.* **128**, 182505–1–10 (2008).
36. V. M. Khomenko, K. Langer, H. Rager and A. Fett, Electronic absorption by Ti^{3+} ions and electron delocalization in synthetic blue rutile, *Phys. Chem. Minerals* **25**, 338–346 (1998).
37. S. Wendt, P. T. Sprunger, E. Lira, G. K. H. Madsen, Z. Li, J. O. Hansen, J. Matthiesen, A. Blekinge-Rasmussen, E. Laegsgaard, B. Hammer, and F. Besenbacher, The Role of Interstitial Sites in the Ti3*d* Defect State in the Band Gap of Titania, *Science* **320**, 1755–1759, (2008).
38. L. Hedin, New method for calculating the one-particle Green's function with application to the electron-gas problem, *Phys. Rev.* **139**, A796–A823 (1965).
39. R. M. Dreizler and E. K. U. Gross, *Density Functional Theory* (Springer Verlag, Heidelberg, 1990).
40. L. Fetter and J. D. Walecka, *Quantum theory of Many Body Systems* (McGraw-Hill, New York, 1981).
41. P. Hohenberg and W. Kohn, Inhomogeneous electron gas, *Phys. Rev.* **136**, B864–B871 (1964).
42. W. Kohn and L. J. Sham, Self-consistent equations including exchange and correlation effects, *Phys. Rev.* **140**, A1133–A1138, (1965).
43. D. M. Ceperley and B. J. Alder, Ground state of the electron gas by a stochastic method, *Phys. Rev. Lett.* **45**, 566–569, (1980).
44. J. P. Perdew and A. Zunger, Self-interaction correction to density-functional approximations for many-electron systems, *Phys. Rev. B* **23**, 5048–5079 (1981).

45. J. P. Perdew, K. Burke and M. Ernzerhof, Generalized gradient approximation made simple, *Phys. Rev. Lett.* **77**, 3865–3868 (1996).
46. C. Adamo and V. Barone, Toward reliable density functional methods without adjustable parameters: The pbe0 model, *J. Chem. Phys.* **110**, 6158–6170 (1999).
47. A. D. Becke, Density-functional thermochemistry. iii. the role of exact exchange, *J. Chem. Phys.* **98**, 5648–5652 (1993).
48. J. P. Perdew and M. Levy, Physical content of the exact Kohn-Sham orbital energies: Band gaps and derivative discontinuities, *Phys. Rev. Lett.* **51**, 1884–1887 (1983).
49. R. D. Mattuck, *A Guide to Feynman Diagrams in the Many-Body Problem* (McGraw-Hill, New York, 1976).
50. L. Hedin and S. Lundqvist Effects of Electron-Electron and Electron-Phonon Interactions on the One-Electron States of Solids, in *Solid State Physics*, Vol. **23**, edited by H. Ehrenreich, F. Seitz and D. Turnbull (Academic Press, New York, 1969) pp. 1–181.
51. G. Strinati, Application of the Green's functions method to the study of the optical properties of semiconductors, *Rivista Nuovo Cimento* **11**, 1–86 (1988).
52. P. M. Echenique, J. M. Pitarke, E. V. Chulkov and A. Rubio, Theory of inelastic lifetimes of low-energy electrons in metals, *Chem. Phys.* **251**, 1–35 (2000).
53. M. S. Hybertsen and S. G. Louie, Electron correlation in semiconductors and insulators: Band gaps and quasiparticle energies, *Phys. Rev. B* **34**, 5390–5413 (1986).
54. R. W. Godby, M. Schlüter and L. J. Sham, Self-energy operators and exchange-correlation potentials in semiconductors, *Phys. Rev. B* **37**, 10159–10175 (1988).
55. S. L. Adler, Quantum theory of the dielectric constant in real solids, *Phys. Rev.* **126**, 413–420 (1962).
56. J. Lindhard, On the properties of a gas of charged particles, *Mat. Fys. Medd. K. Dan. Vidensk. Selsk.* **28**, 1–57 (1954).
57. G. Onida, L. Reining and A. Rubio, Electronic excitations: density-functional versus many-body Green's-function approaches, *Rev. Mod. Phys.* **74**, 601–659 (2002).
58. P. J. Hardmand, G. N. Raikar, C. A. Muryn, G. van der Laan, P. L. Wincott, G. Thornton, D. W. Bullett and P. A. D. M. A. Dale, Valence-band structure of TiO_2 along the Γ-Δ-X and Γ-Σ-M directions, *Phys. Rev. B* **49**, 7170–7177 (1994).
59. S. Eriksen and R. G. Egdell, Electronic excitations at oxygen deficient TiO_2(110) surfaces: A study by EELS, *Surf. Sci.* **180**, 263–278 (1987).
60. R. Heise, R. Courths and S. Witzel, Valence band densities-of-states of TiO_2(110) from resonant photoemission and photoelectron diffraction, *Solid State Commun.* **84**, 599–602 (1992).
61. R. G. Egdell, S. Eriksen and W. R. Flavell, Oxygen deficient SnO_2 (110) and TiO_2 (110): A comparative study by photoemission, *Solid State Commun.* **60**, 835–838 (1986).
62. R. L. Kurtz, R. Stock-Bauer, T. E. Madey, E. Román and J. L. D. Segovia, Synchrotron radiation studies of H_2O adsorption on TiO_2(110), *Surf. Sci.* **218**, 178–200 (1989).
63. R. G. Breckenridge and W. R. Hosler, Electrical properties of titanium dioxide semiconductors, *Phys. Rev.* **91**, 793–802 (1953).
64. F. Arntz and Y. Yacoby, Electroabsorption in rutile TiO_2, *Phys. Rev. Lett.* **17**, 857–860 (1966).

65. A. Frova, P. J. Boddy and Y. S. Chen, Electromodulation of the optical constants of rutile in the UV, *Phys. Rev.* **157**, 700–708 (1967).
66. A. K. Ghosh, F. G. Wakim and R. R. Addiss, Photoelectronic processes in rutile, *Phys. Rev.* **184**, 979–988 (1969).
67. A. Amtout and R. Leonelli, Optical properties of rutile near its fundamental band gap, *Phys. Rev. B* **51**, 6842–6851 (1995).
68. S. Hüfner and G. K. Wertheim, Core-electron splittings and hyperfine fields in transition-metal compounds, *Phys. Rev. B* **7**, 2333–2336 (1973).
69. D. W. Fischer, X-ray band spectra and molecular-orbital structure of rutile TiO_2, *Phys. Rev. B* **5**, 4219–4226 (1972).
70. F. M. F. de Groot, J. Faber, J. J. M. Michiels, M. T. Czyżyk, M. Abbate and J. C. Fuggle, Oxygen $1s$ x-ray absorption of tetravalent titanium oxides: A comparison with single-particle calculations, *Phys. Rev. B* **48**, 2074–2080 (1993).
71. L. A. Grunes, Study of the k edges of $3d$ transition metals in pure and oxide form by x-ray-absorption spectroscopy, *Phys. Rev. B* **27**, 2111–2131 (1983).
72. G. van der Laan, Polaronic satellites in x-ray-absorption spectra, *Phys. Rev. B* **41**, 12366–12368 (1990).
73. F. M. F. de Groot, M. Grioni, J. C. Fuggle, J. Ghijsen, G. A. Sawatzky, and H. Petersen, Oxygen 1s x-ray-absorption edges of transition-metal oxides, *Phys. Rev. B* **40**, 5715–5723 (1989).
74. K. Watanabe, K. Inoue and F. Minami, Resonant phenomena of hyper-raman-scattering of optic phonons in a TiO_2 crystal, *Phys. Rev. B* **46**, 2024–2033 (1992).
75. D. C. Cronemeyer, Electrical and optical properties of rutile single crystals, *Phys. Rev.* **87**, 876–886 (1952).
76. L. Kavan, M. Grätzel, S. E. Gilbert, C. Klemenz and H. J. Scheel, Electrochemical and photoelectrochemical investigation of single-crystal anatase, *J. Am. Chem. Soc.* **118**, 6716–6723 (1996).
77. L. Fleming, C. C. Fulton, G. Lucovsky, J. E. Rowe, M. D. Ulrich, and J. Lüning, Local bonding analysis of the valence and conduction band features of TiO_2, *J. Appl. Phys.* **102**, 033707–1–7 (2007).
78. R. Sanjinés, H. Tang, H. Berger, F. Gozzo, G. Margaritondo, and F. Lévy, Electronic structure of anatase TiO_2 oxide, *J. Appl. Phys.* **75**, 2945–2951 (1994).
79. H. Tang, K. Prasad, R. Sanjinès, P. E. Schmid and F. Lévy, Electrical and optical properties of TiO_2 anatase thin films, *J. Appl. Phys.* **75**, 2042–2047 (1994).
80. M. Cardona and G. Harbeke, Optical properties and band structure of wurtzite-type crystals and rutile, *Phys. Rev.* **137**, A1467–A1476 (1965).
81. J. Pascual, J. Camassel and H. Mathieu, Fine structure in the intrinsic absorption edge of TiO_2, *Phys. Rev. B* **18**, 5606–5614 (1978).
82. N. Hosaka, T. Sekiya, M. Fujisawa, C. Satoko and S. Kurita, Uv reflection spectra of anatase TiO_2, *J. Electron Spec. Rel. Phen.* **78**, 75–78 (1996).
83. N. Hosaka, T. Sekiya, C. Satoko and S. Kurita, Optical properties of single-crystal anatase TiO_2, *J. Phys. Soc. Jpn.* **66**, 877–880 (1997).
84. H. Tang, F. Lévy, H. Berger and P. E. Schmid, Urbach tail of anatase TiO_2, *Phys. Rev. B* **52**, 7771–7774 (1995).
85. M. van Schilfgaarde, T. Kotani and S. Faleev, Quasiparticle self-consistent GW theory, *Phys. Rev. Lett.* **96**, 226402–1–4, (2006).

86. F. Labat, P. Baranek, C. Doman, C. Minot and C. Adamo, Density functional theory analysis of the structural and electronic properties of TiO_2 rutile and anatase polytypes: Performances of different exchange-correlation functionals, *J. Chem. Phys.* **126**, 154703–1–12, (2007).
87. J. Muscat, A. Wander and N. M. Harrison, On the prediction of band gaps from hybrid functional theory, *Chem. Phys. Lett.* **342**, 397–401 (2001).
88. U. Diebold, The surface science of titanium dioxide, *Surf. Sci. Rep.* **48**, 53–229 (2003).
89. L. Chiodo, J. M. García-Lastra, A. Incomino, J. Zhao, H. Petek, and A. Rubio. Self-energy and excitonic effects in the electronic and optical properties of TiO_2 crystalline phases, *Phys. Rev. B* **82**, 045207-1-12 (2010).
90. J. Muscat, V. Swamy and N. Harrison, First-principles calculations of the phase stability of TiO_2, *Phys. Rev. B* **65**, 224112–1–15 (2002).
91. M. Oshikiri, M. Boero, J. Ye, F. Aryasetiawan and G. Kido, The electronic structures of the thin films of $InVO_4$ and TiO_2 by first principles calculations, *Thin Solid Films* **445**, 168–174 (2003).
92. R. Asahi, Y. Taga, W. Mannstadt and A. J. Freeman, Electronic and optical properties of anatase TiO_2, *Phys. Rev. B* **61**, 7459–7465 (2000).
93. M. Calatayud, P. Mori-Sánchez, A. Beltrán, A. Martín Pendás, E. Francisco, J. Andrés and J. M. Recio, Quantum-mechanical analysis of the equation of state of anatase TiO_2, *Phys. Rev. B* **64**, 184113–1–9 (2001).
94. L. Thulin and J. Guerra, Calculations of strain-modified anatase TiO_2 band structures, *Phy. Rev. B* **77**, 195112–1–5, (2008).
95. H. M. Lawler, J. J. Rehr, F. Vila, S. D. Dalosto, E. L. Shirley and Z. H. Levine, Optical to UV spectra and birefringence of SiO_2 and TiO_2: First-principles calculations with excitonic effects, *Phys. Rev. B* **78**, 205108–1–8 (2008).
96. Z.-L. Cai, K. Sendt and J. R. Reimers, Failure of density-functional theory and time-dependent density-functional theory for large extended π systems, *J. Chem. Phys* **117**, 5543–5549 (2002).
97. U. Salzner, J. B. Lagowski, P. G. Pickup and R. A. Poirier, Design of low band gap polymers employing density functional theory – hybrid functionals ameliorate band gap problem, *J. Comput. Chem.* **18**, 1943–1953 (1997).
98. F. De Angelis, A. Tilocca and A. Selloni, Time-dependent dft study of $[Fe(CN)_6]^{4-}$ sensitization of TiO_2 nanoparticles, *J. Am. Chem. Soc.* **126**, 15024–15025 (2004).
99. B. Hammer, L. B. Hansen and J. K. Norskov, Improved adsorption energetics with density functional theory using revised Perdew-Burke-Ernzerhof functionals, *Phys. Rev. B* **59**, 7413–7421 (1999).
100. M. Gratzel, Dye-sensitized solar cells, *J. Photochem. Photobiol. C–Photochem. Rev.* **4**, 145–153 (2003).
101. M. A. L. Marques, C. Ullrich, F. Nogueira, A. Rubio, K. Burke and E. Gross, (eds.), *Time-Dependent Density Functional Theory*, Vol. **706**, (Springer, Berlin, 2006).
102. F. De Angelis, A. Tilocca and A. Selloni, Time-dependent DFT study of [Fe(CN)(6)](4-) sensitization of TiO_2 nanoparticles, *J. Am. Chem. Soc.* **126**, 15024–15025 (2004).

103. F. De Angelis, S. Fantacci, A. Selloni and M. Nazeeruddin, Time dependent density functional theory study of the absorption spectrum of the $[Ru(4,4'\text{-}COO^{-}\text{-}2,2'\text{-}bpy)_2(X)_2]^{4-}$ (X = NCS, Cl) dyes in water solution, *Chem. Phys. Lett.* **415**, 115–120, (2005).
104. M. Nazeeruddin, F. De Angelis, S. Fantacci, A. Selloni, G. Viscardi, P. Liska, S. Ito, B. Takeru and M. Gratzel, Combined experimental and DFT-TDDFT computational study of photoelectrochemical cell ruthenium sensitizers, *J. Am. Chem. Soc.* **127**, 16835–16847 (2005).
105. F. De Angelis, S. Fantacci, A. Selloni, M. K. Nazeeruddin and M. Gratzel, Time-dependent density functional theory investigations on the excited states of Ru(II)-dye-sensitized TiO_2 nanoparticles: The role of sensitizer protonation, *J. Am. Chem. Soc.* **129**, 14156–14157 (2007).
106. F. De Angelis, S. Fantacci and A. Selloni, Alignment of the dye's molecular levels with the TiO_2 band edges in dye-sensitized solar cells: a DFT-TDDFT study, *Nanotechnology* **19**, 424002–1–7 (2008).
107. W. R. Duncan, W. M. Stier and O. V. Prezhdo, *Ab initio* nonadiabatic molecular dynamics of the ultrafast electron injection across the alizarin–TiO_2 interface, *J. Am. Chem. Soc.* **127**, 7941–7951 (2005).
108. W. R. Duncan and O. V. Prezhdo, Theoretical studies of photoinduced electron transfer in dye-sensitized TiO_2, *Ann. Rev. Phys. Chem.* **58**, 143–184 (2007).
109. W. R. Duncan, C. F. Craig and O. V. Prezhdo, Time-domain *ab initio* study of charge relaxation and recombination in dye sensitized TiO_2, *J. Am. Chem. Soc.* **129**, 8528–8543 (2007).
110. O. V. Prezhdo, W. R. Duncan and V. V. Prezhdo, Dynamics of the photoexcited electron at the chromophore-semiconductor interface, *Acc. Chem. Res.* **41**, 339–348 (2008).
111. W. R. Duncan and O. V. Prezhdo, Temperature independence of the photoinduced electron injection in dye-sensitized TiO_2 rationalized by *ab initio* time-domain density functional theory, *J. Am. Chem. Soc.* **130**, 9756–9762 (2008).

Chapter 13

Computational Studies of Tailored Negative-Index Metamaterials and Microdevices

Alexander K. Popov
University of Wisconsin–Stevens Point
apopov@wsp.edu

Thomas F. George
University of Missouri–St. Louis
tfgeorge@umsl.edu

The feasibility of tailoring of optical and nonlinear-optical properties of negative-index nanocomposites with control lasers as well as the feasibility of the design of novel photonic microdevices and all-optical data processing chips are shown and proved with numerical simulations.

13.1. Introduction

Optical negative-index (NI) metamaterials (NIMs) form a novel class of artificial electromagnetic media that promises revolutionary breakthroughs in photonics [1]. Nanostructured metamaterials are expected to play a key role in the development of all-optical data processing chips. Negative refraction does not exists in natural media. Its engineering has become possible only recently with the advent of nanotechnologies and can be implemented to develop a wide variety of devices with enhanced and uncommon functions. Unlike ordinary positive-index (PI) materials (PIMs), the energy flow and wave vector (phase velocity) are counter-directed in NIMs, which determines their extraordinary linear and nonlinear-optical (NLO) properties.

Nonlinear optics in NIMs, especially frequency conversion, still remains a less developed branch of photonics. The feasibility of crafting NIMs with strong NLO responses in the optical wavelength range has been experimentally demonstrated in Ref. [2]. It is well established that local-field enhanced nonlinearities can be attributed to plasmonic nanostructures.

Extraordinary NLO properties of propagation processes in NIMs, such as SHG, three-wave mixing (TWM) and four-wave mixing (FWM) optical parametric amplification (OPA) have been predicted, which are in a stark contrast with their counterparts in natural materials [3–14]. Striking changes in the properties of nonlinear pulse propagation and temporal solitons [15], spatial solitons in systems with bistability [16–18], gap solitons [19], and optical bistability in layered structures including NIMs [20,21] have been revealed. A review of some of the corresponding theoretical approaches is given in Ref. [22]. Frequency-degenerate multi-wave mixing and self-oscillations of counter-propagating waves in ordinary materials have been extensively studied because of their easily achievable phase matching. Phase matching for three-wave mixing (TWM) and four-wave mixing (FWM) of contra-propagating waves that are far from degeneracy seem impossible in ordinary materials and presents a technical challenge in the metamaterials. It became possible only recently due to the advances in nanotechnology [23,24]. The possibility and extraordinary properties of TWM with mirrorless self-oscillations from two co-propagating waves with nearly-degenerate frequencies that fall within an anomalous dispersion frequency domain and gives rise to generation of a far-infrared difference-frequency counter-propagating wave in an anisotropic crystal was proposed in Ref. [25] (and references therein) more than 40 years ago (see also Ref. [26,27]). However, far-infrared radiation is typically strongly absorbed in crystals, which presents an unavoidable strong detrimental factor. For the first time, TWM backward-wave (BW) mirrorless optical parametrical oscillator (BW-MOPO) with all three significantly different optical wavelengths was realized only recently [24]. Phase-matching of counter-propagating waves has been achieved in a submicrometer periodically poled NLO crystal, which has become possible owing to recent advances in nanotechnologies. Both in the proposal [25] and in the experiment [24], *the opposite orientation of wave vectors* was required for mirrorless oscillations due to the fact that a PI crystal was implemented. As outlined, a major technical problem in creating BWMOPO stems from the requirement of phase matching with the opposite orientation of wave vectors in PIMs.

Herein, we show the feasibility and *extraordinary features of coherent nonlinear optical coupling of electromagnetic waves with contra-directed energy flows*, whereas their wave-vectors remain parallel. Hence, distributed feedback becomes possible while *an antiparallel orientation of wave vectors of the coupled waves is not required anymore.* Striking changes in the properties of well-known nonlinear-optical propagation processes that follow from this fact are demonstrated first for SHG and then for three- and four-wave mixing processes. The majority of NIMs realized to date consist of metal-dielectric nanostructures that have highly controllable magnetic and dielectric responses. Significant progress has been achieved recently in the design of bulk, multilayered, negative-index, plasmonic slabs [28–31]. The problem, however, is that these structures introduce strong losses inherent to metals that are difficult to avoid, especially in the visible range of frequencies. Irrespective of their origin, losses constitute a major hurdle to the practical realization of the unique optical applications of these structures. Therefore, developing efficient loss-compensating techniques is of a paramount importance. So far, the most common approach to compensating losses in NIMs is associated with the possibility to embed amplifying centers in the host matrix [1]. The amplification is supposed to be provided through a population inversion between the energy levels of the embedded centers. Herein, *we propose* several *alternative options of compensating losses in NIMs* based on coherent, NLO energy transfer from the control optical field(s) to the negative-phase signal through OPA and demonstrate through the numerical simulations *the possibility to implement such an opportunities to design novel photonic microdevices.* Hence, the major emphasis of the work is placed on the numerical demonstration of the possibility of *compensating losses* in the plasmonic negative-index metamaterials through coherent energy transfer from ordinary control to a negative-index, backward, signal wave. This allows for creation of microscopic all-optical switches, filters, amplifiers and a microscopic BWMOPO at appreciably different frequencies while all the wave vectors of the coupled waves remain co-directed. Such an opportunity makes phase matching much easier, which is offered by the backwardness of electromagnetic waves that is natural to NIMs. Ultimately, the possibilities of producing a robust, all-optically tailored narrow-band transparency, switching, amplification and even mirrorless optical parametric oscillations for contra-propagating entangled right- and left-handed photons in originally strongly absorbing bulk microscopic NI samples of plasmonic metal-dielectric nanostructured composites are demonstrated through numerical simulations. Two options are explored. One is OPA that implements nonlinearities attributed to the building blocks of the NIMs [2]. The other option, FWM [9,10,14], proposes independent engineering of a $\chi^{(3)}$ nonlinearity through embedded, resonant, NLO centers. In the vicinity of the resonances, $\chi^{(3)}$ is exceptionally strong. In addition, optical properties of the composite can be tailored by means of quantum control. It is shown that the outcomes are strongly dependent on the relaxation properties of the coupled optical transitions and on the specific coupling schemes. Comparative analysis of several promising schemes, not considered earlier, is given.

The chapter is organized as follows. Section 13.2 describes specific features of SHG for the model of a transparent double-domain NIM, where the above outlined striking differences are most readily seen. Section 13.3 presents basic analytical approaches and equations employed for the computational studies and numerical experiments presented in the next sections. The feasibility to produce transparency, amplification and generation of entangled counter-propagating right- and left-handed photons is demonstrated in Sec. 13.4. The possibility of creation of a tunable nonlinear-optical negative-index mirror and all-optical switch is shown in Sec. 13.5. Section 13.6 demonstrates the feasibility of tailoring optical and nonlinear-optical properties of the negative-index metamaterials through the independent engineering of the negative refractive index and embedded nonlinearities as well as through the means of quantum control. Section 13.7 summarizes the main results of the work.

13.2. Fundamental Difference in Second-Harmonic Generation in Lossles PIM and Double-Domain NIM

13.2.1. *Poynting and wave-vectors in a lossless NIM*

We consider a traveling electromagnetic wave,

$$\mathbf{E}(\mathbf{r},t) = (1/2)\mathbf{E}_0 \exp[i(\mathbf{k}\cdot\mathbf{r} - \omega t)] + c.c.,$$
$$\mathbf{H}(\mathbf{r},t) = (1/2)\mathbf{H}_0 \exp[i(\mathbf{k}\cdot\mathbf{r} - \omega t)] + c.c.$$

From the equations

$$\nabla\times\mathbf{E} = -\frac{1}{c}\frac{\partial\mathbf{B}}{\partial t}, \mathbf{B} = \mu\mathbf{H}, \nabla\times\mathbf{H} = \frac{1}{c}\frac{\partial\mathbf{D}}{\partial t}, \mathbf{D} = \epsilon\mathbf{E}$$

one finds that

$$\mathbf{k}\times\mathbf{E} = \frac{\omega}{c}\mu\mathbf{H}, \mathbf{k}\times\mathbf{H} = -\frac{\omega}{c}\epsilon\mathbf{E}, \sqrt{\epsilon}E = -\sqrt{\mu}H. \qquad (13.1)$$

Equations (13.2.1) show that the vector triplet **E**, **H**, **k** forms a right-handed system for an ordinary medium with $\epsilon_i > 0$ and $\mu_i > 0$. Simultaneously negative ϵ_i and μ_i result in a left-handed triplet and negative refractive index

$$n = -\sqrt{\mu\epsilon}, \quad k^2 = n^2(\omega/c)^2.$$

We assume here that all indices of ϵ, μ and n are real numbers. The direction of the wave-vector $\mathbf{k}$ with respect to the energy flow (Poynting vector [27,32,33]) depends on the signs of ϵ and μ:

$$\mathbf{S}(\mathbf{r},t) = \frac{c}{4\pi}[\mathbf{E}\times\mathbf{H}] = \frac{c^2\mathbf{k}}{4\pi\omega\epsilon}H^2 = \frac{c^2\mathbf{k}}{4\pi\omega\mu}E^2. \quad (13.2)$$

At $\epsilon_i < 0$ and $\mu_i < 0$, $\mathbf{S}$ and $\mathbf{k}$ become contra-directed, which is in contrast to the electrodynamics of ordinary media and opens opportunities for many revolutionary breakthroughs in photonics.

13.2.2. *Second-harmonic generation in loss-free PIMs and NIMs*

First, consider counter-intuitive effects and unusual characteristics in the spatial distribution and energy exchange between the fundamental and second-harmonic (SH) backward waves beyond the constant-pump approximation. Both semi-infinite and finite-length NIM slabs will be compared with each other and with ordinary PIMs. In order to focus on the basic features of the process, our analysis is based on the solution to equations for the slowly-varying amplitudes of the coupled backward fundamental and ordinary SH waves propagating in lossless NIMs. The Manley–Rowe relations for NIMs will be analyzed and shown to be strikingly different from those in PIMs. The feasibility of a nonlinear-optical mirror converting 100% of the incident radiation into a reflected SH wave is shown for the case of a loss-free NIM.

Since negative refraction index exists only within a certain frequency band, consider a loss-free material, which is NI (left-handed) at the fundamental frequency ω_1 ($\epsilon_1 < 0$, $\mu_1 < 0$), whereas it is PI (right-handed) at the SH frequency $\omega_2 = 2\omega_1$ ($\epsilon_2 > 0$, $\mu_2 > 0$). As mentioned, we assume here that all indices of ϵ, μ and n are real numbers. Thus, the energy-flow $\mathbf{S}_1$ at ω_1 is directed opposite to $\mathbf{k}_1$, whereas $\mathbf{S}_2$ is co-directed with $\mathbf{k}_2$. We assume that an incident flow of fundamental radiation $\mathbf{S}_1$ at ω_1 propagates along the z-axis, which is normal to the surface of a metamaterial. According to (13.3.2), the phase of the wave at ω_1 travels in the reverse direction inside the NIM (Fig. 13.1(a)). Because of the phase-matching requirement, the generated SH radiation also travels backward with energy flow in the same backward direction. This is in contrast with the standard coupling geometry in a PIM (Fig. 13.1(b)).

Following the method of Ref. [4], we assume that a nonlinear response is primarily associated with the magnetic component of the waves. Then the equations for the coupled fields inside a NIM in the approximation of slowly-varying amplitudes acquire the form:

$$dA_2/dz = i\sigma_2 A_1^2 \exp(-i\Delta kz), \tag{13.3}$$

$$dA_1/dz = i\sigma_1 A_2 A_1^* \exp(i\Delta kz). \tag{13.4}$$

Here, $\Delta k = k_2 - 2k_1$; $\sigma_1 = (\epsilon_1\omega_1^2/k_1c^2)8\pi\chi^{(2)}$, $\sigma_2 = (\epsilon_2\omega_2^2/k_2c^2)4\pi\chi^{(2)}$; $\chi^{(2)}$ is the effective nonlinear susceptibility; A_2 and A_1 are the slowly-varying amplitudes of the waves with the phases traveling against the z-axis,

$$H_j(z,t) = A_j \exp[-i(\omega_j t + k_j z)] + c.c., \tag{13.5}$$

$j = \{1,2\}$; $\omega_2 = 2\omega_1$; and $k_{1,2} > 0$ are the moduli of the wavevectors directed against the z-axis. We note that according to Eq. (13.1) the corresponding equations for the electric components can be written in a similar form, with ϵ_j substituted by μ_j and *vice versa*. The factors μ_j were usually assumed to be equal to one in similar equations for PIMs. However, this assumption does not hold for the case of NIMs, and this fact dramatically changes many conventional electromagnetic relations. The Manley–Rowe relations [32,33] for the field intensities and for the energy flows follow from Eqs. (13.2)–(13.4):

$$\frac{k_1}{\epsilon_1}\frac{d|A_1|^2}{dz} + \frac{k_2}{2\epsilon_2}\frac{d|A_2|^2}{dz} = 0, \quad \frac{d|S_1|^2}{dz} - \frac{d|S_2|^2}{dz} = 0. \tag{13.6}$$

The latter equation accounts for the difference in the signs of ϵ_1 and ϵ_2, which brings radical changes to the spatial dependence of the field intensities discussed below.

In order to outline the basic difference between the SHG process in NIMs and PIMs, we assume in our further consideration that the phase matching condition $k_2 = 2k_1$ is fulfilled. The spatially-invariant form of the Manley–Rowe relations follows from Eq. (13.6):

$$|A_1|^2/\epsilon_1 + |A_2|^2/\epsilon_2 = C, \tag{13.7}$$

where C is an integration constant. With $\epsilon_1 = -\epsilon_2$, Eq. (13.7) predicts that the *difference* between the squared amplitudes remains constant through the sample,

$$|A_1|^2 - |A_2|^2 = C, \tag{13.8}$$

as schematically depicted in Fig. 13.1(a). This is in striking difference to the requirement that the *sum* of the squared amplitudes is constant in the analogous case in a PIM, as schematically shown in Fig. 13.1(b).

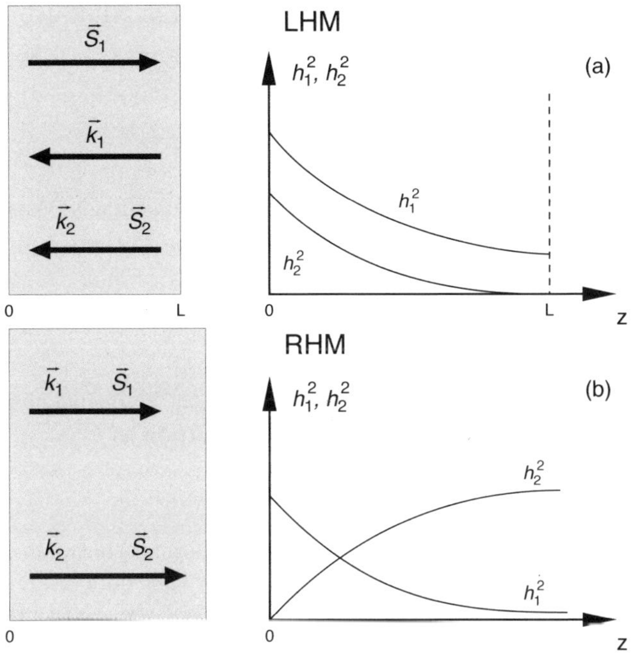

Fig. 13.1. Difference in the phase-matching geometry and in the intensity-distribution for the fundamental and the second harmonic waves, h_1^2 and h_2^2, between a slab of the left-handed metamaterial (a) and a right-handed material (b).

We introduce now the real phases and amplitudes as $A_{1,2} = h_{1,2}\exp(i\phi_{1,2})$. Then the relations for the real amplitudes and phases, which follow from Eqs. (13.3) and (13.4), show that if any of the fields becomes zero at any point, the integral (13.7) corresponds to the solution with the constant phase difference $2\phi_1 - \phi_2 = \pi/2$ over the entire sample. The equations for the slowly-varying amplitudes corresponding to the ordinary coupling scheme in a PIM slab, shown in Fig. 13.1(b), are readily obtained from Eqs. (13.3)–(13.5) by changing the signs of k_1 and k_2 with the known solution

$$h_2(z) = h_{10}\tanh(z/z_0), \tag{13.9}$$

$$h_1(z) = h_{10}/\cosh(z/z_0),\ z_0 = [\kappa h_{10}]^{-1}. \tag{13.10}$$

Here, $\kappa = (\epsilon_2\omega_2^2/k_2c^2)4\pi\chi_{eff}^{(2)}$. The solution has the same form for an *arbitrary* slab thickness with decreasing fundamental and increasing SH squared amplitudes along the z-axis, as shown schematically in Fig. 13.1(b).

Now consider phase-matched SHG in a lossless NIM slab of a finite length L. Equations (13.3) and (13.8) take the form

$$h_1(z)^2 = C + h_2(z)^2, \tag{13.11}$$

$$dh_2/dz = -\kappa[C + h_2(z)^2]. \tag{13.12}$$

Taking into account the *different boundary conditions* in a NIM as compared to a PIM, $h_1(0) = h_{10}$ and $h_2(L) = 0$, the solution to these equations is

$$h_2 = \sqrt{C}\tan[\sqrt{C}\kappa(L - z)], \tag{13.13}$$

$$h_1 = \sqrt{C}/\cos[\sqrt{C}\kappa(L - z)], \tag{13.14}$$

where the integration parameter C depends on the slab thickness L and on the amplitude of the incident fundamental radiation as

$$\sqrt{C}\kappa L = \cos^{-1}(\sqrt{C}/h_{10}). \tag{13.15}$$

Thus, the spatially invariant field intensity difference between the fundamental and SH waves in NIMs depends on the *slab thickness*, which is in strict contrast with the case in PIMs. As seen from Eq. (13.11), the integration parameter $C = h_1(z)^2 - h_2(z)^2$ now represents the deviation of the conversion efficiency $\eta = h_{20}^2/h_{10}^2$ from unity: $(C/h_{10}^2) = 1 - \eta$.

Figure 13.2 depicts the field distribution along the NIM slab with the characteristic conversion length $z_0 = (\kappa h_{10})^{-1}$. One can see from the figure that with an increase in slab length (or intensity of the fundamental wave), the gap between the two plots decreases while the conversion efficiency increases (comparing the main plot and the inset). It can be shown that for the conversion length of 2.5, the NIM slab, which acts as nonlinear mirror, provides about 80% conversion of the fundamental beam into a reflected SH wave.

For a semi-infinite NIM, both waves disappear at $z \to \infty$ due to the entire conversion of the fundamental beam into SH. Hence, in this case $C = 0$. Then Eqs. (13.11) and (13.12) for the amplitudes take the simple form

$$h_2(z) = h_1(z), \tag{13.16}$$

$$dh_2/dz = -\kappa h_2^2. \tag{13.17}$$

Equation (13.16) indicates 100% conversion of the incident fundamental wave into the reflected second harmonic at $z = 0$ in a lossless semi-infinite medium, provided that the phase matching condition $\Delta k = 0$ is fulfilled.

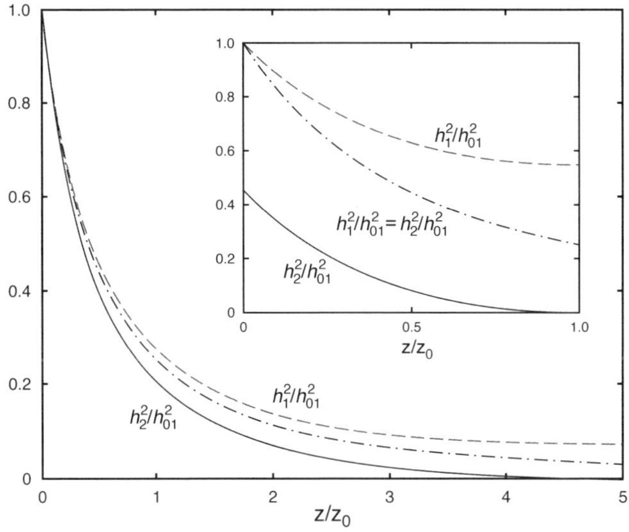

Fig. 13.2. Squared amplitudes for the fundamental wave (dashed line) and SHG (solid line) in a lossless NIM slab of finite length. Inset: The slab has a length equal to one conversion length. Main plot: The slab has a length equal to five conversion lengths. The dash-dot lines show the energy-conversion for a semi-infinite NIM.

The integration of (13.17) with the boundary condition $h_1(0) = h_{10}$ yields

$$h_2(z) = \frac{h_{10}}{(z/z_0) + 1}, \; z_0 = (\kappa h_{10})^{-1}. \tag{13.18}$$

Equation (13.18) describes a *concurrent decrease of both waves of equal amplitudes along the z-axis;* this is shown by the dash-dot plots in Fig. 13.2. For $z \gg z_0$, the dependence is inversely proportional to z. Ultimately, the described features of SHG in a double-domain NIM and spatial dependencies, shown in Fig. 13.2, *are in strict contrast with those for the conventional process of SHG in a PIM*, which are known from various textbooks [compare, for example, with Fig.13.1(b)].

13.3. Backward Waves and Parametric Interaction in a NIM: Master Equations

Consider frequency-mixing nonlinear-optical propagation processes, such as three- and four-wave mixing.

13.3.1. *Coupling geometry and coherent energy transfer from the ordinary control fields to the backward signal in a NIM*

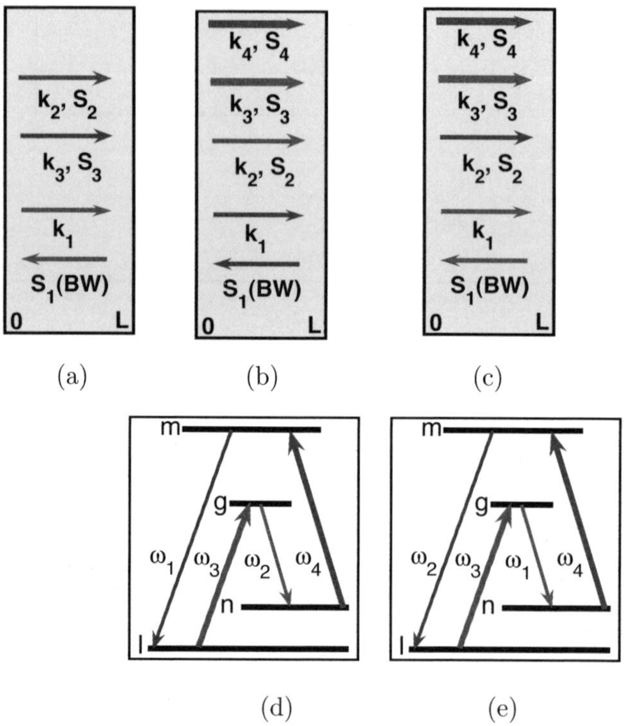

Fig. 13.3. Coupling geometry and alternative schemes of four-wave mixing in the embedded resonant nonlinear-optical centers. (a) Coupling geometry. $\mathbf{S}_1$ — negative-index signal, $\mathbf{S}_3$ — positive-index control field, $\mathbf{S}_2$ — positive-index idler. (b), (c) Coupling geometry for four-wave mixing of the backward and ordinary electromagnetic waves. $\mathbf{S}_1$, $\mathbf{k}_1$ and ω_1 are energy flux, wave-vector and frequency for the backward-wave signal; $\mathbf{S}_2$, $\mathbf{k}_2$ and ω_2 — for the ordinary idler; $\mathbf{S}_{3,4}$, $\mathbf{k}_{3,4}$ and $\omega_{3,4}$ — for the ordinary control fields. (d), (e) Corresponding alternative schemes of quantum controlled four-wave mixing in the embedded resonant nonlinear-optical centers with different ratio of the signal and the idler absorption rates and nonlinear susceptibilities. (d) Shortest-wavelength negative-phase signal, where, depending on the partial relaxation rates, parametric amplification may be assisted by the idler's population-inversion or Raman-type amplification. (e) Longer-wavelength negative-phase signal, where depending on the partial relaxation rates, parametric amplification may be assisted by the signal incoherent amplification attributed to population-inversion or Raman-type gain.

We consider two cases: (1) a NIM slab with a three-wave (TWM) mixing optical nonlinearity $\chi^{(2)}$ attributed to its building blocks (Fig. 13.3(a)), and (2) four-wave mixing (FWM) optical nonlinearity $\chi^{(3)}$ attributed to the embedded nonlinear optical centers (Fig. 13.3(b) and 13.3(c)). In the first case, three coupled optical electromagnetic waves with wave vectors $\mathbf{k}_{1,2,3}$ *co-directed* along the z axis propagate through a slab of thickness L with quadratic, TWM, nonlinearity $\chi^{(2)}$. Only two waves enter the slab — strong control field at ω_3 and weak signal at ω_1 — which then generate a difference-frequency idler at $\omega_2 = \omega_3 - \omega_1$. The idler contributes back to the signal through the similar TWM process, $\omega_1 = \omega_3 - \omega_2$, and thus provides OPA of the signal. The signal is assumed as negative-index, $n(\omega_1) < 0$, and therefore backward wave (BW). This means that the energy flow $\mathbf{S}_1$ is *antiparallel* to $\mathbf{k}_1$ (Fig. 13.3 (a)), in contrasts to the early proposals [25,26] and their recent realization [23,24] of BWOPO in PI materials. The idler and control field are the ordinary waves with parallel $\mathbf{k}_{2,3}$ and $\mathbf{S}_{2,3}$ along the z-axis. Consequently, the control beam enters the slab at $z = 0$, whereas the signal – at $z = L$. In the second FWM case, Fig. 13.3(b) and 13.3(c), the slab is illuminated by two PI control (pump) waves at ω_3 and ω_4. In both cases, all wave-vectors are co-directed along the the z-axis. Due to the parametric interaction, the control and signal fields generate a difference-frequency idler at $\omega_2 = \omega_4 + \omega_3 - \omega_1$ (FWM), which is also assumed to be a PI wave ($n_2 > 0$). The idler contributes back into the wave at ω_1 through the same type of parametric interaction, and thus enables OPA at ω_1 by converting the energy of the control fields into the signal. Thus, all of the coupled waves have their wave-vectors co-directed along z, whereas the energy flow of the signal wave, $\mathbf{S}_1$, is counter-directed to the energy flows of all the other waves, which are codirected with their wave-vectors. Such coupling schemes are in contrast both with the conventional phase-matching scheme for OPA in ordinary materials, where all energy-flows and phase velocities are co-directed, as well as with TWM backward-wave mirrorless OPO [23–26], where both the energy flow and wave-vector of one of the waves are opposite to all others.

13.3.2. *Equations for coupled contrapropagating backward signal and ordinary idler waves*

First, we shall show that magnetic and electric TWM and FWM processes can be treated identically. We consider two alternative types of nonlinearities – electric, $\mathbf{D} = \epsilon\mathbf{E} + 4\pi\mathbf{P}^{NL}$, $\mathbf{B} = \mu\mathbf{H}$, and magnetic,

$\mathbf{B} = \mu\mathbf{H} + 4\pi\mathbf{M}^{NL}$, $\mathbf{D} = \epsilon\mathbf{E}$. Nonlinear polarization and magnetization are sought in the form

$$\mathbf{P}^{NL}(\mathbf{r}, t) = (1/2)\mathbf{P_0}^{NL}(\mathbf{r}) \exp[i(\widetilde{\mathbf{k}} \cdot \mathbf{r} - \omega t)] + c.c.,$$
$$\mathbf{M}^{NL}(\mathbf{r}, t) = (1/2)\mathbf{M_0}^{NL}(\mathbf{r}) \exp[i(\widetilde{\mathbf{k}} \cdot \mathbf{r} - \omega t)] + c.c.$$

Accounting for Eqs. (13.1), one can derive

$$\nabla \times \nabla \times \mathbf{E} = -(\mu/c^2)\partial^2\mathbf{D}/\partial t^2,$$
$$-\Delta\mathbf{E} = \mu(\omega^2/c^2)[\epsilon\mathbf{E} + 4\pi\mathbf{P}^{NL}],$$
$$\nabla \times \nabla \times \mathbf{H} = -(\epsilon/c^2)\partial^2\mathbf{B}/\partial t^2,$$
$$-\Delta\mathbf{H} = \epsilon(\omega^2/c^2)[\mu\mathbf{H} + 4\pi\mathbf{M}^{NL}].$$

For the medium with the electric nonlinearity, the equation for the slowly-varying amplitude $\mathbf{E}_0$ of the wave with the wave-vector along the z-axis takes the form

$$dE_0/dz = i\mu(2\pi\omega^2/kc^2)P_0^{NL} \exp[i(\widetilde{k} - k)z].$$

For the magnetic nonlinearity, the equation is

$$dH_0/dz = i\epsilon(2\pi\omega^2/kc^2)M_0^{NL} \exp[i(\widetilde{k} - k)z].$$

The equations are symmetric and can be converted from one to the other by the replacement $\mu \longleftrightarrow \epsilon$.

For the electric quadratic nonlinearity,

$$P_1^{NL} = \chi_{e1}^{(2)} E_3 E_2^* \exp\{i[(k_3 - k_2)z - \omega_1 t]\},$$
$$P_2^{NL} = \chi_{e2}^{(2)} E_3 E_1^* \exp\{i[(k_3 - k_1)z - \omega_2 t]\},$$

where $\omega_2 = \omega_3 - \omega_1$ and $k_j = |n_j|\omega_j/c > 0$. Then the equations for the slowly-varying amplitudes of the signal and idler in the lossy medium can be given in the form

$$dE_1/dz = i\sigma_{e1}E_2^* \exp[i\Delta kz] + (\alpha_1/2)E_1, \quad (13.19)$$
$$dE_2/dz = i\sigma_{e2}E_1^* \exp[i\Delta kz] - (\alpha_2/2)E_2. \quad (13.20)$$

Here, $\sigma_{ej} = (k_j/\epsilon_j)2\pi\chi_{ej}^{(2)}E_3$, $\Delta k = k_3 - k_2 - k_1$, and α_j are the absorption indices. The depletion of the control (pump) wave E_3 due to the NLO energy conversion is neglected here. Hence, the equation for the control field with the account for absorption takes a standard form and can be solved independently.

For the magnetic type of quadratic nonlinearity,

$$M_1^{NL} = \chi_{m1}^{(2)} H_3 H_2^* \exp\{i[(k_3 - k_2)z - \omega_1 t]\},$$
$$M_2^{NL} = \chi_{m2}^{(2)} H_3 H_1^* \exp\{i[(k_3 - k_1)z - \omega_2 t]\},$$

the equations for the slowly-varying amplitudes are

$$dH_1/dz = i\sigma_{m1} H_2^* \exp[i\Delta kz] + (\alpha_1/2)H_1, \tag{13.21}$$
$$dH_2/dz = i\sigma_{m2} H_1^* \exp[i\Delta kz] - (\alpha_2/2)H_2. \tag{13.22}$$

Here, $\sigma_{mj} = (k_j/\mu_j)2\pi\chi_{mj}^{(2)} H_3$, $H_3 = const$, and the other notations remain the same.

For the electric-type FWM, the equations for the slowly-varying amplitudes are similar:

$$dE_1/dz = i\gamma_1 E_2^* \exp[i\Delta kz] + (\alpha_1/2)E_1, \tag{13.23}$$
$$dE_2/dz = i\gamma_2 E_1^* \exp[i\Delta kz] - (\alpha_2/2)E_2. \tag{13.24}$$

Here, $\gamma_j = (k_j/\epsilon_j)2\pi\chi_j^{(3)} E_3 E_4$ and $\Delta k = k_3 + k_4 - k_1 - k_2$.

We introduce effective amplitudes, $a_{e,m,j}$, and nonlinear coupling parameters, $g_{e,m,j}$, which for the electric and magnetic types of quadratic nonlinearity are defined as

$$a_{ej} = \sqrt{|\epsilon_j/k_j|}E_j, g_{ej} = \sqrt{|k_1 k_2/\epsilon_1\epsilon_2|}2\pi\chi_{ej}^{(2)} E_3,$$
$$a_{mj} = \sqrt{|\mu_j/k_j|}H_j, g_{mj} = \sqrt{|k_1 k_2/\mu_1\mu_2|}2\pi\chi_{mj}^{(2)} H_3,$$

and for FWM as

$$a_j = \sqrt{|\epsilon_j/k_j|}E_j, q_j = \sqrt{|k_1 k_2/\epsilon_1\epsilon_2|}2\pi\chi_j^{(3)} E_3 E_4.$$

The quantities $|a_j|^2$ are proportional to the photon numbers in the energy fluxes. Equations for the amplitudes a_j are identical for all of the types of nonlinearities studied here:

$$da_1/dz = -ig_1 a_2^* \exp(i\Delta kz) + (\alpha_1/2)a_1, \tag{13.25}$$
$$da_2/dz = ig_2 a_1^* \exp(i\Delta kz) - (\alpha_2/2)a_2, \tag{13.26}$$
$$da_3/dz = -(\alpha_3/2)a_3,$$
$$da_4/dz = -(\alpha_4/2)a_4.$$

Here, the last two equations account for absorption of the control fields.

We note the following *three fundamental differences* in Eqs. (13.19)–(13.24) as compared with their counterpart in ordinary, PI materials. First, the signs of σ_1 and γ_1 are opposite to those of σ_2 and γ_2 because $\epsilon_1 < 0$

and $\mu_1 < 0$. Second, the opposite sign appears with α_1 because the energy flow $\mathbf{S_1}$ is against the z-axis. Third, the boundary conditions for the signal are defined at the opposite side of the sample as compared to the idler because the energy flows $\mathbf{S_1}$ and $\mathbf{S_2}$ are counter-directed. Consequently, the sign on the right side of Eq. (13.25) is opposite to that in Eq. (13.26). As will be discussed below, this leads to dramatic changes in the solutions to the equations and in the general behavior of the NLO system. It is convenient to introduce an energy distribution for the backward wave, $T_1(z) = |a_1(z)/a_{1L}|^2$, and for the PI idler, $\eta_2(z) = |a_2(z)/a_{1L}^*|^2$, across the slab. Then the transmission factor for the backward-wave signal at $z = 0$, T_1, and the output idler at $z = L$, η_2, are given by $T_1 = |a_1(0)/a_{1L}|^2$, $\eta_2 = |a_2(L)/a_{1L}^*|^2$. At $a_{1L} = 0$, $a_2(z = 0) = a_{20}$, the slab serves as an NLO mirror with a reflectivity r_1 (output conversion efficiency) at ω_1 given by the equation $r_1 = |a_1(0)/a_{20}^*|^2$. The transmission factor for the PI incident wave is given by $T_2 = |a_2(L)/a_{20}|^2$.

13.3.2.1. *Special cases: Manley–Rowe relations and solutions to the equations for coupled counter-propagating waves*

Consider the special case where the spatial inhomogeneity of the control fields can be neglected. Then Eqs. (13.25)–(13.26) reduce to the the coupled equation for slowly-varying amplitudes a_1 and a_2 with a_3 and a_4 taken constant. For $\alpha_{1,2} = 0$, $g_1 = g_2$, e.g., for off-resonant coupling, one finds with the aid of Eqs. (13.2) and (13.25), (13.26):

$$\frac{d}{dz}\left[\frac{S_{1z}}{\hbar\omega_1} - \frac{S_{2z}}{\hbar\omega_2}\right] = 0, \quad \frac{d}{dz}\left[|a_1|^2 + |a_2|^2\right] = 0.$$

These equations represent the Manley–Rowe relations [32,33], which describe the creation of pairs of entangled counter-propagating photons $\hbar\omega_1$ and $\hbar\omega_2$. The second equation predicts that the *sum* of the terms proportional to the squared amplitudes of the signal and idler remains constant through the sample, which is due to the opposite signs of S_{1z} and S_{2z} and is in contrast to the requirement that the *difference* of such terms is constant in the analogous case in ordinary nonlinear-optical materials.

Taking into account the boundary conditions $a_1(z = L) = a_{1L}$ and $a_2(z = 0) = a_{20}$ (L is the slab thickness) and assuming a_3 and a_4 constant, the solutions to Eqs. (13.25) and (13.26) can be written as

$$a_1(z) = A_1 \exp[(\beta_1 + i\frac{\Delta k}{2})z] + \\ + A_2 \exp[(\beta_2 + i\frac{\Delta k}{2})z], \quad (13.27)$$

$$a_2^*(z) = \kappa_1 A_1 \exp[(\beta_1 - i\frac{\Delta k}{2})z] + \\ + \kappa_2 A_2 \exp[(\beta_2 - i\frac{\Delta k}{2})z], \quad (13.28)$$

where

$$\beta_{1,2} = (\alpha_1 - \alpha_2)/4 \pm iR,\ \kappa_{1,2} = [\pm R + is]/g,$$
$$R = \sqrt{g^2 - s^2}, g^2 = g_2^* g_1, s = (\alpha_1 + \alpha_2)/4 - i\Delta k/2,$$
$$A_1 = \{a_{1L}\kappa_2 - a_{20}^* \exp[(\beta_2 + i\frac{\Delta k}{2})L]\}/D,$$
$$A_2 = -\{a_{1L}\kappa_1 - a_{20}^* \exp[(\beta_1 + i\frac{\Delta k}{2})L]\}/D,$$
$$D = \kappa_2 \exp[(\beta_1 + i\frac{\Delta k}{2})L] - \kappa_1 \exp[(\beta_2 + i\frac{\Delta k}{2})L].$$

For $\Delta k = 0$ and $\operatorname{Im} g = 0$, $(\alpha_1 + \alpha_2)L \ll \pi$ (off-resonance), Eqs. (13.27)–(13.28) reduce to

$$a_1^*(z) \approx \frac{a_{1L}^*}{\cos(gL)} \cos(gz) + \frac{ia_{20}}{\cos(gL)} \sin[g(z - L)],$$
$$a_2(z) \approx \frac{ia_{1L}^*}{\cos(gL)} \sin(gz) + \frac{a_{20}}{\cos(gL)} \cos[g(z - L)].$$

The output amplitudes are then given by

$$a_{10}^* = [a_{1L}^*/\cos(gL)] - ia_{20} \tan(gL),$$
$$a_{2L} = ia_{1L}^* \tan(gL) + [a_{20}/\cos(gL)].$$

For $a_{20} = 0$, the equations for the energy distribution for the backward wave, $T_1(z) = |a_1(z)/a_{1L}|^2$, and for the PI idler, $\eta_2(z) = |a_2(z)/a_{1L}^*|^2$, across the slab take the form

$$T_1(z) = |[\kappa_2 \exp(\beta_1 z) - \kappa_1 \exp(\beta_2 z)]/D|^2, \quad (13.29)$$
$$\eta_2(z) = |[\exp(\beta_1 z) - \exp(\beta_2 z)]/D|^2. \quad (13.30)$$

Then the transmission factor for the backward-wave signal at $z = 0$, T_1, and the output idler at $z = L$, η_2, are given by

$$T_1 = \left|\frac{a_1(0)}{a_{1L}}\right|^2 = \left|\frac{\exp\{-[(\alpha_1/2) - s]\,L\}}{\cos RL + (s/R)\sin RL}\right|^2, \tag{13.31}$$

$$\eta_2 = \left|\frac{a_2(L)}{a_{1L}^*}\right|^2 = \left|\frac{(g/R)\sin RL}{\cos RL + (s/R)\sin RL}\right|^2. \tag{13.32}$$

For $a_{1L} = 0$, $a_2(z = 0) = a_{20}$, the slab serves as an NLO mirror with a reflectivity r_1 (output conversion efficiency) at ω_1 given by an equation identical to Eq. (13.32):

$$r_1 = \left|\frac{a_1(0)}{a_{20}^*}\right|^2 = \left|\frac{(g/R)\sin RL}{\cos RL + (s/R)\sin RL}\right|^2. \tag{13.33}$$

The transmission factor for the PI incident wave is given by

$$T_2 = \left|\frac{a_2(L)}{a_{20}}\right|^2 = \left|\frac{\exp\{-[(\alpha_1/2) - s]\,L\}}{\cos RL + (s/R)\sin RL}\right|^2, \tag{13.34}$$

The above given solutions are useful for a numerical analysis of the basic transmission properties of a NIM slab, their extraordinary features as compared with the ordinary, PI media, and for demonstration of the feasibilities of all-optical control of the transparency and reflectivity of such a slab.

13.3.3. *Density matrix equations and local optical parameters of the medium*

A significant difference between the resonant and near-resonant processes attributed to the embedded centers and off-resonant NLO processes associated with the nonlinearity of the host metamaterial is that all local resonance optical parameters become intensity dependent, and hence their spectral properties may experience a radical change near resonance. In particular, the NLO susceptibilities and, therefore, the parameters γ_1 and γ_2 become complex and differ from each other in the vicinity of the resonances. Hence, the factor g^2 may become negative or complex. This indicates an additional phase shift between the NLO polarization and the generated wave, causing further radical changes in the nonlinear propagation features, which can be tailored. With account for the embedded centers, the macroscopic parameters in Eqs. (13.25) and (13.26) are convenient for calculations with the density matrix technique, which allows one to account for various relaxation and incoherent excitation processes. For the case of Fig. 13.3(d), the power-dependent susceptibility χ_1 responsible for absorption and refraction at ω_1 can be found as

$$P(\omega_1) = \chi_1 E_1,\ P(\omega_1) = N\,\rho_{lm}\,d_{ml} + c.c.,$$

where P is the polarization of the medium oscillating with the frequency ω_1, N is the number density of molecules, d_{ml} is the electric dipole moment of the transition, and ρ_{lm} is the density matrix element. Other polarizations are determined in the same way. The density matrix equations for a mixture of pure quantum mechanical ensembles in the interaction representation can be written in a general form as

$$\begin{aligned} L_{nn}\rho_{nn} &= q_n - i[V,\rho]_{nn} + \gamma_{mn}\rho_{mm}, \\ L_{lm}\rho_{lm} &= L_1\rho_1 = -i[V,\rho]_{lm}, L_{ij} = d/dt + \Gamma_{ij}, \\ V_{lm} &= G_{lm}\cdot\exp\{i[\Omega_1 t - kz]\}, \quad G_{lm} = -\mathbf{E_1}\cdot\mathbf{d_{lm}}/2\hbar, \end{aligned}$$

where $\Omega_1 = \omega_1, -\omega_{ml}$ is the frequency detuning from the corresponding resonance; Γ_{mn} — homogeneous half-widths of the corresponding transition (in the collisionless regime $\Gamma_{mn} = (\Gamma_m + \Gamma_n)/2$); $\Gamma_n = \sum_j \gamma_{nj}$ — inverse lifetimes of levels; γ_{mn} — rate of relaxation from level m to n; and $q_n = \sum_j w_{nj} r_j$ rate of incoherent excitation to state n from underlying levels. The equations for the other elements are written in the same way.

It is necessary to distinguish the open and closed energy-level configurations. In the open case (where the lowest level is not the ground state), the rate of incoherent excitation to various levels by an external source essentially does not depend on the rate of induced transitions between the considered levels. In the closed case (where the lowest level is the ground state), the excitation rate to different levels and velocities depends on the value and velocity distribution at other levels, which are dependent on the intensity of the driving fields. For open configurations, q_i are primarily determined by the population of the ground state and do not depend on the driving fields. Inhomogeneous broadening of the transitions can be accounted for in the final formulas by substituting Ω_i for $\Omega_i' = \Omega_i - \delta_i$, where δ_i is the resonance shift for an individual center embedded in the host material, and then by averaging over the shifts. Equations for the case of Fig. 13.3(e) are obtained by replacing indices $1 \leftrightarrow 2$.

In a steady-state regime, the solution of a set of density-matrix equations can be cast in the following form:

$$\begin{aligned}
&\rho_{ii} = r_i,\ \rho_{lg} = r_3 \cdot \exp(i\Omega_3 t),\ \ \rho_{nm} = r_4 \cdot \exp(i\Omega_4 t),\\
&\rho_{ng} = r_2 \cdot \exp(i\Omega_2 t) + \tilde{r}_2 \cdot \exp[i(\Omega_3 + \Omega_4 - \Omega_1)t],\\
&\rho_{lm} = r_1 \cdot \exp(i\Omega_1 t) + \tilde{r}_1 \cdot \exp[i(\Omega_3 - \Omega_2 + \Omega_4)t],\\
&\rho_{ln} = r_{32} \cdot \exp[i(\Omega_3 - \Omega_2)t] + r_{41} \cdot \exp[i(\Omega_1 - \Omega_4)t].
\end{aligned}$$

The density matrix amplitudes r_i determine the absorption/gain and refraction indexes, and $\tilde{r}_i$ determine the four-wave mixing driving nonlinear polarizations. Then the problem reduces to the set of algebraic equations

$$\begin{aligned}
&P_2 r_2 = iG_2\Delta r_2 - iG_4 r_{42}^* + i r_{32}^* G_3,\\
&d_2\tilde{r}_2 = -iG_4 r_{13}^* + i r_{14}^* G_3,\\
&P_3 r_3 = i\left[G_1\Delta r_1 - G_1 r_{13} + r_{14}G_4\right],\\
&d_1\tilde{r}_1 = -iG_3 r_{42} + i r_{32} G_4\\
&P_{13} r_{13} = -iG_3^* r_1 + i r_3^* G_1,\\
&P_{14} r_{14} = -iG_1 r_4^* + i r_1 G_4^*,\\
&P_{42} r_{42} = -iG_2^* r_4 + i r_2^* G_4,\\
&P_{32} r_{32} = -iG_3 r_2^* + i r_3 G_2^*,\\
&P_3 r_1 = iG_3\Delta r_3,\ P_4 r_4 = iG_4\Delta r_4,
\end{aligned} \tag{13.35}$$

$$\begin{aligned}
&\Gamma_m r_m = -2\,\mathrm{Re}\{iG_4^* r_4\} + q_m,\\
&\Gamma_n r_n = -2\,\mathrm{Re}\{iG_4^* r_4\} + \gamma_{gn} r_g + \gamma_{mn} r_m + q_n,\\
&\Gamma_g r_g = -2\,\mathrm{Re}\{iG_3^* r_3\} + q_g,\\
&\Gamma_l r_l = -2\,\mathrm{Re}\{iG_3^* r_3\} + \gamma_{gl} r_g + \gamma_{ml} r_m + q_l,
\end{aligned} \tag{13.36}$$

where $G_1 = -\mathbf{E_1 d}_{ml}/2\hbar$, $G_2 = -\mathbf{E_2 d}_{gn}/2\hbar$, $G_3 = -\mathbf{E_3 d}_{gl}/2\hbar$, $G_4 = -\mathbf{E_4 d}_{mn}/2\hbar$, $P_1 = \Gamma_{ml} + i\Omega_1$, $P_2 = \Gamma_{ng} + i\Omega_2$, $P_3 = \Gamma_{gl} + i\Omega_3$, $P_4 = \Gamma_{mn} + i\Omega_4$, $P_{32} = \Gamma_{ln} + i(\Omega_3 - \Omega_2)$, $P_{14} = \Gamma_{ln} + i(\Omega_1 - \Omega_4)$, $P_{42} = \Gamma_{gm} + i(\Omega_4 - \Omega_2)$, $P_{13} = \Gamma_{gm} + i(\Omega_1 - \Omega_3)$, $d_2 = \Gamma_{ng} + i(\Omega_3 + \Omega_4 - \Omega_1)$, $d_1 = \Gamma_{lm} + i(\Omega_3 - \Omega_2 + \Omega_4)$, $\Omega_1 = \omega_1 - \omega_{lm}$, $\Omega_3 = \omega_3 - \omega_{lg}$, $\Omega_2 = \omega_2 - \omega_{gn}$, $\Omega_4 = \omega_4 - \omega_{mn}$, $\Delta r_1 = r_l - r_m$, $\Delta r_2 = r_n - r_g$, $\Delta r_3 = r_l - r_g$, and $\Delta r_4 = r_n - r_m$. For a closed scheme, Eq. (13.36) must be replaced by

$$r_l = 1 - r_n - r_g - r_m.$$

13.3.3.1. *Nonlinear susceptibilities and energy level populations for the case where each level is coupled to only one driving field: Open and closed schemes*

With the aid of the solution of a set of Eqs. (13.35) for the off-diagonal elements of the density matrix up to first order in perturbation theory with respect to the weak fields, the equations for the susceptibilities can be expressed as [34]

$$\frac{\chi_i}{\chi_i^0} = \frac{\Gamma_i \Delta r_i}{P_i \Delta n_i}(i = 3, 4), \frac{\chi_i}{\chi_i^0} = \frac{\Gamma_i R_i}{P_i \Delta n_i}(i = 1, 2).$$

Here, χ_i^0 is a resonance value of the susceptibility for all fields turned off, where $R_{1,2}$ are given below:

$$\tilde{\chi}_2 = -iN\frac{d_{ml}d_{lg}d_{gn}d_{nm}/8\hbar^3}{d_2(1 + v_5^* + g_5^*)} \times \left[\left(\frac{\Delta r_3}{P_3 P_{13}^*} + \frac{\Delta r_4}{P_4 P_{14}^*}\right) + \frac{R_1^*}{P_1^*}\left(\frac{1}{P_{13}^*} + \frac{1}{P_{14}^*}\right)\right],$$

$$\tilde{\chi}_1 = -iN\frac{d_{ml}d_{lg}d_{gn}d_{nm}/8\hbar^3}{d_1(1 + v_7^* + g_7^*)} \times \left[\left(\frac{\Delta r_3}{P_3 P_{32}} + \frac{\Delta r_4}{P_4 P_{42}}\right) + \frac{R_2^*}{P_2^*}\left(\frac{1}{P_{32}} + \frac{1}{P_{42}}\right)\right],$$

$$R_2 = \Delta r_2(1 + g_7 + v_7) \times \frac{1 - v_3\dfrac{(1 + v_7 - g_8)\Delta r_4}{(1 + g_7 + v_7)\Delta r_2} - g_3\dfrac{(1 + g_7 - v_8)\Delta r_3}{(1 + g_7 + v_7)\Delta r_2}}{(1 + g_2 + v_2) + [g_7 + g_2(g_7 - v_8) + v_7 + v_2(v_7 - g_8)]},$$

$$R_1 = \Delta r_1(1 + v_5 + g_5) \times \frac{1 - \dfrac{g_1(1 + g_5 - v_6)\Delta r_3}{(1 + v_5 + g_5)\Delta r_1} - v_1\dfrac{(1 + v_5 - g_6)\Delta r_4}{(1 + v_5 + g_5)\Delta r_1}}{(1 + g_4 + v_4) + [v_5 + v_4(v_5 - g_6) + g_5 + g_4(g_5 - v_6)]},$$

$$g_1 = |G_3|^2/P_{13}P_3^*, g_2 = |G_3|^2/P_{32}^*P_2, g_3 = |G_3|^2/P_{42}^*P_4^*,$$
$$g_4 = |G_3|^2/P_{13}P_1, g_5 = |G_3|^2/P_{14}d_2^*, g_6 = |G_3|^2 P_{13}d_2^*,$$
$$g_7 = |G_3|^2/P_{42}^*d_1^*, g_8 = |G_3|^2/P_{32}^*d_1^*, v_1 = |G_4|^2/P_{14}P_4^*,$$
$$v_2 = |G_4|^2/P_{42}^*P_2, v_3 = |G_4|^2/P_{42}^*P_4^*, v_4 = |G_4|^2/P_{14}P_1,$$
$$v_5 = |G_4|^2/P_{13}d_2^*, v_6 = |G_4|^2/P_{14}d_2^*, v_7 = |G_4|^2/P_{32}^*d_1^*,$$
$$v_8 = |G_4|^2/P_{42}^*d_1^*.$$

The populations are described by the formulas below.

OPEN CONFIGURATION:

$$\Delta r_3 = \frac{(1+æ_4)\Delta n_3 + b_1 æ_4 \Delta n_4}{(1+æ_3)(1+æ_4) - a_1 æ_3 b_1 æ_4},$$
$$\Delta r_4 = \frac{(1+æ_3)\Delta n_4 + a_1 æ_3 \Delta n_3}{(1+æ_3)(1+æ_4) - a_1 æ_3 b_1 æ_4},$$
$$\Delta r_2 = \Delta n_2 - b_2 æ_4 \Delta r_4 - a_2 æ_3 \Delta r_3,$$
$$\Delta r_1 = \Delta n_1 - a_3 æ_3 \Delta r_3 - b_3 æ_4 \Delta r_4,$$

$$r_m = n_m + (1-b_2)æ_4 \Delta r_4,$$
$$r_g = n_g + (1-a_3)æ_3 \Delta r_3,$$
$$r_n = n_n - b_2 æ_4 \Delta r_4 + a_1 æ_3 \Delta r_3,$$
$$r_l = n_l - b_1 æ_4 \Delta r_4 + a_3 æ_3 \Delta r_3,$$

where

$$æ_3 = æ_3^0 \Gamma_{lg}^2/|P_3|^2, \quad æ_4 = æ_4^0 \Gamma_{mn}^2/|P_4|^2,$$
$$æ_3^0 = 2(\Gamma_l + \Gamma_g - \gamma_{gl})|G_3|^2/\Gamma_l \Gamma_g \Gamma_{lg},$$
$$æ_4^0 = 2(\Gamma_m + \Gamma_n - \gamma_{mn})|G_4|^2/\Gamma_m \Gamma_n \Gamma_{mn},$$

$$a_1 = \frac{\gamma_{gn} a_2}{\Gamma_n - \gamma_{gn}} = \frac{\gamma_{gn}\Gamma_l a_3}{\Gamma_n(\Gamma_g - \gamma_{gl})} = \frac{\gamma_{gn}\Gamma_l}{\Gamma_n(\Gamma_l + \Gamma_g - \gamma_{gl})},$$
$$b_1 = \frac{\gamma_{ml}\Gamma_n b_2}{\Gamma_l(\Gamma_m - \gamma_{mn})} = \frac{\gamma_{ml} b_3}{\Gamma_l(\Gamma_l - \gamma_{ml})} = \frac{\gamma_{ml}\Gamma_n}{\Gamma_l(\Gamma_m + \Gamma_n - \gamma_{mn})}.$$

CLOSED CONFIGURATION:

In this case, the populations of levels are given by the equations

$$\begin{aligned}
\Gamma_m r_m &= w_m r_l - 2\,\mathrm{Re}\left\{iG_4^* r_4\right\},\\
\Gamma_g r_g &= w_g r_l - 2\,\mathrm{Re}\left\{iG_3^* r_3\right\},\\
\Gamma_n r_n &= w_n r_l + 2\,\mathrm{Re}\left\{iG_4^* r_4\right\}\\
&\quad + \gamma_{gn} r_g + \gamma_{mn} r_m,\\
r_l &= 1 - r_m - r_g - r_n,
\end{aligned}$$

whose solution is

$$\begin{aligned}
&r_l = n_l(1+\ae_4)(1+\ae_3)/\beta,\\
&r_g = (1+\ae_4)[n_l(1+\ae_3) - \Delta n_3]/\beta,\\
&r_n = \{n_m(1+\ae_4)(1+\ae_3)\\
&\quad + [\Delta n_4(1+\ae_3) + \Delta n_3\gamma_2\ae_3/\Gamma_n](1+b\ae_4)\ \}/\beta,\\
&r_m = \{n_m(1+\ae_4)(1+\ae_3)\\
&\quad + [\Delta n_4(1+\ae_3) + \Delta n_3\gamma_2\ae_3/\Gamma_n]b\ae_4\ \}/\beta,
\end{aligned}$$

$$\begin{aligned}
&\Delta r_4 = r_n - r_m = \left[\Delta n_4(1+\ae_3) + \Delta n_3\gamma_2\ae_3/\Gamma_n\right]/\beta,\\
&\Delta r_3 = r_l - r_g = \Delta n_3(1+\ae_4)/\beta.
\end{aligned}$$

Here,

$$\begin{aligned}
\Delta n_3 &= n_l - n_g, \Delta n_4 = n_n - n_m,\\
n_m &= n_l w_m/\Gamma_m, n_g = n_l w_g/\Gamma_g, n_n = n_l w_n{}'/\Gamma_n,\\
n_l &= (1 + w_m/\Gamma_m + w_g/\Gamma_g + w_n{}'/\Gamma_n)^{-1},\\
w_n{}' &= w_n + w_g\gamma_{gn}/\Gamma_n + w_m\gamma_{mn}/\Gamma_n,\\
b &= \Gamma_n/(\Gamma_m + \Gamma_n - \gamma_4).
\end{aligned}$$

$$\begin{aligned}
&\ae_3 = \frac{(2|G_3|^2}{\Gamma_3\Gamma_g)(\Gamma_3^2/|P_3|^2)}, \ae_4 = \frac{2|G_4|^2(\Gamma_m + \Gamma_n - \gamma_4)}{\Gamma_m\Gamma_n\Gamma_4(\Gamma_4^2/|P_4|^2)},\\
&\beta = (1+\ae_4)[1 - \Delta n_4 + 2(n_l + n_m)\ae_3]\\
&\quad + (1+2b\ae_4)[\Delta n_4(1+\ae_3) + \Delta n_3\gamma_2\ae_3/\Gamma_n].
\end{aligned}$$

The remaining notations are as before.

13.4. Laser-Induced Transparency, Amplification and Generation of the Backward Wave

The fundamental difference between the spatial distribution of the signal in ordinary and NI slabs is explicitly seen at $\alpha_j = \Delta k = 0$. Then, Eq. (13.31) reduces to

$$T_{10} = 1/[\cos(gL)]^2. \tag{13.37}$$

Equation (13.37) shows that the output signal and idler experience a sequence of geometrical resonances at $gL \to (2j+1)\pi/2$, (j = 0, 1, 2, ...), as functions of the slab thickness L and of the intensity of the control field (factor g). Such behavior is in stark contrast to that in an ordinary medium, where the signal would grow exponentially as $T_1 \propto \exp(2gL)$. The resonances indicate that strong absorption of the left-handed wave and of the idler can be turned into transparency, amplification and even *cavity-free self-oscillation* when the denominator tends to zero. The reflection factor, r_{10}, and conversion factor, η_{2L}, experience a similar resonance increase. Self-oscillations would provide for the generation of *entangled counter-propagating left-handed, $\hbar\omega_1$, and right-handed, $\hbar\omega_2$, photons without a cavity.* Similar behavior is characteristic for distributed-feedback lasers and is equivalent to a great extension of the NLO coupling length. It is known that even weak amplification per unit length may lead to lasing, provided that the corresponding frequency coincides with a high-quality cavity or feedback resonances. Numerical simulations described below in subsection 13.4.1 show that absorption and phase mismatch $\Delta k = k_3 - k_2 - k_1$ may essentially change the properties of the spatial distribution and the output values of the signal. However, the basic cardinal difference between the field distributions and transmission properties of an ordinary and NIM slab of the same optical thickness is explicitly seen with the numerical examples shown in Fig. 13.4. Figure 13.4(a) displays "geometrical" resonances where amplification may exceed the oscillation threshold, which provides mirrorless OPO. It contrasts with the exponential dependence depicted in Fig. 13.4(b) which is computed for a positive index slab. Figures 13.4(c)–13.4(f) display the differences in the corresponding spatial distributions.

13.4.1. *Effect of absorption and phase mismatch on the laser-induced transparency resonances*

The fact that the waves decay towards opposite directions causes a specific strong dependence of the entire propagation process and, consequently, of

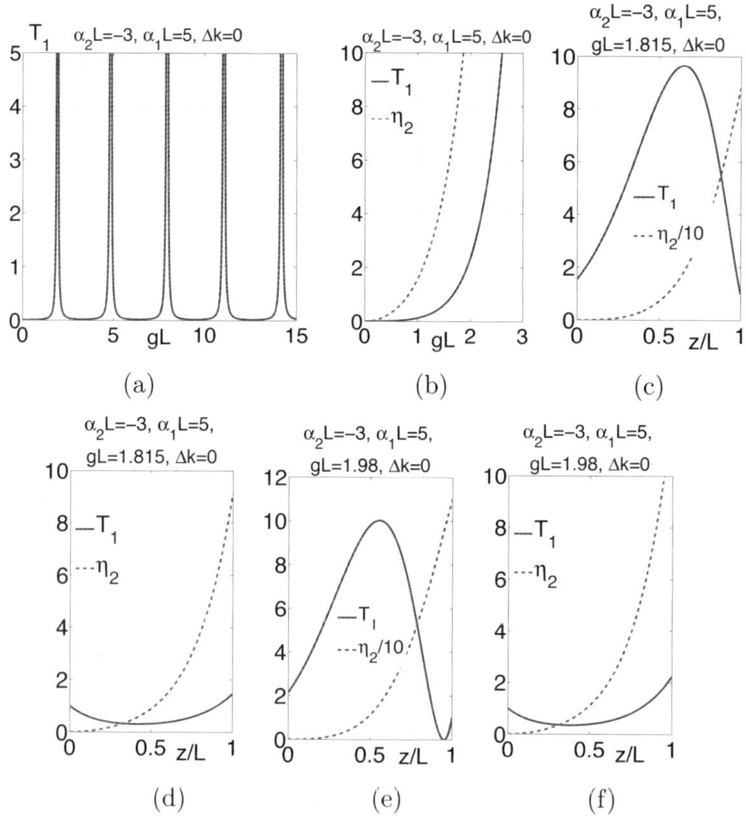

Fig. 13.4. (a) Transmitted negative-index signal, $T_1(z = 0)$, and generated idler, $\eta_2(z = L)$. (b) $T_1(z = L)$ and generated idler, $\eta_2(z = L)$ for positive-index slab of the same optical thickness $\alpha_1 L$ and $\alpha_2 L$. $\alpha_3 L = 0$. (c–f) The difference in the distribution of the fields across the negative-index, (c) and (e), and positive-index, (d) and (f), slabs. Panels (c) and (d) correspond to the left and (e) and (f) to the right slopes of the first peak in panel (a). Here, the materials are absorptive at the frequency of the signal and amplifying for the idler, $\Delta k = 0$.

the transmission properties of the slab on the ratio of the decay rates. A typical plasmonic NIM slab absorbs about 90% of light at the frequencies which are in the NI frequency-range. Such absorption corresponds to $\alpha_1 L \approx 2.3$. As outlined above, the transparency exhibits an extraordinary resonance behavior as a function of the intensity of the control field and the NIM slab thickness, which occurs due to the backwardness of the light waves in NIMs. Basically, such resonances are narrow, and the sample remains

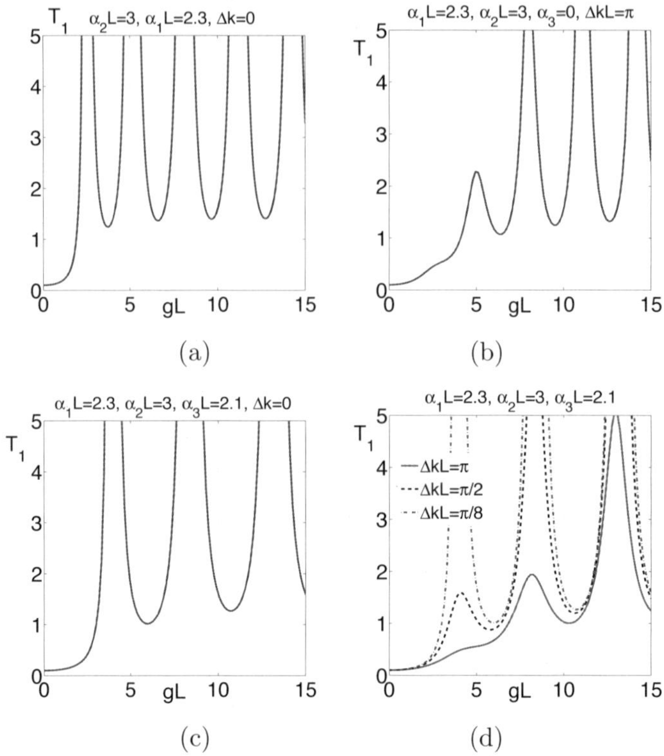

Fig. 13.5. Effect of inhomogeneity of the control field across the slab caused by its absorption on the transmission resonances for different phase mismatches.

opaque anywhere beyond the resonance intensity of the control field. If the nonlinear susceptibility varies within the negative-index frequency domain, this translates into relatively narrow-band filtering. Alternatively, the slab would become transparent within the broad range of the slab thickness and the control field intensity if the transmission in all of the minima is around 1 or more. The results of the numerical simulations presented below (see also Ref. [12]) show that such robust transparency can be achieved thorough the appropriate adjustment of the absorption indices at the frequencies of the coupled fields. The model used in Ref. [12] assumes that the slab is transparent for the control field, which is often not the case in the real world. In order to demonstrate the major effects of absorption of all three fields and of the phase mismatch, we adopt the model where the dependence of the local optical and NLO parameters on the intensity of the control field can

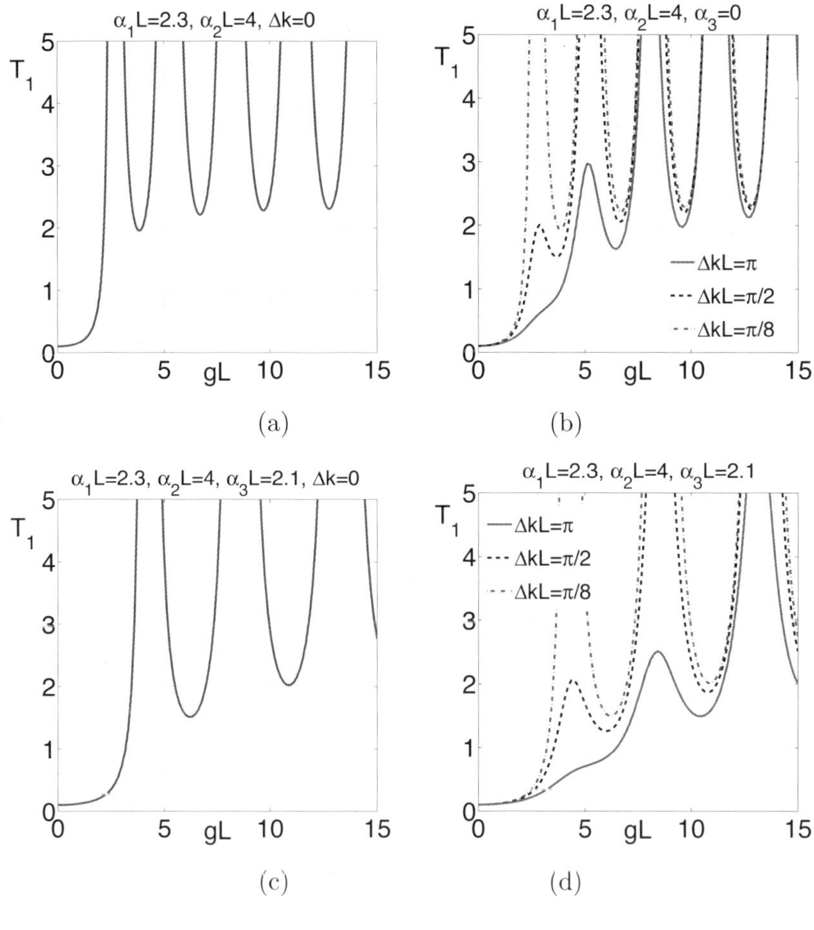

Fig. 13.6. Changes in transparency with increase of the idler absorption. At a given ratio of the absorption indices, transparency does not fall below 100% in the transmission minima.

be neglected and the parameter g is real. Such a model is relevant to, e.g., off-resonant quadratic and cubic nonlinearities attributed to the structural elements of metal-dielectric nanocomposites [2]. The results will be used in Sec. 13.6 for optimization of transparency achievable through embedded resonant FWM centers with power-dependent optical parameters.

Figures 13.5–13.7 present numerical simulations of the dependence of the transparency on the ratio of the absorption indices at the frequencies of the coupled waves and on the phase mismatch. Figures 13.5–13.6 show the

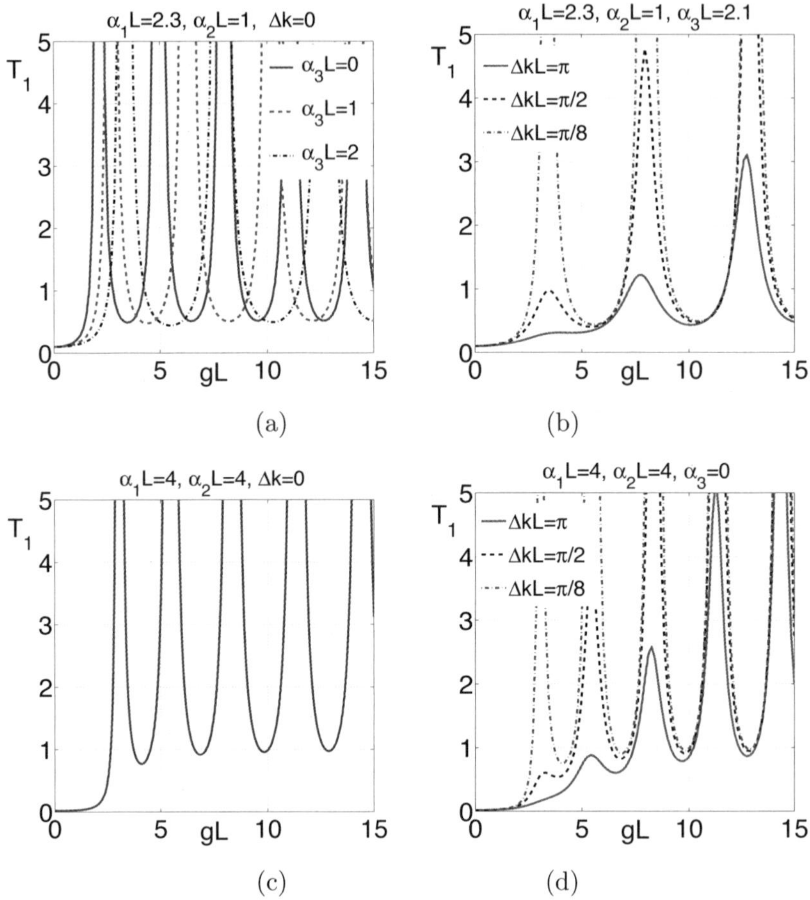

Fig. 13.7. Transmission of the negative-index slab for $\alpha_2 L \leqslant \alpha_1 L$. The transmission in minima does not exceed 100%.

feasibility of achieving robust transparency and amplification in a NIM slab at the signal frequency through a wide range of the control field intensities by the appropriate adjustment of the absorption indices $\alpha_2 \geq \alpha_1$. It is seen that the transmissions does not drop below 1 for $\alpha_2 > \alpha_1$. Figure 13.7 proves that larger absorption for the idler is advantageous for robust transmission of the signal, which is counterintuitive. The increase of the idler's absorption is followed by the relatively small shift of the resonances to larger magnitudes of gL. Oscillation amplitudes grow sharply near the resonances, which indicates cavity-less generation. Phase mismatch causes the

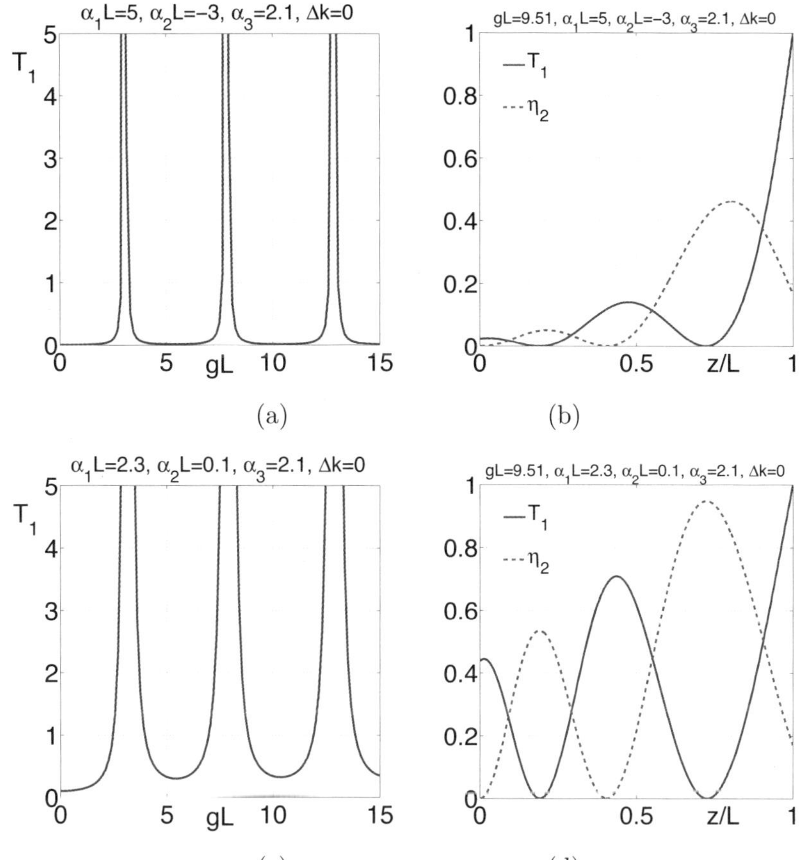

Fig. 13.8. Transmission resonances and distribution of the signal and the idler inside the negative-index slab in the vicinity of second transmission minimum.

decrease of maxima of the first resonances. Inhomogeneity of the control field due to its absorption causes the decrease of next maxima and their insignificant shift to the larger intensities.

The distribution of the signal and the idler inside the slab would also dramatically change with the ratio of the depletion rates (Fig. 13.8). Unless optimized, the signal maximum inside the slab may appear much greater than its output value at $z = 0$. The spatial distributions of the signal and the idler also experience a strong dependence on phase mismatch. Such dependencies are in strong contrast with their counterparts in PI materials and are determined by the backwardness of the coupled waves that is inherent to NIMs.

A qualitative explanation of the described dependencies revealed through numerical simulations is as follows. Besides the factor g, the local NLO energy conversion rate for each of the waves is proportional to the amplitude of another coupled wave and depends on the phase mismatch Δk. Hence, the fact that the waves decay in opposite directions causes a specific, strong dependence of the entire propagation process and, consequently, of the transmission properties of the slab on the ratio of their decay rates. Since the idler and the control field are absorbed toward the back facet of the slab and the signal experiences absorption in the opposite direction, the maximum of the signal for the given parameters is located somewhere inside the slab. A change in the slab optical thickness or in the intensity of the control fields leads to significant changes in the distributions of the signal and idler along the slab. Hence, the simulations suggest a general procedure of optimization and control of the output signal and slab transparency without a change in its composition and structure.

13.5. Nonlinear-Optical Negative-Index Mirror

Alternatively, for $a_{1L} = 0$, $a_2(z = 0) = a_{20}$, where only a positive-index signal at ω_2 and control field at ω_3 enter a slab at $z = 0$, and the generated difference-frequency electromagnetic wave at $\omega_1 = \omega_3 - \omega_2$ falls into the negative-index frequency domain, the slab serves as a generator of a backward wave at ω_1, i.e., as a NLO mirror. Basically, the reflected wave has a different frequency, and the reflectivity may significantly exceed 100%. Results of numerical simulations of such process and properties of the mirror are presented below. Figure 13.9 displays reflectivity resonances which may exceed the self-oscillation threshold. Transmission minima depend on the ratio of absorption rates, whereas reflectivity minima remain robust, which is in stark contrast with the process investigated in the preceding Sec. 13.4. Alternatively, phase mismatch causes decrease of the reflectivity maxima and increase the minima (Fig. 13.10). Reflectivity becomes relatively robust against phase mismatch with increase of intensity of the control field. It drops dramatically in the range of small phase mismatch and then remains relatively robust within the range of greater phase mismatch (Figs. 13.11 and 13.12). The outlined properties of the NLO mirror are determined by the interplay of several processes which have a strong effect on the NLO coupling of contrapropagating waves as indicated in Sec. 13.4 and, consequently, on their distribution inside the slab (Figs. 13.13 and 13.14).

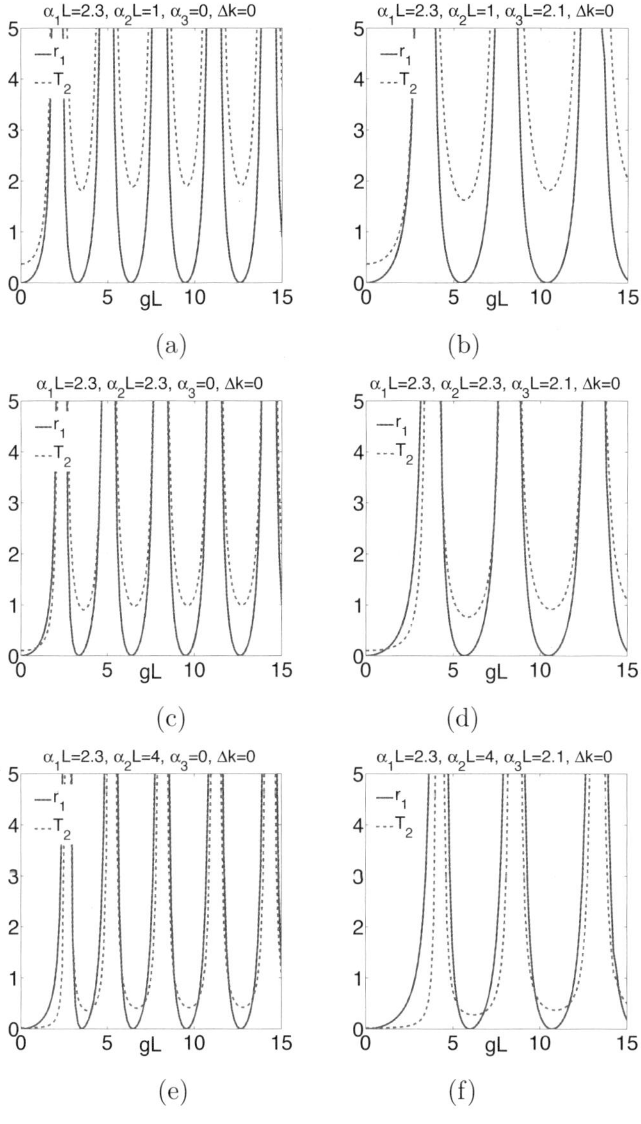

Fig. 13.9. Effect of absorption on reflectivity and transmittance of the NLO mirror. Here, reflectivity can be switched from zero to magnitudes exceeding 100%.

Ultimately, the simulations show the possibility to tailor and switch the reflectivity of such a mirror over the wide range by changing intensity of the control field.

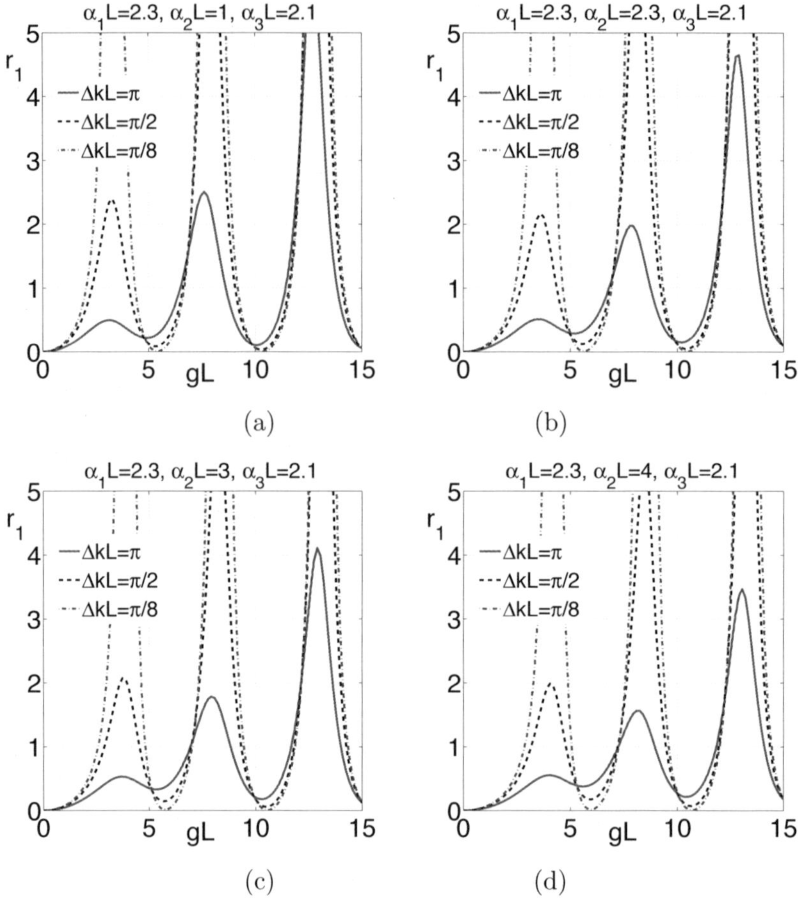

Fig. 13.10. Effect of of absorption on the reflectivity and transmittance of the NLO mirror at different phase mismatches.

Only rough estimations can be made regarding $\chi^{(2)}$ attributed to metal-dielectric nanostructures. Assuming $\chi^{(2)} \sim 10^{-6}$ ESU ($\sim 10^3$ pm/V), which is on the order of that for $CdGeAs_2$ crystals, and a control field of $I \sim 100$ kW focused on a spot of $D \sim 50$ μm in diameter, one can estimate that the typical required value of the parameter $gL \sim 1$ can be achieved for a slab thickness in the *microscopic* range of $L \sim 1\mu$m, which is comparable with that of the multilayer NIM samples fabricated to date [28,30].

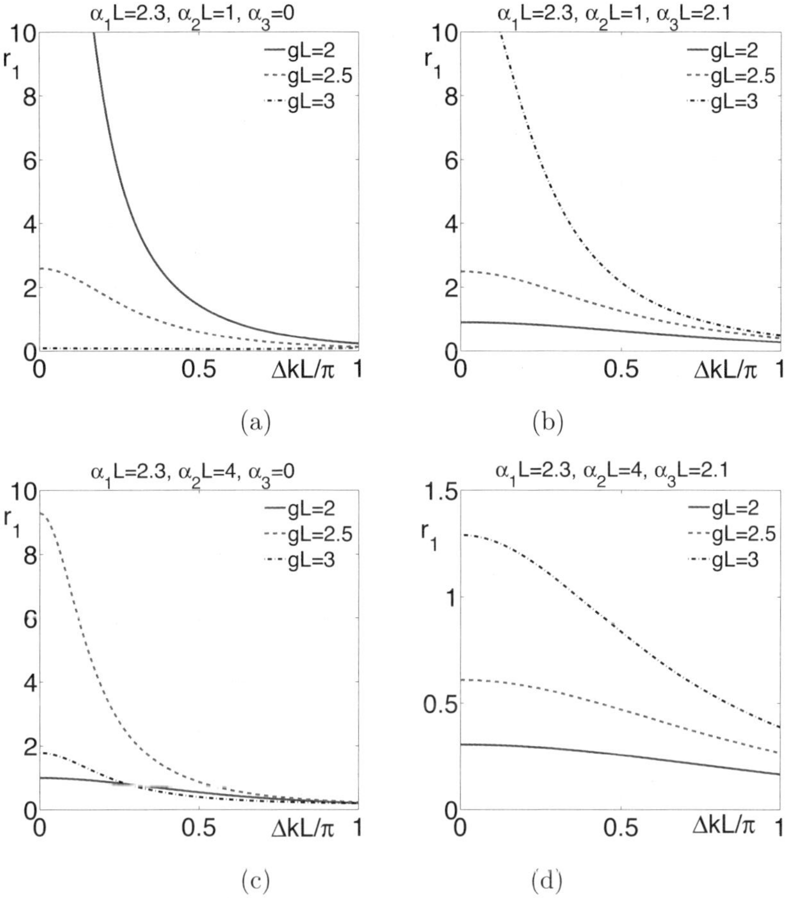

Fig. 13.11. Effect of phase mismatch on the reflectivity and transmittance of the NLO mirror at different intensities of the control field.

13.6. Embedded Nonlinearity

The above described features allow us to propose and optimize the feasibility of independently engineering the NI and the resonantly-enhanced higher-order ($\chi^{(3)}$) NLO response of a composite metamaterial with embedded NLO centers (ions or molecules) (Fig. 13.3(d,e)). The sample is illuminated by two control fields, E_3 and E_4, so that the amplification of the NI signal, E_1, and the generation of the counter-propagating PI idler, E_2, occur due to the FWM process $\omega_1 + \omega_2 = \omega_3 + \omega_4$ (Fig. 13.3(b,c)). The

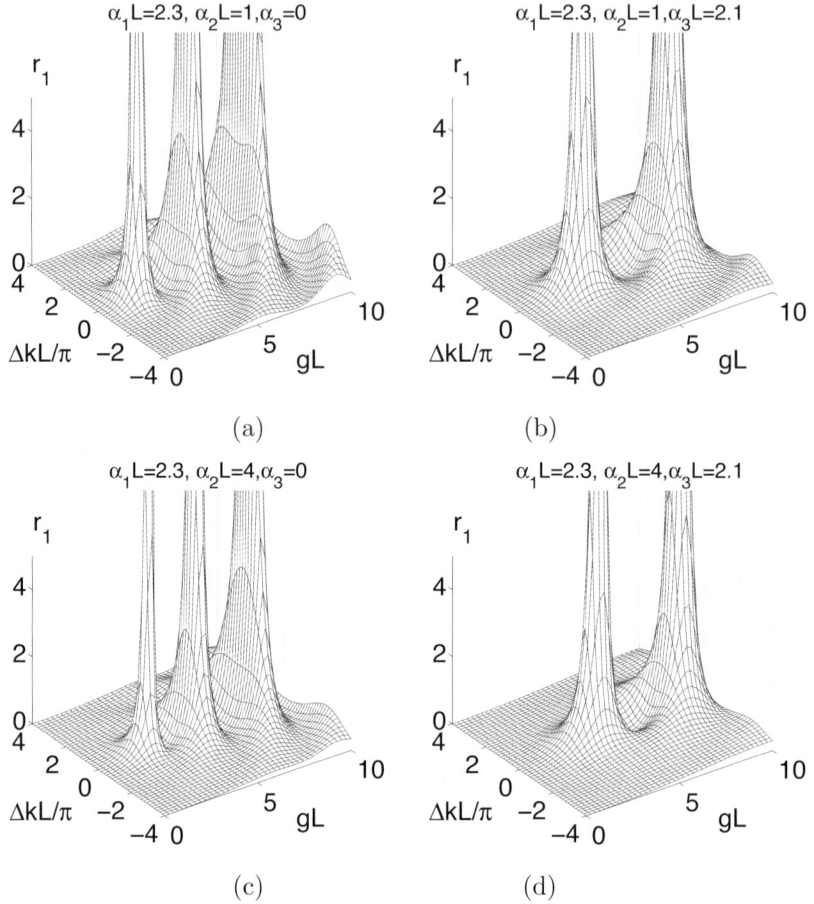

Fig. 13.12. Reflectivity vs. intensity of the control field and phase mismatch for different absorption indices for the coupled waves.

transmission factor for the signal, $T_1(z = 0)$ can be computed as described in subsection 13.3.2. Due to resonant or near-resonant coupling, all local parameters here become strongly dependent on the intensity of the control fields and can be tailored by the means of quantum control. The schemes Fig. 13.3(b,d) and (c,e) provide for different relations between local linear and NLO parameters at ω_1 and ω_2 and for their dependencies on the control fields. Linear and NLO local parameters attributed to the embedded centers are calculated by the density-matrix method, as described in subsection 13.3.3.1. The strength of the control fields is represented by the

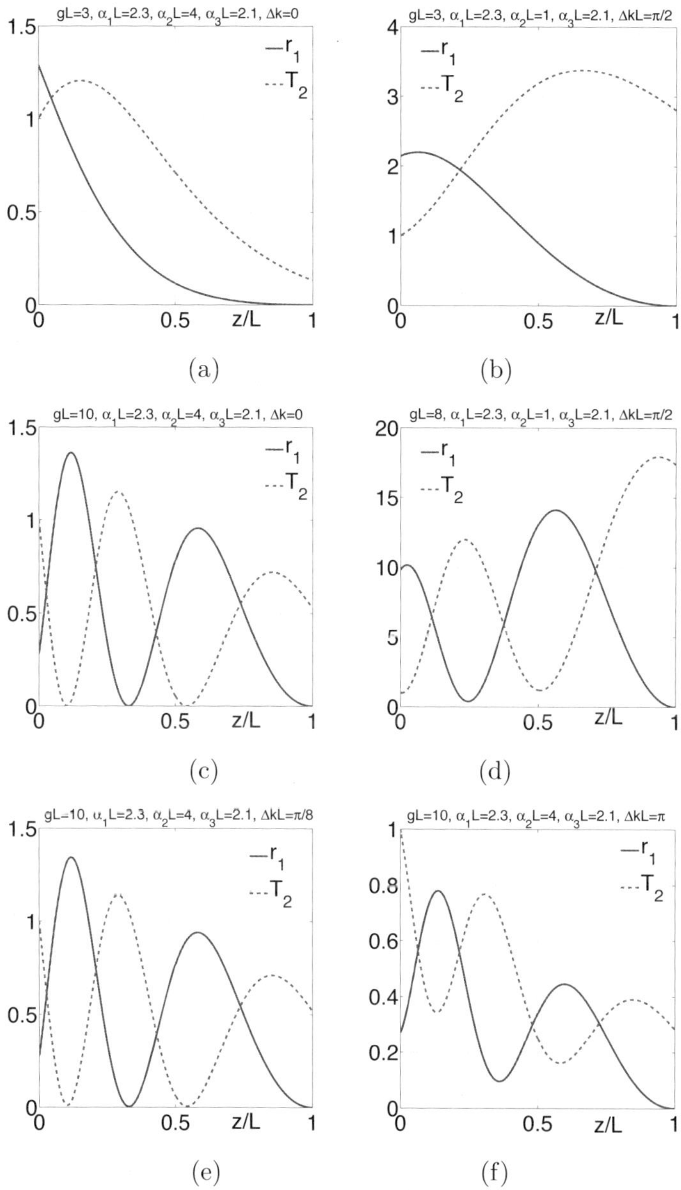

Fig. 13.13. Intensity distribution for the ordinary and backward waves inside the slab for different absorption index, phase mismatch and intensity of the control field.

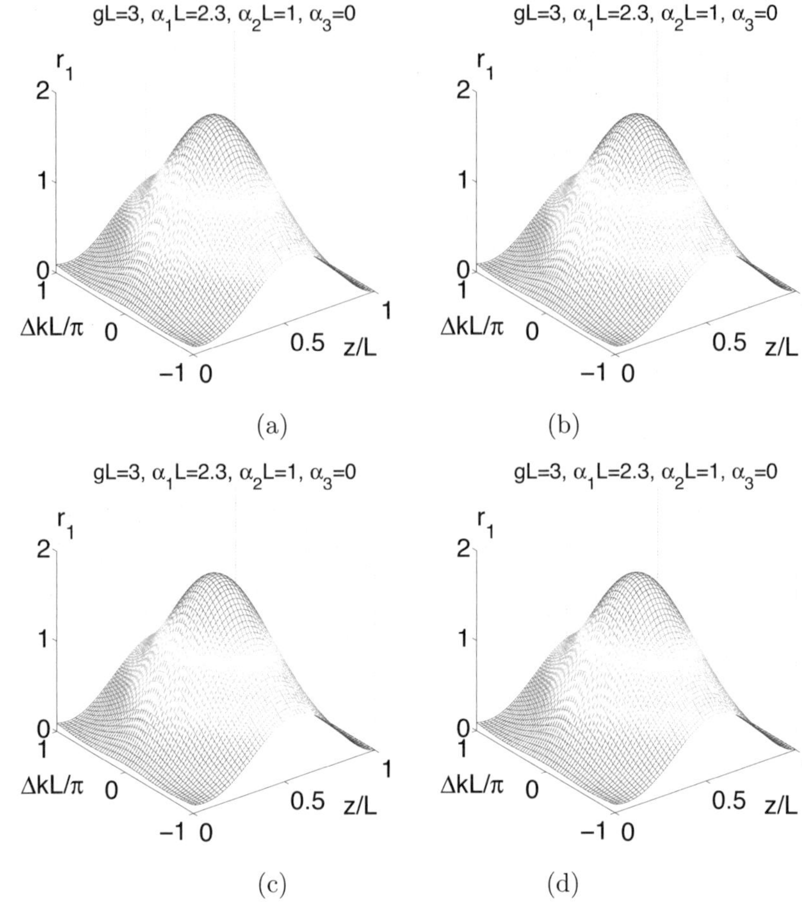

Fig. 13.14. Dependence of distribution of generated backward wave inside the slab on phase mismatch.

coupling Rabi frequencies $G_3 = E_3 d_{lg}/2\hbar$ and $G_4 = E_4 d_{nm}/2\hbar$, where d_{ij} are electrodipole transition matrix elements. The quantities α_{10} and α_{20} denote the value of fully-resonant absorption introduced by the embedded centers at the frequencies of the corresponding transitions with all driving fields turned off. In this section, the absorption of the host material in the slab at ω_1 is assumed fixed at 90% and is equal to 88% at ω_2. Hence, the absorption indices, attributed to the host slab at the frequencies of the signal and the idler, are taken $\alpha_{h1}L$=2.3, and $\alpha_{h2}L$=2.1 for both schemes Fig. 13.3(b,d) and (c,e). For scheme Fig. 13.3(e), the electrical linear and nonlinear polarizations, Eq. (13.19), are calculated as

$$P_1(z,t) = (1/2)\{P_{01}^L \exp(ik_1 z) + P_{01}^{NL} \exp[i(k_3 + k_4 - k_2)z]\} \exp(-i\omega_1 t) + c.c.$$
$$= N(\rho_{ng} d_{gn} + \rho_{gn} d_{ng});$$
$$P_2(z,t) = (1/2)\{P_{02}^L \exp(ik_2 z) + P_{02}^{NL} \exp[i(k_3 + k_4 - k_1)z]\} \exp(-i\omega_2 t) + c.c.$$
$$= N(\rho_{ml} d_{lm} + \rho_{lm} d_{ml}).$$

Here, ρ_{ij} are the density matrix elements, and d_{ij} are the transition dipole elements. For scheme Fig. 13.3(d), they are calculated in a similar way. Effective linear, $\chi_{1,2}$, and NLO, $\chi_{1,2}^{(3)}$, susceptibilities dependent on the intensities of the driving control fields E_3 and E_2 are defined as

$$P_{01}^L = \chi_1 E_1, \quad P_{01}^{NL} = \chi_1^{(3)} E_3 E_4 E_2^*;$$
$$P_{02}^L = \chi_2 E_2, \quad P_{02}^{NL} = \chi_2^{(3)} E_3 E_4 E_1^*.$$

The linear susceptibilities determine the intensity-dependent contributions to absorption and to the refractive indices of the composite attributed to the embedded centers, while the NLO susceptibilities determine the FWM. Here, $\omega_1 + \omega_2 = \omega_3 + \omega_4$, and $k_j = |n_j|\omega_j/c > 0$.

13.6.1. *Fully-resonant control fields*

The results of numerical simulations for two examples of fully-resonant control fields, $\Omega_3 = \omega_3 - \omega_{gl} = 0$ and $\Omega_4 = \omega_4 - \omega_{mn} = 0$, are presented in Figs. 13.15 and 13.16. Relaxation properties of the model are taken as follows: energy level relaxation rates $\Gamma_n = 20$, $\Gamma_g = \Gamma_m = 120$; partial transition probabilities $\gamma_{gn} = 50$, $\gamma_{mn} = 90$, (all in 10^6 s^{-1}); homogeneous transition half-widths $\Gamma_{lg} = 1$, $\Gamma_{lm} = 1.9$, $\Gamma_{ng} = 1.5$, $\Gamma_{nm} = 1.8$ (all in 10^{11} s^{-1}); $\Gamma_{gm} = 5$, $\Gamma_{ln} = 0.5$ (all in 10^9 s^{-1}); and $\lambda_1 = 756$ nm, $\lambda_2 = 480$ nm.

Fig. 13.15(a,b) depicts the dependence of the transmission of the signal on the thickness of the doped slab and on its resonance frequency for the coupling schemes of Fig. 13.3(b,d). The plots display the appearance of transmission and amplification that may exceed the oscillation threshold. For the given fields and relaxation parameters, the power-dependent energy level populations are $r_l \approx 0.42$, $r_n \approx 0.2$, $r_g \approx 0.19$, $r_m \approx 0.19$ and, hence, no population-inversion or Raman-type gain is involved in the coupling. Figure 13.15 proves the feasibility of compensating losses, producing narrow-band transparency, amplification, and mirrorless generation.

A similar dependence for the alternative option Fig. 13.3(c,e) is shown in Fig. 13.16(a,b). The plots also prove the possibilities of compensating

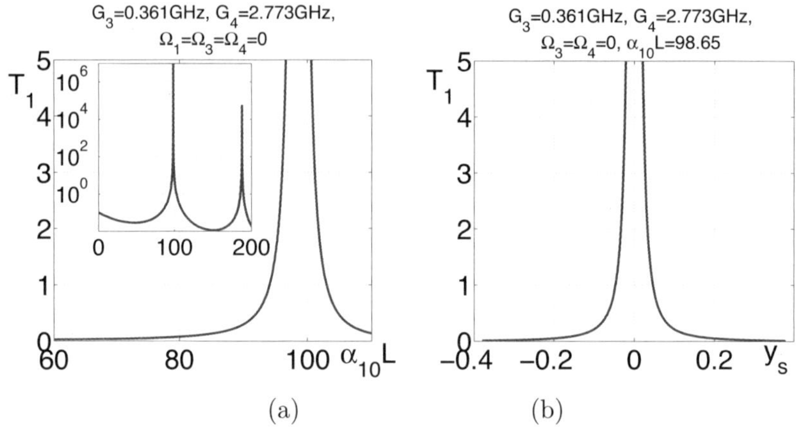

Fig. 13.15. Transmission of the negative-index signal in the vicinity of the higher frequency transition (Fig. 13.3(d)), tailored by the fully-resonant control fields. (a) Transmission of the signal vs. optical thickness of the slab, $\alpha_{10}L$. $\omega_1 = \omega_{ml}$. (b) Transmission of the signal at $\alpha_{10}L$=98.65 vs. signal resonance detuning. $y_s = (\omega_1 - \omega_{ml})/\Gamma_{ln}$, $\omega_2 = \omega_3 + \omega_4 - \omega_1$, and the coupling Rabi frequencies for the control fields are G_3=0.361 GHz, G_4=2.773 GHz.

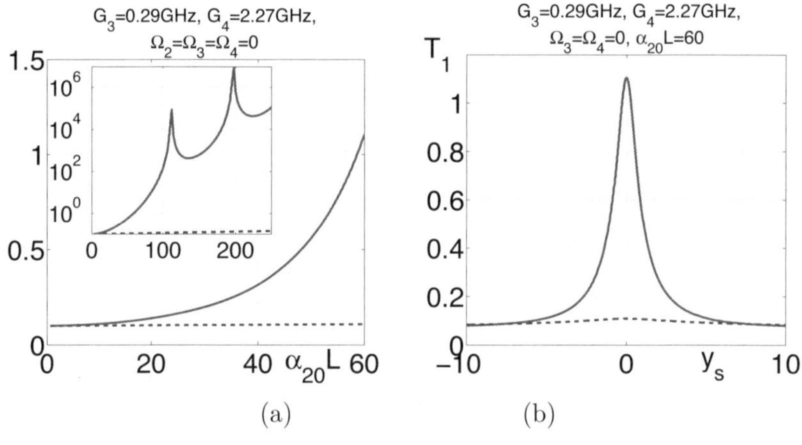

Fig. 13.16. Resonant coupling in the scheme Fig. 13.3(e). (a) Transmission of the signal vs. optical thickness of the slab, $\alpha_{20}L$. $\omega_1 = \omega_{gn}$. (b) Transmission of the signal at $\alpha_{20}L$=60 vs. signal resonance detuning. $y_s = (\omega_1 - \omega_{gn})/\Gamma_{ln}$, $\omega_2 = \omega_3 + \omega_4 - \omega_1$, where the coupling Rabi frequencies for the control fields are G_3=0.29 GHz, G_4=2.27 GHz. The dashed line shows transmission at $\chi^{(3)} = 0$.

strong losses in NIMs through independently engineered embedded resonant nonlinearity. For the given fields and relaxation parameters, the power-dependent energy level populations are $r_l \approx 0.48$, $r_n \approx 0.18$, $r_g \approx 0.17$, $r_m \approx 0.17$.

Hence, counterintuitively, the simulations show that one can produce transparency of the initially strongly lossy NIM by introducing an additional strong absorption at the frequencies in the vicinity of transition ml through the embedded centers.

13.6.2. *Quasi-resonant control fields*

Here, we simulate the case, where the idler corresponds to a higher-frequency transition from the ground state, and the signal corresponds to a lower-frequency transition between the excited states (Fig. 13.4(e)). No incoherent amplification is possible here for the idler, and the dependence of the idler and the signal absorption indices on the control fields changes cardinally. First, the scheme with relatively fast quantum coherence relaxation rates and the case where only a two-photon, Raman-like resonance for the signal holds is considered; all other one-photon frequency offsets are on the order of several tens of the optical transition widths. Then the scheme with the same quantum coherence relaxation rates, but with higher partial spontaneous transition rates is considered, in which case population inversion at the coupled optical transitions is impossible. Finally, we consider the scheme with longer quantum coherence lifetimes, which still does not allow population inversion at the optical transitions nor Raman-like amplification. The fact that all involved optical transitions are absorptive determines essentially different features of the overall loss-compensation technique in such composites in each proposed scheme. In all of the schemes outlined above, the linear and nonlinear local parameters can be tailored through quantum control by varying the intensities and frequency-resonance offsets for combinations of the two control driving fields.

First, consider the model with slightly different relaxation parameters $\gamma_{mn} = 70 \times 10^6$ s^{-1}, $\Gamma_{ln} = 1 \times 10^{10}$ s^{-1}. All other parameters are the same as in the preceding subsection. The changes in absorption, amplification, and refractive indices as well as in the magnitudes and signs of NLO susceptibilities caused by the control fields depend on the population redistribution over the coupled levels, which in turn strongly depends on the partial transition probabilities. Figure 13.17 depicts such modifi-

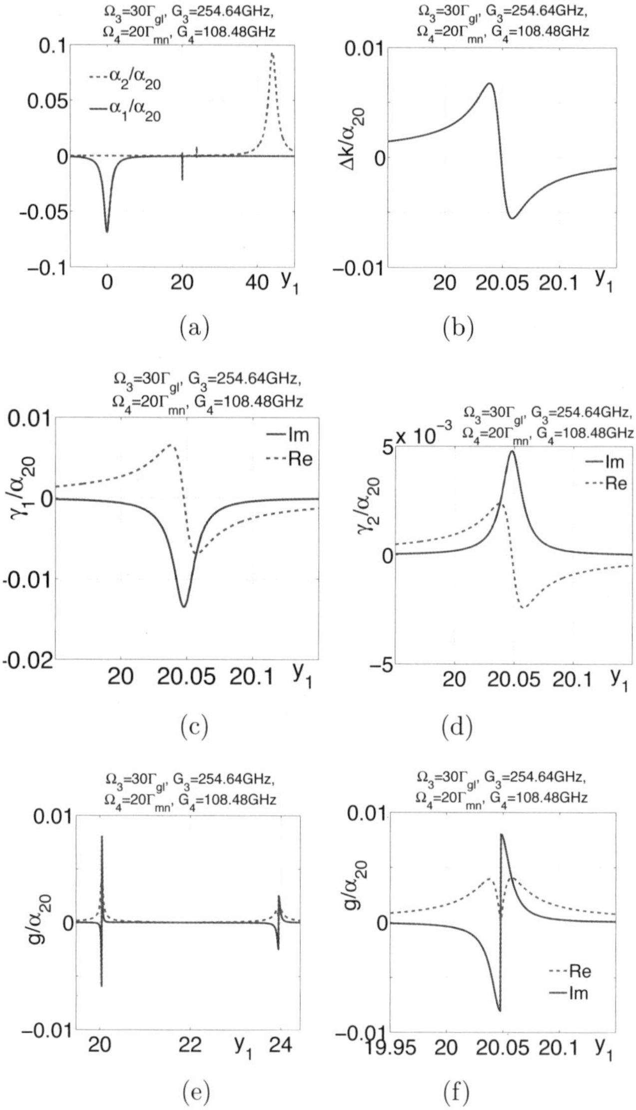

Fig. 13.17. Nonlinear spectral structures in local optical quantities produced by the control fields. $y_1 = (\omega_1 - \omega_{gn})/\Gamma_{gn}$, $\omega_2 = \omega_3 + \omega_4 - \omega_1$. (a) Absorption/gain indices for the signal and the idler; (b) Phase mismatch; (c-f) Four-wave mixing coupling parameters. Coupling Rabi frequencies and resonance frequency offsets for the control fields are: $G_3 = 254.64$ GHz, $\Omega_3 = 30\Gamma_{gl}$, $G_4 = 108.48$ GHz, $\Omega_4 = 20\Gamma_{mn}$.

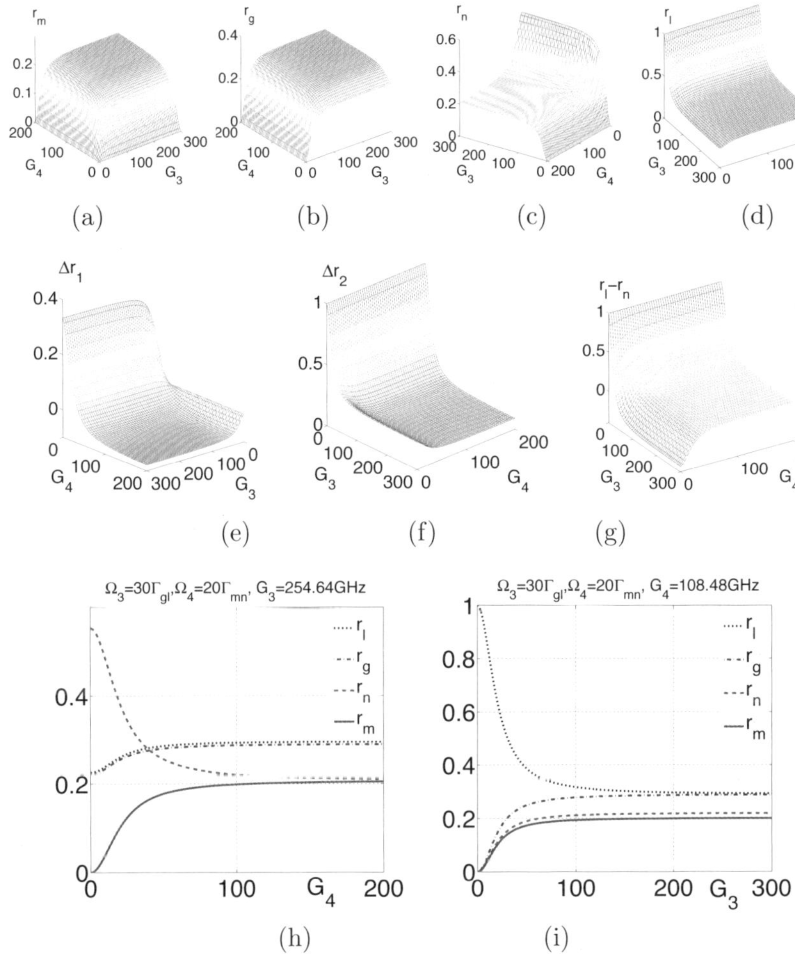

Fig. 13.18. (a)–(g) Difference of the energy-level populations and their dependence on the Rabi frequency of the control fields G_3 and G_4 (given in GHz). $\Omega_3 = 30\Gamma_{gl}$, $\Omega_4 = 20\Gamma_{mn}$. (h) G_3=254.64 GHz, (i) G_4=108.48 GHz.

cations at the given resonance offsets and intensities of the control fields. Figure 13.17(a) displays the modified absorption/gain indices. The nonlinear spectral structures are caused by the modulation of the probability amplitudes, which exhibits itself as an effective splitting of the energy levels coupled with the driving fields. Figure 13.17(b) shows the contribution to the phase mismatch associated with one such spectral structure.

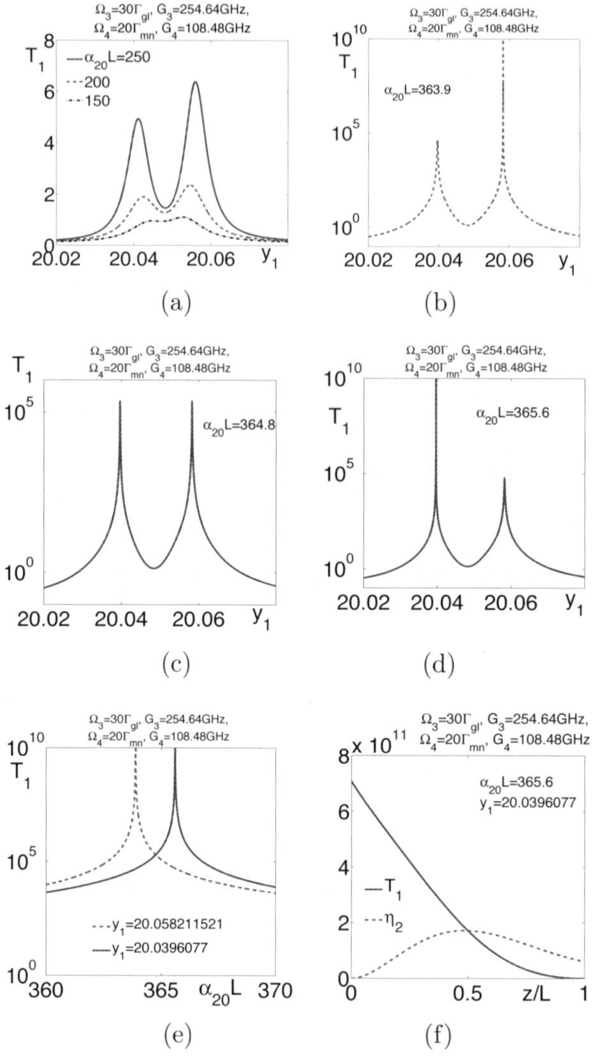

Fig. 13.19. Dependence of the transmission of the slab on the resonance frequency offset $y_1 = (\omega_1 - \omega_{gn})/\Gamma_{gn}$ for different optical densities of the slab, (a–d), on the resonant optical density of the slab, (e), and the distribution of the signal and the idler along the slab, (f). G_3=254.64 GHz, $\Omega_3 = 30\Gamma_{gl}$, G_4=108.48 GHz, $\Omega_4 = 20\Gamma_{mn}$.

Figures 13.17(c,d) indicate that the real and imaginary parts of the NLO susceptibilities become commensurate for the given susceptibility, but may exceed their counterparts for the idler by several times. This occurs due to the fact that different population differences contribute in different ways to the NLO susceptibilities [34], and driving fields cause significant redistributions of the level populations (Fig. 13.18). At the given partial probabilities of spontaneous transition between the levels, population inversions at the signal transition become possible (Figs. 13.18(e,g–i)). However, for the given frequency offsets of the control fields, corresponding amplification contributes negligibly to the energy conversion (Fig. 13.17(a)). Alternatively, two-photon, Raman-like amplification at $\Omega_1 \approx 20.05\Gamma_{gn}$ shown in Fig. 13.17(a) supports coherent, parametric energy conversion from the control fields to the signal. Figures 13.19(a–d) display the spectral properties of the output signal at $z = 0$ for one of the resonances in the vicinity of the signal frequency offset $\omega_1 - \omega_{gn} \approx 20.05\Gamma_{gn}$ at different optical densities of the slab at ω_{ml} attributed to the impurity centers. The density of the embedded centers and the slab thickness, and hence, the additional resonant optical thickness of the slab contributed by these impurities, may vary as shown in the panels. Actual quasi-resonant absorption/gain indices depend on the intensities and frequency offsets of the control fields, as shown in Fig. 13.17(a). Besides the features imposed by the counter-propagation of the coupled waves, the output magnitudes of the signal at $z = 0$ and the idler at $z = L$ and their distributions inside the slab are determined by the interplay of several contributing linear and nonlinear processes. They include the phase mismatch, absorption of the signal and the idler, and the parametric gain g, which are all controlled by the driving fields E_3 and E_4. The dependence of the overall optimized output signal on the density of the impurities and on the slab thickness (on the resonant optical thickness of the slab) is depicted in Fig. 13.19(e). Such a behavior is determined by the radically different distributions of the idler, which propagates from left to right, and the signal propagating from right to left (Fig. 13.19(f)). Figures 13.19(b–f) indicate the possibility of mirrorless self-oscillation. Figure 13.20 shows the role of partial spontaneous transitions between the energy levels. Here, $\gamma_{mn} = 9 \times 10^7$ sec^{-1}, which makes both population inversion and two-photon gain impossible (Fig. 13.20(a)). At the indicated Rabi frequencies and frequency offsets for the driving control fields, the energy-level populations are: $r_l \approx 0.4$, $r_g \approx 0.2009$, $r_n \approx 0.2031$, $r_m \approx 0.2$. The magnitude of the FWM coupling parameters appear comparable with those depicted in Fig. 13.17(c-h). However, the absence of one- and two-photon amplification that would support energy-conversion processes, like in Ref. [10] and in Fig. 13.17(a), dramatically decreases the achievable amplification and increases the required optical thickness of the slab (Fig. 13.20(g,h)).

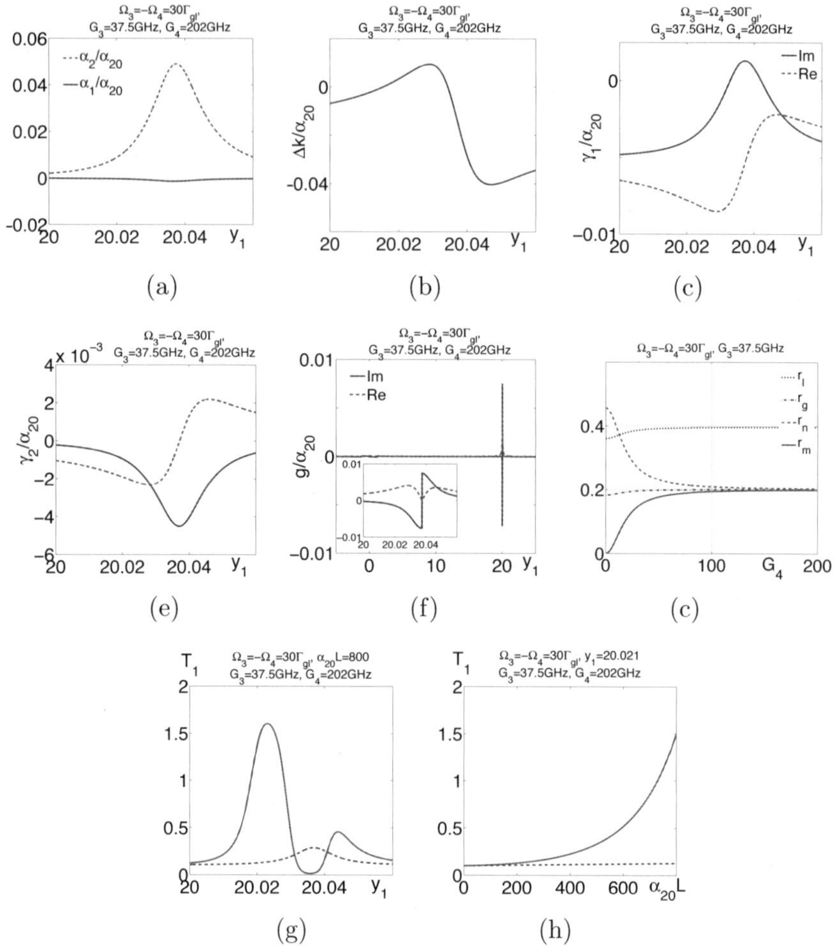

Fig. 13.20. Energy-conversion in the scheme with neither population inversion nor two-photon gain possible [$\gamma_{mn} = 9 \times 10^7$ sec^{-1}); all other relaxation parameters are the same as in the previous case. $y_1 = (\omega_1 - \omega_{gn})/\Gamma_{gn}$, $\omega_2 = \omega_3 + \omega_4 - \omega_1$. (a) Absorption indices for the signal and the idler; (b) Phase mismatch; (c–f) Four-wave mixing coupling parameters; (g,h) Transmission factor, where the dashed line shows transmission at $g = 0$. The coupling Rabi frequencies and resonance frequency offsets for the control fields are: G_3=37.5 GHz, G_4=202 GHz, $\Omega_3 = -\Omega_4 = 30\Gamma_{gl}$.

Figure 13.21 shows that, even in such cases, the optimized magnitude of the required control field intensities and the slab optical density can be substantially reduced for centers with lower coherence relaxation rates and quasi-resonant coupling. Here, quantum nonlinear interference effects play

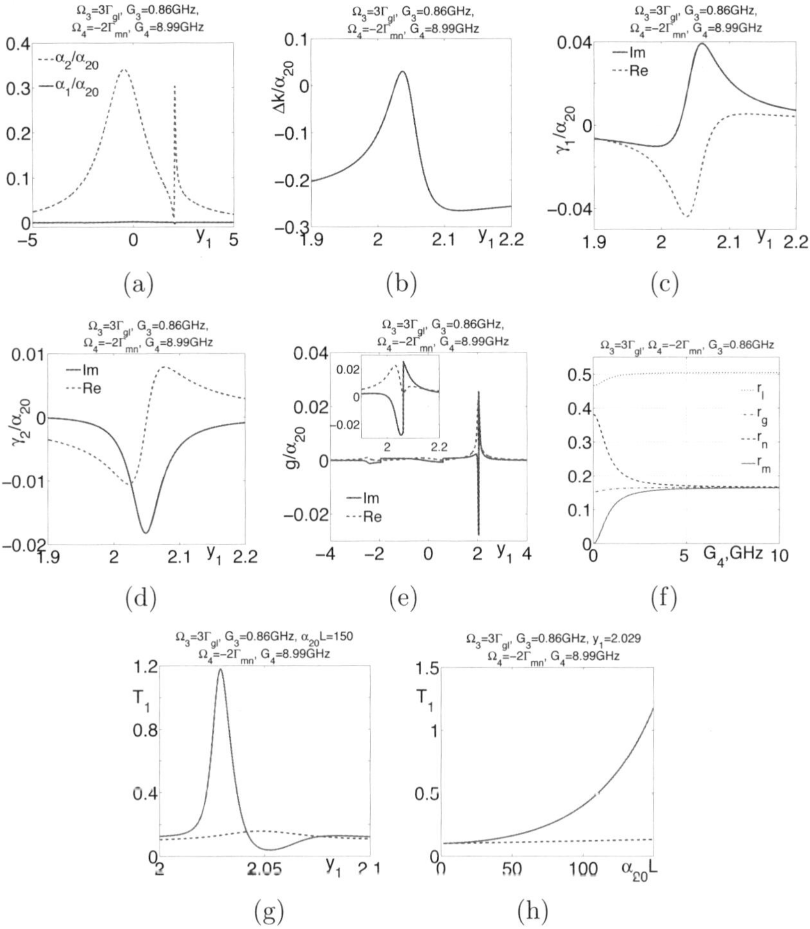

Fig. 13.21. Quasi-resonant coupling at lower quantum coherence relaxation rates with neither population inversion nor two-photon gain possible [Γ_{gl}=1.8, Γ_{mn}=1.9, Γ_{gn}=1, Γ_{ml}=1.5, $\Gamma_{mg} = 5\times10^{-2}$, $\Gamma_{nl} = 5\times10^{-3}$ (in 10^{11} sec^{-1}); all other relaxation parameters are the same as in the previous case]. $y_1 = (\omega_1 - \omega_{gn})/\Gamma_{gn}$, $\omega_2 = \omega_3 + \omega_4 - \omega_1$. (a) Absorption indices for the signal and the idler; (b) Phase mismatch; (c–e) Four-wave mixing coupling parameters; (f) Energy level populations; (g,h) Transmission factor, where the dashed line shows transmission at $g = 0$. The coupling Rabi frequencies and resonance frequency offsets for the control fields are: G_3=0.86 GHz, $\Omega_3 = 3\Gamma_{gl}$; G_4=8.99 GHz, $\Omega_4 = -2\Gamma_{mn}$.

an important role [34]. At the indicated Rabi frequencies and frequency offsets for the driving control fields shown in Fig. 13.21, the energy-level populations are: $r_l \approx 0.504$, $r_g \approx 0.165$, $r_n \approx 0.167$, $r_m \approx 0.164$.

The requirements for the proposed all-optical control of transmission and reflectivity of a doped metamaterial slab are as follows: For the given transitions, the magnitude G=1 GHz corresponds to the control field intensities of $I \sim 1$ W/(0.1mm)^2. At a resonance absorption cross-section $\sigma_{40} \sim 10^{-16}$ cm^2, which is typical for transitions with oscillator strength of about one, and a concentration of the embedded centers $N \sim 10^{19}$ cm^{-3}, one estimates $\alpha_{10} \sim 10^3$ cm^{-1} and the required slab thickness in the microscopic range $L \sim (1-100)\mu$m. The contribution by the impurities to the refractive index is estimated as $\Delta n < 0.5(\lambda/4\pi)\alpha_{40} \sim 10^{-3}$, which essentially does not change the negative refractive index.

13.7. Conclusions

In conclusion, extraordinary properties of coherent nonlinear-optical propagation processes, such as second-harmonic generation and three- and four-wave mixing processes in negative-index metamaterials, are demonstrated by the means of numerical experiments. In particular, the feasibility of compensation of strong losses in negative-index metamaterials, e.g., in metal-dielectric nanostructured composites, are numerically demonstrated. This is the key problems in determining numerous revolutionary applications of such a novel class of electromagnetic metamaterials. All-optical tailoring of transparency and reflectivity is shown to be possible through coherent nonlinear-optical energy transfer from the ordinary control electromagnetic wave to the negative-index, backward wave. Backwardness of traveling electromagnetic waves is intrinsic to the negative-index metamaterials. It shown that besides the nonlinearity attributed to the building blocks of the negative-index host, a strong nonlinear optical response of the composite can be provided by the embedded resonant four-level nonlinear-optical centers. This opens the way to independent nanoengineering and adjustment of the negative index and nonlinearity of metamaterials. In addition, the opportunity for quantum control over the local optical parameters of the metamaterial in this case, which employs constructive and destructive quantum interference tailored by two auxiliary driving control fields is shown. Such a possibility is proven with the aid of a realistic numerical model. Among the possible applications of the proposed technique are a novel class of the miniature frequency-tunable narrow-band filters, nonlinear-optical mirrors, quantum switchers, amplifiers, and cavity-free microscopic optical parametric oscillators that allow generation of entangled counter-propagating left- and right-handed photons, and all-optical data processing microchips. The unique unparalleled features of the underlying processes are revealed, such as the strongly-resonant behavior with respect to the material thickness, the density of the embedded resonant centers and the intensities of the control fields, the feasibility of negating the linear phase mismatch introduced by the host material, and the role of absorption or, conversely, the supplementary nonparametric amplification of the idler.

Acknowledgments

The authors thank S. A. Myslivets, V. M. Shalaev and V. V. Slabko for collaboration and fruitful discussions of the results of this paper and S. A. Myslivets for help with numerical simulations. This material is based upon work supported by by the U. S. Army Research Laboratory and by the U. S. Army Research Office under grant number W911NF-0710261.

References

1. V. M. Shalaev, Optical negative-index metamaterials, *Nat. Photonics* **1**, 41–48 (2007).
2. M. W. Klein, M. Wegener, N. Feth and S. Linden, Experiments on second- and third-harmonic generation from magnetic metamaterials, *Opt. Express* **15**, 5238–5247 (2007); erratum:*ibid*, **16**, 8055 (2008).
3. V. M. Agranovich, Y. R. Shen, R. H. Baughman and A. A. Zakhidov, Linear and nonlinear wave propagation in negative refraction metamaterials, *Phys. Rev. B* **69**, 165112(7) (2004).
4. I. V. Shadrirov, A. A. Zharov and Yu. S. Kivshar, Second-harmonic generation in nonlinear left-handed metamaterials, *J. Opt. Soc. Am. B* **23**, 529–534 (2006).
5. A. K. Popov, V. V. Slabko and V. M. Shalaev, Second-harmonic generation in left-handed metamaterials, *Laser Phys. Lett.* **3**, 293–297 (2006).
6. A. K. Popov and V. M. Shalaev, Negative-index metamaterials: Second-harmonic generation, Manley–Rowe relations and parametric amplifications, *Appl. Phys. B* **84**, 131–137 (2006).
7. M. Scalora, G. D'Aguanno, M. Bloemer, M. Centini, N. Mattiucci, D. de Ceglia and Yu. S. Kivshar, Dynamics of short pulses and phase-matched second-harmonic generation in negative-index materials, *Opt. Express* **14**, 4746–4756 (2006).
8. A. K. Popov and V. M. Shalaev, Compensating losses in negative-index metamaterials by optical parametric amplification, *Opt. Lett.* **31**, 2169–2171 (2006).
9. A. K. Popov, S. A. Myslivets, T. F. George and V. M. Shalaev, Four-wave mixing, quantum control and compensating losses in doped negative-index photonic metamaterials, *Opt. Lett.* **32**, 3044–3046 (2007).

10. A. K. Popov, S. A. Myslivets and V. M. Shalaev, Resonant nonlinear optics of backward waves in negative-index metamaterials, *Appl. Phys. B: Lasers and Optics* **96**, 315–323 (2009).
11. A. K. Popov, S. A. Myslivets and V. M. Shalaev, Microscopic Mirrorless Negative-index Optical Parametric Oscillator, *Opt. Lett.* **34**, 1165–1167 (2009).
12. A. K. Popov and S. A. Myslivets, Transformable broad-band transparency and amplification in negative-index films, *Appl. Phys. Lett.* **93**, 191117-1-3 (2008).
13. N. M. Litchinitser and V. M. Shalaev, Loss as a route to transparency, *Nat. Photonics* **3**, 75–79 (2009).
14. A. K. Popov, S. A. Myslivets and V. M. Shalaev, Coherent nonlinear optics and quantum control in negative-index metamaterials, *J. Opt. A: Pure Appl. Opt.* **11**, 114028-1-13, (2009)
15. N. Lazarides and G. P. Tsironis, Coupled nonlinear Schrödinger field equations for electromagnetic wave propagation in nonlinear left-handed materials, Phys. Rev. E **71**, 036614-1-4 (2005)
16. P. Tassin, L. Gelens, J. Danckaert, I. Veretennicoff, G. Van der Sande, P. Kockaert and M. Tlidi, Dissipative structures in left-handed material cavity optics, *Chaos* **17**, 037116-1-11 (2007)
17. P. Kockaert, P. Tassin, G. Van der Sande, I. Veretennicoff and M. Tlidi, Negative diffraction pattern dynamics in nonlinear cavities with left-handed materials, *Phys. Rev. A* **74**, 033822-1-8 (2006).
18. A. D. Boardman, N. King, R. C. Mitchell-Thomas, V. N. Malnev and Y. G. Rapoport, Gain control and diffraction-managed solitons in metamaterials, *Metamaterials* **2**, 145–154 (2008)
19. G. D'Aguanno, N. Mattiucci, M. Scalora and M. J. Bloemer, Bright and Dark Gap Solitons in a Negative Index Fabry-Perot Etalon, *Phys. Rev. Lett.* **93**, 213902(1) (2004).
20. N. M. Litchinitser, I. R. Gabitov, A. I. Maimistov and V. M. Shalaev, Negative Refractive Index Metamaterials in Optics, *Prog. in Optics* **51**, 1–68 (2007).
21. N. M. Litchinitser, I. R. Gabitov, A. I. Maimistov and V. M. Shalaev, Effect of an optical negative index thin film on optical bistability, *Opt. Lett.* **32**, 151–153 (2007).
22. A. I. Maimistov and I. R. Gabitov, Nonlinear optical effects in artificial materials, *Eur. Phys. J.* Special Topics, **147**, 265–286 (2007).
23. J. B. Khurgin, Mirrorless magic, *Nat. Photonics* **1**, 446–447 (2007).
24. C. Canalias and V. Pasiskevicius, Mirrorless optical parametric oscillator, *Nat. Photonics* **1**, 459–462 (2007).
25. S. E. Harris, Proposed backward wave oscillations in the infrared, *Appl. Phys. Lett.* **9**, 114–117 (1966).
26. K. I. Volyak, A. S. Gorshkov, Investigations of a reverse-wave parametric oscillator, *Radiotekhnika i Elektronika (Radiotechnics and Electronics)* **18** 2075–2082 (1973).
27. A. Yariv, *Quantum Electronics*, 2nd Edn. (Wiley, New York, 1975), Ch. 18.

28. C. M. Soukoulis and M. Kafesaki, Weakly and strongly coupled optical metamaterials, Invited talk at Nanometa 2009, 2nd European Topical Meeting on Nanophotonics and Metamaterials (Seefeld, Tirol, Austria, 2009).
29. N. Katsarakis, G. Konstantinidis, A. Kostopoulos, R. S. Penciu, T. F. Gundogdu, M. Kafesaki, E. N. Economou, Th. Koschny and C. M. Soukoulis, Magnetic response of split-ring resonators in the far-infrared frequency regime, *Opt. Lett.* **30**, 1348–1350 (2005).
30. X. Zhang, Optical Bulk Metamaterials, Plenary talk at Nanometa 2009, 2nd European Topical Meeting on Nanophotonics and Metamaterials (Seefeld, Tirol, Austria, 2009).
31. J. Valentine, S. Zhang, T. Zentgraf, E. Ulin-Avila, D. A. Genov, G. Bartal and X. Zhang, Three-dimensional optical metamaterial with a negative refractive index, *Nature* **455**, 376–378 (2008).
32. J. M. Manley and H. E. Rowe, Some general properties of nonlinear elements–Part I. General energy relations, *Proc. IRE* **44**, 904–913 (1956).
33. J. M. Manley and H. E. Rowe, General energy relations in nonlinear reactances, *Proc. IRE* **47**, 2115–2116 (1959).
34. A. K. Popov, S. A. Myslivets and T. F. George, Nonlinear interference effects and all-optical switching in optically-dense inhomogeneously-broadened media, *Phys. Rev. A* **71**, 043811-1-13 (2005).

Chapter 14

Nanoscale Resolution in the Near and Far Field Intensity Profile of Optical Dipole Radiation

Xin Li[*] and Henk F. Arnoldus[†]

Mississippi State University

[*]xl121@msstate.edu

[†]hfa1@msstate.edu

Jie Shu

Rice University

js35@rice.edu

When an electric dipole moment rotates, the flow pattern of the emitted energy exhibits a vortex structure in the near field. The field lines of energy flow swirl around an axis which is perpendicular to the plane of rotation of the dipole. This rotation leads to an apparent shift of the dipole when viewed from the far field. The shift is of the order of the spatial extend of the vortex, which is about a fraction of an optical wavelength. We also show that when an image of the radiation is formed on an observation plane in the far field, the rotation of the field lines in the near field leads to a shift of the dipole image.

14.1. Introduction

When light is emitted by a localized source, it appears as if the light travels along straight lines from the source to the observer, when viewed from the far field (many wavelengths from the source). These light rays are the flow lines of energy, and they are usually referred to as optical rays. The rays are the orthogonal trajectories of the wave fronts. In the geometrical optics limit of light propagation certain terms in Maxwell's equations can be neglected under the assumption that the wavelength of the light is small compared to other relevant distances. It can then be shown [1] that in a homogeneous medium, like the vacuum, the rays are straight lines, running from the source to the far field, and they coincide with the field lines of energy flow. However, when the light is detected within a fraction of a wavelength from the source, or with nanoscale precision at a larger distance, the geometrical optics limit does not apply, and we have to consider the exact solution of Maxwell's equations at all distances. The flow lines of energy will in general be curves. Far away from the source, each curve will asymptotically approach a straight line, but, when measured with nanoscale resolution, these asymptotic lines will not coincide with the optical rays. The rays run radially out from the source, but an asymptote of a field line of energy flow could be displaced with respect to this direction.

When the dimension of a localized source of radiation is small compared to the wavelength of the emitted light, then the source is in first approximation an electric dipole, located at the center of the source, and when viewed from outside the source, the radiation is identical to the radiation emitted by a point dipole. Also, radiation emitted by atoms and molecules is usually electric dipole radiation, and since atoms and molecules are much smaller than the wavelength of the light they emit, we can consider them as point sources. Since electric dipole radiation is the most elementary type of radiation, we shall consider the nanoscale structure of this type of radiation. We shall consider the spatial distribution of the energy flow in the near field, and we shall show that the curving of the field lines in the near field has an effect on the observable intensity profile in the far field, provided that the measurement is carried out with sub-wavelength resolution.

14.2. Electric Dipole Radiation

When the current density of a localized source oscillates harmonically with angular frequency ω, for instance when the source is placed in a laser beam oscillating at the same frequency ω, then the induced electric dipole moment has the form

$$\mathbf{d}(t) = d_o \operatorname{Re}(\boldsymbol{\varepsilon} e^{-i\omega t}) \ , \tag{14.1}$$

with $d_o > 0$ and $\boldsymbol{\varepsilon}$ a unit vector, normalized as $\boldsymbol{\varepsilon} \cdot \boldsymbol{\varepsilon}^* = 1$. The radiated electric field will also have a harmonic time dependence, and can be written as

$$\mathbf{E}(\mathbf{r},t) = \mathrm{Re}[\mathbf{E}(\mathbf{r})e^{-i\omega t}] \ , \tag{14.2}$$

with $\mathbf{E}(\mathbf{r})$ the complex amplitude. A similar expression holds for the magnetic field $\mathbf{B}(\mathbf{r},t)$. When the dipole is located at the origin of coordinates, the complex amplitudes of the electric and magnetic fields are given by [2]

$$\mathbf{E}(\mathbf{r}) = \frac{d_o k_o^3}{4\pi\varepsilon_o} \left\{ \boldsymbol{\varepsilon} - (\boldsymbol{\varepsilon} \cdot \hat{\mathbf{r}})\hat{\mathbf{r}} + [\boldsymbol{\varepsilon} - 3(\boldsymbol{\varepsilon} \cdot \hat{\mathbf{r}})\hat{\mathbf{r}}] \frac{i}{q}\left(1 + \frac{i}{q}\right) \right\} \frac{e^{iq}}{q} \ , \tag{14.3}$$

$$\mathbf{B}(\mathbf{r}) = -\frac{d_o k_o^3}{4\pi\varepsilon_o} \boldsymbol{\varepsilon} \times \hat{\mathbf{r}} \left(1 + \frac{i}{q}\right) \frac{e^{iq}}{q} \ , \tag{14.4}$$

where $k_o = \omega / c$ is the wave number in free space and $\hat{\mathbf{r}}$ is a unit vector which is directed from the location of the dipole to the field point, represented by **r**. We have also introduced the dimensionless variable $q = k_o r$, which represents the distance between the field point and the dipole. In this way, a distance of 2π in terms of q corresponds to a distance of one optical wavelength in terms of *r*. The properties of the electric dipole moment enter the expressions for $\mathbf{E}(\mathbf{r})$ and $\mathbf{B}(\mathbf{r})$ only through the complex-valued unit vector $\boldsymbol{\varepsilon}$ (apart from an overall d_o).

14.3. Energy Flow and the Poynting Vector

In an electromagnetic field the energy flow is determined by the Poynting vector, defined as

$$\mathbf{S}(\mathbf{r},t) = \frac{1}{\mu_o} \mathbf{E}(\mathbf{r},t) \times \mathbf{B}(\mathbf{r},t) \ , \tag{14.5}$$

for propagation in vacuum. If d*A* is an infinitesimal surface element at the position **r**, with unit normal $\hat{\mathbf{n}}$, then the power flowing through d*A* is equal to $\mathrm{d}P = \mathbf{S}\cdot\hat{\mathbf{n}}\mathrm{d}A$. Therefore, at position **r** the energy flows into the direction of the Poynting vector **S**. For time-harmonic fields, as in Eq. (14.2), the Poynting vector simplifies to

$$\mathbf{S}(\mathbf{r}) = \frac{1}{2\mu_o}\mathrm{Re}\,[\mathbf{E}(\mathbf{r})\times\mathbf{B}(\mathbf{r})^*] \ , \tag{14.6}$$

and here terms that oscillate at twice the optical frequency ω have been dropped, since these average to zero in an experiment, on a time scale of an optical cycle. The Poynting vector in Eq. (14.6) only involves the complex amplitudes, rather than the fields themselves, and $\mathbf{S}(\mathbf{r})$ is independent of time *t*.

With expressions (14.3) and (14.4) the Poynting vector for the radiation emitted by a dipole can be evaluated immediately. We obtain

$$\mathbf{S}(\mathbf{r}) = \frac{3P_o}{8\pi r^2}\left\{\zeta(\theta,\phi)\hat{\mathbf{r}} - \frac{2}{q}\left(1+\frac{1}{q^2}\right)\mathrm{Im}\,[(\hat{\mathbf{r}}\cdot\boldsymbol{\varepsilon})\boldsymbol{\varepsilon}^*]\right\} \ , \tag{14.7}$$

where θ and ϕ are the angles of the observation point **r** in a spherical coordinate system. The function $\zeta(\theta,\phi)$ is given by

$$\zeta(\theta,\phi) = 1 - (\hat{\mathbf{r}}\cdot\boldsymbol{\varepsilon})(\hat{\mathbf{r}}\cdot\boldsymbol{\varepsilon}^*) \ , \tag{14.8}$$

and

$$P_o = \frac{ck_o^4}{12\pi\varepsilon_o}d_o^2 \ , \tag{14.9}$$

is the power emitted by the dipole. The unit vector $\hat{\mathbf{r}}$ is in spherical coordinates

$$\hat{\mathbf{r}} = (\mathbf{e}_x\cos\phi + \mathbf{e}_y\sin\phi)\sin\theta + \mathbf{e}_z\cos\theta \ , \tag{14.10}$$

and this determines the dependence of $\zeta(\theta,\phi)$ on θ and ϕ, given the dipole moment vector $\boldsymbol{\varepsilon}$.

When we take d*A* as part of a sphere, centered around the origin, and with radius *r*, then $\mathrm{d}A = r^2\mathrm{d}\Omega$, and the radiated power per unit solid angle becomes

$$\frac{dP}{d\Omega} = r^2 \mathbf{S}(\mathbf{r}) \cdot \hat{\mathbf{r}} \ . \tag{14.11}$$

With Eq. (14.7) this becomes

$$\frac{dP}{d\Omega} = \frac{3P_o}{8\pi} \zeta(\theta,\phi) \ , \tag{14.12}$$

since $(\hat{\mathbf{r}} \cdot \boldsymbol{\varepsilon})\boldsymbol{\varepsilon}^* \cdot \hat{\mathbf{r}}$ is real. The emitted power per unit solid angle is independent of r, and this may give the impression that the power flows radially outward, as in the geometrical optics limit of light propagation. We shall see below that this is usually not the case.

When vector $\boldsymbol{\varepsilon}$ in Eq. (14.1) is real, the dipole moment is $\mathbf{d}(t) = d_o \boldsymbol{\varepsilon} \cos(\omega t)$. This corresponds to a linear dipole moment, oscillating back and forth along an axis through vector $\boldsymbol{\varepsilon}$. For this case, the Poynting vector becomes

$$\mathbf{S}(\mathbf{r}) = \frac{3P_o}{8\pi r^2} \zeta(\theta,\phi) \hat{\mathbf{r}} \ , \tag{14.13}$$

since $(\hat{\mathbf{r}} \cdot \boldsymbol{\varepsilon})\boldsymbol{\varepsilon}^*$ is real. Then $\mathbf{S}(\mathbf{r})$ is proportional to $\hat{\mathbf{r}}$ at any field point, and hence the power flow is exactly in the radial direction. The field lines of $\mathbf{S}(\mathbf{r})$, which are the field lines of energy flow, are straight lines, running from the location of the dipole to infinity.

14.4. Elliptical Dipole Moment

In its most general state of oscillation, the dipole moment $\mathbf{d}(t)$ traces out an ellipse in a plane [3,4]. We take this plane to be the xy-plane, and we parametrize vector $\boldsymbol{\varepsilon}$ as

$$\boldsymbol{\varepsilon} = -\frac{1}{\sqrt{\beta^2+1}} (\beta \mathbf{e}_x + i\mathbf{e}_y) \ , \tag{14.14}$$

with β real. The parametrization is chosen such that for $\beta = \pm 1$, vector $\boldsymbol{\varepsilon}$ reduces to the standard spherical unit vectors

$$\mathbf{e}_{\pm 1} = -\frac{1}{\sqrt{2}}(\pm \mathbf{e}_x + i\mathbf{e}_y) \;, \tag{14.15}$$

with respect to the z-axis. With $\boldsymbol{\varepsilon}$ from Eq. (14.14), the dipole moment becomes

$$\mathbf{d}(t) = -\frac{d_o}{\sqrt{\beta^2+1}}[\beta \mathbf{e}_x \cos(\omega t) + \mathbf{e}_y \sin(\omega t)] \;. \tag{14.16}$$

As time progresses, vector $\mathbf{d}(t)$ follows the contour of an ellipse, as shown in Fig. 14.1. For $\beta = \pm 1$ the ellipse reduces to a circle. When β is positive, the rotation is counterclockwise as in the figure, which is the positive direction with the z-axis, as given by the right-hand rule. For $\beta < 0$ the rotation is in the opposite direction. For $\beta = 0$, vector $\boldsymbol{\varepsilon}$ becomes $-i\mathbf{e}_y$, and the oscillation is linear along the y-axis. In the limit $\beta \to \pm\infty$ we have $\boldsymbol{\varepsilon} \to \mp \mathbf{e}_x$, and the oscillation is linear along the x-axis.

For an elliptical dipole in the xy-plane, the function $\zeta(\theta,\phi)$ from Eq. (14.8) becomes

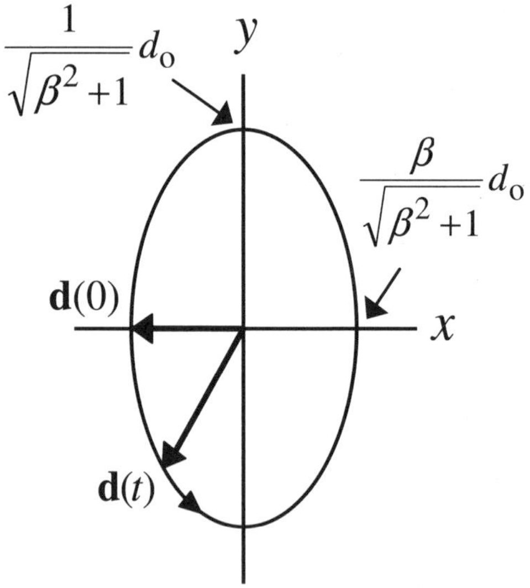

Fig. 14.1. In its most general state of oscillation, the dipole moment $\mathbf{d}(t)$ rotates along an ellipse. The parameter β (positive in the figure) determines the lengths of the major and minor axes, as shown in the figure, and the sign of β determines the direction of rotation.

$$\zeta(\theta,\phi)=1-\frac{1}{2}\sin^2\theta\left[1+\frac{\beta^2-1}{\beta^2+1}\cos(2\phi)\right] . \qquad (14.17)$$

This function is, apart from an overall constant, the emitted power per unit solid angle into the direction (θ,ϕ), according to Eq. (14.12). For the Poynting vector, Eq. (14.7), we find

$$\mathbf{S}(\mathbf{r})=\frac{3P_o}{8\pi r^2}\left\{\zeta(\theta,\phi)\hat{\mathbf{r}}+\frac{2}{q}\left(1+\frac{1}{q^2}\right)\frac{\beta}{\beta^2+1}\mathbf{e}_\phi\sin\theta\right\} , \qquad (14.18)$$

with

$$\mathbf{e}_\phi = -\mathbf{e}_x\sin\phi+\mathbf{e}_y\cos\phi . \qquad (14.19)$$

The term proportional to $\hat{\mathbf{r}}$ is responsible for the radial power outflow, and the term with $\mathbf{e}_\phi$ gives a rotation in the flow of energy around the z-axis. Close to the dipole this term is of order r^{-5}, whereas the radial term is of order r^{-2}. Therefore, in the near field the rotation will dominate the energy flow pattern, even though this rotation does not contribute to the power outflow at any distance from the dipole. In the limit of a linear dipole, $\beta=0$ or $\beta\to\infty$, the term with $\mathbf{e}_\phi$ vanishes, and the Poynting vector only has a radial component, corresponding to power flowing radially outward from the dipole, without any swirling around the z-axis.

14.5. Field Lines of the Poynting Vector

In order to investigate in detail the energy flow out of a dipole, we consider the field lines of the Poynting vector $\mathbf{S}(\mathbf{r})$. Expression (14.18) for $\mathbf{S}(\mathbf{r})$ determines a vector field in space, and a field line of $\mathbf{S}(\mathbf{r})$ is a curve for which at any point along the curve the vector $\mathbf{S}(\mathbf{r})$ is on its tangent line. When $\mathbf{r}$ is a point on the field line, then we can parametrize such a field line as $\mathbf{r}(u)$, with u a dummy variable. A field line is determined only by the direction of $\mathbf{S}(\mathbf{r})$, and not its magnitude, and hence a field line is a solution of

$$\frac{d\mathbf{r}}{du}=f(\mathbf{r})\mathbf{S}(\mathbf{r}) , \qquad (14.20)$$

with $f(\mathbf{r})$ any positive function of $\mathbf{r}$. Equation (14.19) is an autonomous differential equation, since the variable u does not appear on the right-hand side. The variable u itself has no physical significance.

We now use spherical coordinates (q,θ,ϕ) to represent a point $\mathbf{r}$, so that $\mathbf{r}=(q/k_o)\hat{\mathbf{r}}$, with $\hat{\mathbf{r}}$ given by Eq. (14.10). For a point on a field line, q, θ and ϕ then become functions of u, and when we write out Eq. (14.20) we obtain

$$\frac{\mathrm{d}q}{\mathrm{d}u}=k_o f(\mathbf{r})\hat{\mathbf{r}}\cdot\mathbf{S}(\mathbf{r}) \ , \tag{14.21}$$

$$q\frac{\mathrm{d}\theta}{\mathrm{d}u}=k_o f(\mathbf{r})\mathbf{e}_\theta\cdot\mathbf{S}(\mathbf{r}) \ , \tag{14.22}$$

$$q\sin\theta\frac{\mathrm{d}\phi}{\mathrm{d}u}=k_o f(\mathbf{r})\mathbf{e}_\phi\cdot\mathbf{S}(\mathbf{r}) \ , \tag{14.23}$$

with

$$\mathbf{e}_\theta=(\mathbf{e}_x\cos\phi+\mathbf{e}_y\sin\phi)\cos\theta-\mathbf{e}_z\sin\theta \ . \tag{14.24}$$

For the function $f(\mathbf{r})$ we take $8\pi r^2/(3P_o k_o)$, and with Eq. (14.18) we then find the set of equations

$$\frac{\mathrm{d}q}{\mathrm{d}u}=\zeta(\theta,\phi) \ , \tag{14.25}$$

$$\frac{\mathrm{d}\theta}{\mathrm{d}u}=0 \ , \tag{14.26}$$

$$\frac{\mathrm{d}\phi}{\mathrm{d}u}=\frac{2}{q^2}\left(1+\frac{1}{q^2}\right)\frac{\beta}{\beta^2+1} \ , \tag{14.27}$$

for the coordinates q, θ and ϕ as functions of u for points on a field line.

From Eq. (14.26) it follows that θ is constant along a field line, and let us indicate this constant by θ_o. Any field line starts at the location of the dipole, and we then see that it remains on a cone which has an angle

θ_o with the z-axis. Then we can replace θ by θ_o on the right-hand side of Eq. (14.25), and divide Eq. (14.27) by Eq. (14.25). This yields

$$\frac{\mathrm{d}\phi}{\mathrm{d}q}=\frac{2}{q^2}\left(1+\frac{1}{q^2}\right)\frac{1}{\zeta(\theta_o,\phi)}\frac{\beta}{\beta^2+1} \;. \tag{14.28}$$

When we see ϕ as a function of q, rather than u, this is a nonlinear first-order equation for the function $\phi(q)$. The equation is separable, and is most easily solved by considering q as a function of ϕ. The result is [5]

$$q(\phi)=\frac{1}{[(1+\frac{1}{4}A^2)^{\frac{1}{2}}+\frac{1}{2}A]^{\frac{1}{3}}-[(1+\frac{1}{4}A^2)^{\frac{1}{2}}-\frac{1}{2}A]^{\frac{1}{3}}} \;, \tag{14.29}$$

where A is a function of ϕ, defined as

$$A=\frac{3}{8}\left(\beta-\frac{1}{\beta}\right)\sin^2\theta_o\,[\sin(2\phi)-\sin(2\phi_o)]$$

$$-\frac{3}{2}\left(\beta+\frac{1}{\beta}\right)\left(1-\frac{1}{2}\sin^2\theta_o\right)(\phi-\phi_o) \;. \tag{14.30}$$

We see from Eq. (14.28) that for q large we have $\mathrm{d}\phi/\mathrm{d}q\to 0$, and therefore ϕ approaches a constant ϕ_o at a large distance. This is angle ϕ_o in the expression for A, and this ϕ_o serves as the integration constant. Angle ϕ is now the free parameter, and it follows from the derivation in Ref. [5] that, given ϕ_o, angle ϕ has to be taken in the range

$$-\infty<\phi<\phi_o \quad , \quad \beta>0 \;, \tag{14.31}$$

$$\phi_o<\phi<\infty \quad , \quad \beta<0 \;. \tag{14.32}$$

Therefore, each field line is determined by a choice of θ_o and ϕ_o, which are the values of θ and ϕ at a large distance.

We indicate by $\bar{x}=k_o x$, $\bar{y}=k_o y$ and $\bar{z}=k_o z$ the dimensionless Cartesian coordinates of a field point. We then have along a field line

$$\bar{x}(\phi) = q(\phi)\sin\theta_o \cos\phi \ , \tag{14.33}$$

$$\bar{y}(\phi) = q(\phi)\sin\theta_o \sin\phi \ , \tag{14.34}$$

$$\bar{z}(\phi) = q(\phi)\cos\theta_o \ , \tag{14.35}$$

with ϕ the free variable, and $q(\phi)$ given by Eq. (14.29). Figure 14.2 shows several field lines for $\theta_o = \pi/4$ and $\theta_o = 3\pi/4$ and for different values of ϕ_o. For the figure we took $\beta = 1$, which corresponds to a circular dipole moment, rotating counterclockwise in the *xy*-plane. The field lines swirl around the *z*-axis with the same orientation as the rotation of the dipole moment. The field line pattern has a vortex structure, with the dipole at the center of the vortex. All field lines wind around the *z*-axis, while remaining on a cone. The scale in the figure is such that 2π corresponds to one optical wavelength, and we see that the spatial extend of the vortex is well below a wavelength. Figure 14.3 shows field lines for $\beta = 1$, so for the same dipole as in Fig. 14.2, but now for different values of θ_o. Angle ϕ_o is the same for all field lines, and is taken as $\pi/2$. Therefore, all field lines approach asymptotically a straight line parallel to the positive *y*-axis.

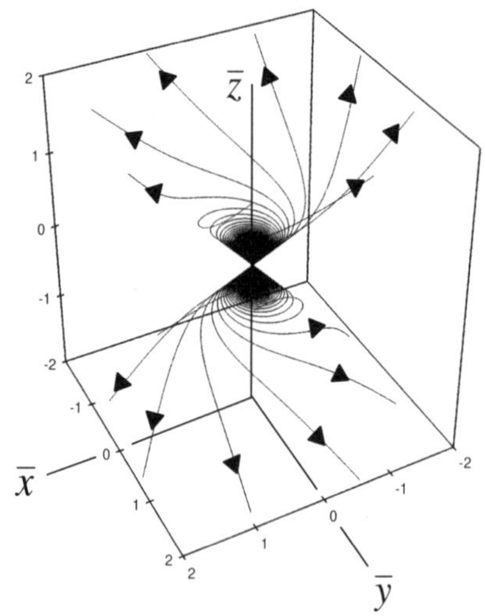

Fig. 14.2. The figure shows several field lines of the Poynting vector for a circular dipole, with a dipole moment that rotates counterclockwise in the *xy*-plane. Each field line lies on a cone which makes an angle of 45° with the *z*-axis.

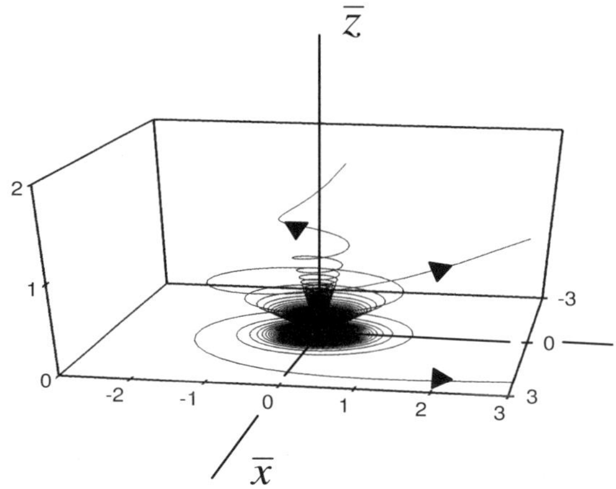

Fig. 14.3. The figure shows field lines for the same dipole as in Fig. 14.2, but now for different values of θ_o, and with $\phi_o = \pi/2$ for all field lines.

The result for the field lines depends parametrically on the value of β. For $\beta = 1$ the field lines show a vortex structure near the location of the dipole, as illustrated in Figs. 14.2 and 14.3, and the dimension of this vortex is a fraction of a wavelength. For $\beta \to 0$ and $\beta \to \infty$ the ellipse reduces to a line, and the oscillation of the dipole moment becomes linear. As shown above, for a linear dipole the field lines are straight lines emanating from the dipole. In order to see the transition from the vortex pattern for a circular dipole to the straight-line pattern for a linear dipole we have graphed field lines for three values of β in Fig. 14.4. It is seen from the figure that when β decreases from unity to zero, the size of the vortex diminishes, until it reaches a point for $\beta = 0$. For $\beta = 1$, the field lines bend around the z-axis over an extend of about a wavelength from the dipole. For smaller values of β, the field line becomes straight already in the very neighborhood of the dipole.

14.6. Field Lines in the Far Field

At a distance of many wavelengths from the dipole, each field line approaches a straight line, as can be seen from the figures above. This line, however, does not appear to come exactly from the location of the dipole, but it is slightly displaced with respect to the radial direction, as

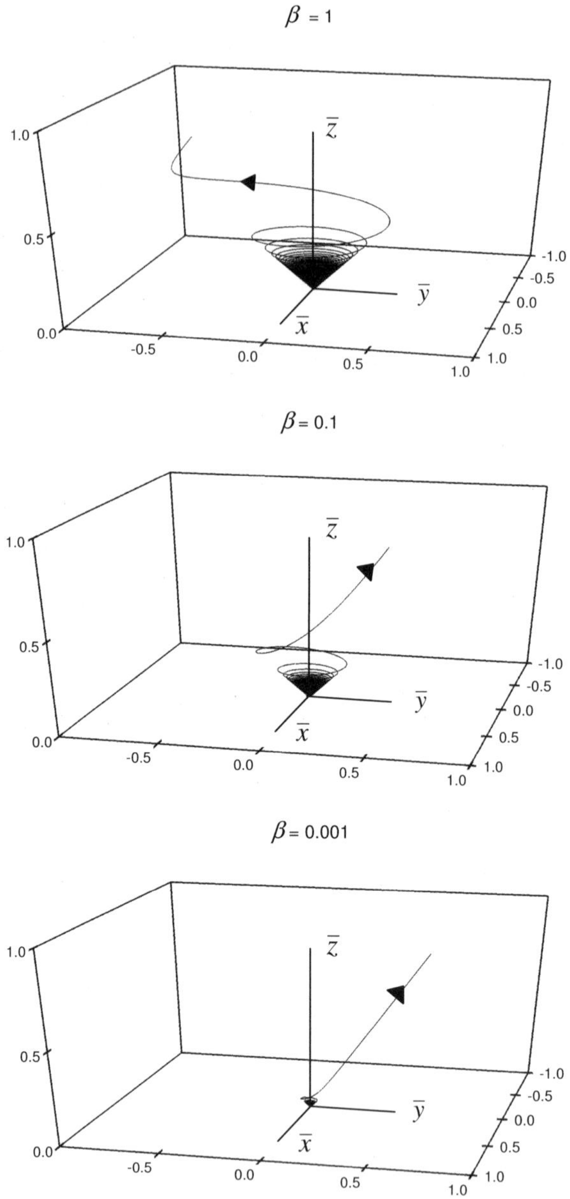

Fig. 14.4. Shown are field lines for dipoles with different values of β. For each, $\theta_o = \pi/4$ and $\phi_o = \pi/2$.

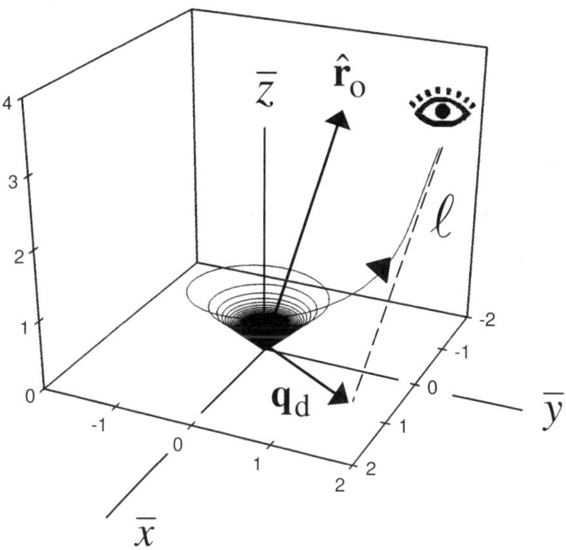

Fig. 14.5. Due to the rotation of a field line near the source, it appears as if the source is displaced when viewed from the far field.

illustrated in Fig. 14.5. For an observer, indicated by the eye in the figure, the field line of energy flow seems to come from a point in the xy-plane that does not coincide with the position of the source, and hence it appears as if the source is displaced with respect to its actual position. This apparent displacement, indicated by the displacement vector $\mathbf{q}_d$, is a result of the rotation of the field lines near the source, and as such this near field effect should be observable in the far field.

In order to compute this displacement, we consider q as the independent variable along a field line, rather than ϕ, even though in the explicit solution (14.29) it is the other way around. For q large, ϕ approaches the value ϕ_o, and the value of θ along a field line is θ_o at any point. Therefore, for q large we can expand ϕ in an asymptotic series as $\phi(q) = \phi_o + c_1/q + c_2/q^2 ...$, with the coefficients c_1, c_2, ... to be determined. In expression (17) for $\zeta(\theta,\phi)$ we set $\theta = \theta_o$ and for ϕ we substitute the asymptotic expansion. This gives $\zeta(\theta_o,\phi) = \zeta(\theta_o,\phi_o)$ $+ \ddots (1/q)$. We use this to expand the right-hand side of Eq. (14.28), which yields

$$\frac{\mathrm{d}\phi}{\mathrm{d}q}=\frac{1}{q^2}\frac{1}{\zeta(\theta_o,\phi_o)}\frac{2\beta}{\beta^2+1}+\cdots\left(\frac{1}{q^3}\right) . \tag{14.36}$$

The right-hand side of Eq. (14.36) is an asymptotic series in q, and term-by-term integration gives

$$\phi(q)=\phi_o-\frac{1}{q}Y(\theta_o,\phi_o;\beta)+\ldots , \tag{14.37}$$

where we have set

$$Y(\theta_o,\phi_o;\beta)=\frac{1}{\zeta(\theta_o,\phi_o)}\frac{2\beta}{\beta^2+1} , \tag{14.38}$$

which is a constant along a field line.

Equations (14.33)–(14.35) give the Cartesian coordinates of a point on a field line. We now view q as the independent variable, and we expand $\cos\phi$ and $\sin\phi$ in Eqs. (14.33) and (14.34) with the help of Eq. (14.37). We then obtain

$$\cos\phi(q)=\cos\phi_o+\frac{1}{q}Y(\theta_o,\phi_o;\beta)\sin\phi_o+\ldots , \tag{14.39}$$

$$\sin\phi(q)=\sin\phi_o-\frac{1}{q}Y(\theta_o,\phi_o;\beta)\cos\phi_o+\ldots . \tag{14.40}$$

For a point on a field line, q is the dimensionless distance between this point and the origin of coordinates. When the terms represented by ellipses in Eq. (14.37) are omitted, parameter q loses this significance. Therefore, we shall write t instead of q. Equations (14.33)–(14.35) then become

$$\bar{x}=\sin\theta_o[t\cos\phi_o+Y(\theta_o,\phi_o;\beta)\sin\phi_o] , \tag{14.41}$$

$$\bar{y}=\sin\theta_o[t\sin\phi_o-Y(\theta_o,\phi_o;\beta)\cos\phi_o] , \tag{14.42}$$

$$\bar{z} = t\cos\theta_o \ . \tag{14.43}$$

For a given observation direction (θ_o, ϕ_o), these Cartesian coordinates are linear functions of t, and hence Eqs. (14.41)–(14.43) are the parameter equations of a straight line. This is line ℓ in Fig. 14.5, which is the asymptote of the field line in the direction (θ_o, ϕ_o). Vector $\hat{\mathbf{r}}_o$ in Fig. 14.5 is the unit vector in the (θ_o, ϕ_o) direction, which follows from Eq. (14.10) by replacing (θ, ϕ) by (θ_o, ϕ_o). When we introduce the vector

$$\mathbf{q}_d = Y(\theta_o, \phi_o; \beta)\sin\theta_o(\mathbf{e}_x \sin\phi_o - \mathbf{e}_y \cos\phi_o) \ , \tag{14.44}$$

then Eqs. (14.41)–(14.43) can be written in vector form as

$$\mathbf{q} = \mathbf{q}_d + t\hat{\mathbf{r}}_o \ , \tag{14.45}$$

where $\mathbf{q} = k_o\mathbf{r}$ is a point on the line ℓ. According to Eq. (14.43), $t = 0$ gives $\bar{z} = 0$, so this corresponds to the intersection point of ℓ and the xy-plane. The position vector of this point is $\mathbf{q}_d$, as follows from Eq. (14.45), and therefore vector $\mathbf{q}_d$ is the displacement vector of the source as shown in Fig. 14.5. It represents the apparent position of the source in the xy-plane, when viewed from the far field. From Eqs. (14.10) and (14.44) we see that $\mathbf{q}_d \cdot \hat{\mathbf{r}}_o = 0$, so $\mathbf{q}_d$ is perpendicular to $\hat{\mathbf{r}}_o$.

14.7. The Displacement

The magnitude of the displacement vector $\mathbf{q}_d$ is given by

$$q_d = \frac{2|\beta|}{\beta^2 + 1} \frac{\sin\theta_o}{\zeta(\theta_o, \phi_o)} \ , \tag{14.46}$$

with $\zeta(\theta_o, \phi_o)$ given by Eq. (14.17). This source displacement depends on the observation angles θ_o and ϕ_o, and on the parameter β of the ellipse. For observation along the z-axis we have $\sin\theta_o = 0$ and the displacement is zero. As a function of θ_o, the displacement is maximum for $\theta_o = \pi/2$, so for observation along the *xy*-plane. In the *xy*-plane and for a given β, the displacement depends on the location in the *xy*-plane, e.g., on the angle ϕ_o. We find that the displacement is maximum for $\cos(2\phi_o) = 1$ when $|\beta| > 1$, and for $\cos(2\phi_o) = -1$ when $|\beta| < 1$. From Fig. 14.1 we then see that this corresponds to an observation along the major axis of the ellipse in both cases. In this direction, the displacement is given by

$$q_d = \begin{cases} 2|\beta| \ , & |\beta| > 1 \\ \dfrac{2}{|\beta|} \ , & |\beta| < 1 \end{cases} , \tag{14.47}$$

for a given β. For a circular dipole we have $|\beta| = 1$, and the maximum displacement is $q_d = 2$. For $\beta \to 0$ or $\beta \to \infty$ the eccentricity of the ellipse increases and the dipole becomes more linear. We then have $q_d > 2$ and the displacement can grow without bounds when the dipole approaches a linear dipole.

14.8. Intensity in the Image Plane

The observation direction (θ_o, ϕ_o) can be represented by a unit vector $\hat{\mathbf{r}}_o$, as in Fig. 14.5. We now consider an observation plane, shown in Fig. 14.6, which is a plane perpendicular to $\hat{\mathbf{r}}_o$, and a distance r_o away from the dipole. The origin of coordinates in this plane is represented by the vector $\mathbf{r}_o$. We then define λ and μ axes in this plane, such that the axes run into the direction of the the spherical unit vectors $\mathbf{e}_{\theta_o}$ and $\mathbf{e}_{\phi_o}$, as shown in the figure. Therefore, λ and μ are the Cartesian coordinates of a point in this plane, and the position vector $\mathbf{r}$ of a point in this plane can be written as

$$\mathbf{r} = \mathbf{r}_o + \lambda \mathbf{e}_{\theta_o} + \mu \mathbf{e}_{\phi_o} \ . \tag{14.48}$$

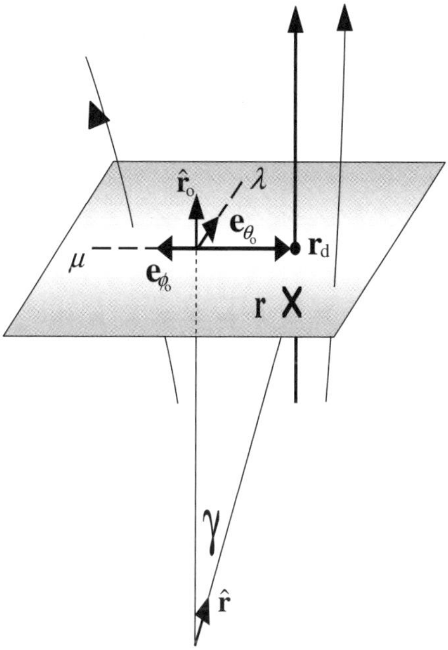

Fig. 14.6. The figure shows the coordinate system for the image plane and several field lines of the Poynting vector. Point **r** in the plane, represented by **X**, has Cartesian coordinates (λ, μ).

The dimensionless displacement vector (14.44) is a vector in this plane, and can be expressed as

$$\mathbf{q}_d = -\mathbf{e}_{\phi_o} Y(\theta_o, \phi_o; \beta) \sin\theta_o \ . \qquad (14.49)$$

The corresponding vector $\mathbf{r}_d = \mathbf{q}_d / k_o$ is shown in the figure. Therefore, the displacement vector is along the μ-axis. The asymptote ℓ of the field line for this direction (θ_o, ϕ_o) goes through the point $\mathbf{r}_d$, and when the distance r_o is sufficiently large, the field line through this point is perpendicular to the observation plane.

When an image is formed on the observation plane, or image plane, it may be expected that the major contribution comes from the field lines in the neighborhood of $\mathbf{r}_d$, since these cross the plane almost perpendicularly. In that case, the image in this plane would be

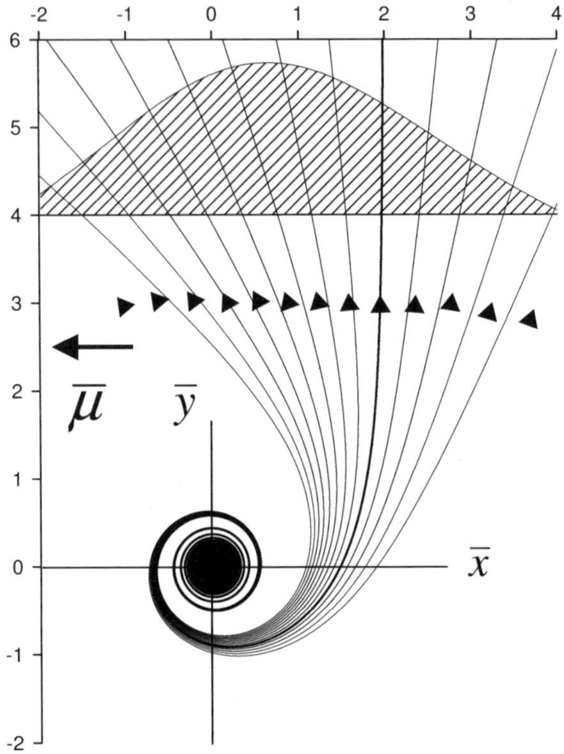

Fig. 14.7. Field lines of the Poynting vector in the xy-plane for a dipole moment which rotates in the xy-plane are shown. They swirl around the origin numerous times, and are then incident upon the image plane. The intensity distribution on the image plane shows a maximum in the neighborhood of the central field line for this direction, which goes through the point $\bar{x} = 2$ on the image plane.

shifted over $\mathbf{r}_\mathrm{d}$ with respect to the origin, and this shift would then be the same as the virtual displacement of the dipole in the xy-plane. However, since a bundle of field lines passing through this plane determines the image, rather than the single field line from Fig. 14.5, the image of the dipole is not necessarily exactly located at the position $\mathbf{r}_\mathrm{d}$. Figure 14.7 shows a bundle of field lines and the corresponding intensity distribution over a plane (defined below), and we observe that indeed the maximum of the image does not coincide with the displacement of the corresponding field line for this case ($\mathbf{q}_\mathrm{d} = 2\mathbf{e}_x$). It should also be noted that the shift in the xy-plane is defined through the extrapolation of the asymptote ℓ of the field line in the observation direction, as in Fig. 14.5, and, from an experimental point of view, this may not be a directly observable shift.

In order to investigate this issue in detail, we now consider the intensity distribution over the image plane. Since $\hat{\mathbf{r}}_o$ is the unit normal vector at all points in the image plane, the intensity (power per unit area) at point $\mathbf{r}$ is

$$I(\mathbf{r}_o;\lambda,\mu)=\mathbf{S}(\mathbf{r})\cdot\hat{\mathbf{r}}_o \ . \tag{14.50}$$

This intensity depends on r_o, the distance between the source and the image plane, the observation direction (θ_o,ϕ_o), and the coordinates λ and μ of point $\mathbf{r}$ in the plane. In addition, it depends parametrically on the parameter β of the ellipse.

We introduce dimensionless coordinates $\overline{\lambda}=k_o\lambda$ and $\overline{\mu}=k_o\mu$ in the observation plane, and the dimensionless distance between the origin of the plane and the dipole is $q_o=k_or_o$. For point $\mathbf{r}$ in the plane we then have $q=k_or$, and the relation to its Cartesian $(\overline{\lambda},\overline{\mu})$ is

$$q=\sqrt{q_o^2+\overline{\lambda}^2+\overline{\mu}^2} \ . \tag{14.51}$$

Putting everything together yields [6]

$$I(\mathbf{r}_o;\lambda,\mu)=I_o\left(\frac{q_o}{q}\right)^3\left\{1-\frac{1}{q^2}\frac{1}{1+\beta^2}\right.$$

$$\times\left[\beta^2(\overline{\rho}\cos\phi_o-\overline{\mu}\sin\phi_o)^2+(\overline{\rho}\sin\phi_o+\overline{\mu}\cos\phi_o)^2\right]$$

$$\left.-\frac{1}{q_oq}\left(1+\frac{1}{q^2}\right)\frac{2\beta}{\beta^2+1}\overline{\mu}\sin\theta_o\right\} , \tag{14.52}$$

for the intensity distribution in an observation plane for the case of an elliptical dipole moment, rotating in the *xy*-plane. In Eq. (14.52) we use the abbreviation

$$\bar{\rho} = q_o \sin\theta_o + \bar{\lambda}\cos\theta_o \ , \tag{14.53}$$

and the overall factor I_o is given by

$$I_o = \frac{3P_o}{8\pi r_o^2} \ . \tag{14.54}$$

The first two lines in Eq. (14.52) come from the angular dependence of the emitted power, which is accounted for by the function $\zeta(\theta,\phi)$ in Eq. (14.12) for $\mathrm{d}P/\mathrm{d}\Omega$. The appearance in Eq. (14.52) of this angular dependence is not through the function $\zeta(\theta,\phi)$, since here we consider the power flow through a plane, whereas $\mathrm{d}P/\mathrm{d}\Omega$ refers to the power flow through a sphere. The third line in Eq. (14.52) originates in the possible rotation of the field lines near the source. This is most easily seen from the fact that this term changes sign with the sign of β. The overall factor of $1/(q_o q)$ indicates that this term vanishes rapidly in the far field, as compared to the remaining terms in braces. We shall see below, however, that a finite and observable effect of this term survives in the far field.

14.9. Linear Dipole

Let us first consider $\beta = 0$, corresponding to a linear dipole, oscillating along the y-axis. The intensity distribution (52) on an image plane then becomes

$$I(\mathbf{r}_o;\lambda,\mu) = I_o\left(\frac{q_o}{q}\right)^3\left[1 - \frac{1}{q^2}(\bar{\rho}\sin\phi_o + \bar{\mu}\cos\phi_o)^2\right] \ . \tag{14.55}$$

For a linear dipole, all field lines are straight, and any dependence of the intensity on $\mathbf{r}_o$, λ and μ comes from the angular dependence of $\mathrm{d}P/\mathrm{d}\Omega$ and from the fact that we cut through $\mathrm{d}P/\mathrm{d}\Omega$ with a plane. As an illustration, let us consider an observation plane perpendicular to the y-axis at the positive side. We then have $\theta_o = \pi/2$, so $\bar{\rho} = q_o$, and $\phi_o = \pi/2$. We introduce angle γ as the observation direction for a point in the image plane, as seen from the site of the dipole (see Fig. 14.6). We then have $\cos\gamma = q_o / q$, and the intensity becomes

$$I(\mathbf{r}_o;\lambda,\mu) = I_o \cos^3\gamma \sin^2\gamma \ . \tag{14.56}$$

Since the right-hand side of Eq. (14.56) only depends on angle γ, the intensity distribution in the image plane is circularly symmetric around the origin. At the origin of the image plane we have $\gamma = 0$ and therefore $I = 0$, which corresponds to the well-known fact that no radiation is emitted along the dipole axis for a linear dipole. The intensity has a maximum for $\cos\gamma = \sqrt{3/5}$, corresponding to an angle of $\gamma = 39^o$, and this defines a ring in the observation plane. The radius of this ring is $q_o\sqrt{2/3}$. The intensity profile is shown in Fig. 14.8.

14.10. Rotating Dipole and the Far Field

The intensity distribution in Fig. 14.8 scales with the distance q_o between the dipole and the observation plane, which is a reflection of the fact that the field lines of the Poynting vector are straight for a linear dipole. When the image plane moves further away, the picture remains the same, apart from a scale factor. We now consider the effect of the rotation of the field lines near the source for a rotating dipole moment on the intensity profile in the far field. We first look at a circular dipole for which $\beta = \pm 1$. The intensity on the image plane, given by Eq. (14.52), simplifies for $\beta = \pm 1$ to

$$I(\mathbf{r}_o;\lambda,\mu) = I_o\left(\frac{q_o}{q}\right)^3\left[1 - \frac{1}{2q^2}(\bar{\rho}^2 + \bar{\mu}^2)\right.$$

$$\left. - \frac{\beta}{q_o q}\left(1 + \frac{1}{q^2}\right)\bar{\mu}\sin\theta_o\right] . \tag{14.57}$$

There is no dependence on the observation angle ϕ_o, as could be expected for a circular dipole moment in the *xy*-plane.

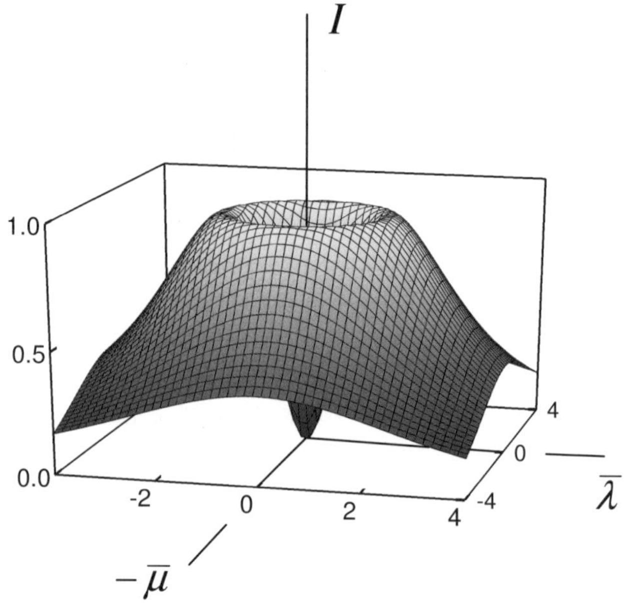

Fig. 14.8. The figure shows the intensity distribution on an image plane for a linear dipole. The plane is perpendicular to the dipole axis.

We are looking for effects in the far field that are due to the rotation of the field lines near the source. The displacement from Sec. 14.7 was maximum for observation in the *xy*-plane, e.g. $\theta_o = \pi/2$, so here we consider the same situation in order to find the maximum effect. From Eq. (14.53) we then have $\bar{\rho} = q_o$, and from Eq. (14.57) we see that the dependence on the coordinate λ in the image plane only enters through the λ dependence of q in Eq. (14.51). Therefore, the intensity only depends on λ as λ^2, and consequently there is an extremum at $\lambda = 0$. With $\theta_o = \pi/2$ and $\lambda = 0$, Eq. (14.57) becomes

$$I(\mathbf{r}_o;0,\mu) = I_o\left(\frac{q_o}{q}\right)^3\left[\frac{1}{2} - \frac{\beta}{q_o q}\left(1+\frac{1}{q^2}\right)\bar{\mu}\right], \qquad (14.58)$$

which is the intensity along the μ axis in the image plane. To find the extrema, we set $\partial I/\partial\bar{\mu} = 0$. This yields

$$\frac{3}{2}\overline{\mu}=-\frac{\beta}{q_{\mathrm{o}}q}\left[1+q_{\mathrm{o}}^{2}-3\overline{\mu}^{2}\left(1+\frac{2}{q^{2}}\right)\right] . \tag{14.59}$$

In the far field we have $q_{\mathrm{o}} >> 1$, and Eq. (14.59) becomes

$$\frac{3}{2}\overline{\mu}=-\frac{\beta}{q_{\mathrm{o}}q}(q_{\mathrm{o}}^{2}-3\overline{\mu}^{2}) . \tag{14.60}$$

We are looking for a possible shift in the intensity profile due to the rotation of the field lines. Such a shift has to remain finite for $q_{\mathrm{o}} \to \infty$, as can be seen from Fig. 14.7, so at such an extremum we must have $\overline{\mu}/q_{\mathrm{o}} \to 0$ for q_{o} large. If we indicate by $\overline{\mu}_{\mathrm{p}}$ the solution of Eq. (14.60), we find for the location of the peak in the intensity profile

$$\overline{\mu}_{\mathrm{p}}=-\frac{2}{3}\beta . \tag{14.61}$$

The magnitude of this shift is 2/3 in dimensionless coordinates, and since 2π corresponds to one wavelength, the shift is equal to a wavelength divided by 3π. Apparently, this is a nanoscale shift for optical radiation, but it is a shift that persists in the far field. Figure 14.9 shows the intensity distribution for this case.

For Fig. 14.9 we took $q_{\mathrm{o}}=4$ (and $I_{\mathrm{o}}=1$), which is $2/\pi$ times a wavelength. This is just enough to justify the far-field approximation $q_{\mathrm{o}} >> 1$. For larger values of q_{o}, the background becomes very large since it scales with q_{o}, and it may not be possible experimentally to resolve the shift of the peak. In an experiment, the dipole is set in oscillation with a laser beam. A circular dipole moment in the *xy*-plane is induced by a circularly polarized laser beam, propagating along the *z*-axis, and the sign of β is determined by the helicity of the laser. In Eq. (14.57), the first line is the broad background and the second line is due to the rotation of the field lines. The background does not depend on the sign of β, so if we would measure the intensity for left- and right-handed helicity then the difference of the profiles would be twice the second line of Eq. (14.57), and any resulting difference profile would be due entirely to the rotation of the field lines. This experiment was performed recently [7] and such an asymmetric profile was found indeed.

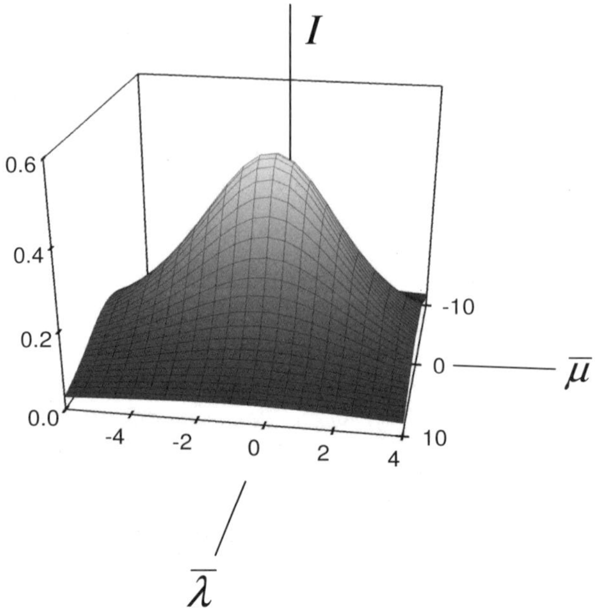

Fig. 14.9. The figure illustrates the nanoscale shift of the intensity profile for a circular dipole. The image plane is the same as in Fig. 14.8.

For an elliptical dipole, similar calculations show [6] that the profile is more complicated. It can have maxima, minima and saddle points. In particular, the maximum shown in Fig. 14.9 can become a minimum, and the effect of the rotation of the field lines can be a moving hole rather than a moving peak.

14.11. Intensity in the Near Field

With very precise near field techniques, involving fiber tip probes, intensities can be measured in the near field with nanoscopic precision [8,9]. In this approach the existence of the vortex could be verified experimentally by measuring the intensity directly at the location of the vortex, rather than through the observation of the displacement of the maximum of the intensity profile in the far field. Figure 14.10 shows the intensity distribution on an image plane very close to a circular dipole, for the case where the image plane is perpendicular to the y-axis. Near

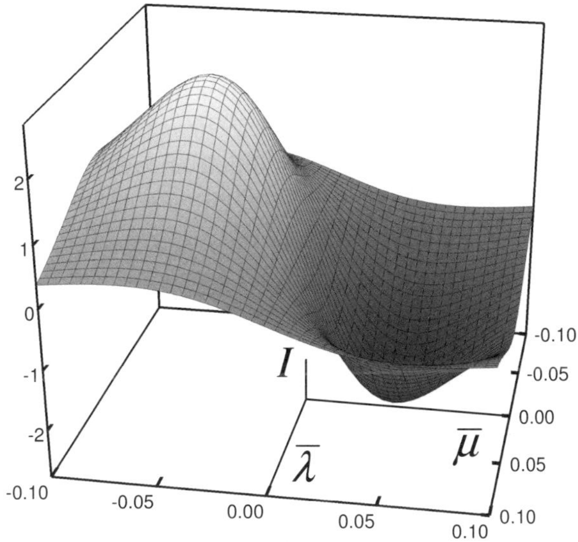

Fig. 14.10. The figure shows an intensity profile in the near field of a circular dipole.

the positive peak, the field lines pass through the image plane in the outward direction. The negative peak indicates that in this region of the observation plane the field lines are inward. This image is a direct consequence of the spiraling of the field lines near the source. A field line first passes through the image plane in the outward direction, and then spirals back inwards near the peak with negative intensity. Although it may not be possible to measure a negative intensity directly in an experiment, the figure nevertheless illustrates the winding of the field lines in an appealing manner.

14.12. Conclusions

When an electric dipole moment rotates in the *xy*-plane, the field lines of energy flow swirl numerous times around the *z*-axis, while remaining on the surface of a cone. Far away from the dipole, the field lines approach asymptotically a straight line, reminiscent of an optical ray. This line is displaced with respect to the radial direction, and this leads to an apparent shift of the dipole location, when viewed from far away. This displacement depends on the observation direction and on the eccentricity of the ellipse. In order to define an image of the dipole, we have considered the intensity distribution of the emitted radiation over an image plane. For a circular dipole, the image is a single peak when viewed in the plane of rotation of the dipole. It was shown that in the far field the peak of this intensity profile is shifted with respect to the central direction, and that this shift is due to the rotation of the field lines near the source. In this fashion, the near field curving of the field lines affects the image in the far field, and this makes the vortex near the dipole observable at a macroscopic distance. The shift is of the order of a fraction of a wavelength, as is the dimension of the vortex.

References

1. M. Born and E. Wolf, *Principles of Optics*, 6th edn. (Pergamon, 1980), Chapter 3.
2. J. D. Jackson, *Classical Electrodynamics*, 3rd edn. (Wiley, 1991), p. 411.
3. I. V. Lindell, *Methods for Electromagnetic Field Analysis* (Oxford U. Press, 1992), Section 1.4.
4. L. Mandel and E. Wolf, *Optical Coherence and Quantum Optics* (Cambridge U. Press, 1995), p. 469.
5. J. Shu, X. Li and H. F. Arnoldus, Energy flow lines for the radiation emitted by a dipole, *J. Mod. Opt.* **55**, 2457–2471 (2008).
6. J. Shu, X. Li and H. F. Arnoldus, Nanoscale shift of the intensity distribution of dipole radiation, *J. Opt. Soc. Am. A* **26**, 395–402 (2009).
7. D. Haefner, S. Sukhov and A. Dogariu, Spin Hall effect of light in spherical geometry, *Phys. Rev. Lett.* **102**, 123903 1–4 (2009).
8. K. G. Lee, H. W. Kihm, J. E. Kihm, W. J. Choi, H. Kim, C. Ropers, D. J. Park, Y. C. Yoon, S. B. Choi, D. H. Woo, J. Kim, B. Lee, Q. H. Park, C. Lienau and D. S. Kim, Vector field microscopic imaging of light, *Nat. Photonics* **1**, 53–56 (2007).
9. Y. Ohdaira, T. Inoue, H. Hori and K. Kitahara, Local circular polarization observed in surface vortices of optical near-fields, *Opt. Exp.* **16**, 2915–2921 (2008).

Chapter 15

Laser-Induced Femtosecond Magnetism

G. P. Zhang
Indiana State University
gpzhang@indstate.edu

Thomas F. George
University of Missouri–St. Louis
tfgeorge@umsl.edu

This chapter presents an overview of laser-induced femtosecond magnetism. We first review some basic elements in femtomagnetism and then explain the roles of exchange interaction and spin-orbit coupling. Finally, we summarize our current understanding.

15.1. Introduction

Magnetism is one of the most exciting research fields in modern physics, with applications ranging from magnetic storage, giant magneto-resistance to magnetic sensors. To manipulate the electron spins, one normally relies on a magnetic or a thermal field. A magnetic field can realign the spin along one particular direction. The process involves the domain motion and domain rotation. A thermal field can demagnetize a ferromagnetic sample by raising the spin temperature of the sample above its Curie temperature and then destroying the long-range magnetic ordering. In nearly all the traditional investigations in magnetism, one rarely is concerned about the time scale. This is true even in the conventional magneto-optical Kerr (MOKE) recording, where the laser pulse is used as a heating source to warm up the spin. For a small data processing, the speed is not a major problem, but as the disk becomes much larger (2 TB, 2009), one has to think about how quickly a few hundred GB file can be saved to a disk within a reasonable time frame. This is very important to the current technology.

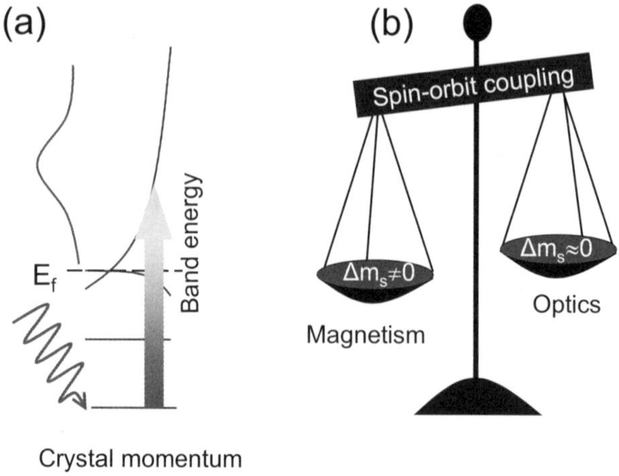

Fig. 15.1. (a) Upon laser excitation, an initial charge distribution is created. The laser energy selects an energy window. (b) Two competing requirements ($m_s \approx 0$ versus $m_s \neq 0$) for optical and magnetic excitations. The delicate balance between them is mediated by the spin-orbit coupling.

15.2. Fundamentals in Femtosecond Magnetism

Femtosecond magnetism or femtomagnetism refers to the laser-induced femtosecond magnetization process in ferromagnets [1]. In essence, it explores the ultrafast short time scale, offered by the laser pulse, on which to manipulate the spin [2]. This discovery has attracted an enormous attention [3–11].

15.2.1. *Optical and thermal excitations*

Different from thermal excitations where nearly all the electrons excited are around the Fermi surface, the laser excites electrons around and away from the Fermi surface [2]. Excited electrons peak at the laser energy, and there is a distribution around this peak. Figure 15.1(a) shows what happens when a laser hits a particular material. Within the first few hundred femtoseconds, the electrons are highly excited and lack well-defined statistics, so the temperature is not well defined [12]. During the excitation, the electron orbital angular momentum must be changed, according to optical selection rules (see below for more discussion). Its total angular momentum must be changed by ±1. For instance, if the electron is originally in a p orbital, the final state may be in an s or d orbital. The helicity of the light will determine how the magnetic angular momentum is changed: 0 for the linearly-polarized light or ± 1 for right or left circularly polarized light [13]. Therefore, if the helicity of the light does not affect the dynamics of the system, the magnetic orbital momentum is not changed, but this does not mean that the total angular momentum is not changing. In thermal excitations, the temperature is well defined, and it does not have an "equivalent" selection rule since all the interactions (such as spin-orbit coupling, spin–phonon interaction and so on) are hidden in the Bose–Einstein distribution.

15.2.2. *Necessity of symmetry breaking*

A big challenge in femtomagnetism is that the optical transition conserves the spins if the spin-orbit coupling is ignored (see Fig. 15.1(b)). In order to affect the spin system, the electron must find a way to couple the orbital dynamics to the spin dynamics. Spin-orbit coupling is a possibility, as it plays a role as a messenger to exchange the orbital and spin momenta. The role is very different from the exchange interaction on the spin. Such interaction does not bring in the "time dependence" of the magnetization change. In other words, the magnetic moments at $t = 0$ and $t \neq 0$ remain the same. The only time dependence from the exchange interaction is when we increase the lattice temperature, which is on a much longer time scale. On the other hand, the spin-orbit coupling brings in the real time dependence into the system. This is done through transferring the momentum from the orbital to spin system or *vice versa* [14,15]. Finally, it is important to point out that the spin-orbit coupling in the ground state and excited states may be different because both the electron spin and orbital momenta are normally different in the ground and excited states.

15.2.3. *Selection rule*

The dipole selection rule is about the quantum number, not about the expectation value of an operator [16]. In other words, even if two states have zero angular momentum or $\langle L_z \rangle = 0$, the transition is still possible. Here L_z is the angular momentum operator along the z direction. This appears to confuse many researchers. To be more convincing, we use a concrete example below.

Assume that we have two states ψ_1 and ψ_2 as

$$\psi_1 = p_1 Y_{10} + p_2 Y_{11} + p_3 Y_{1-1} \tag{15.1}$$

$$\psi_2 = d_1 Y_{2-2} + d_2 Y_{2-1} + d_3 Y_{20} + d_4 Y_{21} + d_5 Y_{22}, \tag{15.2}$$

where the radial dependence is not included, and p_i and d_i are coefficients to ensure the orthogonality and normalization of the wavefunctions. Y_{lm} is a spherical harmonic, with l and m being the angular and magnetic angular quantum numbers, respectively. We choose p_i and d_i such that both $\langle\psi_1|L_z|\psi_1\rangle$ and $\langle\psi_2|L_z|\psi_2\rangle$ are exactly zero. The transition matrix element for the z component is

$$\langle\psi_2|\hat{z}|\psi_1\rangle = p_1 d_3\langle Y_{20}|\hat{z}|Y_{10}\rangle + p_2 d_4\langle Y_{21}|\hat{z}|Y_{11}\rangle + p_3 d_2\langle Y_{2-1}|\hat{z}|Y_{1-1}\rangle. \quad (15.3)$$

One of the simplest choices to have zero orbital momentum for both states is $|p_i| = 1/\sqrt{3}$ and $|d_i| = 1/\sqrt{5}$, where the sign of those coefficients can be either positive or negative. Inserting these coefficients into the transition matrix element, we find

$$\langle\psi_2|\hat{z}|\psi_1\rangle = \pm\frac{2}{15} \pm \frac{1}{5\sqrt{3}} \pm \frac{1}{5\sqrt{3}}, \quad (15.4)$$

where $\pm$ represents any possible combinations of plus and minus signs. It is obvious that the transition moment is nonzero, even though the expectation value of both the orbital momenta is zero.

15.2.4. *Orbital momentum quenching*

In solids, due to the translation symmetry, or in molecules, due to the molecular symmetry, the degeneracy of the orbital momentum is lifted. If we sum over all those orbitals, we get a zero orbital momentum[a]. We should point out: (1) Zero orbital momentum in the ground state does not necessarily mean zero orbital moment in the excited states. For instance, for Gd in the ground state, the orbital momentum is quenched, but optical excitation is still possible [17]. (2) An optical transition relies on the off-diagonal orbital momentum change, not only on the diagonal orbital momentum or the expectation value of the orbital momentum. To be more precise, even if the expectation values of the orbital momentum for the ground state (GS) and excited state (EX) are zero, or $\langle GS|L_z|GS\rangle = 0$ and $\langle EX|L_z|EX\rangle = 0$, but this does not necessarily mean that the matrix $\langle GS|L_z|EX\rangle$ is zero.

[a]The quenching also depends on the electron filling. For instance, if a p orbital is split into three orbitals with $m_l = 0, \pm 1$ but only two orbitals $m_l = -1$ and $m_l = 1$ are occupied, then a complete quenching is not possible.

15.3. Exchange Interaction

15.3.1. *Static properties*

Magnetism originates from the exchange interaction between electron spins. The sign and magnitude of the exchange interaction determine whether a material is ferromagnetic, antiferromagnetic or ferrimagnetic, and how strong the magnetization is. A rigorous theoretical description is very difficult, because it involves the electron–electron interaction or many-body wavefunctions. In insulators, if the kinetic energy of the electrons is small, we can ignore it to some extent. Only the potential energy is left. If we keep the most important exchange terms, we arrive at the Heisenberg model.

$$H = -\sum_{i,j} J_{ij} \mathbf{S}_i \cdot \mathbf{S}_j, \tag{15.5}$$

where J_{ij} is the exchange integral between sites or atoms i and j, and $\mathbf{S}$ is the spin operator at these two sites. Note that the exchange integral formally has four indices for the "exchange charge." It condenses into two indices when two of those indices are the same. To be more precise,

$$\mathcal{J}_{ij;kl} = \int\int d\tau_1 d\tau_2 \phi_i^*(\mathbf{r}_1)\phi_j^*(\mathbf{r}_2)v(\mathbf{r}_1 - \mathbf{r}_2)\phi_k(\mathbf{r}_1)\phi_l(\mathbf{r}_2), \tag{15.6}$$

where $\mathcal{J}_{ij;kl}$ includes both Coulomb and exchange interactions.

One important note here is that the wavefunctions ϕ are not regular basis functions. As pointed out by Anderson long ago [18], they may or may not be orthogonal to each other. Under the Hartree-Fock approximation, they are computed iteratively by solving Fock matrices. In the exact many-body theory, these wavefunctions are only "coefficients" of the true many-body wavefunctions, where "coefficients" refers to a product of coefficients and basis functions (a series of Slater determinants). The potential term $v(\mathbf{r}_1 - \mathbf{r}_2)$ is not a regular bare Coulomb term, either, except if the system is just an atom, in which case the screening effect must be considered. This potential must include the influence from the atoms or crystal field and screening of other electrons. If in Eq. (15.6), we let $i = l$ and $j = k$, then we have only two independent indices, which are close to Eq. (15.5), but not exactly same. A further assumption is that ϕ_i is localized on atom i and ϕ_j on atom j. Then we arrive at Eq. (15.5). Once ϕ is localized on a particular atom, it is easy to see that the vibration of the nucleus will enter the exchange integral, though if the localization of the wavefunction is not good, then the effect will be smaller. This is important when one compares metals with insulators. Secondly, the vibration of the nucleus also affects the exchange integral through the potential $v(\mathbf{r}_1 - \mathbf{r}_2)$. These two contributions represent part of how much the vibration can affect the exchange interaction or magnetic ordering.

In a simple two-atom system, the exchange interaction is reduced if the distance between these two atoms increases. Unfortunately, even in the simplest molecules such as H_2, the exchange integral has to be found numerically, though an approximate analytical expression is possible [19]. Ma *et al.* recently fitted the numerical exchange interaction [20,21] as

$$J_{ij} = J_0 \left(1 - \frac{r_{ij}}{r_c}\right)^3 \Theta(r_c - r_{ij}), \tag{15.7}$$

where $r_c = 3.75$ Å is the cut-off distance for Fe, r_{ij} is the distance between atoms i and j, and Θ is the Heaviside step function. While such fitting is not unique, it provides some idea how the exchange interaction changes with the distance.

However, in the general demagnetization theory such as spin wave or magnon formalism, J is always kept constant, but the spin is assumed to deviate from its original orientation (such a deviation is initiated by electrons and nuclei). Doing so is equivalent to the conversion

$$J(\mathbf{R}_i, \mathbf{R}_j)\mathbf{S}_i \cdot \mathbf{S}_j \quad \longrightarrow \quad J(\mathbf{R}_i(0), \mathbf{R}_j(0))(\mathbf{S}_i + \Delta\mathbf{S}_i) \cdot (\mathbf{S}_j + \Delta\mathbf{S}_j), \tag{15.8}$$

where $\mathbf{R}$ is the position of a nucleus, and $\mathbf{R}(0)$ is its equilibrium position. Then, one converts $\Delta\mathbf{S}$ to a magnon operator. The temperature or phonon effect is added back through the Bose–Einstein distribution function $1/[\exp(\beta E) - 1]$, where $\beta = 1/k_B T$, k_B is the Boltzmann constant, T is temperature, and E is the magnon energy. Because of this treatment, the phonon effect is hidden.

In metals, the kinetic energy becomes comparable to the potential energy. Rigorously speaking, the Heisenberg model is not applicable, and the Hubbard model is often used. In this model, the interaction term is on the same atom and is not of the Heisenberg type. The spin wave excitation over different atoms in the Hubbard model is through the kinetic energy term, not the spin-spin interaction as in the Heisenberg model.

15.3.2. *Longer-time scale dynamics: Magnon–phonon interaction*

It is very instructive to consider the following permutation relation for the Heisenberg model:

$$[\mathbf{S}_z, \sum_{ij} J_{ij}\mathbf{S}_i \cdot \mathbf{S}_j] = 0, \tag{15.9}$$

where $\mathbf{S}_z = \sum_i \mathbf{S}_{zi}$. This means that in the Heisenberg model, $\mathbf{S}_z$ is a conserved quantity, which is also true for the magnon Hamiltonian. An important consequence is that $\mathbf{S}_z(t) = \mathbf{S}_z(0)$ will be time-independent, where t is time. In other words, in the traditional spin wave theory, even though demagnetization is possible, this is a static process and one can not tell when and how long the demagnetization takes place.

In real experiments, the time-dependence comes from the change of the exchange interaction. Since we are considering the vibrations of atoms or phonon effects (all the electronic degrees of freedom are integrated over), we expand J around its equilibrium position $\mathbf{R}_i(0)$ of atom i as [20]

$$J(\mathbf{R}_i, \mathbf{R}_j) = J(0) + \left(\frac{\partial J}{\partial \mathbf{R}_{i(j)}}\right)_0 \Delta\mathbf{R}_{i(j)} + \frac{1}{2}\left(\frac{\partial^2 J}{\partial^2 \mathbf{R}_i \mathbf{R}_j}\right)_0 \Delta\mathbf{R}_i \Delta\mathbf{R}_j + ..., \tag{15.10}$$

where $\Delta\mathbf{R}_i$ is the lattice displacement for atom i, and a similar term appears for j. Different from the equilibrium calculation (such as magnon spectrum), here the first order is not zero. If the displacement is quantized, it can be written as $(a + a^\dagger)$, where a is the phonon annihilation operator. $\frac{\partial J}{\partial \mathbf{R}_i}$ is exchange part of electron–phonon interaction to first order. Using the data from on bcc Fe and hcp Co [20], we can estimate the magnitude of $\frac{\partial J}{\partial \mathbf{R}_i}$[a]. For Fe, if we assume a lattice displacement of 1% of the lattice constant, we have an exchange energy change of 0.125 mRy, or 0.0017 eV or 2352 femtoseconds. For hcp Co, we find 0.2 mRy, or 0.00272 eV or 1470.5 fs. These values set the lower limit of the magnon time scale, since it does not take into account the phonon time scale.

One important question is how the magnon dynamics affects the demagnetization process. First of all, even though J is changing with time as a result of the lattice vibration, it does not mean that the magnetization is

[a]Figure 2 in Ref. 20 does not allow for an accurate determination. For bcc Fe, we choose a first point at 0.92 a (a is a lattice constant, not to be confused with the phonon operator a in the main text) with exchange of $J = 2$ mRy, and the second point at 0.96 a with exchange of 1.5 mRy. The first-order derivative of J is -12.5 mRy/a. For hpc Co, we find -20 mRy/a.

changing, which is clear from the permutation in Eq. (15.9). It only means that the spin can fluctuate from atom to atom, but when we sum over all the spins, we get the exact same spin as what we start with. This is due to the law of spin conservation [19]. Second, it is necessary to introduce some mechanism to break spin conservation and couple the electron spin to other subsystems. One way is to use the magnetic field, and another way is to introduce spin-orbit coupling (SOC). Since the magnon time scale is already on the phonon time scale, its effect on magnetization will be in the nanosecond regime. With SOC, the spin couples to the orbital momentum. The demagnetization proceeds by transferring some momentum to orbital part.

In the Hubbard model as well as the Stoner model, since the Hamiltonian permutes with total $\mathbf{S}_z$, there is a spin dynamics but without real time-dependent demagnetization. Therefore, the Stoner excitation itself can not induce time-dependent demagnetization. It must rely on the same mechanism as discussed above.

15.3.3. *Dynamics on shorter time scales*

One of the perplexing issues in femtomagnetism is the role of exchange interaction on the shorter time scale. We start with the Kohn–Sham equation (in Ry atomic units),

$$[-\nabla^2 + V_{Ne} + V_{ee} + V_{xc}^{\sigma}]\psi_{i\mathbf{k}}^{\sigma}(r) = E_{i\mathbf{k}}^{\sigma}\psi_{i\mathbf{k}}^{\sigma}(r). \tag{15.11}$$

The terms on the left-hand side represent the kinetic energy, electron-nuclear attraction, Coulomb and exchange interactions, respectively. $\psi_{i\mathbf{k}}^{\sigma}(r)$ is the Bloch wavefunction of band i at crystal momentum $\mathbf{k}$ with spin σ, and $E_{i\mathbf{k}}^{\sigma}$ is the band energy. SOC is included using a second-variational method in the same self-consistent iteration. We use E_{ik} to represent the eigenvalue with the SOC. We choose two different exchange-correlation functionals, the local density approximation (LDA) and the general gradient approximation (GGA). Then, we perform a self-consistent calculation. The number of k points is huge, $104 \times 104 \times 104$, which gives us a well-converged result. We employ the full-potential augmented planewave method (FLAPW). We use a much higher planewave cut-off K_{max}, i.e., $RK_{max} = 9.5$, where R is the muffin-tin atomic radius. We also use the exact angular momentum for the spin-orbit coupling calculation to eliminate any possible truncation errors which may occur if one uses a smaller subspace for the SOC calculations.

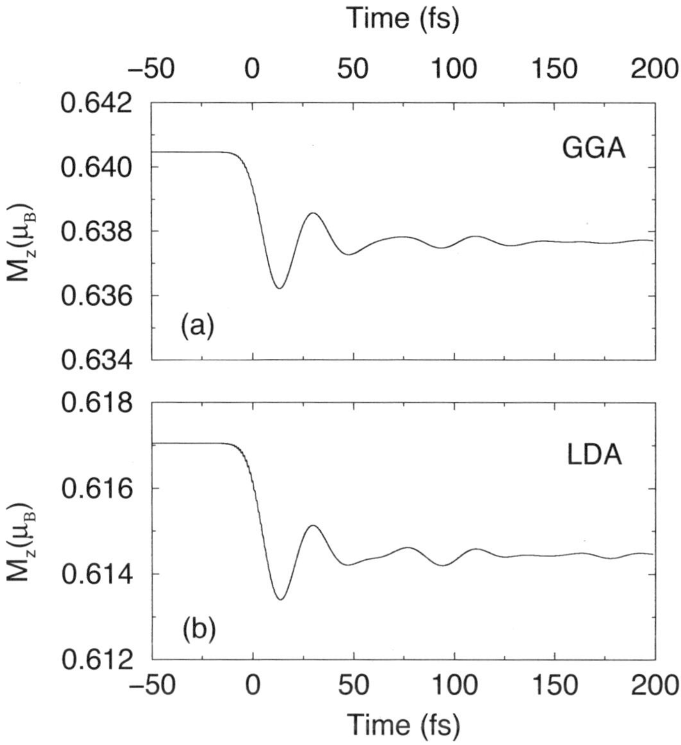

Fig. 15.2. (a) Spin moment change as a function of time. The results are computed within the GGA. (b) Same as (a), but the calculation is done with the LDA.

The dynamical simulation is done by solving the Liouville equation [22 24],

$$i\hbar\frac{\partial\rho_{i,j;\mathbf{k}}}{\partial t} = (E_{i\mathbf{k}} - E_{j\mathbf{k}})\rho_{i,j;\mathbf{k}} - e\mathbf{F}(t)\cdot\sum_m\left(\mathbf{r}_{i,m;\mathbf{k}}\rho_{m,l;\mathbf{k}} - \rho_{i,m;\mathbf{k}}\mathbf{r}_{m,j;\mathbf{k}}\right), \tag{15.12}$$

where $\mathbf{F}(t)$ is the laser field, $\mathbf{r}_{i,m;\mathbf{k}}$ is the transition matrix between states i and m at point $\mathbf{k}$, and $\rho_{i,j;\mathbf{k}}$ refers to the density matrix element between states i and j at point $\mathbf{k}$. Once we have obtained the density matrix at any time step, we can compute the spin moment as

$$M_z(t) = \sum_{\mathbf{k}} Tr(\rho_{\mathbf{k}}S^z_{\mathbf{k}}), \tag{15.13}$$

where $S^z_{\mathbf{k}}$ is the spin matrix.

Our results are shown in Fig. 15.2, where GGA is (a) and LDA is (b). Here the laser field amplitude is 0.05V/Å and the energy is 2.0 eV. The pulse duration is 12 fs. One static difference is that the net magnetic moment in GGA is larger than the LDA results. However, the dynamics evolution is very similar. A small difference is seen at 50 fs.

15.4. Spin-Orbit Coupling

The electric field affects the orbital momentum of the electron, but not the spin, since the field has no spin dependence. Therefore, in order for the electric field to play a role in the demagnetization, we must invoke a mechanism that can couple the electric field to the spin. SOC plays roles on two different time scales: one is on the electron time scale, and the other is on the phonon time scale.

15.4.1. *Static properties*

Spin-orbit coupling is a relativistic effect. When electrons are orbiting around a nucleus, it sees an effective magnetic field from the nucleus, so that the orbital degree of freedom of the electron couples to its spin degree of freedom. The most important effect of SOC in femtomagnetism is that it provides a way to couple the orbital momentum to the spin momentum, so that the light electric field can indirectly affect the spin momentum. Different from the magnetic field, which under time-reversal symmetry is odd, the SOC's symmetry is even. This is because both the spin and momentum reverse their directions.

For a spherical potential, SOC can be written as (SI units)

$$\lambda = \frac{1}{2m^2c^2}\frac{1}{4\pi\epsilon_0}\frac{1}{r}\frac{dV}{dr}\mathbf{S}\cdot\mathbf{L}, \tag{15.14}$$

where m is the mass of the electron, c is the speed of light, r is electron radius, and V is the potential. For a nonspherical potential, there will be two extra terms, corresponding to the polar and azimuthal directions. It is easy to see that SOC contains two parts: one is the angular part $\mathbf{S}\cdot\mathbf{L}$, and the other is the radial part. To have a rough idea of how large these terms are, we first estimate the value of $\mathbf{S}\cdot\mathbf{L}$ by introducing the basis function for the total angular momentum $\mathbf{J}$ (not to be confused with the exchange interaction above) according to Cohen–Tannoudji, Diu and Laloë [16]. For the s state, the angular momentum is zero, so there is no SOC. Given the quantum number $j = l \pm s$, $\mathbf{S}\cdot\mathbf{L} = ls$ or $\mathbf{S}\cdot\mathbf{L} = -s(l+1)$, excluding $l = 0$.

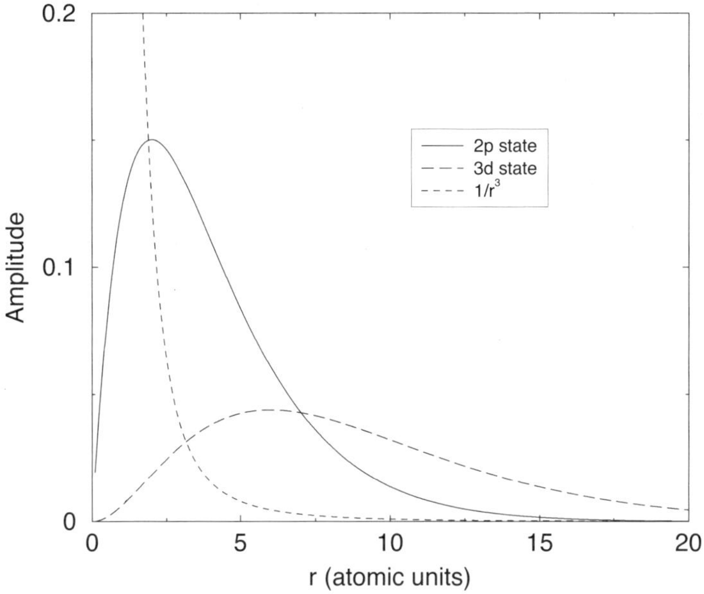

Fig. 15.3. A comparison of $2p$ and $3d$ states. $1/r^3$ is from the spherical potential.

It is clear that the larger l is, the larger λ is, but as seen below, this is only part of the story.

We first define a universal spin-orbit coupling constant ξ_0,

$$\xi_0 = \frac{1}{4\pi\epsilon_0}\frac{e^2}{2m^2c^2}\frac{\hbar^2}{a_0^3} = 0.0007245222 \text{ eV}, \tag{15.15}$$

where $-e$ is the electron charge, a_0 is the Bohr radius, ϵ_0 is the permittivity of free space, m is the electron mass, and $\hbar$ is the Planck constant over 2π. For the radial contributions, we need to use real wavefunctions. We take hydrogen atom as an example, where V is just the Coulomb interaction. Cohen–Tannoudji, Diu and Laloë [16] worked out the coupling for $2p$ states. If we represent it by the universal SOC constant, we have

$$\xi_{2p} = \xi_0/24 = 3.02 \times 10^{-5} \text{ eV}. \tag{15.16}$$

For the $3d$ orbital, we can write down a similar term as

$$\xi_{3d} = \xi_0 \int_0^\infty \frac{1}{r^3} R_{32}^2 r^2 dr = \xi_0/405 = 1.79 \times 10^{-6} \text{ eV}, \tag{15.17}$$

where R_{32} is the hydrogen $3d$ wavefunction. It is interesting to see that the $3d$ radial contribution is 17 times smaller than $2p$'s. The reason is that the

$3d$ state is much more extended than the $2p$ states (see Fig. 15.3), while the spherical potential has a dependence of $1/r^3$. The overlap between the potential and the wavefunction is much larger for the $2p$ state.

For other elements, one important modification is that the radius of the atomic orbital is contracted by a_0/Z, where Z is the atomic number. Since ξ_0 is proportional to $1/a_0^3$, ξ_0 is increased to $Z^3\xi_0$. And also the Coulomb potential increases to $ZV(r)$. Collecting all the terms together, from Eq. (15.15), we see a power of 4 increase in the spin-orbit coupling constant, or

$$\xi_0 \longrightarrow Z^4\xi_0 \tag{15.18}$$

This qualitatively explains why heavy elements have strong SOC, but this result is different from what Baym found [25]. He showed that the spin-orbit coupling is linearly proportional to Z, which is incorrect. He made his estimation from the integral

$$\xi_{nl} \sim \int r^2 dr |\phi_{nl}|^2 \frac{e^2}{r^3} \sim \frac{Ze^2}{a_0^3} \quad \text{(incorrect)}, \tag{15.19}$$

where he ignored the spatial dependence of the wavefunction ϕ_{nl} on Z.

If we take Ni as an example, ξ_{3d} for $3d$ state becomes $28^4 \times 1.79 \times 10^{-6} = 1.09958$ eV. If we approximate $\mathbf{L}\cdot\mathbf{S} = ls$ (see above) and use bulk nickel's orbital moment ($\langle l_z \rangle = 0.06\hbar$) and spin moment ($\langle s_z \rangle = 0.64\hbar$) as l and s, respectively, we find the spin-orbit coupling is 0.0422 eV. This value is smaller than the accepted value of 0.07 eV, but the order of magnitude is correct. This underestimation indicates that the electron delocalization, filling and correlation effect may affect the value of SOC in bulk nickel.

For a hydrogen-like atom, the generic spin-orbit coupling is

$$\xi_{nl} = \xi_0 \frac{Z^4}{n^3 l(l+1/2)(l+1)} \tag{15.20}$$

which was first worked out by Di Bartolo [26] in 1967. The reader can directly use this expression to estimate the strength of the spin-orbit coupling. It confirms our above observation that the spin-orbit coupling becomes smaller for larger quantum n and l, but it gets a boost from the atomic number Z.

15.4.2. *How is the magnetic moment changed?*

The role of spin-orbit coupling in femtomagnetism differs from that of the regular magnetic field. The magnetic field only shifts the spin up and down the energy levels in the diagonal matrix elements of the Hamiltonian, but the SOC shifts not only the diagonal matrix elements of the Hamiltonian but also its off-diagonal elements, where the latter is due to spin mixing. As a result, all the band states consist of both spin-up and spin-down parts of the wavefunction.

When a laser excites the system, the optical transition is still between the same spin subspace of the wavefunctions. Since the spin mixing in the ground and excited states is different, the spin moment changes. This can be easily seen if we consider two states,

$$\psi_1 = \frac{1}{\sqrt{3}}(\sqrt{2}|\uparrow\rangle + |\downarrow\rangle) \tag{15.21}$$

$$\psi_2 = \frac{1}{\sqrt{2}}(|\uparrow\rangle + |\downarrow\rangle). \tag{15.22}$$

Since the spin for state ψ_2 $\langle s_z \rangle = 0$ and for state ψ_1 is $1/3\ \mu_B$, any transition from state 1 to 2 leads to a reduction of the magnetic moment. However, the optical transition matrix element is between the spin-up (spin-down) parts of the wavefunctions. There will be no crossing transition. If we want to see a spin flip from a positive spin moment to a negative spin moment (for instance, from $1/3\ \mu_B$ to $-1/3\ \mu_B$), then these two states should look like,

$$\psi_1 = \frac{1}{\sqrt{3}}(\sqrt{2}|\uparrow\rangle + |\downarrow\rangle) \tag{15.23}$$

$$\psi_2 = \frac{1}{\sqrt{3}}(|\uparrow\rangle + \sqrt{2}|\downarrow\rangle). \tag{15.24}$$

It is also clear from these equations that for a single optical transition, the maximum spin change is smaller than $2\mu_B$. In other words, one can not optically flip a down spin to an upper spin through interacting with the light field only once. This is because if one state is completely polarized along $+z$ with $s_z = \hbar/2$ and the other along $-z$ with $s_z = -\hbar/2$, the transition becomes forbidden. If the system interacts with the laser field several times, then it is possible to have a spin flip from a spin-down state to a spin-up state since the system can undergo multiple steps of transition, although the probability of such flipping depends on the details of electronic structures and laser fields.

The level of spin-up and spin-down mixing in the wavefunction is jointly determined by SOC and exchange interaction. By checking the spin moment difference between two states, it is possible to estimate the maximum possible change in spin. However, the exact amount of the spin change is also determined by the optical transition probabilities, which in turn are determined by the external laser field and the transition matrix element between two states.

15.4.3. *Time scale determined by spin-orbit coupling*

This time scale of femtomagnetism through SOC mechanism is determined by the spin-orbit coupling magnitude. The exchange interaction time scale is much shorter, which is the reason why recent experiments on rare earth compounds show an extremely long time scale for the inverse Faraday effect [27], even though the laser pulse duration is very much shorter. Physically, when the light field excites the electron system's orbital momentum, the spin can not respond quickly since the magnitude of the SOC is small, and it takes time for the spin to feel the effect. In the highly-excited states, the SOC may change, in comparison to the ground states. If the excited states' wavefunction turns to be more delocalized, the overlap matrix see Eq. (15.17) becomes smaller, as does the spin-orbit coupling. This leads to a very long decay time if one uses a strong laser field. Another possible channel is the restriction of the band structure. For instance, in half-metallic CrO_2, the optical transition is dominated by one spin channel transition. Since the single channel does not allow a substantial spin moment change as seen above, the reduction in the magnetic moment is smaller and must proceed through high-order spin-orbit coupling. This results in a slower process, in comparison to metals such as nickel.

15.5. Discussion

To summarize our current understanding, we review the fundamental process of femtomagnetism as follows. We caution that each item below assumes that the laser pulse is very short (less than 50 fs), so that it does not blur the intrinsic dynamics of the system. Otherwise, a clear distinction between different processes is difficult if not impossible.

- In the beginning of the laser radiation, the electron is highly excited. The electron excitation has two effects: (i) the electron orbital moment is changed, and (ii) the Coulomb and exchange interactions between electrons lead to individual excitations (such as excitons on a time scale of 100 fs) and collective excitations (such as plasmons on a time scale of 10 fs).

- Since the laser electric field does not couple to the spin, spin-orbit coupling must be invoked. With a changed orbital momentum, the spin experiences a new effective magnetic field, different from the magnetic field in the ground state. How quickly the spin is going to respond depends on the strength of the SOC and the exchange interaction. How large the spin changes depends on the band structure and laser field (intensity, duration and energy). If the conduction band has a smaller spin polarization than the valence band, the excitation is likely to lead to demagnetization. Here, we purposely avoid using the total angular momentum conservation, since in many systems, the total angular momentum is not a good quantum number.
- Due to the electron–phonon interaction, the electron transfers energy to the lattice system, but since the nuclei are much heavier than electrons, they only move on a 1 picosecond time scale. The exact value of this time scale depends a particular system. Note that the electron–phonon interaction does not affect spin.
- The spin–phonon interaction is a result of the spin-orbit coupling with the moving nucleus. The time scale is determined jointly by the phonons and the strength of the spin-orbit coupling. This normally appears on a few hundred picosecond time scale.

Acknowledgments

This work was supported by the U. S. Department of Energy under Contract No. DE-FG02-06ER46304 and also by a Promising Scholars grant from Indiana State University. We acknowledge part of the work as done on Indiana State University's high-performance computers. This research used resources of the National Energy Research Scientific Computing Center at Lawrence Berkeley National Laboratory, which is supported by the Office of Science of the U.S. Department of Energy under Contract No. DE-AC02-05CH11231. Initial studies used resources of the Argonne Leadership Computing Facility at Argonne National Laboratory, which is supported by the Office of Science of the U.S. Department of Energy under Contract No. DE-AC02-06CH11357.

References

1. E. Beaurepaire, J.-C. Merle, A. Daunois and J.-Y. Bigot, Ultrafast spin dynamics in ferromagnetic nickel, *Phys. Rev. Lett.* **76**, 4250–4253 (1996).

2. G. P. Zhang, W. Hübner, E. Beaurepaire and J.-Y. Bigot, Laser-induced ultrafast demagnetization: Femtomagnetism, a new frontier? *Topics Appl. Phys.* **83**, 245–290 (2002).
3. J.-Y. Bigot, M. Vomir and E. Beaurepaire, Coherent ultrafast magnetism induced by femtosecond laser pulses, *Nature Phys.* **5**, 515–520 (2009).
4. G. P. Zhang, W. Hübner, G. Lefkidis, Y. Bai and T. F. George, Paradigm of the time-resolved magneto-optical Kerr effect for femtosecond magnetism, *Nature Phys.* **5**, 499–502 (2009).
5. G. P. Zhang and W. Hübner, Laser-induced ultrafast demagnetization in ferromagnetic metals, *Phys. Rev. Lett.* **85**, 3025 (2000); W. Hübner and G. P. Zhang, Ultrafast spin dynamics in nickel, *Phys. Rev. B* **58**, R5920-R5930 (1998); G. P. Zhang and W. Hübner, Femtosecond spin dynamics in the time domain, *J. Appl. Phys.* **85**, 5657–5659 (1999).
6. P. M. Oppeneer and A. Liebsch, Ultrafast demagnetization in Ni: Theory of magneto-optics for non-equilibrium electron distributions, *J. Phys.: Condens. Matt.* **16**, 5519–5530 (2004).
7. A. Vernes and P. Weinberger, Formally linear response theory of pump-probe experiments, *Phys. Rev. B* **71**, 165108-1-12 (2005).
8. I. Radu, G. Woltersdorf, M. Kiessling, A. Melnikov, U. Bovensiepen, J.-U. Thiele and C. H. Back, Laser-induced magnetization dynamics of lanthanide-doped permalloy thin films, *Phys. Rev. Lett.* **102**, 117201-1-4 (2009).
9. B. Koopmans, M. van Kampen, J. T. Kohlhepp and W. J. M. de Jonge, Ultrafast magneto-optics in nickel: Magnetism or optics? *Phys. Rev. Lett.* **85**, 844–847 (2000).
10. G. M. Müller, J. Walowski, M. Djordjevic, G. X. Miao, A. Gupta, A.V. Ramos, K. Gehrke, V. Moshnyaga, K. Samwer, J. Schmalhorst, A. Thomas, A. Hütten, G. Reiss, J. S. Moodera and M. Münzenberg, Spin polarization in half-metals probed by femtosecond spin excitation, *Nature Mater.* **8**, 56–61 (2009).
11. J. Hohlfeld, C. D. Stanciu and A. Rebei, Athermal all-optical femtosecond magnetization reversal in GdFeCo, *Appl. Phys. Lett.* **94**, 152504-1-3 (2009).
12. L. Guidoni, E. Beaurepaire and J.-Y. Bigot, Magneto-optics in the ultrafast regime: Thermalization of spin populations in ferromagnetic films, *Phys. Rev. Lett.* **89**, 017401-1-4 (2002).
13. C. D. Stanciu, F. Hansteen, A. V. Kimel, A. Kirilyuk, A. Tsukamoto, A. Itoh and Th. Rasing, All-optical magnetic recording with circularly polarized light, *Phys. Rev. Lett.* **99**, 047601-1-4 (2007).
14. G. P. Zhang and T. F. George, Total angular momentum conservation in laser-induced femtosecond magnetism, *Phys. Rev. B* **78**, 052407-1-4 (2008).
15. G. P. Zhang, Laser-induced orbital and spin excitations in ferromagnets: Insights from a two-level system, *Phys. Rev. Lett.* **101**, 187203-1-4 (2008).
16. C. Cohen–Tannoudji, B. Diu and F. Laloë, *Quantum Mechanics* (Wiley, Asia Pte. Ltd., 2005), p. 1223.
17. A. F. Bartelt, A. Comin, J. Feng, J. R. Nasiatka, T. Eimüller, B. Ludescher, G. Schütz, H. A. Padmore, A. T. Young and A. Scholl, Element-specific spin and orbital momentum dynamics of Fe/Gd multilayers, *Appl. Phys. Lett.* **90**, 162503-1-3 (2007).

18. P. W. Anderson, Theory of magnetic exchange interactions: Exchange in insulators and semiconductors, in *Solid State Physics, Advances in Research and Applications*, Vol. **14**, ed. F. Seitz and D. Turnbull (Academic, New York, 1963), pp. 99–214.
19. P. W. Ma, C. H. Woo and S. L. Dudarev, Large-scale simulation of the spin-lattice dynamics in ferromagnetic iron, *Phys. Rev. B* **78**, 024434-1-12 (2008).
20. R. F. Sabiryanov and S. S. Jaswal, Magnons and magnon-phonon interactions in iron, *Phys. Rev. Lett.* **83**, 2062–2064 (1999).
21. S. Moran, C. Ederer and M. Fähnle, *Ab initio* electron theory for magnetism in Fe: Pressure dependence of spin-wave energies, exchange parameters, and Curie temperature, *Phys. Rev. B* **67**, 012407-1-4 (2003).
22. G. P. Zhang, Y. Bai and T. F. George, An energy- and crystal momentum-resolved study of laser-induced femtosecond magnetism, *Phys. Rev. B* (accepted, 2009).
23. G. P. Zhang, Y. Bai, W. Hübner, G. Lefkidis and T. F. George, Understanding laser-induced ultrafast magnetization in ferromagnets, *J. Appl. Phys.* **103**, 07B113-1-3 (2008).
24. T. Hartenstein, G. Lefkidis, W. Hübner, G. P. Zhang, and Y. Bai, Time-resolved and energy-dispersed spin excitation in ferromagnets and clusters under the influence of femtosecond laser pulses, *J. Appl. Phys.* **105**, 07D305-1-3 (2009).
25. G. Baym, *Lectures on Quantum Mechanics* (Benjamin, New York, 1969), pp. 462–465.
26. D. Di Bartolo, *Optical Interactions in Solids* (John Wiley & Sons, INC. New York, 1968), pp. 143–145.
27. K. Vahaplar, A. M. Kalashnikova, A. V. Kimel, D. Hinzke, U. Nowak, R. Chantrell, A. Tsukamoto, A. Itoh, A. Kirilyuk and Th. Rasing, Ultrafast path for optical magnetization reversal via a strongly nonequilibrium state, *Phys. Rev. Lett.* **103**, 117201-1-4 (2009).

Chapter 16

Gas-Dispersed Materials as an Active Medium of Chemical Lasers

Renat R. Letfullin
Rose-Hulman Institute of Technology
letfullin@rose-hulman.edu

Thomas F. George
University of Missouri–St. Louis
tfgeorge@umsl.edu

A promising avenue in the development of high-energy pulsed chemical HF/DF lasers and amplifiers is the utilization of a photon-branched chain reaction initiated in a two-phase active medium, i.e., a medium containing a laser working gas and ultradispersed passivated metal particles. These particles are evaporated under the action of IR laser radiation, resulting in the appearance of free atoms, their diffusion into the gas, and the development of a photon-branching chain process, which involves photons as both reactants and products. The key obstacle here is the formation a relatively-large volume (in excess of 10^3 cm^3) of the stable active medium and filling this volume homogeneously for a short time with a sub-micron monodispersed metal aerosol, which has specified properties. In this chapter, results are presented for an extensive study of laser initiation of a photon-branched chain reaction in a gas-dispersed H_2-F_2 medium.

16.1. Introduction

Theoretical investigations [1–9] have been carried out for the problem of photon branching in a two-phase (consisting of a gas and disperse particles) chemically-active media. It has been shown that the introduction of a certain concentration of ultradisperse particles into the active medium of a chemical laser based on a chain reaction, and further their (particles) evaporation by external infrared (IR) radiation, can result in the generation of coherent radiation of energy considerably higher than the energy of an initiating input pulse at the same frequency. New laser photons generated in the cavity can also participate directly in the laser-chemical reactions, facilitating the formation of active centers of a chain reaction (free atoms). If the number of emitted photons exceeds the number of those used in the initiation process, one can speak of a new type of branched chain reaction: photon-branched chain reaction (PBCR). The PBCR has practical interest because of its possible use in constructing super-high energy pulsed HF/DF-lasers and amplifiers without external energy sources.

The first experimental results to achieve a photon-branched chain reaction in a two-phase medium were published in Ref. [10]. The authors showed that during the injection of a fine-disperse aluminum powder with particle diameters of 0.3 μm covered by a thin oxide film in a mixture of $F_2 + H_2 + O_2 + He$, the system remains stable. They observed the formation of free aluminum atoms under the action of radiation of a chemical laser with intensities as high as 10^7 W/cm^2, with demonstration of the chemical mixture ignition. Comparison of the output energy in the presence and absence of disperse particles led to the conclusions that about 15% of the output energy is due to a PBCR.

The successes of the first experiments have given a positive answer to the question of the feasibility of a photon-branched chain chemical process in a two-phase active medium of an HF laser, and it has confirmed the accuracy of the theoretical models of the chemical and vibration-relaxation kinetics. We have developed a relatively-simple HF laser kinetic model [1,2], which includes the most important physical-chemical kinetics features under the assumption of rotational-translational non-equilibrium, making it possible to calculate the specific energy of the coherent radiation, the pulse duration, the effective chain length of the reaction between fluorine and hydrogen, and the chemical and engineering efficiencies of the laser. This model was supplemented with an electrodynamical model [2–9] of the pulsed laser based on a solution of the wave equation. The wave approach for the description of a pulsed chemical HF-laser amplifier based on a PBCR in an unstable telescopic cavity enables the detection of interesting optical effects in an active two- phase medium of H_2-F_2-O_2-He and Al particles, and also of the new properties of the laser. These effects, the optimal parameters and the output performances of the laser, and also some characteristics of an active medium, are presented in Table 16.1.

It has been shown [3,5–9] that the laser-chemical reaction can be ignited in an initial small focal volume of an active medium, and then spreads out of this minimal volume spontaneously in the auto-wave regime, without external power sources, to subsequently fill the whole volume of the laser with a highly-intense electromagnetic field with self-supporting cylindrical photon-branching zones formed by the paths of the rays inside the unstable telescopic cavity. The ignition of an auto-wave PBCR under the condition of external signal focusing reduces strongly the input pulse energy necessary for initiation down to ~10^{-8} J and, thereby, enables a huge value of the energy gain of about 10^{11}. In a system of this kind, there is no danger of self-excitation, even for large values of the gain, because the medium is not inverted in advance and the energy is reserved in the form of the free energy of reactants, instead of through working laser transitions.

A parametrical study [7–9] of a pulsed chemical laser based on a PBCR initiated in a gaseous disperse medium, composed of H_2-F_2-O_2-He and Al particles, by focused external IR radiation has been conducted. We see (Table 16.1) that both the observed effects of auto-wave spreading of a PBCR under the condition of external signal focusing and the effect of a huge laser energy gain of 10^{11} give the pulsed chemical HF laser the important properties of autonomy and compactness. We have determined the performances of the main laser units: the minimal linear size of the cavity, the minimal energy of the master oscillator and its power source, and the dimensions of the system of gas cylinders. We have also obtained the optimal composition of the reactants and the pressure of the laser working mixture. For example, at a general pressure of the working gases of P = 2.3 bar, the optimal parameters of the dispersed component (Al particles with radius r_0 = 0.09 μm and concentration $N_0 = 1.4\times10^9$ cm^{-3}) and the composition of the working

Table 16.1. Main properties and parameters of the HF-laser amplifier based on a PBCR.[a]

Observed effects and properties of the laser	**Calculated optimal laser parameters**	**Output characteristics**
Effect of auto-wave spreading of a PBCR throughout the volume of the active medium as self-supporting cylindrical zones of photon branching.	Ratio of the reagents as H_2:F_2 = 1:2. The range of general pressures of the mixture is P = 1–2.3 bar	E_{out} = 1.5×2.3 kJ E_{sp} = 200–730 J/L $P_{max} \sim 10^{11}$ W.
Effect of the giant laser energy gain of $k_{amp} = E_{out}/E_{in} \sim 10^{11}$, where E_{out} and E_{in} are the output and input energies, respectively.	Ranges of Al particle radii and concentrations are, respectively: r_0 = 0.09–0.4 μm and $N_0 = 10^9$–10^7 cm^{-3}.	$t^L \approx 800$ ns $\tau_{1/2}^L = 100$ ns.
Autonomous generation: a self-contained laser can be initiated by a small sub-microjoule master oscillator powered by an accumulator.	Performances of the master oscillator: input signal intensity, $I_0 = 10^4$ W/cm^2; energy of the input pulse, E_{in} = 2.18×·10^{-8} J; and duration of the initiating pulse, τ_{in} = 250 ns.	$k_{amp} = E_{out}/E_{in} \approx 10^7$–$10^{11}$.
Property of compactness, which is determined by the performances of the main laser units: minimal linear sizes of the cavity, ~ 13×14×14 cm^3; minimal energy of the master oscillator, $E_{in} = 2.18\times10^{-8}$ J; and minimal initially-excited volume, $V_{in}^{min} = 10^{-7}$ cm^3. The volume of the active medium is about 2L.	Geometric parameters of the unstable telescopic cavity are: number of passes, N_p = 4; range of cavity lengths, 13–50 cm; range of magnification coefficients, β= 1.5–2; and diameter of the coupling aperture, $d \leq 3$ mm.	The profile of the output pulse in the near-field zone is a torus, and in the far-field zone the profile is Gaussian-like.

[a] In the above table, r_0 and N_0 are the radius and concentration of the dispersed particles; I_0 and τ_{in} are the intensity and duration of the initiating radiation; E_{out} and E_{sp} are the full (i.e., for the whole volume of the laser active medium) and specific (i.e., per unit volume) energies of the output pulse of radiation; P_{max} is the maximum generated power; k_{amp} is the energy gain factor; t^L is the duration of the output pulse; $\tau_{1/2}^L$ is the output pulse duration at half of the maximum power; and I_{thr} is the threshold intensity for optical breakdown of the active medium.

mixture of H_2:F_2 = 1:2, the HF laser ensures an output energy up to ~ 1.5 kJ in a pulse from a rather small volume of about 2 L of the active medium. A specific design has been proposed for a self-contained compact pulsed HF laser with small linear sizes of the unstable telescopic cavity (~ 13×14×14 cm^3, where the diameter of the input (large) mirror is 14 cm), which can be initiated by a small sub-microjoule master oscillator powered by an accumulator. The wave studies of the spatio-temporal behavior of the output laser profile of the output laser radiation showed that the profile of the output pulse in the near-field zone is a torus, whereas in the far-field zone the profile is Gaussian-like.

For a full-scale demonstration of the effect of photon-branching and design of a chemical laser based on a PBCR, it is necessary to investigate the permissible range of the concentration and size of disperse particles in the laser active medium, its dynamics and aging; and to develop the technique of preparing the gas-disperse medium of required properties. In this chapter, we present an overview of the optimal and permissible parameters of the disperse component of the HF-active medium and give recommendations for the experimental realization of PBCR in a chemically-active gas-disperse medium.

16.2. Initiation of a Photon-Branched Chain Reaction in Aerosol

16.2.1. *Kinetic scheme*

The initiation of a PBCR in an active medium of a H_2-F_2-laser amplifier, containing passivated fine aluminum particles covered by an oxide film with a thickness ~ 5–10 nm, occurs by the mechanism offered in Refs. [1,2,8] for a pulsed HF-generator. These aluminum particles are evaporated under the action of laser radiation, resulting in the appearance of free Al atoms. These atoms diffuse into the gas, and the photon-branching process begins. The initiation can be described by the following scheme:

$$\text{Al (solid)} + n\hbar\omega \rightarrow \text{Al (gas)}$$
$$\text{Al (gas)} + F_2\text{ (gas)} \rightarrow \text{AlF} + \text{F}.$$

After the appearance of free F atoms, the atoms then diffuse into the gas and the chain reaction starts, leading to the formation of vibrationally-excited HF molecules and, further, to generation of new photons:

$$\begin{aligned} F + H_2 &\rightarrow HF^* + H \\ H + F_2 &\rightarrow HF^* + F \\ HF^* &\rightarrow HF + \hbar\omega. \end{aligned}$$

Some of the emitted photons are absorbed, which results in the formation of new free Al atoms, while other emitted photons contribute to the net laser energy gain. The progressive growth of photons and free atoms takes place over time, and all of the particles (photons and free F and Al atoms) participate in active chain reactions. A reaction of this type is called a PBCR. The kinetic scheme and rate constants of the elementary processes in an H_2-F_2-laser medium for a PBCR as well as the electrodynamics of the laser of this type are discussed, for example, in Ref. [8].

The specific power P_L of the laser radiation generated as a result of the $(u,j\text{-}1) \rightarrow (l,j)$ transition in vibrationally-excited HF molecule is described by the following expression deduced on the basis of the two-level model:

$$P_L = h\omega_{ul}\frac{2j+1}{4j}\left[\frac{n_u}{M_{j-1}\tau} - \frac{2j-1}{2j+1}\frac{n_l}{M_j\tau}\right].$$

Here, $M_j\tau$ is the collective relaxation time from all the levels, in the rotational reservoir model [8], into the given level j, where M_j is a dimensionless coefficient multiplying the characteristic rotational relaxation time τ. Calculations show τ to be given by $\tau = (\pi\Delta\nu_L)^{-1}$, where $\Delta\nu_L = \sum\gamma_M P_M$ is the homogeneous half-width of the HF line, and P_M is the partial pressure of the Mth component. Here, $\gamma_{H_2} = \gamma_{O_2} = 0.02$ cm^{-1} atm^{-1}, $\gamma_{F_2} = 0.035$ cm^{-1} atm^{-1}, $\gamma_{HF} = 0.09$ cm^{-1} atm^{-1}, and $\gamma_{He} = 0.0055$ cm^{-1} atm^{-1}. The power density can affect the gain of the active medium $\alpha(t)$ through the relation $P_L = \alpha(t)I(r)$, where the intracavity distribution of intensity $I(r)$ is determined by the solution of the wave equation with the given boundary conditions on the mirrors [8].

The advantages of the using fine disperse particles of metals or other elements as the laser-active media are:

- high, in principle, concentrations of free atoms;
- short time for producing high concentration of free atoms;
- uniformity of filling large active volumes with free atoms;
- universality (ability to use various atoms);
- high specific surface area of the particles, exceeding the area of the walls (of, for example, a discharge chamber), which increases the rate of depopulation of the lower active level.

The properties of an aerosol as an active medium of a pulsed chemical HF laser based on a photon-branched reaction should satisfy a number of requirements, which can be divided into two independent groups. These are: (1) the requirements imposed by the conditions for sustaining the photon-branched chain reaction in an H_2-F_2 mixture with fine disperse particles; and (2) those imposed by the aerosol optics.

16.2.2. *Laser evaporation of particles*

The formation of active centers of the chain reaction is achieved through the fast reaction of a fluorine molecule with an evaporated aluminum atom from the surface of a disperse particle under the action of external IR-laser radiation:

$$\mathrm{Al\ (gas)} + F_2 \rightarrow \mathrm{AlF} + \mathrm{F}.$$

The characteristic time of this reaction is $\leq 10^{-9}$ s at the concentration $[F_2] \geq 10^{18}$ cm^{-3}. Laser heating and evaporation of the disperse component can be described within the approximation of free molecular flow, which is valid for particles whose size does not exceed the mean free path of molecules in a laser mixture (see, for example, Ref. [8]):

$$4\pi r_0^2 \frac{dr_0}{dt}\rho_0 = -\eta 4\pi r_0^2 \bar{V}_s(T_s)\rho_s(T_s),$$

$$\frac{4}{3}\pi r_0^3 \rho_0 C_s \frac{dT_s}{dt} = \sigma_{abs} I -$$

$$-(T_s - T_g)\sum_M C_M 4\pi r_0^2 n_m \bar{V}_M(T_g) +$$

$$+L_{evap} 4\pi r_0^2 \frac{dr_0}{dt}\rho_0 .$$

Here, C_s and L_{evap} are, respectively, the specific heat and heat of evaporation of aluminum; I is the radiation intensity inside the cavity; σ_{abs} is the cross section of absorption of laser radiation by an Al particle; T_s and T_g are the temperatures of the aluminum particles and the laser-active medium, respectively; C_M, V_M and n_M are, respectively, the molar specific heat, average velocity and concentration of the molecules of the Mth component of the laser mixture; η is the accommodation coefficient; V_s is the average velocity of the vapor molecules; and $\rho_s(T_s)$ is the saturated vapor density of aluminum at T_s.

16.2.3. *Optimal concentration of particles*

Formation of active centers of the reaction chain and their diffusing into the medium is thus possible around each fine particle in the IR laser radiation field. The kinetic scheme and rate constants of the elementary processes in a H_2-F_2-laser medium for a PBCR have been reviewed in Ref. [8]. We will consider below only the features of a laser-chemical process involving disperse particles.

The kinetic equation describing the formation of active centers of a chain can be written in the approximation of quasi-steady-state concentrations of the F and H atoms:

$$\frac{dn_a}{dt} = -4\pi r_0^2 \frac{dr_0}{dt}\frac{\rho_0}{m_{Al}N_A} N_0 + W - \beta n_a .$$

Here, $n_a = n_F + n_H$ is the total concentration of the active centers; r_0 and ρ_0 are the radius and density of Al particles; N_0 and m_{Al} are the particle concentration and atomic mass of aluminum; and N_A is Avogadro's number. The first term on the right-hand side of this equation takes into account the nucleation of active centers of the chain as a result of the interaction of F_2 molecules with evaporated Al atoms. The second term, W, describes the formation of the active centers for the dissociation of molecules, and the last term, βn_a, is the rate of losses of the centers.

The distribution of free atoms in the active volume can become sufficiently homogeneous in a time τ_h, which is limited by the duration of the laser-chemical process (~ 1 μs), provided the condition $R_{ev} \le 2R_{diff}$ is satisfied, where R_{ev} is the average distance between the particles, $R_{diff} \sim \sqrt{D\tau_h}$, and D is the diffusion coefficient of the active centers in the laser medium. This sets the lower limit to the required concentration of disperse particles as $N_0 \approx \left(1/R_{ev}^3\right)$. Optimization of the aerosol parameters from the point of view of the laser-chemical process under the conditions of initiation of a pulsed HF amplifier was reported in Ref. [8]. Calculations showed that the optimal concentration of aluminum particles is $N_{Al} = 1.4\times10^9$ cm^{-3} for the radius $r_0 = 0.09$ μm.

16.2.4. *Optimal particle size*

The optimal particle size is governed, for the selected concentration, by the aerosol optics, so that the attenuation coefficient of laser radiation representing scattering and absorption in a disperse medium does not exceed the local active-medium gain α. Calculations of the optical characteristics of spherical disperse particles at a given radiation wavelength λ were based on the Mie diffraction theory and the single-scattering approximation [11].

The Mie formalism requires the use of two dimensionless input parameters, $\rho = 2\pi r_0/\lambda$ and $\delta = \rho m$, where m *is* the relative value of the complex refractive index of the aerosol at the wavelength λ. Computer calculations of the dependences of the attenuation K_{att}, scattering K_{sc} and absorption K_{abs} coefficients at the HF-laser wavelength (λ = 3.3 μm) in the size range r_0 = 0.5–1 μm of aluminum particles (m = 3.2–29.5i at

300 K) were performed in Refs. [2,8] and plotted in Fig. 16.1. It is evident from Fig. 16.1(a) that, for particle radii $r_0 > 0.5$ μm, the scattering coefficient K_{sc} is considerably greater than the absorption coefficient K_{abs} , i.e., lasing is impossible in such a strong-scattering medium. Figure 1(b) demonstrates, on a much larger scale, the region of intersection of the $K_{sc}(r_0)$ and $K_{abs}(r_0)$ curves. The absorption of IR radiation predominates over the scattering by aluminum particles, $K_{abs}(r_0) \geq K_{sc}(r_0)$, if $r_0 \leq 0.15$ μm. It follows that the aerosol optics imposes stringent conditions on the maximum permissible particle size in a two-phase active medium of the laser, which should satisfy the condition $\rho = 2\pi r_0/\lambda < 1$. An aluminum aerosol with the parameters $r_0 = 0.09$–0.4 μm and $N_{Al} = 10^7$-10^9 cm^{-3} can be recommended as a model two-phase active medium of an HF laser based on a PBCR.

16.2.5. *Lifetime of a disperse component*

The main shortcoming of lasers with two-phase active media is a fast degradation of the disperse component and the consequent short lifetime of the active medium with specified properties. Continuous variation of the properties of the disperse phase with time results in a deterioration of the output characteristics of a laser or incomplete quenching of the laser action.

The first attempt to model degradation of the disperse phase in a laser-active medium was made for the operating conditions in a pulsed chemical HF laser [12]. The degradation processes are practically the same for all types of lasers with disperse media. We adopted an approximate coagulation model [13] and assumed the Stokes law. In the first approximation, we studied the time dependences of the overall characteristics of an aerosol in a laser-active medium, including the average size and total concentration of disperse particles, and we also calculated the formation time and lifetime of a two-phase active medium of an HF laser with given parameters.

The processes occurring in a gaseous medium containing a disperse aerosol may be modeled on the assumption of coagulation, precipitation and (in the presence of electric charges) electrostatic scattering of

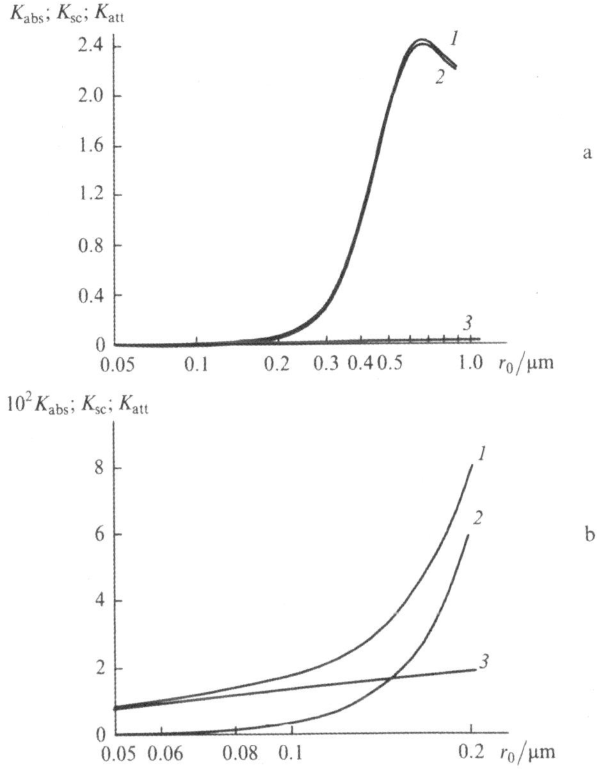

Fig. 16.1. Dependencies of the attenuation K_{att} (1), scattering K_{sc} (2), and absorption K_{sc} (3) coefficients at the HF-laser wavelength (λ = 3.3 μm) for the size r_0 = 0.5–1 μm of aluminum particles (m = 3.2–29.5i at 300 K) (a), with the region of intersection of curves 2 and 3 shown on an enlarged scale (b).

disperse particles. Aerosol coagulation was described by the log-normal quasi-self-sustaining model of Ref. [14], based on an approximation in which a log-normal model [15] and high-current sectional model, conserving the square of the particle volume [16], are adopted. Here, the particle size distribution is described by the standard log-normal function.

The precipitation of particles was accounted for on the basis of the following model assumptions: the active medium was taken to be ideally mixed, continuous, incompressible, and characterized by a finite viscosity. The particle shape was assumed to be spherical. All these assumptions were in order to apply the Stokes law to aerosol precipitation. The concentration of the disperse component remaining in the medium after particle precipitation was calculated from

$$N_{prec} = N_0 \frac{\exp(-vt)}{H},$$

where H is the transverse size of the active medium of the HF laser, assumed to be equal to the diameter of the larger cavity mirror.

An analysis of the influence of the induced electrostatic charge of fine disperse particles of a conducting aerosol on the rate of its degradation was made within the overall framework of the problem of the lifetime of a two-phase active medium. The formation of a metal aerosol is practically always accompanied by its charging, for example, as a result of thermionic and photoemission of electrons. The concentration of the aerosol remaining after electrostatic scattering of the particles with the same charge and the same size can be found from expression

$$N_{esc} = \frac{N_0}{1 + 4\pi q^2 B N_0 t},$$

where B is average mobility of the particles.

Specific calculations were carried out for a standard active mixture of the composition H_2:F_2:O_2:He = 100:400:40:210 torr, into which an aluminum aerosol was injected in advance. At the initial time $t = 0$, the aerosol was assumed to be quasi-monodisperse with the average initial radius $r_0 = 0.05$ μm and concentration $N_0 = 2\times10^9$ cm^{-3}.

Fast coagulation, with the constant $\beta = 5\times10^{-10}$ cm^{3}s^{-1} (Fig. 16.2) and sticking probability practically equal to unity for a submicron aerosol, results in continuous enlargement of the particles with a certain size distribution. The time for the establishment of a self-sustaining log-normal particle size distribution in the investigated two-phase active medium is about 25 s. Figure 16.2 shows the dependencies of coagulation and electrostatic scattering constants on the particle size. We can see that an increase in the average charge alters slightly the coagulation constant;

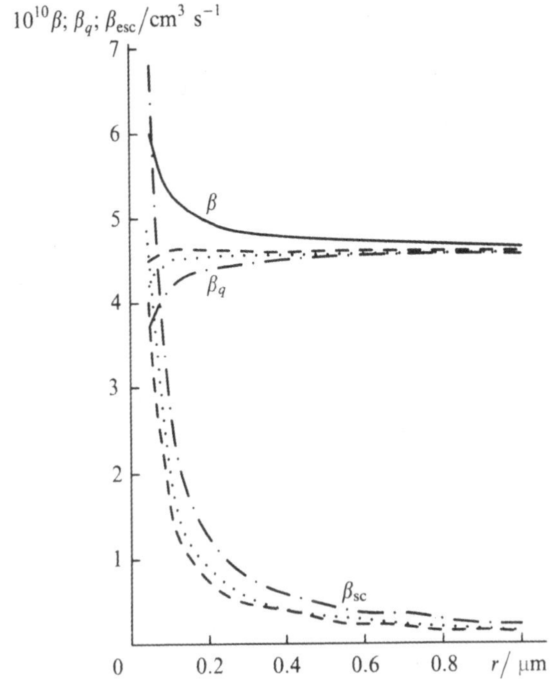

Fig.16.2. Dependencies of the coagulation constant of neutral particles β, the coagulation constant of charged particles β_q, and aerosol electrostatic scattering constant β_{esc} on the particle radius r for the average induced unipolar charge $q = e$ (dashed curve), 1.1e (dotted curve) and 1.3e (chain curve).

this effect is significant only for small ($r < 0.1$ μm) particles. For an aerosol with $r = 0.05$ μm, the constants of these processes are comparable, and the rate of electrostatic scattering of the charged particles exceeds the rate of coagulation of the neutral particles only if the average charge is $q \geq 1.3e$. Therefore, at the first instant, such electrostatic scattering of submicron charged particles competes with their coagulation, which results in rapid loss of the particles from the volume of the active medium.

It follows from Fig. 16.3(a) that during the first 50 s there is a strong fall of the concentration of the charged aerosol with the average unit charge per particle down to the lowest value, ~ 10^8 cm^{-3}, needed to sustain the laser-chemical reaction. An increase in the coagulation time results in continuous spreading of the initial particle size distribution in the direction of larger particles and in a fall of the aerosol concentration, because electrostatic scattering is accelerated by the coagulation and precipitation of the particles.

A comparison of the change, during the selected time step, in the concentration of the particles remaining in the medium after electrostatic scattering with the change in the total concentration of the particles in the medium undergoing coagulation and precipitation shows (Fig. 16.3(b)) that, beginning from the particle radius $r \sim 0.2$ μm, such coagulation and precipitation in an aerosol predominate strongly over electrostatic scattering, irrespective of the particle charge. When the selected time step is relatively large, $t \geq 1000$ s, so that the particles reach micron sizes, particle precipitation begins to predominate over coagulation, as demonstrated in Ref. [13]. Therefore, in describing the gradual degradation of two-phase active media of lasers, it is necessary to take into account all three processes – electrostatic scattering, coagulation and precipitation – which compete in different ways during different time intervals.

If the criteria for optical transparency of the active medium to the laser radiation are assumed to be $I_{abs}/I_0 << 1$ and $I_{sc}/I_0 << 1$ (I_{abs}, I_{sc} and I_0 are the intensities of the absorbed, scattered and incident radiations), we can estimate the maximum permissible lifetime of a two-phase active medium of the investigated laser. The strongest scattering of the $\lambda = 3.3$ μm radiation corresponds to the aluminum particle radius $r = 0.6$ μm and, consequently, our selected criteria can be satisfied if the final size distribution $n(r)$ excludes almost completely particles of this radius.

It follows from Fig. 16.4, which gives the distributions $n(r)$ calculated at different instants in time, taking account of all the processes resulting in the aging of a neutral aluminum aerosol in the active medium of an HF laser, that the condition $n(r = 0.6\ \mu\text{m})/N_0 << 1$ is well satisfied for $t = 250$ s. For a longer time interval, for example $t = 300$ s, numerical calculations give an aluminum particle concentration $\sim 3\times10^3$ cm^{-3} with the impermissible radius $r = 0.6$ μm. The dashed curve in Fig. 16.4 represents the particle size distribution calculated by ignoring precipitation. We can see that the proportion of the large particles in the size distribution increases and, consequently, the lifetime of an aerosol with the specified parameters decreases further by 50 s.

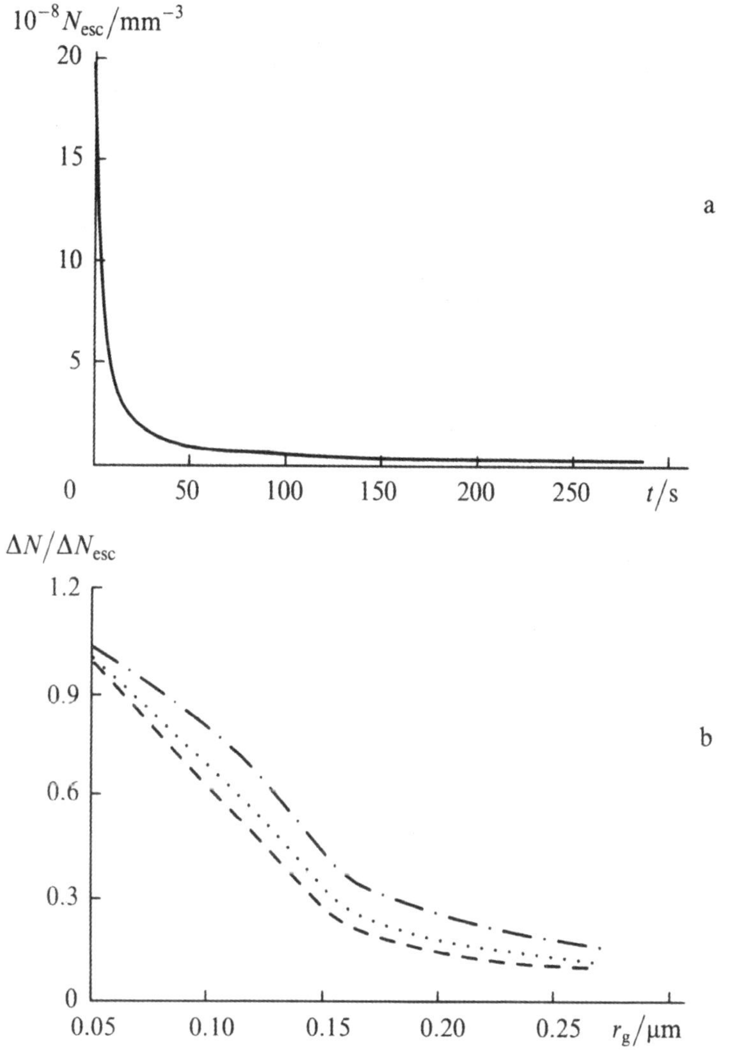

Fig. 16.3. Time dependencies of the concentration N_{esc} of a unipolarly-charged aerosol with the average unit elementary charge per particle under electrostatic scattering conditions (a) and of the ratio of the change in the total particle concentration N in the medium, which experiences coagulation and precipitation, to the change in the concentration N_{esc} of the particles which remain in the medium after electrostatic scattering, plotted as a function of the average size r_g of the particles participating in these processes, calculated for $q = e$ (dashed curve), 1.1e (dotted curve), and 1.3e (chain curve) (b).

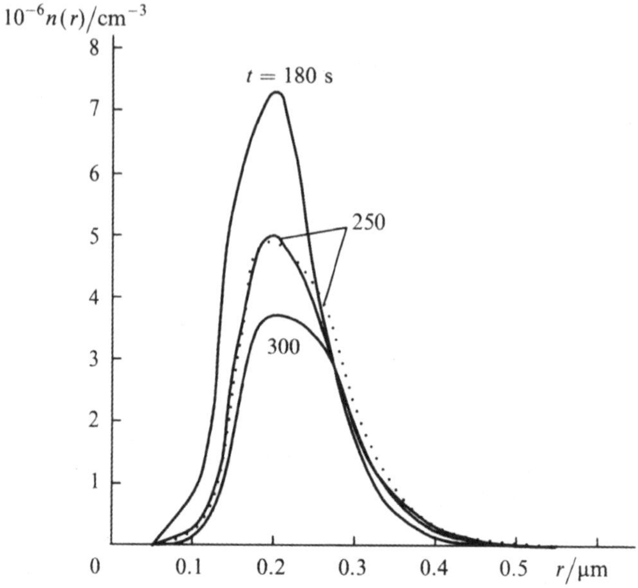

Fig. 16.4. Particle size distribution $n(r)$ at different limiting time moments, calculated by taking account of coagulation and precipitation (continuous curves) and ignoring precipitation (dotted curve) of an aluminum aerosol in the active medium of a pulsed HF laser.

Thus, the above analysis of the processes resulting in degradation of the disperse component of two-phase laser-active media allows us to draw the following main conclusions:

- The lifetime of two-phase active media with specified properties is determined by electrostatic scattering, coagulation and precipitation of the aerosol when the real particle size distribution function is taken into account.
- An analysis of the optics and laser-chemical kinetics makes it possible to recommend an aluminum aerosol with the parameters r_0 = 0.09–0.4 μm and N = 10^7–10^9 cm^{-3} for a two-phase active medium of the HF laser.
- The calculated lifetime of the HF mixture with disperse Al particles, characterized by the properties given above, is 250 s.

16.2.6. *Aerosol-evaporation reactor cavity*

In the previous sections, we have determined the aerosol properties satisfying the laser-active medium and their dynamics over time. We have considered, also, the first experiments on the formation of a gaseous disperse medium of a pulsed chemical H_2-F_2 laser. However, there remains an unsolved problem of generating a relatively-large volume (in excess of 10^3 cm^3) of the active medium of such a laser amplifier, and filling this volume homogeneously with a submicron monodisperse metal aerosol which has specified properties. An analysis of the aerosol optics and laser-chemical kinetics of the relevant PBCR shows that the ranges of the permissible parameters of the disperse component in which oscillation is possible in an HF laser are as follows: the radius of passivated aluminum particles should be r_0 = 0.09–0.4 μm, and their concentration $N_0 = 10^9$–10^7 cm^{-3}.

The simplest method for the formation of a two-phase active medium is atomization of a previously-prepared fine powder. However, when this method is used, it is impossible to ensure the optimal parameters of the dispersion, size and concentration of the particles, and also the homogeneity of the filling (because, for example, of particle agglomeration). Moreover, coagulation, gravitational precipitation and electrostatic scattering of the injected particles result in continuous variation of the parameters of the disperse component and, consequently, in deterioration of the output characteristics of the laser amplifier or even suppression of lasing. An analysis of the degradation processes in a two-phase active medium of an HF laser shows that, within these ranges of the parameters, the lifetime of the active mixture containing disperse aluminum particles is τ_{coag} = 250 s.

Experiments [10] have demonstrated stability of a fluorine-hydrogen mixture with an injected highly-disperse component (Al particles with r_0 = 0.2–1 μm and concentration $N_0 = 10^5$ cm^{-3}) against spontaneous self-inflammation, and that a chain reaction can be initiated by evaporating such particles by IR laser radiation. On the other hand, these experiments reveal that, in order to achieve lasing and a full-scale PBCR, it is necessary to employ improved methods for the formation of a gas-disperse medium with the required parameters and for homogeneous filling of large working volumes in a short time. For example, in a time interval $t < \tau_{coag} = 250$ s, it is necessary to form an aerosol with specified properties and to inject a gaseous mixture into the active volume of the laser. The time needed to fill this volume with an H_2-F_2 mixture and to ensure its homogeneous mixing amounts to 3–4 min in the experiments and, consequently, no more than 10 s remains for the formation of the aerosol.

A solution to this problem, proposed below, involves adoption of a fundamentally-new method for the preparation of a two-phase active medium of a pulsed HF laser in an aerosol reactor coupled structurally to an unstable telescopic cavity. The use of such a closed HF oscillator-amplifier system, based on a PBCR and containing a device for the generation of a homogeneous aerosol, reduces considerably the time needed for the formation of a two-phase active medium and ensures that the disperse component has the necessary parameters. On the other hand, this system requires an external pulsed energy source, and it is promising primarily for experimental research on the implementation of a PBCR and investigation of this reaction. The method may be of practical interest also for the improvement of metal vapor lasers and for the nascent new aerosol laser technologies [17].

The basic design of the proposed laser amplifier with a two-phase active medium is shown in Fig. 16.5. This laser amplifier consists of an unstable telescopic cavity (1), which is structurally coupled to an aerosol reactor (2). The front cavity mirror has an aperture for coupling to a master oscillator (3). The aerosol reactor is in the form of a demountable cylindrical chamber made of heat-resistant quartz glass. Plasma deposition or some other method is used to deposit, on the inner wall of this chamber, an aluminum film (4) of the required thickness. This metal film can extend over the whole length of the aerosol reactor (Fig. 16.5(a)), or it may form a closed ring of limited length at the centre of the internal wall of the quartz glass chamber (Fig. 16.5(b)). The aerosol chamber variant with a limited length of the evaporated film is preferable from the point of view of preventing contamination of the end walls of the chamber by the metal, which is not permissible when the chamber is used inside the laser cavity to generate a two-phase active medium.

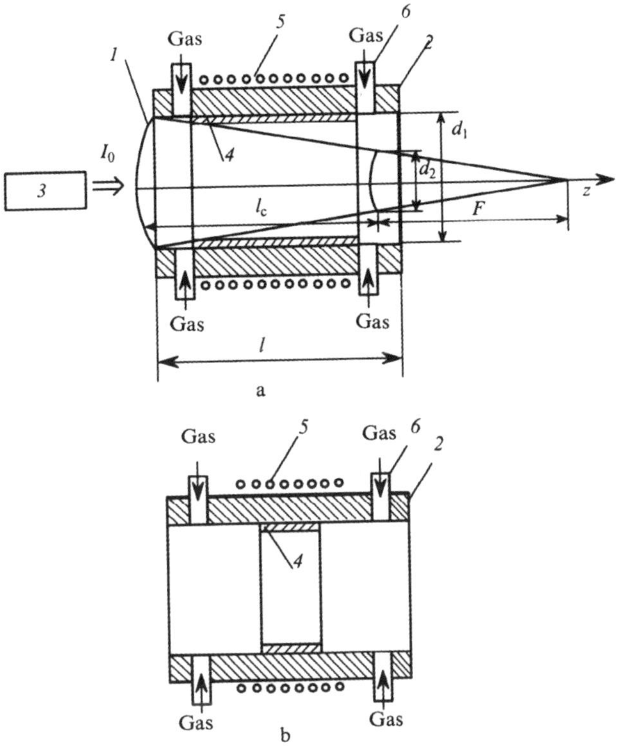

Fig. 16. 5. Simplified basic designs of a laser with an aerosol-evaporation reactor (a) and an aerosol reactor with a limited length of the evaporated film (b): $d_{1,2}$ are the cavity mirror diameters; l_c is the cavity length; F is the focal length of the output mirror; l is the reactor length; (1) unstable telescopic cavity; (2) aerosol reactor; (3) master oscillator; (4) evaporated film; (5) solenoid; and (6) gas-supply pipes.

The metal film is subjected to a homogeneous electromagnetic field generated in a solenoid (5) and evaporated, so that it mixes with the atmosphere of a low-pressure carrier gas (helium). The reactor contains also a power supply and a capacitor bank. The linear dimensions of the chamber are governed by the cavity parameters.

Let us now consider the formation of a two-phase laser-active medium in the aerosol reactor. A bank of capacitors with a calculated capacitance C is discharged over a short time interval by closing the circuit. The discharge excites the solenoid, which has the following parameters: its length is l_L, the number of turns is N_L, the wire cross section is S_L, and the resistance of the primary winding is R_1. A homogeneous magnetic flux Φ is generated inside a closed region surrounded by the primary winding whose inductance is L_1. This flux is sufficient for the evaporation, over a short time interval, of the metal film, where the film acts also as a secondary winding whose resistance is R_2. The metal vapor, expanding in the surrounding cold gaseous He, then condenses to submicron metal particles.

The metal can be prevented from reaching the end faces of the chamber, and the region with aluminum vapor can be filled homogeneously if the film is evaporated under the conditions of laminar flow of the carrier gas through pipes (6) placed at the ends of the chamber (2). A small amount of oxidant (O_2) injected into the reactor after condensation of the metal vapor produces an oxide film on the surfaces of the aluminum particles. Such an oxide film is 5–10 nm thick and prevents spontaneous self-inflammation of the fluorine-hydrogen working mixture of the laser. The necessary average size of the resultant particles and their concentration are ensured by varying the thickness of the initial metal film and by suitable selection of the electric parameters of the apparatus. The aerosol chamber variant with a limited length of the evaporated film (Fig. 16.5(b)) has an additional advantage in that it is possible to vary the aerosol parameters by altering the evaporated film length.

In Refs. [2–9] we proposed a theoretical model for the design of a pulsed HF laser system based on a PBCR, initiated in an aerosol medium in an unstable telescopic cavity. The following parameters of the proposed laser design were assumed in the actual calculations (Fig. 16.5):

- length of the unstable telescopic cavity $l_c = 49.5$ cm;
- diameter of the input mirror $d_1 = 14$ cm;
- diameter of the output mirror d2 = 6 cm;
- focal length of the output mirror F = 53 cm;
- diameter of the input aperture coupling the cavity to the master oscillator $d_0 = 1.6$ mm.

The linear dimensions of the aerosol reactor were correspondingly selected as follows:

- length l = 50 cm;
- diameter of the quartz glass chamber d = 15 cm;
- wall thickness $H = 1$ cm.

An aluminum film of thickness h_0 = 350 nm was assumed to be deposited over the whole length of the internal wall of the chamber, and evaporation of this film was ensured by selecting suitable parameters of the electric circuit of the aerosol reactor. The working volume of the chamber was assumed to be filled with helium at a pressure varied from 300 to 600 torr.

The evaporation of an aluminum film with a given mass $m_0 = 0.1$ g requires an energy $\Delta E = L_{ev} m_0 \approx 1088$ J. Calculations show that a solenoid with a primary winding made of copper wire 4 mm in diameter and containing N_L = 50 turns, in combination with a capacitor of capacitance C = 10 μF subjected to a voltage U_c = 50 kV, can ensure evaporation of this film over a time $\tau_{ev} = 1/(2\gamma) \approx 0.217$ ms. The same results can be obtained for a lower voltage across the capacitor, for example U_c = 18 kV, if selection is made of a solenoid with a primary winding consisting of wires with a small diameter (1.5 mm) when the number of turns is $N_L = 4$. The energy stored in the capacitor is then E_C = 1620 J, which is sufficient to evaporate the film and to compensate for the energy lost in heating the wires in the primary winding and for the heat lost to the ambient gas and to the chamber walls. The variant with a small number of primary-winding turns in the solenoid corresponds to an aerosol reactor with an evaporated film of finite length (Fig. 16.5(b)), which is preferable for intracavity use.

Discharge of a capacitor which stores this energy results in slight heat dissipation in the primary winding of the solenoid as well. Consequently, the primary winding temperature rises by $\Delta T = E_C/C_p m_{Cu}$ = 40°C, where C_p and m_{Cu} are the specific heat and mass of a copper wire in the primary winding. Film evaporation is accompanied also by the loss of heat to the ambient gas and to the quartz-glass chamber walls. Let us ignore the energy lost as a result of heat conduction in the gas, and estimate the upper limit to the heat lost to the chamber walls from the film being evaporated. The depth Δh of penetration of heat into quartz glass over the course of diffusive heat exchange with the heated film during the total evaporation time $\tau_{ev} \approx 0.217$ ms is $\Delta h = \sqrt{2\chi\tau_{ev}} \approx 2\times10^{-5}$ m, where χ is the thermal diffusivity of quartz glass. The energy lost through the chamber walls in the case of a four-turn solenoid does not exceed $\Delta Q = \Delta m_1 C_{p1} \Delta T_1 \approx 500$ J, where Δm_l is the mass of the heated chamber wall, C_{p1} is the specific heat of quartz glass, and ΔT_1 is the temperature drop at the film-glass interface. Evaporation of the film occurs in a diffusive regime without causing film rupture and without formation of a large-sized fraction, since the pressure exerted by the magnetic field on an element of the film surface of length dl and of area dS is

$$\frac{dF}{dS} = \frac{I_2}{dS}\left[dl \times B\right] = 0.26 \text{ bar} .$$

Here, $B = \dfrac{\mu_0 N_L I_1}{dl}$ is the magnetic field strength; $\mu_0 = 1.26\times10^{-6}$ H/m is the magnetic constant; and I_1 and I_2 are, respectively, the currents in the primary and secondary windings. This pressure is much less than that needed to rupture an aluminum film in the solid or liquid state (and in the case of liquids, this pressure amounts to hundreds of bars).

The newly-formed aluminum vapor expands adiabatically inside the reactor and experiences a condensation 'jump' accompanied by the formation of clusters with critical radius r_{cr} and high concentration (Table 16.2) in a short time of to $t \approx 1$ μs. The evolution of these clusters at various ambient gas pressures is illustrated in Fig. 16.6 and Table 16.2.

Table 16.2. Results of numerical calculations of the formation of a two-phase active medium and of the output characteristics of an HF laser. [a]

P_{He}/torr	r_{cr}/μm ×10^{-4}	r_0/μm	N_0/cm^{-3} ×10^7	t^*/μs	E_{out}/J	E_{sp}/J L^{-1}	P_{max}/W ×10^{10}	τ_p/ns	k_{amp} ×10^4
300	9.4	0.276	5	134	1076	141.23	1.9	736	7.1
400	9.5	0.198	10	80.5	1465	192.26	2.0	716	9.7
600	11	0.158	13	57.6	1485	194.92	1.5	737.6	9.9

[a] Here, P_{He} is the buffer (carrier) gas pressure; r_{cr} and r_0 are the critical and final radii of the condensed particles; N_0 is the particle concentration; t^* is the time needed for the formation of particles with the final radius r_0; E_{out} and E_{sp} are the total and specific energies of the output radiation pulses; P_{max} is the maximum output power; τ_p is the duration of the output pulse; and k_{amp} is the energy gain.

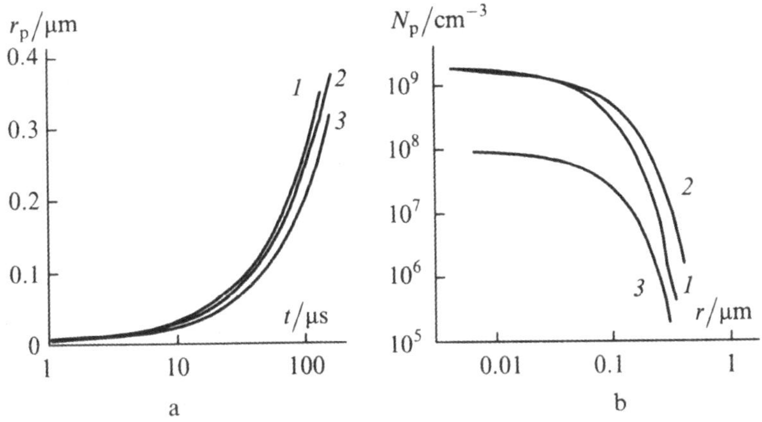

Fig. 16.6. (a) Evolution of clusters in time and (b) distribution of the concentration of the particles formed by the condensation of Al vapor, described in terms of the particle size $N_p(r_p)$, calculated for helium buffer (carrier) gas pressures P_{He} = 600 torr (1), 400 torr (2), and 300 torr (3).

It follows from our calculations that the time t^* needed for the formation, over the course of condensation, of particles with final radius r_0 is ~ 50–150 μs, which corresponds to the time of attainment of the balance between the masses of the evaporated film on the one hand, and of the clusters and vapor on the other. The growth of these particles in the range of radii from r_{cr} to $r_0 \approx 0.1$ μm occurs at a practically constant and maximum concentration N_0 (Fig. 16.6b), which makes it easier to satisfy the condition on r_0 for a two-phase active medium of an HF laser. For large particles, $0.1\ \mu m < r_0 \leq r_{max} = 0.5\ \mu m$, the particle (cluster) concentration N_0 rapidly falls to ~10^5 cm^{-3}, and the attenuation of the generated IR radiation is then slight. It follows that condensation of the aluminum vapor in 50–150 μs produces a polydisperse aerosol with a distribution of the particle radii close to a 'step' (Fig. 16.6(b)). The width of this step matches the particle radius range of $r_{cr} \approx 1\ nm \leq r_0 \leq 0.1\ \mu m$.

A fast reaction of aluminum with oxygen added to helium occurs on the surfaces of the growing clusters: $4Al + 3O_2 \rightarrow 2Al_2O_3$. The result is a dense protective Al_2O_3 film, 5–10 nm thick, on the particle surfaces. This film prevents spontaneous self-inflammation of the working mixture of the laser during the dark reactions. It was demonstrated experimentally in Ref. [10] that mixing of passivated aluminum particles

~ 0.5 μm in diameter and with a concentration ~ 10^5 cm, with a laser-active mixture H_2:F_2:O_2:He = 50:114:11:635 torr, does not increase the rate of accumulation of the HF molecules, and such a mixture has the required long-term stability.

The output characteristics of the HF laser with an aerosol reactor, whose cavity is filled with the working mixture H_2:F_2:O_2:He = 76:228:23:578 torr, are listed in Table 16.2 and also presented in Fig. 16.7. These numerical results were obtained on the basis of our laser model described, for example, in Ref. [8]. A control radiation beam of intensity I_0 = 5 MW/cm^2 with pulses of τ_{in} = 150 ns duration is injected into the laser cavity as shown in Fig. 16.5. The diameter of the aperture coupled to the master oscillator is d_0 = 1.6 mm. The formation of smaller particles over the course of condensation improves the output energy characteristics of the laser.

Reduction of the particle size is possible if the vapor condenses at a high pressure of the ambient He gas. However, as we discussed in subsection 16.2.4, there is a lower limit to the radius, r_{min} = 0.09 μm, at which the efficiency of absorption of the control radiation is still sufficiently high ($K_{abs} \geq 10^{-2}$), and particles are evaporated efficiently and

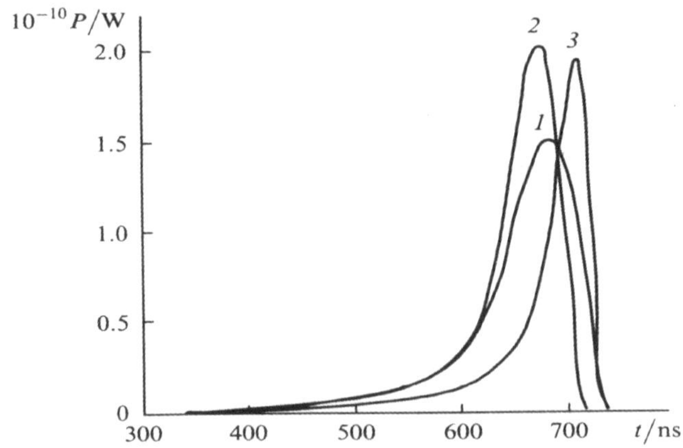

Fig. 16.7. Calculated time dependencies of the output power P of a laser with an active mixture of the H_2:F_2:O_2:He = 76:228:23:578 torr composition when the input signal of intensity I_0 = 5 MW cm^{-2} is in the form of pulses of duration τ_{in} = 150 ns: r_p = 0.158 μm (*1*), 0.198 μm (2), and 0.276 μm (3); N_p = 1.3×10^8 cm^{-3} (1), 10^8 cm^{-3} (2), and 5×10^7 cm^{-3} (3).

then react with fluorine. Further reduction in the particle radius reduces the output power of the laser, since such reduction leads to a very small coefficient of attenuation ($K_{ext} \leq 10^{-3}$) of the radiation by the particles, and the medium is not heated very strongly. It follows from Fig. 16.6(b) that the optimal variant of the particle parameters ensuring the maximum laser output power and short duration of the output pulses corresponds to vapor condensation in an aerosol reactor where the initial pressure is 400 torr. Then at a time of $t^* \approx 80$ μs, an aluminum aerosol with final particle radius $r_0 = 0.198$ μm is formed, and the concentration of such particles is $N_{Al} = 10^9$ cm^{3}.

It therefore follows from the calculations carried out on the basis of our model (see, for example, Ref. [8]) that the proposed method of formation of a two-phase active medium of a laser in a short time ~ 0.217 ms directly in the laser cavity should make it possible to construct a powerful pulsed chemical HF laser based on a photon-branched chain reaction. In particular, the calculations show that such a laser has extremely high output energy characteristics (see Fig. 16.7 and Table 16.2): maximum power per pulse is $P_{max} = 2\times10^{10}$ W; the pulse duration is $\tau_p \approx 740$ ns; the total output energy per pulse is $E_{out} = 1.5$ kJ; the specific output energy is $E_{sp} \approx 200$ J/L; and the energy gain is $k_{amp} \approx 10^5$.

Thus the proposed design of a pulsed chemical laser with a two-phase active medium, which includes an aerosol reactor coupled structurally to an unstable telescopic cavity, should make it possible to reduce considerably the time of formation of such an active medium and to ensure the necessary parameters of the disperse component. Our simulations show that at a total time of ~ 0.217 ms, it should be possible to form in such an aerosol reactor a polydisperse aerosol distributed homogeneously over the cavity volume, characterized by a 'step' partcle-size distribution function and by the maximum concentration of particles whose radii are in the range 1 nm $\leq r_0 \leq 0.1$ μm.

The size of the condensing particles and their concentration can be varied by altering the pressure of the gas inside the chamber, as well as the thickness of the film being evaporated and the electrical performances of the aerosol reactor. The proposed method of formation of a two-phase active medium should make it possible to generate HF laser pulses as a result of a photon-branched chain reaction, and the output characteristics of the laser should be extremely high.

16.2.7. *Beam stability of a chemically-active aerosol*

The off-resonant initiation mechanism of a PBCR by laser evaporation of ultradisperse metal particles injected into a laser-active medium imposes restrictions from below and above on the radiation intensity of the master oscillator. On the one hand, the intensity of the input radiation should be sufficient for effective evaporation of submicron metal particles, and on the other hand, it should be below the threshold intensity of the optical breakdown of an H_2-F_2 laser gas-disperse active medium. The kinetics of plasma formation in the laser infrared radiation field has been studied for gas-disperse fluorine-containing media in Ref. [8].

The limiting factors for maximum values of the output energy of HF/DF laser based on a PBCR are only the beam stability of the laser active medium and the cavity mirrors. The thresholds for optical breakdown are:

- Beam stability of metallic mirrors from molybdenum and copper: at a pulse duration of 100 ns, $E_{thr} \sim 100$ J/cm^2, i.e., $I_{thr} \sim 10^9$ W/cm^2 (E_{thr} and I_{thr} are the threshold energy and intensity).
- Optical breakdown threshold of the laser active medium is $I_{thr} \sim 10^{10}$ W/cm^2 ($\tau_{thr} \sim 10$ ns).
- Optical breakdown threshold of raw air is $I_{thr} \sim 2 \cdot 10^9$ W/cm^2.
- Optical breakdown threshold of purified air is $I_{thr} \geq 10^{10}$ W/cm^2.

16.3. Conclusions

The results presented in this chapter lead to the following main conclusions:

- An analysis of the aerosol optics, based on the Mie diffraction theory, and the laser-chemical kinetics of a PBCR have been used to determine the range of the maximal permissible parameters of the disperse component, which is capable of supporting the operation of a pulsed chemical HF laser with a two-phase active medium, with particle radius r_0 = 0.09–0.4 μm and particle concentration $N_0=10^9$–10^7 cm^{-3}.

- In this range of aerosol parameters, the lifetime of the active medium of the HF laser with the disperse Al particles is governed by the coagulation, precipitation and electrostatic scattering of particles occurring in the time range of ~10^3 s.
- The first experiments demonstrated the stability of a fluorine-hydrogen mixture with an injected fine disperse component (Al particles of radius r_0=0.2–1 μm with concentration N_0= 10^5 cm^{-3}) under the conditions when spontaneous combustion is possible. It was also shown that a chain reaction could be initiated by the evaporation of particles under the action of IR radiation. The corresponding numerical calculations let us conclude that in some experiments, conducted in Ref. [13], the photon-branched chain reaction has been observed for the first time. For a full-scale demonstration of the effect of the photon-branched chain reaction, it is necessary to improve the technique of preparing the gas-disperse medium as discussed above.
- The optical breakdown of a chemically-active aerosol at the wavelength of the generated radiation (λ_g = 3.3 μm) is characterized by the threshold intensity I_{thr} = 1–3×10^{10} W/cm^2.

References

1. N. G. Basov, A. S. Bashkin, V. I. Igoshin, A. N. Oraevsky and V A. Shcheglov, *Chemical Lasers* (Springer-Verlag, Berlin, 1990).
2. V. I. Igoshin and R. R. Letfullin, High power laser amplifier based on an auto-wave photon-branched chain reaction in an unstable telescopic cavity, *Quantum Electron.* **27**, 487–491 (1997).
3. V. I. Igoshin and R. R. Letfullin, Diffractive focusing of an input pulse and giant energy gain in a laser based on an auto-wave photon-branched chain reaction, *Quantum Electron.* **29**, 37–42 (1999).
4. R. R. Letfullin and S. P. Sannikov, Laser amplifier based on a photon-branched chain reaction in an aerosol evaporation reactor cavity, *Quantum Electron.* **29**, 43–48 (1999).

5. V. I. Igoshin and R. R. Letfullin, Ultra-high energy gain in small volumes of the active medium in a laser based on an auto-wave photon-branched chain reaction, *Quantum Electron.* **30**, 1049–1054 (2000).
6. R. R. Letfullin and T. F. George, Theory of energy gain in a laser-amplifier based on a photon-branched chain reaction: Auto-wave amplification mode under the condition of input signal focusing, *J. Appl. Phys.* **88**, 3824- 3831 (2000).
7. R. R. Letfullin and T. F. George, Self-contained compact pulsed laser based on an auto-wave photon-branched chain reaction, *Appl. Phys. B* **71**, 813–818 (2000).
8. R. R. Letfullin, V. I. Igoshin and T. F. George, Chemical laser based on a photon-branched chain reaction, in *Modern Topics in Chemical Physics*, eds. by T. F. George, X. Sun and G. P. Zhang (Research Signpost, Trivandrum, India, 2002), pp. 183–285.
9. R. R. Letfullin, Compact pulsed HF laser on an auto-wave photon-branched chain reaction, *J. Russian Laser Res.* **23**, 252–273 (2002).
10. M. A. Azarov, V. A., Drozdov, V. I Igoshin, Yu. Kurov, A. L. Petrov, Yu. Pichugin and G. A. Troshchinenko, Formation and experimental investigation of a gaseous disperse medium of a pulsed chemical H_2-F_2 laser initiated by IR radiation, *Quantum Electron.* **27**, 953–956 (1997).
11. H. C. van de Hulst, *Light Scattering by Small Particles* (Wiley, New York, 1957).
12. A. N. Bekrenev, V. I. Igoshin, R. R. Letfullin and K. G. Melikhov, Degradation of the disperse component of the active medium of a pulsed chemical HF laser, *Quantum Electron.* **27**, 221–223 (1997).
13. R. R. Letfullin, K. G. Melikhov, V. I. Igoshin and L. A. Mitlina L. A., Lifetime of a two-phase active medium of a pulsed chemical HF laser, *Quantum Electron.* **28**, 887–892 (1998).
14. E. Otto, F. Stratmann F., H. Fissan, S. Vemury and S. E. Pratsinis, Quasi-self-preserving log-normal size distributions in the transition regime *Part. Part. Syst. Charact.* **11**, 359–366 (1994).
15. F. Stratmann and E. R. Whitby, Numerical solution of aerosol dynamics for simultaneous convection. Diffusion and external forces, *J. Aerosol Sci.*, **20**, 437–440 (1989).
16. J. D. Landgrebe and S. E. Pratsinis, A discrete-sectional model for particulate by gas-phase chemical reaction and aerosol coagulation in the free-molecular regime, *J Colloid Interface Sci.* **139**, 63–86 (1990).
17. R. R. Letfullin and V. I. Igoshin, Multipass optical reactor for laser processing of disperse materials, *Quantum Electron.* **25**, 684–689 (1995).

Chapter 17

Transport Coefficients in ^3He-^4He Mixtures

Sahng-Kyoon Yoo
Green University

Chung-In Um
Korea University
cium@korea.ac.kr

Thomas F. George
University of Missouri–St. Louis
tfgeorge@umsl.edu

The calculation of transport coefficients, such as thermal conductivity and first viscosity, in dilute ^3He-^4He mixtures is reviewed. Besides phonon ^3He scattering processes, the effects of three-phonon processes, which are the dominant scattering mechanism for liquid ^4He at low temperatures, are included.

17.1. Introduction

The transport properties of dilute ^3He-^4He mixtures at low temperatures come from the interactions among the ^3He quasiparticles, phonon and roton excitations of superfluid ^4He. In these mixtures, the interaction between ^3He quasiparticles is modified due to the presence of ^4He background. At the same time, since the scattering or absorption processes of phonons are largely limited by ^3He quasiparticles, the phonon contribution to the transport coefficients are different from the pure ^4He case.

Bardeen, Baym and Pines (BBP) [1] deduced the nature of the ^{3}He quasiparticle interactions in dilute ^{3}He-^{4}He mixtures and showed that the effective interaction is weakly attractive. Using this interaction, Baym [2–4] calculated the first viscosity, thermal conductivity and the attenuation of first sound at very low temperatures by considering the quasiparticle–quasiparticle and phonon-quasiparticle scattering processes. However, their results for the thermal conductivity and viscosity were much larger at high temperatures for which the phonon contributions begin to play a role.

This fact clearly means that some of the phonon scattering processes may be omitted in the consideration of transport. It is well known that the phonon spectrum of pure ^{4}He is not normal but anomalous [5], which means that the phonon absorption or emission (three-phonon processes, 3PP) are possible and dominant below 0.6 K. We have previously calculated the transport coefficients in both thin [6] and bulk [7] liquid ^{4}He. It is reasonable to include 3PP effects in order to obtain the transport coefficients of dilute ^{3}He-^{4}He mixtures.

In this chapter, we review the investigation of the 3PP effects on superfluid and dilute ^{3}He-^{4}He mixtures [8–10]. We first study the 3PP rates by using the phonon energy spectrum, and then the dilute ^{3}He-^{4}He mixtures, mainly for the case of low ^{3}He concentration in which 3PP effects appear due to the decrease in ^{3}He-^{3}He and ^{3}He-phonon interactions. The first viscosity [8–10] and thermal conductivity [9,10] are discussed.

17.2. Elementary Excitation of ^{4}He and 3PP Scattering Rates

Landau and Khalatnikov [11,12] developed a kinetic theory to take into account the interaction between elementary excitations in superfluid ^{4}He. In the long-wavelength limit (phonon), the energy spectrum is given by

$$\epsilon(q) = c_0 \hbar q(1 - \gamma q^2 - \delta q^4 + \cdots). \tag{17.1}$$

The positivity of γ results in the normal energy spectrum. However, the viscosity of ^{4}He, obtained from the normal energy spectrum, was not in agreement with the experimental measurements. Maris and Messey [5] proposed that the coefficient γ should be negative and the consequent energy spectrum is anomalous. Using this spectrum, Maris obtained a better fit with experiments. In the anomalous spectrum, 3PP as the lowest-order processes are allowed, while in case of a normal spectrum, the lowest-order processes are (four-phonon processes, 4PP). Afterwards, many workers continued the studies of this topic [13–15].

In this section, we study the 3PP rates in superfluid ^{4}He. Among the various spectra, we take two of those suggested by Greywall [16] which is obtained by fitting the specific heat measurement under pressure. The phonon energy spectrum is

$$\epsilon_q = sq(1 + \alpha_2 q^2 + \alpha_4 q^4 + \alpha_6 q^6) \tag{17.2}$$

where $s = 237.0$ m/s for $P = 0$ atm (298.9 m/s for $P = 10$ atm),

$$\alpha_2 = 1.30 - 0.065\,P,$$

$$\alpha_4 = -10.25\,\alpha_2 - 108.5\left(\frac{s_4}{s} - 1\right) - 28.44\left(\frac{s_5}{s} - 1\right),$$

$$\alpha_6 = 25.0\,\alpha_2 - 434.0\left(\frac{s_4}{s} - 1\right) - 177.8\left(\frac{s_5}{s} - 1\right),$$

$$s_4 = 247.0 + 2.86\,P,\ s_5 = 242.0 + 2.20\,P, \tag{17.3}$$

with s as the velocity of first sound and P as the pressure.

Note that Maris [17] obtained another one given by

$$\epsilon_q = sq\left[1 + \gamma q^2 \frac{1-(q/q_A)^2}{1-(q/q_D)^2}\right] \tag{17.4}$$

Where $s = 238.3$ *m/s*, $\gamma = 10\times10^{37}$ (*cgs units*) and $q_A = 0.542\,A^{-1}$, $q_B = 0.332\,A^{-1}$.

With these spectra, the 3PP does not change the total momentum and total energy. Therefore, two kinds of processes are possible: (1) $\vec{\mathbf{q}} \leftrightarrow \vec{\mathbf{q}}' + \vec{\mathbf{q}}''$ and (2) $\vec{\mathbf{q}}' \leftrightarrow \vec{\mathbf{q}} + \vec{\mathbf{q}}''$.

Before we go to the mixture case, we first calculate the eigenvalue matrix equation representing various relaxations resulting from 3PP small angle processes, using the quantum hydrodynamic Hamiltonian [18]. In the long-wavelength limit, 3PP interaction is given by

$$V_3 = \int d\vec{\mathbf{r}}\left[\frac{1}{2} m_4\, \vec{\mathbf{v}}_{\mathbf{s}}(\vec{\mathbf{r}})\, \rho_4(\vec{\mathbf{r}})\, \vec{\mathbf{v}}_{\mathbf{s}}(\vec{\mathbf{r}}) + \frac{1}{6}\frac{\partial}{\partial n_4}\left(\frac{m_4 s^2}{n_4}\right) \rho_4(\vec{\mathbf{r}})^3\right], \tag{17.5}$$

where m_4 is the ^{4}He mass, and $\vec{\mathbf{v}}_{\mathbf{s}}$ and ρ_4 are the local superfluid density and local density variation of ^{4}He from equilibrium density n_4, which are small quantities. These variations can be expanded in terms of phonon annihilation and creation operators b_q and b_q^+ as

$$\rho_4(\vec{\mathbf{r}}) = \sum_{\vec{\mathbf{q}}} \left(\frac{q^2 n_4}{2m_4 \epsilon_q V}\right)^{\frac{1}{2}} \left(b_q\, e^{i\vec{\mathbf{q}}\cdot\vec{\mathbf{r}}} + b_q^+\, e^{-i\vec{\mathbf{q}}\cdot\vec{\mathbf{r}}}\right), \tag{17.6}$$

and

$$\vec{\mathbf{v}}_{\mathbf{s}}(\vec{\mathbf{r}}) = \sum_{\vec{\mathbf{q}}} \left(\frac{\epsilon_q}{2m_4 n_4 V}\right)^{\frac{1}{2}} \hat{q}\left(b_q\, e^{i\vec{\mathbf{q}}\cdot\vec{\mathbf{r}}} + b_q^+\, e^{-i\vec{\mathbf{q}}\cdot\vec{\mathbf{r}}}\right). \tag{17.7}$$

We then obtain the matrix element for 3PP as

$$< \vec{\mathbf{q}}', \vec{\mathbf{q}}'' \mid V_3 \mid \vec{\mathbf{q}} > = \left[\frac{T(q,q',q'')}{V}\right]^{1/2} \delta_{\vec{\mathbf{q}}' + \vec{\mathbf{q}}'', \vec{\mathbf{q}}}\ , \tag{17.8}$$

where

$$T(q,q',q'') = \left(\frac{1}{8m_4 n_4}\right)\left[\left(\frac{\epsilon_q \epsilon_{q''}}{\epsilon_{q'}}\right)^{\frac{1}{2}} q'\, \hat{q}\cdot\hat{q}'' + \left(\frac{\epsilon_q \epsilon_{q'}}{\epsilon_{q''}}\right)^{\frac{1}{2}} q''\, \hat{q}\cdot\hat{q}' + \left(\frac{\epsilon_{q'} \epsilon_{q''}}{\epsilon_q}\right)^{1/2} q\, \hat{q}'\cdot\hat{q}'' + s^2(2u-1)\frac{q\, q' q''}{\left(\epsilon_q \epsilon_{q'} \epsilon_{q''}\right)^{1/2}}\right]^2. \tag{17.9}$$

Here, u is the Grüneisen constant defined by

$$u = \frac{n_4}{s}\frac{\partial s}{\partial n_4}. \tag{17.10}$$

The 3PP collision integral is given by

$$I_{3PP}(\vec{\mathbf{q}}) = \frac{1}{2}\sum_{\vec{\mathbf{q}}',\vec{\mathbf{q}}''} 2\pi \mid < \vec{\mathbf{q}}', \vec{\mathbf{q}}'' | V_3 | \vec{\mathbf{q}} > |^2\, \delta(\varepsilon_f - \varepsilon_i)\left[n_{q'} n_{q''}(1+n_q) - n_q(1+n_{q'})(1+n_{q''})\right]$$
$$+ \sum_{\vec{\mathbf{q}}',\vec{\mathbf{q}}''} 2\pi \mid < \vec{\mathbf{q}}' | V_3 | \vec{\mathbf{q}}, \vec{\mathbf{q}}'' > |^2\, \delta(\varepsilon_f - \varepsilon_i)\left[n_{q'}(1+n_{q''})(1+n_q) - n_q n_{q''}(1+n_{q'})\right], \tag{17.11}$$

where ε_i and ε_f are the phonon energy of initial and final states, respectively. The first term represents the first' process ($\vec{\mathbf{q}} \leftrightarrow \vec{\mathbf{q}}' + \vec{\mathbf{q}}''$), and the second term the second process ($\vec{\mathbf{q}}' \leftrightarrow \vec{\mathbf{q}} + \vec{\mathbf{q}}''$).

If the phonons are in the equilibrium state, i.e., $n_q = n_q^0$, the collision integral vanishes by detailed balancing, where n_q^0 is the equilibrium distribution function. Let us consider a small variation from the equilibrium phonon distribution as

$$n_q = n_q^0 + \delta n_q. \tag{17.12}$$

Then the collision integral can be rewritten within the first order of δn_q as

$$\begin{aligned} I_{3PP}(\vec{\mathbf{q}}) = \\ &-\frac{1}{2(2\pi)^2}\int_{(1)} dq' d\Omega_{q'}\, q'^2\, T(q,q',q'')\delta(\varepsilon_f \\ &\qquad - \varepsilon_i)\left[\delta n_q\left(1 + n_{q'}^0 + n_{q''}^0\right) - \delta n_{q'}\left(n_{q''}^0 - n_q^0\right)\right. \\ &\qquad \left. - \delta n_q\left(n_{q'}^0 - n_q^0\right)\right] \\ &-\frac{1}{(2\pi)^2}\int_{(2)} dq' d\Omega_{q'}\, q'^2\, T(q,q',q'')\delta(\varepsilon_f - \varepsilon_i)[\delta n_q\left(n_{q''}^0 - n_{q'}^0\right) \\ &\qquad - \delta n_{q'}\left(1 + n_q^0 - n_{q''}^0\right) \\ &\qquad\qquad + \delta n_{q''}\left(n_q^0 - n_{q'}^0\right). \end{aligned} \tag{17.13}$$

Due to momentum conservation, $\vec{\mathbf{q}}'' = \vec{\mathbf{q}} - \vec{\mathbf{q}}'$ in the first process, and $\vec{\mathbf{q}}'' = \vec{\mathbf{q}}' - \vec{\mathbf{q}}$ in the second process are denoted by (1) and (2), respectively, in Eq. (17.13).

The variation of distribution function from the equilibrium state can be expanded in spherical harmonics $Y_{lm}(\Omega_q)$ as

$$\delta n_q = \vec{q}\, \bar{n}_q^0 \sum_{l,m} \Phi_{lm}(q) Y_{lm}(\Omega_q), \tag{17.14}$$

where for simplicity we define

$$\bar{n}_q^0 \equiv \frac{\partial n_q^0}{\partial \varepsilon_q}\ . \tag{17.15}$$

and Ω_q is the solid angle of $\vec{\mathbf{q}}$. Using the addition theorem, the spherical harmonics for $\vec{\mathbf{q}}'$ and $\vec{\mathbf{q}}''$ are transformed to those for $\vec{\mathbf{q}}$ as follows:

$$Y_{lm}(\Omega_{q'}) \rightarrow P_l(\cos\theta)Y_{lm}(\Omega_q), \tag{17.16}$$

$$Y_{lm}(\Omega_{q''}) \rightarrow P_l(\cos\theta')\,Y_{lm}(\Omega_q), \tag{17.17}$$

where P_l is the Legendre polynomial, and θ and θ' are the angles between momenta $(\vec{\mathbf{q}},\vec{\mathbf{q}}')$ and $(\vec{\mathbf{q}},\vec{\mathbf{q}}'')$, respectively. This transform procedure is explained well in Ref. [19].

After performing the angle integration, we obtain the collision integral as

$$I_{3PP}(\vec{\mathbf{q}}) = -q\bar{n}_q^0 \sum_{l,m} Y_{lm}(\Omega_q)\Bigg\{\frac{1}{2(2\pi)}\int_{(1)} dq'q'^2 \frac{T(q,q',q'')}{B(q,q',q'')}\Big[(1+n_{q'}^0 + n_{q''}^0)\Phi_{lm}(q) - \frac{q'\bar{n}_{q'}^0}{q\bar{n}_q^0}\left(n_{q''}^0 - n_q^0\right)P_l(\cos\theta)\Phi_{lm}(q') - \frac{q''\bar{n}_{q''}^0}{q\bar{n}_q^0}\left(n_{q'}^0 - n_q^0\right)P_l(\cos\theta')\Phi_{lm}(q'')\Big] + \frac{1}{2\pi}\int_{(2)} dq'q'^2 \frac{T(q,q',q'')}{B(q,q',q'')}\Big[(n_{q''}^0 - n_{q'}^0)\Phi_{lm}(q) - \frac{q'\bar{n}_{q'}^0}{q\bar{n}_q^0}\left(1+n_q^0 + n_{q''}^0\right)P_l(\cos\theta)\Phi_{lm}(q') + \frac{q''\bar{n}_{q''}^0}{q\bar{n}_q^0}\left(n_q^0 - n_{q'}^0\right)P_l(\cos\theta')\Phi_{lm}(q'')\Big]\Bigg\}, \tag{18}$$

where $B(q,q',q'')$, originating from the delta function representing energy conservation through 3PP, is defined by

$$B(q,q',q'') = \left|\frac{\partial(\varepsilon_f-\varepsilon_i)}{\partial x}\right| = \left|\frac{\partial\varepsilon_{q''}}{\partial q''}\frac{\partial q''}{\partial x}\right|, \tag{17.19}$$

with $x \equiv \cos\theta$ and $q'' = (q^2 + q'^2 - 2qq'x)^{1/2}$.

Since we want to represent the collision integral in the form

$$I_{3PP}(\vec{\mathbf{q}}) = -q\bar{n}_q^0 \sum_{l,m} \frac{Y_{lm}(\Omega_q)\Phi_{lm}(q)}{\tau_l}, \tag{17.20}$$

we obtain the eigenvalue equation for each l, where τ_l is the relaxation time with l. Since the eigenvalue equation does not change with different m, we suppress the index m.

The eigenvalue equation for a given l is

$$\begin{aligned}\frac{1}{2(2\pi)}\int_{(1)} dq' q'^2 \frac{T(q,q',q'')}{B(q,q',q'')}&[(1+n^0_{q'}+n^0_{q''})\Phi_l(q)\\ &-\frac{q'\bar{n}^0_{q'}}{q\bar{n}^0_q}\left(n^0_{q''}-n^0_q\right)P_l(\cos\theta)\Phi_l(q')\\ &-\frac{q''\bar{n}^0_{q''}}{q\bar{n}^0_q}\left(n^0_{q'}-n^0_q\right)P_l(\cos\theta')\Phi_l(q'')]\\ &=\mu_l\Phi_l(q),\end{aligned} \tag{17.21}$$

where the eigenvalues are given by $\mu_l = \tau_l^{-1}$. We cannot solve this integral equation analytically and therefore should use a numerical method. As we replace the integral on the left-hand side by a sum over a finite set of points, a matrix eigenvalue equation is obtained as

$$\mathrm{M_l}\Phi_l(q) = \mu_l\Phi_l(q), \tag{17.22}$$

where $\mathrm{M_l}$ denotes the collision matrix with a given l symbolically. The diagonal elements of $\mathrm{M_l}$ consists of terms containing Φ_l, while the off-diagonal elements come from the remaining terms involving $\Phi_l(q')$ and $\Phi_l(q'')$. Then the number of eigenfunctions and eigenvalues is equal to the matrix size.

From the condition of energy conservation through 3PP, we can show that the lowest eigenvalue with $l = 0$ is zero. This can be seen as follows. Since 3PP does not change the total energy,

$$\int \varepsilon_q I_{3PP}(\vec{\mathbf{q}})\, d\vec{\mathbf{q}} = 0. \tag{17.23}$$

Let the variation of phonon distribution be only φ_{S_0}. Then the eigenfunction with the lowest eigenvalue μ_{S_0}, which has no node, becomes

$$\int \varepsilon_q q\, \bar{n}^0_q \mu_{S_0} Y_{00}(\Omega_q)\varphi_{S_0} d\vec{\mathbf{q}} = 0. \tag{17.24}$$

Since the sign of integrand except μ_{S_0} is that of φ_{S_0}, in order to make the above equation valid, μ_{S_0} should be zero. Through a similar argument, we can see that the lowest eigenvalue with $l = 1$ also is zero using the condition of momentum conservation. Therefore, we focus our attention on the $l = 2$ case, in which the lowest eigenvalue is finite.

Maris [17] also transformed the integral eigenvalue problem into a matrix one. In his numerical calculations, he replaced the integral by a sum over the set of points $j\Delta\varepsilon_q$, where $j = 1, \cdots, j_m$, and he took a suitable choice of $\Delta\varepsilon_q$ between $0.6\, k_B T$ and $1.25\, k_B T$. This choice corresponds to about $10 \leq j_m \leq 20$. He argued that the results were found to be independent of the details of the mesh to better than 1%, and that the spectrum of eigenvalues is discrete.

In our numerical calculation, we use $j\Delta q$ instead of $j\Delta\varepsilon_q$. The lowest eigenvalue is for $l = 2$, i.e., μ_2, which is the wide-angle scattering rate characterizing viscosity, since $\Delta\theta = \pi/2$, where $\Delta\theta$ is the angular distance between the maximum and minimum phonon distributions. For a given l, where $\Delta\theta = \pi/l$, in order to see the effect of the phonon energy spectrum on the wide-angle scattering rate, we evaluate the rates using both Greywell's spectrum and Maris' spectrum. For Greywell's with $P = 0$ atm and $P = 10$ atm, we show the wide-angle scattering rates in Fig. 17.1. Note also that when Greywell's spectrum is used, the value of wide-angle scattering rates is larger than the case of Maris, which is also shown in the figure.

17.3. ^{3}He-Phonon Scattering in Dilute ^{3}He-^{4}He Mixtures

In this section we review Baym and Ebner's theory for the ^{3}He-phonon interaction which considers two processes: phonon absorption by ^{3}He and phonon-^{3}He elastic scattering. in the spirit of Landau and Khalatnikov [11,12], Baym and Ebner [2] derived this interaction as determined by thermodynamic and Galilean-invariance arguments. The energy of a slowly-moving quasiparticle of momentum $\vec{\mathbf{p}}$ in superfluid ^{4}He at rest is simply

$$\epsilon_p = \epsilon_0 + \frac{p^2}{2m^*}, \tag{17.25}$$

where ϵ_0 is the chemical potential μ_3 of ^{3}He in liquid ^{4}He in the limit of zero concentration, and m* is the effective mass of ^{3}He, which increases under pressure.

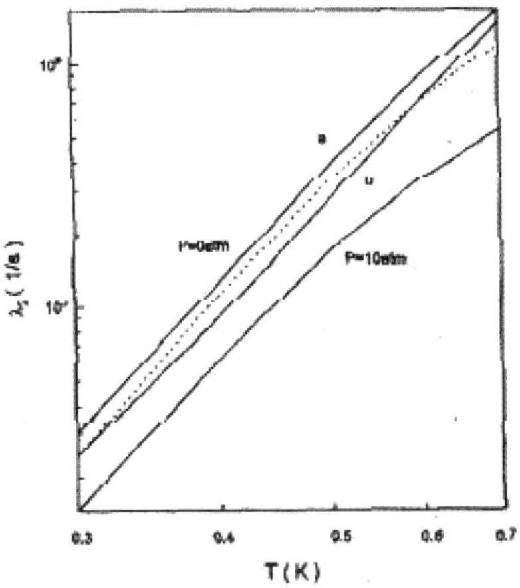

Fig. 17.1. The wide-angle scattering rates as a function of T: (a) Greywall's spectrum and (b) Maris' spectrum. The dotted line represents the result using Greywall's spectrum with the artificial cut-off momentum $q_c = 0.35$ A^{-1}. The case of $P = 10$ atm is also shown using Greywall's spectrum.

The long-wavelength and low-frequency deviation of ^{4}He from equilibrium may be described by a density variation $\rho_4(\vec{\mathbf{r}}, t)$ from the mean density, and by a superfluid velocity $\vec{\mathbf{v}}_{\mathbf{s}}(\vec{\mathbf{r}}, t)$. As one knows how the quasiparticle energy in Eq. (17.25) depends on the local ^{4}He density and superfluid velocity, the interaction between the long-wavelength phonon and the quasiparticle can be found.

From the requirements of Galilean invariance, the dependence of ϵ_p on $\vec{\mathbf{v}}_{\mathbf{s}}$ is determined. Suppose that there is a uniform superfluid velocity in the liquid ^{4}He. Then in the frame of $\vec{\mathbf{v}}_{\mathbf{s}} = 0$, the energy of a quasiparticle of momentum $\vec{\mathbf{p}}$ is given by Eq. (17.25). In the laboratory frame, the momentum required to create this quasiparticle is $\vec{\mathbf{p}} + m_3\vec{\mathbf{v}}_{\mathbf{s}}$, where m_3 is the ^{3}He atomic mass, while the energy required to create it is

$$\epsilon_p + \vec{\mathbf{p}} \cdot \vec{\mathbf{v}}_{\mathbf{s}} + \frac{m_3 v_s^2}{2}. \tag{17.26}$$

The energy in the laboratory frame is

$$\epsilon_p[\vec{\mathbf{v}}_{\mathbf{s}}] = \epsilon_0 + \frac{p^2}{2m^*} + \frac{\delta m}{m^*}\vec{\mathbf{p}} \cdot \vec{\mathbf{v}}_{\mathbf{s}} - \frac{\delta m}{m^*}\frac{m_3 v_s^2}{2}. \tag{17.27}$$

where $\delta m = m^* - m_3$, and the square bracket $[\vec{\mathbf{v}}_\mathbf{s}]$ denotes the $\vec{\mathbf{v}}_\mathbf{s}$ dependence of ϵ_0 and m^* on the ^{4}He density. The dependence of m^* on ρ_4 can be neglected since it leads to corrections of order $(v_f/s)^2 \sim 0.01$, where v_f is the Fermi velocity of the ^{3}He quasiparticles. The dependence of ϵ_0 on ρ_4 can be written up to second order as

$$\epsilon_0[\rho_4] = \epsilon_0 + \left(\frac{\partial \epsilon_0}{\partial n_4}\right)_{n_3} \rho_4 + \frac{1}{2}\left(\frac{\partial^2 \epsilon_0}{\partial n_4^2}\right)_{n_3} \rho_4^2. \tag{17.28}$$

The ^{3}He quasiparticle sees a long-wavelength phonon as a uniform motion of ^{4}He with a local velocity v_s and a local density $n_4 + \rho_4$. Thus, the change in the energy of a quasiparticle of momentum $\vec{\mathbf{p}}$, due to the presence of a long-wavelength phonon, is

$$\delta\epsilon_p(\vec{\mathbf{r}},t) = \left(\frac{\partial \epsilon_0}{\partial n_4}\right)\rho_4(\vec{\mathbf{r}},t) + \left(\frac{\delta m}{m^*}\right)\vec{\mathbf{p}}\cdot\vec{\mathbf{v}}_\mathbf{s}(\vec{\mathbf{r}},t) + \frac{1}{2}\left(\frac{\partial^2 \epsilon_0}{\partial n_4^2}\right)\rho_4^2(\vec{\mathbf{r}},t) - \frac{1}{2}\left(\frac{\delta m}{m^*}\right) m_3 v_s^2(\vec{\mathbf{r}},t), \tag{17.29}$$

where $\vec{\mathbf{r}}$ is the position of the quasiparticle.

Since ϵ_0 is the ^{3}He chemical potential at zero concentration, one can have [1]

$$\left(\frac{\partial \epsilon_0}{\partial n_4}\right)_{n_3} = \left(\frac{m_4 s^2}{n_4}\right)(1+\alpha), \tag{17.30}$$

where m_4 is the ^{4}He atomic mass and α is the fractional excess molar volume of ^{3}He in ^{4}He. Thus,

$$\left(\frac{\partial^2 \epsilon_0}{\partial n_4^2}\right) = \frac{m_4 s^2}{n_4}\left[(1+\alpha)\left(2\frac{n_4}{s}\frac{\partial s}{\partial n_4} - 1\right) + n_4 \frac{\partial \alpha}{\partial n_4}\right]. \tag{17.31}$$

The amplitude for phonon-quasiparticle scattering can be obtained by quantizing the interaction in Eq. (17.29) using Eqs. (17.6) and (17.7) for $\rho_4(\vec{\mathbf{r}})$ *and* $\vec{\mathbf{v}}_\mathbf{s}(\vec{\mathbf{r}})$. Let the first two terms in Eq. (17.29) be V_1 and the last two terms be V_2. Then, V_1 corresponds to absorption or emission of a phonon by a quasiparticle. Since the position and momentum of the quasiparticle do not commute, the $\vec{\mathbf{p}}\cdot\vec{\mathbf{v}}_\mathbf{s}(\vec{\mathbf{r}})$ term must be symmetrized as $\frac{1}{2}[\vec{\mathbf{p}}\cdot\vec{\mathbf{v}}_\mathbf{s}(\vec{\mathbf{r}}) + \vec{\mathbf{v}}_\mathbf{s}(\vec{\mathbf{r}})\cdot\vec{\mathbf{p}}]$. The matrix element for a quasiparticle of momentum $\vec{\mathbf{p}}$ absorbing a phonon of momentum

$\vec{\mathbf{q}}$, and ending up with momentum $\vec{\mathbf{p}}''$ is

$$<\vec{\mathbf{p}}''|V_1|\vec{\mathbf{p}},\vec{\mathbf{q}}>=\delta_{\vec{\mathbf{p}}'',\vec{\mathbf{p}}+\vec{\mathbf{q}}}\left(\frac{qn_4}{2m_4sV}\right)^{1/2}\left[\frac{\partial\epsilon_0}{\partial n_4}+\frac{\delta m\, s}{n_4 q}\left(\epsilon_{p''}-\epsilon_p\right)\right]. \quad (17.32)$$

Similarly, V_2 corresponds to the scattering of a phonon by a quasiparticle. The matrix element for the scattering process is

$$<\vec{\mathbf{p}}',\vec{\mathbf{q}}'|V_2|\vec{\mathbf{p}},\vec{\mathbf{q}}>=\delta_{\vec{\mathbf{p}}'+\vec{\mathbf{q}}',\vec{\mathbf{p}}+\vec{\mathbf{q}}}\frac{(qq')^{1/2}n_4}{2m_4sV}\left[\frac{\partial^2\epsilon_0}{\partial n_4^2}-\frac{\delta m}{m^*}\frac{m_3s^2}{n_4^2}\hat{q}\cdot\hat{q}'\right]. \quad (17.33)$$

The total amplitude $<\vec{\mathbf{p}}',\vec{\mathbf{q}}'|T|\vec{\mathbf{p}},\vec{\mathbf{q}}>$ for the phonon-quasiparticle scattering is the sum of the matrix elements for the five processes in Fig. 17.2(b)–(f). The first three processes, i.e., (b), (c), (d), are analogues to those in nonrelativistic scattering of photon by free electrons. Their contribution to the total scattering amplitude is denoted by $<\vec{\mathbf{p}}',\vec{\mathbf{q}}'|T_d|\vec{\mathbf{p}},\vec{\mathbf{q}}>$.

The two processes (e) and (f) in the figure are the scattering of a phonon by a quasiparticle through the exchange of a virtual phonon; they constitute the "phonon-induced" interaction between a phonon and a quasiparticle, analogous to that between electrons in a metal. In order to calculate their contribution, denoted by $<\vec{\mathbf{p}}',\vec{\mathbf{q}}'|T_{ph}|\vec{\mathbf{p}},\vec{\mathbf{q}}>$, the three-phonon matrix element, Eq. (17.8), is needed. The scattering amplitude due to phonon exchange is simply evaluated on the energy shell when the scattering is elastic. Then we find that the total phonon-quasiparticle scattering amplitude is simplified as

$$\begin{aligned}<\vec{\mathbf{p}}',\vec{\mathbf{q}}'|T|\vec{\mathbf{p}},\vec{\mathbf{q}}> &= <\vec{\mathbf{p}}',\vec{\mathbf{q}}'|T_d|\vec{\mathbf{p}},\vec{\mathbf{q}}> + <\vec{\mathbf{p}}',\vec{\mathbf{q}}'|T_{ph}|\vec{\mathbf{p}},\vec{\mathbf{q}}> \\ &= \delta_{\vec{\mathbf{p}}'+\vec{\mathbf{q}}',\vec{\mathbf{p}}+\vec{\mathbf{q}}}\frac{sq}{2n_4V}\cdot \\ &\quad\times\left[n_4\frac{\partial\alpha}{\partial n_4}+\left(1+\alpha+\frac{\delta m}{m_4}\right)\frac{m_4}{m^*}\left(1+\alpha-\frac{m_3}{m_4}\right)\cos\theta\right].\end{aligned} \quad (17.34)$$

The rate at which phonons are scattered from $\vec{\mathbf{q}}$ to $\vec{\mathbf{q}}'$ by the ^{3}He is given by

$$2\sum_{\vec{\mathbf{p}},\vec{\mathbf{p}}'}2\pi\,\delta\left(\epsilon_p+sq-\epsilon_{p'}-sq'\right)|<\vec{\mathbf{p}}',\vec{\mathbf{q}}'|T|\vec{\mathbf{p}},\vec{\mathbf{q}}>|^2[1+ n(\vec{\mathbf{q}})]f(\vec{\mathbf{p}})[1-f(\vec{\mathbf{p}})], \quad (17.35)$$

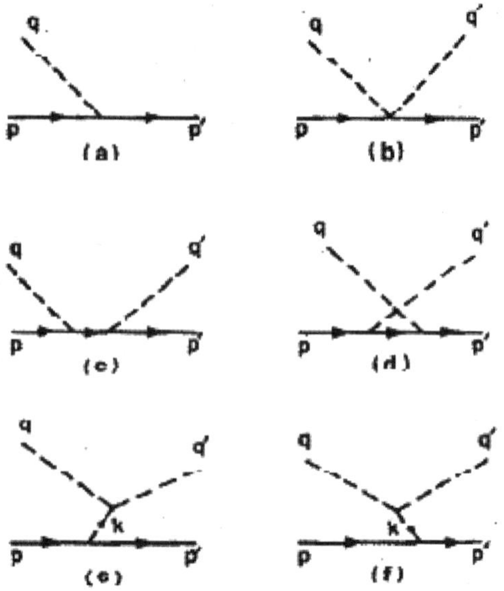

Fig. 17.2. Interactions of phonons (dashed line) with quasiparticles (solid line). (a) is phonon absorption; (b), (c) and (d) are three processes leading to the scattering of phonons by quasiparticles; and (e) and (f) are "phonon-induced" scattering of phonons by quasiparticles.

where the initial factor of 2 comes from ^{3}He spin states, and $f(\vec{\mathbf{p}})$ is the quasiparticle distribution function. For the purpose of calculating the phonon contribution to thermal conductivity and viscosity, it is sufficient to assume ^{3}He to be in thermal equilibrium and to neglect the exclusion principle for $\vec{\mathbf{p}}$. Then multiplying Eq. (17.35) by $\delta(\cos\theta - \hat{q}\cdot\hat{q}')$ and summing over all $\vec{\mathbf{q}}'$, one can find finally that the rate at which phonons of momentum $\vec{\mathbf{q}}$ are scattered by angle θ, divided by $1 + n(\vec{\mathbf{q}})$, is

$$\Gamma_{sc}(q,\theta) = \frac{n_3 s q^4}{8\pi n_4^2}\left[n_4 \frac{\partial \alpha}{\partial n_4} + \left(1 + \alpha + \frac{\delta m}{m_4}\right)\frac{m_4}{m^*}\left(1 + \alpha - \frac{m_3}{m_4}\right)\cos\theta\right]^2. \tag{17.36}$$

This scattering rate has a q^4-dependence characteristic of Rayleigh scattering, and thus a T^4-dependence. Consequently, it plays an important role in limiting phonon-transport phenomena at high temperatures.

At very low temperatures, the attenuation of first sound is due to the viscosity of ^{3}He. Since very long-wavelength phonons are simply first sound waves, ultrasonic attenuation is the dominant mechanism for limiting transport by such phonons. At a very long wavelength, only V_1 is significant for absorption. However, in calculating the absorption rate, the quasiparticle interaction is important, and the description of quasiparticles as plane wave states is inappropriate. The reason is that since $v_f < s$, energy and momentum conservation forbid the absorption of a phonon by a noninteracting quasiparticle. It is convenient, therefore, to write the interaction Eq. (17.29) between ^{3}He and the phonons in terms of the ^{3}He particle density operator $\rho_3(\vec{\mathbf{r}})$ and the ^{3}He particle current operator $\vec{\mathbf{j}}_3(\vec{\mathbf{r}})$. Then, the interaction becomes

$$\int d\vec{\mathbf{r}}\left\{\left[\frac{\partial\epsilon_0}{\partial n_4}\rho_4(\vec{\mathbf{r}}) + \frac{1}{2}\frac{\partial^2\epsilon_0}{\partial n_4^2}\rho_4^2(\vec{\mathbf{r}}) - \frac{1}{2}\frac{\delta m}{m^*}m_3 v_s^2(\vec{\mathbf{r}})\right]\rho_3(\vec{\mathbf{r}}) + \delta m\vec{\mathbf{j}}_3(\vec{\mathbf{r}})\cdot\vec{\mathbf{v}}_s(\vec{\mathbf{r}})\right\}. \tag{17.37}$$

The matrix element for the absorption of a phonon of momentum $\vec{\mathbf{q}}$, with ^{3}He going from the initial state $|i>$ to the finial state $|f>$, is

$$< f|V_1|i,\vec{\mathbf{q}} >= \left(\frac{qn_4}{2m_4 sV}\right)^{\frac{1}{2}}\int d\vec{\mathbf{r}}\, e^{i\vec{\mathbf{q}}\cdot\vec{\mathbf{r}}}\left[\frac{\partial\epsilon_0}{\partial n_4} < f|\rho_3(\vec{\mathbf{r}})|i > + \frac{s\delta m}{n_4}\hat{q}\cdot < f|\vec{\mathbf{j}}_3(\vec{\mathbf{r}})|i >\right]. \tag{17.38}$$

From the conservation of ^{3}He atoms, or quasiparticles, we can replace $\hat{q}\cdot< f|\vec{\mathbf{j}}_3(\vec{\mathbf{r}})|i >$ by $(\varepsilon_f - \varepsilon_i) < f|\rho_3(\vec{\mathbf{r}})|i >= sq < f|\rho_3(\vec{\mathbf{r}})|i >$, where ε_i and ε_f are the energies of the states $|i>$ and $|f>$. The absorption rate is found by squaring Eq. (17.38) and summing over all $|f>$ consistent with energy conservation. Averaging over a thermal ensemble of initial states with the integral form of the delta-function, one can find

$$\Gamma_{abs}(q) = \frac{-qm_4 s^3}{n_4}\left(1 + \alpha + \frac{\delta m}{m_4}\right)^2 Im < \rho_3\rho_3 > (q, sq). \tag{17.39}$$

Here, $< \rho_3\rho_3 >$ is the Fourier transform of the retarded ^{3}He density-density response function which can be evaluated by solving the ^{3}He kinetic equation including ^{3}He-^{3}He collisions by means of a relaxation-time approximation that includes conservation of the quasiparticle number, momentum and energy. The result for $\omega \gg v_f q$ and $\hbar\omega \ll k_B T$ is

$$< \rho_3 \rho_3 > (q, \omega) = \frac{q^2 n_3}{m^* \omega^2 + 4iq^2 \omega\eta / 3n_3 (1 + \tau_\eta \omega^2)}. \tag{17.40}$$

The ^{3}He first viscosity η_3 is given in terms of the ^{3}He relaxation time τ_η for the viscosity by

$$\eta_3 = P_f(T)\tau_\eta, \tag{17.41}$$

where $P_f(T)$ is the pressure of an ideal Fermi gas of the same effective mass and number density as ^{3}He in the mixture.

Then, the phonon absorption rate by ^{3}He becomes

$$\Gamma_{abs}(q) = \frac{1}{\tau_a(q)} = \frac{4m_4 P_f(T)}{3m^{*2} n_4} \frac{(1+\alpha+\frac{\delta m}{m_4})^2 q^2 \tau_\eta \left[1+(\frac{\hbar s q}{2\pi k_B T})^2\right]}{\left[1+(\frac{\hbar s q}{2\pi k_B T})^2\right]^2 + s^2 q^2 \tau_\eta{}^2}. \tag{17.42}$$

In deriving this equation, τ_η is replaced by the more general expression $\tau_\eta \left[1 + (\frac{\hbar s q}{2\pi k_B T})^2\right]^{-1}$, which can be shown by direct consideration of the collision integral [20].

17.4. Transport Coefficients of Dilute ^{3}He-^{4}He Mixtures

17.4.1. *First viscosity*

The phonon Boltzmann equation in the mixtures has the following form:

$$\frac{\partial n_{\vec{\mathbf{q}}}}{\partial t} + \frac{\partial \varepsilon_q}{\partial \vec{\mathbf{q}}} \cdot \frac{\partial n_q}{\partial \vec{\mathbf{r}}} - \frac{\partial \varepsilon_q}{\partial \vec{\mathbf{r}}} \cdot \frac{\partial n_q}{\partial \vec{\mathbf{q}}} = I_{3-ph} + I_{ph-ph}. \tag{17.43}$$

When a superfluid flow with velocity $\vec{\mathbf{v}}_s$ exists, the phonon energy is given by

$$\varepsilon_q = sq + \vec{\mathbf{q}} \cdot \vec{\mathbf{v}}_{\mathbf{s}}, \tag{17.44}$$

where the sound velocity s may depend on position and time. I_{3-ph} and I_{ph-ph} are the ^{3}He-phonon and phonon-phonon collision operators, respectively. I_{3-ph} includes both scattering, I_{sc}, and absorption or emission of phonons, I_{abs}, by the ^{3}He quasiparticle. For the phonon-phonon interactions, we consider only 3PP, which dominate below 0.6 K. We neglect boundary scattering when we calculate the viscosity.

The rate at which ^{3}He-phonon scattering changes the phonon distribution $n_{\vec{\mathbf{q}}}$ is linearized to first order of deviation of the phonon distribution about global equilibrium as

$$I_{sc} = -\int \frac{d\Omega_{q'}}{2\pi} \Gamma_{sc}(q,\theta) \left[\delta n_q - \delta n_{q'} - \frac{\partial n_q^0}{\partial \varepsilon_{q'}} (\vec{\mathbf{q}} - \vec{\mathbf{q}}') \cdot (\vec{\mathbf{v}}_{\mathbf{s}} - \vec{\mathbf{v}}_{\mathbf{3}})\right], \tag{45}$$

where $\Gamma_{sc}(q,\theta)$ is given by Eq. (17.36), and v_3 is the drift velocity of ^{3}He quasiparticles. Since $q = q'$ in Eq. (17.45) due to the elasticity of the scattering, the eigenfunctions of the scattering operator are simply spherical harmonics. Similar to Eq. (17.14) in Sec. 17.2, we express the deviation of the phonon distribution from global equilibrium as

$$\delta n_q = q \frac{\partial n_q^0}{\partial \varepsilon_{q'}} \Phi(\vec{q}) = q \frac{\partial n_q^0}{\partial \varepsilon_{q'}} \sum_{l=0}^{\infty} \Phi_l(q)$$

$$= q \frac{\partial n_q^0}{\partial c_{q'}} \sum_{l=0}^{\infty} \sum_{m=-l}^{l} Y_{lm}(\Omega_q) \Phi_{lm}(q)\,. \tag{46}$$

Then making use of the addition theorem for spherical harmonics, the rate becomes

$$I_{sc} = q \frac{\partial n_q^0}{\partial \varepsilon_{q'}} \left[\frac{\hat{q} \cdot (\vec{\mathbf{v}}_3 - \vec{\mathbf{v}}_s)}{\tau_1(q)} + \sum_{l=1}^{\infty} \frac{\Phi_l(q)}{\tau_l(q)}\right], \tag{17.47}$$

where

$$\tau_l(q)^{-1} = \int_{-1}^{1} d(\cos\theta)[1 - P_l(\cos\theta)]\, \Gamma_{sc}(q,\theta), l \geq 1. \tag{17.48}$$

The net effect of the τ_l terms in Eq. (17.48) is to urge the phonon distribution to relax about the local ^{3}He velocity $\vec{\mathbf{v}}_3$. Because $\Gamma_{sc}(q,\theta)$

contains terms only up to $P_2(\cos\theta)$, all of the $\tau_l(q)^{-1}$ for $l \geq 3$ are equal to

$$\tau_0{}^{-1} = \int_{-1}^{1} d(\cos\theta)\, \Gamma_{sc}(q,\theta). \tag{17.49}$$

Now we consider the contribution I_{abs} to I_{3-ph} due to absorption or emission of phonons by the ^{3}He quasiparticles, which is the mechanism of attenuation of ultrasound at low temperatures. The net absorption rate was given by Eq. (17.42) in previous section. When ^{3}He is in local equilibrium at temperature T_3' and velocity $\vec{\mathbf{v}}_3$, the net rate of absorption $I_{abs}(q)$ can be written as

$$I_{abs} = -\frac{(n_q - n_q^{(3)})}{\tau_a(q)}, \tag{17.50}$$

where

$$n_q^{(3)} = \frac{1}{exp\left[\beta_3'(\varepsilon_{q'} - \vec{\mathbf{q}}\cdot\vec{\mathbf{v}}_3)\right]-1}, \tag{17.51}$$

and β_3' is $\frac{1}{k_B T_3'}$. Then the final phonon Boltzmann equation becomes

$$\frac{\partial n_{\vec{\mathbf{q}}}}{\partial t} + \frac{\partial \varepsilon_q}{\partial \vec{q}} \cdot \frac{\partial n_q}{\partial \vec{r}} - \frac{\partial \varepsilon_q}{\partial \vec{r}} \cdot \frac{\partial n_q}{\partial \vec{q}} = -q\frac{\partial n_q^0}{\partial \varepsilon_q}\left[\frac{\Phi(\vec{\mathbf{q}}) - \beta^{-1}\delta(\beta_3' s) - \hat{q}\cdot(\vec{\mathbf{v}}_s - \vec{\mathbf{v}}_3)}{\tau_a(q)} + \frac{\hat{q}\cdot(\vec{\mathbf{v}}_s - \vec{\mathbf{v}}_3)}{\tau_1(q)} + \sum_{l=1}^{\infty}\frac{\Phi_l(q)}{\tau_l(q)} + \frac{\hat{q}\cdot(\vec{\mathbf{v}}_s - \vec{\mathbf{v}}_n)}{\tau_1{}'(q)} + \sum_{l=0}^{\infty}\frac{\Phi_l(q)}{\tau_1{}'(q)}\right], \tag{17.52}$$

where the last two terms on the right-hand side represent the 3PP collision integral.

The stress tensor is determined from the equation of motion for the total momentum. The total momentum is given by

$$g = \sum_{\vec{\mathbf{p}}} \vec{\mathbf{p}}\, f_p + \sum_{\vec{\mathbf{q}}} \vec{\mathbf{q}}\, n_q + m_4 n_4 \vec{\mathbf{v}}_s, \tag{17.53}$$

The phonon momentum can be written in linear order as

$$\sum_{\vec{\mathbf{q}}} \vec{\mathbf{q}}\, n_q = \rho_{ph}(\vec{\mathbf{v}}_{\mathbf{n}} - \vec{\mathbf{v}}_{\mathbf{s}}), \tag{17.54}$$

where

$$\rho_{ph} = -\sum_q \frac{q^2}{3} \frac{\partial n_q^0}{\partial \varepsilon_q}. \tag{17.55}$$

Summing over all phonon states after multiplying Eq. (17.52) by $\vec{\mathbf{q}}$, we find the linearized equation

$$\frac{\partial}{\partial t} \sum_{\vec{\mathbf{q}}} \vec{\mathbf{q}}\, n_q + \frac{\partial}{\partial \vec{\mathbf{r}}} \sum_{\vec{\mathbf{q}}} s\hat{q}\vec{\mathbf{q}}\, n_q + \rho_{ph} \frac{\partial s}{\partial \vec{r}} = \sum_{\vec{\mathbf{q}}} \vec{\mathbf{q}}(I_{abs} + I_{sc} + I_{3PP}). \tag{17.56}$$

The Boltzmann equation for ^{3}He is

$$\frac{\partial f_p}{\partial t} + \frac{\partial \epsilon_p}{\partial \vec{\mathbf{p}}} \cdot \frac{\partial f_p}{\partial \vec{\mathbf{r}}} - \frac{\partial \epsilon_p}{\partial \vec{\mathbf{r}}} \cdot \frac{\partial f_p}{\partial \vec{\mathbf{p}}} = I_{3-3} + I_{3-ph}, \tag{17.57}$$

where I_{3-3} and I_{3-ph} denote the collision integrals due to the ^{3}He-^{3}He interaction and ^{3}He-phonon interaction, respectively. Multiplying Eq. (17.57) by the ^{3}He quasiparticle momentum, $\vec{\mathbf{p}}$, and summing over all quasiparticle states, we have

$$\frac{\partial}{\partial t} \sum_{\vec{\mathbf{p}}} \vec{\mathbf{p}}\, f_p + \frac{\partial}{\partial \vec{\mathbf{r}}} \sum_{\vec{\mathbf{p}}} \frac{\vec{\mathbf{p}}\vec{\mathbf{p}}}{m} f_p + n_3 \frac{\partial \epsilon_0}{\partial \vec{\mathbf{r}}} = \sum_{\vec{\mathbf{p}}} \vec{\mathbf{p}}(I_{3-3} + I_{3-ph}). \tag{17.58}$$

From Eqs. (17.56), (17.57) and the equation of motion for superfluid

$$m_4 \frac{\partial \vec{\mathbf{v}}_{\mathbf{s}}}{\partial t} + \frac{\partial \mu_4}{\partial \vec{\mathbf{r}}} = 0, \tag{17.59}$$

we can derive the equation of motion for $\vec{\mathbf{g}}$ as

$$\frac{\partial \vec{\mathbf{g}}}{\partial t} + n_3 \frac{\partial \epsilon_0}{\partial \vec{\mathbf{r}}} + n_4 \frac{\partial \mu_4}{\partial \vec{\mathbf{r}}} + \rho_{ph} s \frac{\partial s}{\partial \vec{\mathbf{r}}} + \frac{\partial}{\partial \vec{\mathbf{r}}} \cdot \left(\sum_{\vec{\mathbf{q}}} s\hat{q}\, \vec{\mathbf{q}} n_q + \sum_{\vec{\mathbf{p}}} \frac{\vec{\mathbf{p}}\vec{\mathbf{p}}}{m} f_p \right) = 0. \tag{17.60}$$

To first order with fixed $\vec{\mathbf{v}}_{\mathbf{s}}$, the local equilibrium distributions obey

$$\sum_{\vec{\mathbf{p}}} \epsilon_p \left(\delta f_p - \delta f_p^{le}\right) + \sum_{\vec{\mathbf{q}}} \varepsilon_q \left(\delta n_q - \delta n_q^{le}\right) = 0, \qquad (17.61)$$

where δf_p^{le} and δn_q^{le} are the local equilibrium deviations from global equilibrium for the impurities and phonons, respectively. At the temperatures and concentrations of interest, the ^{3}He particles tend to come into local equilibrium at temperature T_3' due to rapid collisions between themselves. The common local equilibrium temperature T thus differs from the initial T_3'.

The local equilibrium distribution, n_q^{le}, obeys

$$\frac{\partial}{\partial \vec{\mathbf{r}}} \sum_{\vec{\mathbf{q}}} s\hat{q}\vec{\mathbf{q}}\, n_q^{le} = -s\rho_{ph} \frac{\partial s}{\partial \vec{\mathbf{r}}} + S_{ph} \frac{\partial T}{\partial \vec{\mathbf{r}}}, \qquad (17.62)$$

where the equilibrium phonon entropy per unit volume is given as

$$S_{ph} = \frac{s^2 \rho_{ph}}{T}. \qquad (17.63)$$

Similarly,

$$\frac{\partial}{\partial \vec{r}} \sum_{\vec{\mathbf{p}}} \frac{\vec{\mathbf{p}}\vec{\mathbf{p}}}{m} f_p^{le} = S_3 \frac{\partial T}{\partial \vec{\mathbf{r}}}, + n_3 \frac{\partial \epsilon_0 - \mu_3^{le}}{\partial \vec{\mathbf{r}}}, \qquad (17.64)$$

where S_3 is the ^{3}He entropy defined as

$$S_3 = \frac{\frac{5}{2} P_f(T) + (\epsilon_0 - \mu_3) n_3}{k_B T}. \qquad (17.65)$$

From Eqs. (17.60) and (17.62)–(17.65) together with the local thermodynamic Gibbs-Duhem relation, the equation of motion for $\vec{\mathbf{g}}$ can be written as

$$\frac{\partial \vec{\mathbf{g}}}{\partial t} + \frac{\partial}{\partial \vec{\mathbf{r}}} \cdot \vec{\Pi} = 0. \qquad (17.66)$$

Here, Π_{ij} is the stress tensor expressed as

$$\Pi_{\mathrm{ij}} = \sum_{\vec{\mathbf{p}}} \frac{p_i p_j}{m} \left(f_p - f_p^{le}\right) + \sum_{\vec{\mathbf{q}}} \frac{q_i q_j}{q} \left(n_q - n_q^{le}\right) + \delta_{ij}\left[P + n_4\left(\mu_4 - \mu_4^{le}\right)\right], \qquad (17.67)$$

where P is the local thermodynamic pressure, and μ_4^{le} is the ^{4}He chemical potential. From this stress tensor, we can deduce the viscosity coefficient.

Now we solve the phonon Boltzmann equation. We assume that the variations take place in the z-direction and the fluid moves in the x-direction. Furthermore, all longitudinal variations such as $\vec{\mathbf{v}}_s, \delta P, \delta s$ and $\delta\beta$ are neglected. We consider n_q and $\vec{\mathbf{v}}_3$ in terms of deviations from equilibrium proportional to $\exp(i\vec{\mathbf{k}}\cdot\vec{\mathbf{r}} - i\omega t)$. Then the phonon Boltzmann equation can be transformed into

$$(\omega - s\hat{q}\cdot\vec{\mathbf{k}})\Phi(\vec{\mathbf{q}}) \\ = -i\left[\hat{q}\cdot\vec{\mathbf{v}}_3\left(\frac{1}{\tau_a}+\frac{1}{\tau_1}\right) + \frac{\Phi(\vec{\mathbf{q}})}{\tau_a} + \sum_{l=1}^{\infty}\Phi_l(\vec{\mathbf{q}})\left(\frac{1}{\tau_l}+\frac{1}{\tau_{l'}}\right)\right]. \quad (17.68)$$

Let us consider the hydrodynamic limit, i.e., ω and $\vec{\mathbf{k}} \to 0$. Then

$$\Phi(\vec{\mathbf{q}}) = \Phi_{11}(Y_{11} - Y_{1,-1}) = -A\,\hat{q}\cdot\vec{\mathbf{v}}_3, \quad (17.69)$$

where

$$A = \frac{\tau_a{}^{-1}+\tau_1{}^{-1}}{\tau_a{}^{-1}+\tau_1{}^{-1}+\tau_1{}'^{-1}}, \quad (17.70)$$

and all other terms which are different from $l = 1$ and $m = \pm 1$ are zero. Using the zeroth-order solution, we find the first-order equation in ω and $\vec{\mathrm{k}}$ as

$$(\omega - s\hat{q}\cdot\vec{\mathbf{k}})(-A\,\hat{q}\cdot\vec{\mathbf{v}}_3) = \\ -i\left[\hat{q}\cdot\vec{\mathbf{v}}_3\left(\frac{1}{\tau_a}+\frac{1}{\tau_1}\right) + \frac{\Phi(\vec{q})}{\tau_a} + \sum_{l=1}^{\infty}\Phi_l(\vec{\mathbf{q}})\left(\frac{1}{\tau_a}+\frac{1}{\tau_l}+\frac{1}{\tau_{l'}}\right)\right]. \quad (17.71)$$

Since

$$s\hat{q}\cdot\vec{\mathbf{k}}\,\hat{q}\cdot\vec{\mathbf{v}}_3 = -(6\pi/45)^{1/2}(Y_{11} - Y_{1,-1})skv_3, \quad (17.72)$$

comparing all the coefficients of the spherical harmonics, we find that

$$\Phi(\vec{\mathbf{q}}) = \Phi_{11}(Y_{11} - Y_{1,-1}) + \Phi_{21}(Y_{21} - Y_{2,-1}) = -\,\hat{q}\,\cdot$$

$$\vec{\mathbf{v}}_3(i\omega\tau_K + 1) + is\,\hat{q}\cdot\vec{\mathbf{k}}\,\hat{q}\cdot\vec{\mathbf{v}}_3\tau_v, \tag{17.73}$$

where

$$\tau_K = \frac{A^2}{\tau_a{}^{-1}+\tau_1{}^{-1}+\tau_1{}'^{-1}}, \tau_v = \frac{A}{\tau_a{}^{-1}+\tau_2{}^{-1}+\tau_2{}'^{-1}}. \tag{17.74}$$

From the phonon stress tensor, the viscosity due to phonons is given by

$$\eta_{ph} = \frac{1}{5}s^2\rho_{ph} < \tau_v >, \tag{17.75}$$

where

$$< \tau_v >= -\frac{1}{\rho_{ph}}\sum_{\vec{\mathbf{q}}}\frac{q^2}{3}\frac{\partial n_q^0}{\partial \varepsilon_q}\tau_v(q) \tag{17.76}$$

is the average lifetime for the phonon viscosity. Therefore, the total viscosity of a dilute ^{3}He-^{4}He mixture is

$$\eta_{tot} = \eta_3 + \eta_{ph} = P_f(T)\tau_\eta + \frac{1}{5}s^2\rho_{ph} < \tau_v > \tag{17.77}$$

in the hydrodynamic limit, where η_3 is the ^{3}He viscosity. The average lifetime $< \tau_v >$ can be calculated numerically.

From the parameters given in Ref. [8], we first numerically calculate the rate $\tau_2{}'^{-1}(q)$ of 3PP. However, since the 3PP rates of pure ^{4}He are different from those in mixtures, $\tau_2{}'^{-1}(q)$ can be obtained as follows. First, by taking the initial $\tau_2{}'^{-1}(q)$ as the diagonal element of M_2, we find a trial $\tau_v{}^0$ using Eq. (17.74) with $\vec{\mathbf{v}}_3 = 0$, and in turn obtain a trial Φ_2. Second, inserting the trial Φ_2 into Eq. (17.22) with $l = 2$, we obtain a new $\tau_2{}'^{-1}(q)$. Then we obtain the optimized $\tau_2{}'^{-1}(q)$ iteratively.

The scattering rate of phonons by quasiparticles can be written as

$$\Gamma_{sc}(q,\theta) = \frac{xsq^4}{8\pi n_4}(-1.01 + 0.69\cos\theta)^2, \tag{17.78}$$

which we see is proportional to x. To get the absorption rate, we must determine the ^{3}He-^{3}He characteristic time τ_η for the viscosity in the degenerate (quantum) regime and classical regime. For the former case, we calculate τ_η from the theory of a Fermi liquid using the BBP interaction [1,21]. τ_η is proportional to T^{-2}. In the classical regime, τ_η may be obtained by solving the classical Boltzmann equation for the viscosity to the lowest order with the proper effective quasiparticle interaction. The classical τ_η can also be determined from an empirically chosen η_3. Although some workers estimated τ_η using the latter procedure, their results are much smaller than those from the Baym–Ebner theory. In their estimation, 3PP were ignored completely. Considering 3PP, we have obtained η_3 for high temperatures, and then obtained τ_η as $5 \times 10^{-12}/xT$, which is smaller than other estimates. This reduction corresponds to the increase of the absorption rate and the decrease of thermal conductivity and viscosity. Lacking a solution in the intermediate region, we properly interpolate the two limiting regions.

Figure 17.3 represents the total viscosities at various ^{3}He concentrations. All viscosities meet around $T \cong 0.3$ K, and are in good agreement with those of Kuenhold *et al.* [23] obtained through hydrodynamic measurements utilizing capillary flow. The decrease of viscosity at high temperatures is due to the fact that the scattering and absorption rates of phonons by ^{3}He quasiparticles strongly depends on the ^{3}He concentration. Note that as Fig. 17.3 is taken from Ref. [8], it shows the discrepancy at $x = 0.005$. In Ref. [9], this discrepancy was resolved completely by reconsidering the 3PP as described above (see also Fig. 3.5 of Ref. [10]).

However, a complete microscopic calculation of the quasiparticle effective interaction is needed with the inclusion of other processes in connection with phonons and rotons. For the phonon-phonon processes, the temperature-dependent excitation spectrum will give a better result than the zero-temperature spectrum.

The effect of 3PP on the viscosity in dilute ^{3}He-^{4}He mixtures decreases the total viscosity and gives good agreement with experiment by a proper reduction of the ^{3}He viscosity. It is essential that 3PP are considered in order to explain the viscosities at various ^{3}He concentrations consistently.

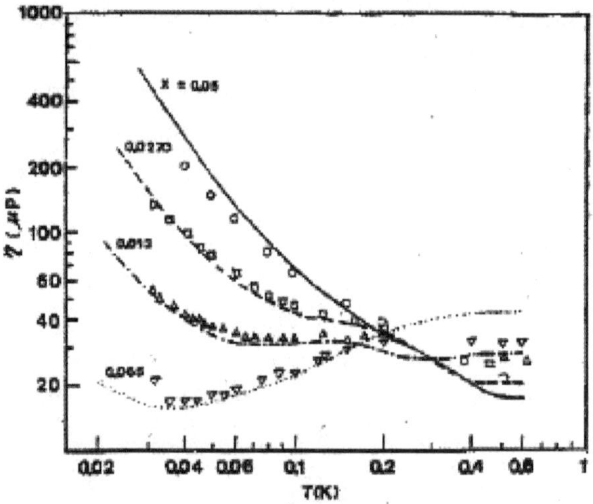

Fig. 17.3. Calculated viscosities (dotted, dash-dotted, dashed, and solid lines) at various concentrations of ^{3}He [8]. The experimental data (Δ, ▽, □ and ○) are taken from Ref. [22]. For $x \geq 0.013$, the results are in good agreement.

17.4.2. *Thermal conductivity*

Now we present the theory for thermal conductivity in the presence of 3PP based on the theory of Callaway [23]. Consider a tube with diameter d containing dilute ^{3}He-^{4}He mixtures. When there is a temperature gradient along the x-axis, heat may be transferred by phonon diffusion, and at higher temperatures by the drifts of phonons and ^{3}He quasiparticles.

The deviation of the phonon distribution from the equilibrium distribution can be expressed by an expression in terms of Legendre polynomials $\mathrm{P_l}(\cos\theta)$ as

$$\delta n_q = q\frac{\partial n_q}{\partial \varepsilon_q}\Sigma_l\,\varphi_l(q)\mathrm{P_l}(\cos\theta). \qquad (17.79)$$

The phonon Boltzmann equation is then written as

$$\frac{s^2}{T}\frac{dT}{dx} = \frac{\varphi_l}{\tau_b} + \frac{\varphi_l + v_3}{\tau_a} + \frac{\varphi_l + v_3}{\tau_1} + \frac{\varphi_l + v_n}{\tau_1{}'} = \frac{\varphi_l}{\tau_{all}} + v_3\left(\frac{1}{\tau_a} + \frac{1}{\tau_1}\right) + \frac{v_n}{\tau_1{}'}. \qquad (17.80)$$

In this equation, the second equality is given since four-phonon relaxation processes are considered, with four corresponding scattering times: boundary scattering (τ_b), phonon absorption by ^{3}He (τ_a), phonon scattering with ^{3}He (τ_1), and 3PP (τ_1'). Here τ_{all} is defined by

$$\frac{1}{\tau_{all}} \equiv \frac{1}{\tau_b} + \frac{1}{\tau_a} + \frac{1}{\tau_1} + \frac{1}{\tau_1{}'}. \qquad (17.81)$$

From Eq. (17.80), we see that the phonons scattered diffusively by the boundary relax to global equilibrium, while the phonons taking part in the phonon-^{3}He interaction relax to a drifting state with velocity v_3. Through 3PP, phonons also relax to a drifting with v_n.

The energy flux density is

$$q_x = \int \varepsilon_q \, \delta n_q s \cos\theta d\vec{\mathbf{q}}. \qquad (17.82)$$

We substitute Eq. (17.79) for δn_q into Eq. (17.82), and due to the orthogonality of P_l, only a term containing P_1 remains. Thus, the energy flux density becomes

$$q_x = \int s^2 q^2 \frac{\partial n_q}{\partial \varepsilon_q} \varphi_l {P_l}^2(\cos\theta) \, d\vec{\mathbf{q}} \ , \qquad (17.83)$$

where we have taken the phonon energy as simply $\varepsilon_q = sq$.

We assume that the coefficient $\varphi_l(q)$ has the form

$$\varphi_l = \tau_{eff} \frac{s^2}{T} \frac{dT}{dx}, \qquad (17.84)$$

where τ_{eff} is the effective scattering time characterizing the thermal conductivity. Then, from Eq. (17.83), the thermal conductivity can be written as

$$\kappa = -\int s^4 q^2 {P_l}^2(\cos\theta) \, \frac{\tau_{eff}(q)}{T} \frac{\partial n_q}{\partial \varepsilon_q} d\vec{\mathbf{q}}. \qquad (17.85)$$

In order to evaluate the thermal conductivity, we must obtain τ_{eff} as a function of known scattering times. To do this, we take v_3 and v_n as

$$v_3 = -\alpha \frac{s^2}{T}\frac{dT}{dx}, v_n = -\beta \frac{s^2}{T}\frac{dT}{dx}, \qquad (17.86)$$

where α and β are constants. Substituting Eq. (17.86) into (17.80), we arrive at

$$\varphi_l = \tau_{all}\left\{1 + \alpha\left(\frac{1}{\tau_a} + \frac{1}{\tau_1}\right) + \frac{\beta}{\tau_1{}'}\right\}\frac{s^2}{T}\frac{dT}{dx}. \qquad (17.87)$$

Comparing Eq. (17.87) with (17.84), we obtain the expression for τ_{eff} as

$$\tau_{eff} = \tau_{all}\left\{1 + \alpha\left(\frac{1}{\tau_a} + \frac{1}{\tau_1}\right) + \frac{\beta}{\tau_1{}'}\right\}. \qquad (17.88)$$

Since 3PP do not change the total momentum, we can write

$$\Sigma_{\vec{\mathbf{q}}}\vec{\mathbf{q}}\, I_{3PP}(\vec{\mathbf{q}}) = 0. \qquad (17.89)$$

From this constraint on the 3PP collision integral, we obtain the relation between v_n and φ_l as

$$v_n = -\frac{\langle \varphi_l/\tau_1{}' \rangle}{\langle 1/\tau_1{}' \rangle}. \qquad (17.90)$$

Then the effective scattering time becomes

$$\tau_{eff} = \tau_{all}\left[1 + \left\{\frac{\alpha}{\beta}\right\}\left(\frac{1}{\tau_a} + \frac{1}{\tau_1}\right) + \frac{1}{\tau_1{}'}\right] \times \left\{\frac{\langle \tau_{all}/\tau_1' \rangle}{\langle 1/\tau_1' \rangle - \langle \tau_{all}/\tau_1'^2 \rangle - \left(\frac{\alpha}{\beta}\right)\langle \tau_{all}(\frac{1}{\tau_a}+\frac{1}{\tau_1})/\tau_1' \rangle}\right\}.$$

This expression contains the ratio of the ^{3}He drift velocity to the phonon drift velocity, α/β. When this ratio is equal to zero, the expression for τ_{eff} is the same as that of Callaway [23].

In this calculation, the boundary scattering rate $\tau_b{}^{-1}$ is taken simply as s/d, where d is the tube diameter. To compare with experimental data [24], we take $d = 26$ cm. The 3PP scattering rate

${\tau_1}'^{-1}(q)$ characterizing thermal conductivity can be obtained the same way as ${\tau_2}'^{-1}$ by using Eq. (17.83). The only difference is that $l = 1$, with τ_{eff}^0 instead of τ_v^0.

In the first step of iteration, the ${\tau_1}'^{-1}(q)$ obtained from Eq. (17.22) have negative values in the high momentum region, which means that the phonon variation goes far away from equilibrium in that region. However, the dominant part of the phonon distribution is within the momentum range with positive collision rates at the temperatures considered. Therefore, we set all the negative values equal to zero except for replacing the first negative one by the value equal to the last positive one. This makes the range of positivity enlarged and reduces the discontinuity in the resulting Φ_1 caused by setting the negative values as zero.

For the successive step we follow the same procedure for the negative part. We find ${\tau_1}'^{-1}(q)$ to be saturated successfully. Moreover, the momentum range of positive values becomes larger until a certain enlarged range which covers almost all the dominant part of the phonon distribution. Even though ${\tau_1}'^{-1} = 0$ is taken as a trial 3PP rate instead of the diagonal elements of the collision matrix, the saturated 3PP rates are identical. This indicates that our iterative procedure is meaningful. We emphasize that the explicit momentum dependence of the 3PP rates can be estimated, and the smaller range of 3PP rates explains the Gaussian factor introduced by Bowley and Sheard [25].

The thermal conductivities for $x = 0.0001$ and $x = 0.0132$ at $P = 0$ atm are shown in Fig. 17.4. According to the experimental situations, it may be possible that the ratio v_3/v_n is not zero, in which case the thermal conductivity becomes higher than that of the case of $\frac{v_3}{v_n} = 0$, and when $\frac{v_3}{v_n} = 1$, only the contribution of boundary scattering survives. It is possible that the discrepancy at higher temperatures for $x = 0.0001$ is related to the situation of nonzero v_3.

The effect of 3PP on the thermal conductivity of dilute mixtures is apparently important when the effective ^3He-^3He interaction is relatively small: small ^3He concentrations and relatively high temperatures. Our theory based on the matrix elements is very successful in fitting the experimental data.

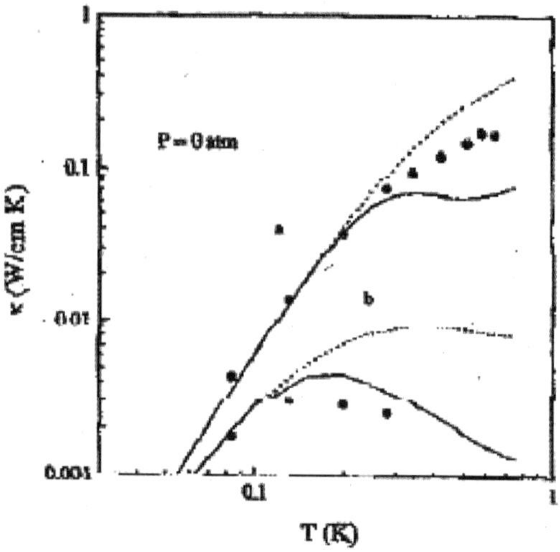

Fig. 17.4. Thermal conductivities for (a) $x = 0.0001$ and (b) $x = 0.0132$ at $P =$ 0 atm, where the solid lines represent the present results, and the dotted lines are those in which 3PP are neglected. The solid circles indicate experimental data [24].

17.5. Summary

In this chapter, we have reviewed the transport coefficients, such as first viscosity and thermal conductivity, of dilute ^{3}He-^{4}He at low temperatures. The 3PP are now known to be dominant scattering mechanisms below 0.6 K, as the phonon spectrum of superfluid ^{4}He is anomalous. We have shown that the inclusion of 3PP is very essential for understanding the transport phenomena of dilute mixtures.

By calculating the 3PP rates in mixtures which are different from the pure ^{4}He case based on the collision matrix and iteration method, we have resolved most existing discrepancies with experiments. Furthermore, it should be noted that our theory, as an extension of Baym and Saam's theory, does not need to adjust the phonon-^{3}He scattering rate and the rate of phonon absorption by ^{3}He. In other words, we have confirmed rigorously that the phonon-^{3}He interaction (BBP [1]) works quantitatively as well as qualitatively. Our results have been discussed in detail in a recent review [26].

For the viscosity, 3PP effects are important for temperatures greater than 0.1 K, and the results show excellent agreements with experiments which have been obtained through hydrodynamic measurements. In the calculation, the value of the ^{3}He viscosity, η_3, has been revised by correcting τ_η of a previous theory, and shown to decrease the viscosities. We also have shown similar agreements and better explanations about the experimental data on thermal conductivity than any previous theory.

However, even though our theory is very successful in fitting the experiments, there are some difficulties in obtaining 3PP scattering rates. Such rates in dilute mixtures must be studied further in order to obtain purely theoretical results without having to take into account the experimental data. In order to study the high temperature case, the temperature-dependent phonon excitation spectra which we have derived should be used instead of the zero-temperature ones, together with the higher-order phonon processes. For relatively high ^{3}He concentrations, a study to find an exact effective ^{3}He-^{4}He interaction in mixtures is needed, since phonon absorption processes by ^{3}He become important.

References

1. J. Bardeen, G. Baym and D. Pines, Interactions between He^3 atoms in dilute solutions of He^3 in superfluid He^4, *Phys. Rev. Lett.* **17**, 372–375 (1966); Effective Interaction of He^4 atoms in dilute solutions of He^3 in He^4 at low termperatures, *Phys. Rev.* **156**, 207–221 (1967).
2. G. Baym and C. Ebner, Phonon-quasiparticle interactions in dilute solutions of He^3 in superfluid He^4. I. Phonon thermal conductivity and ultrasonic attenuation, *Phys. Rev.* **164**, 235–244 (1967).
3. G. Baym and W. F. Saam, Phonon-quasiparticle interactions in dulte solutions of He^3 in superfluid He^4 II. Phonon Boltzmann equation and first viscosity, *Phys. Rev.* **171**, 172–178 (1968).
4. G. Baym, W. F. Saam and C. Ebner, Phonon-quasiparticle interactions in dulute solutions of He^3 in superfluid He^4. III. Attenuation of first sound above 0.2°K, *Phys. Rev.* **173**, 306–313 (1968).
5. H. Maris and W. E. Massey, Phonon dispersion and the propagation of sound in liquid helium-4 below 0.6 K, *Phys. Rev. Lett.* **25**, 220–222 (1970).
6. C. I. Um, W. H. Kahng, K. H. Yeon, S. T. Choh and A. Isihara, Temperature variation of sound velocity in liquid He-II, *Phys. Rev. B* **29**, 5203–5206 (1984); C. I. Um, C. W. Jun, W. H. Kahng and T. F. George, Thermal conductivity and viscosity via phonon-phonon, phonon-roton, and roton-roton scattering in thin ^{4}He films, *Phys.*

Rev. B **38**, 8838–8849 (1988); C. W. Jun, C. I. Um and T. F. George, Coefficients of the second viscosity in thin liquid-helium films, *Phys. Rev. B* **43**, 2748–2755 (1991).

7. A. Isihara, C. I. Um, C. W. Jun, W. H. Kahng and S. T. Choh, Effects of the anomalous energy dispersion on the attenuation coefficient of first sound in liquid ^{4}He, *Phys. Rev. B* **37**, 7348–7351 (1988); C. I. Um, C. W. Jun, W. H. Kahng and T. F. George, Coefficient of first viscosity via three-phonon processes in bulk liquid helium, *Phys. Rev. B* **38**, 8834–8837 (1988).
8. C. I. Um, S. K. Yoo, S. Y. Lee, T. F. George and L. N. Pandey, First viscosity of dilute ^{3}He-^{4}He mixtures below 0.6 K, *J. Low Temp. Phys.* **94**, 145–160 (1994).
9. C. I. Um, S. Y. Lee, S. K. Yoo, T. F. George and L. N. Pandey, Thermal conductivity and viscosity in dilute ^{3}He-^{4}He mixtures below 0.6 K, *J. Low Temp. Phys.* **109**, 495–510 (1997).
10. S. Y. Lee, The effects of three-phonon processes on the transport properties of superfluid ^{4}He and dilute ^{3}He-^{4}He mixtures at low temperatures, Ph.D. thesis, Korea University (1995).
11. L. D. Landau and I. M. Khalatnikov, The theory of viscosity of helium-II. Collisions of elementary excitations in helium-II, *Sov. Phys.-JETP* **19**, 637 (1949).
12. I. M. Khalatnikov, *Sov. Phys.-JETP* **22**, 687; **23**, 8, 21, 169, 253, 265 (1952).
13. R. W. Whitworth, Experiments on the flow of heat in liquid helium below 0.7 K, *Proc. R. Soc. Lond.* A **246**, 390–405 (1958); B. M. Abraham, Y. Eckstein, J. B. Ketterson and M. Kuchnir, Three-phonon process and the propagation of sound in liquid helium-4, *Phys. Rev. Lett.* **19**, 690–692 (1967); B. M. Abraham, Y. Eckstein, J. B. Ketterson, M. Kuchnir and J. Vignos, Sound propagation in liquid ^{4}He, *Phys. Rev.* **181**, 347–373 (1969).
14. H. J. Maris, Attenuation and velocity of sound in superfluid helium, *Phys. Rev. Lett.* **28**, 277–280 (1972).
15. R. C. Dynes and V. Narayanamurti, Evidence for upward of 'anomalous' dispersion in the excitation spectrum of He-II, *Phys. Rev. Lett.* **33**, 1195–1197 (1974); Measurement of anomalous dispersion and the excitation spectrum of He-II, *Phys. Rev. B* **12**, 1720–1730 (1975); W. R. Junker and C. Elbaum, Pressure dependence of the low-momentum phonon dispersion relation in liquid ^{4}He, *Phys. Rev. B* **15**, 162–172 (1977); R. J. Donnelly, J. A. Donnelly and R. N. Hills, Specific heat and dispersion curve for helium-II, *J. Low Temp. Phys.* **44**, 471–489 (1981); W. G. Stirling and K. A. Anderson, Neutron scattering studies of the excitations of liquid ^{4}He, *J. Phys. C: Cond. Matt.* **6A**, 63–70 (1994).
16. D. S. Greywall, Specific heat and phonon dispersion of liquid ^{4}He, *Phys. Rev. B* **18**, 2127–2144 (1978); Erratum: Specific heat and phonon dispersion of liquid ^{4}He, *ibid,* **21**, 1329–1331 (1980); Thermal-conductivity measurements in liquid ^{4}He below 0.7K, *ibid,* **23**, 2152–2168 (1981).
17. H. J .Maris, Hydrodynamics of superfluid helium below 0.6 K. 1. viscosity of the normal fluid, *Phys. Rev. A* **8**, 1980–1987 (1973).

18. I. M. Khalatnikov, *An Introduction to the Theory of Superfluidity* (W. A. Benjamin, New York, 1965).
19. S. K. Yoo, Studies on the ground state properties and transport phenomena of ^{3}He-^{4}He mixtures, Ph.D. thesis, Korea University (1994).
20. A. A. Abrikosov and I. M. Khalatnikov, The theory of a Fermi liquid (the properties of liquid ^{3}He at low temperatures), *Rept. Progr. Phys.* **22**, 329–367 (1959).
21. K. H. Bennemann and I. B. Ketterson, *The Physics of Liquid and Solid Helium* (Wiley, New York, 1976), Part II, Chap. 1.
22. K. A. Kuenhold, D. B. Crum and R. E. Sarwinski, The viscosity of dilute solutions of ^{3}He in ^{4}He at low temperatures, *Phys. Lett. A* **41**, 13–14 (1972).
23. J. Callaway, Model for lattice thermal conductivity at low temperatures, *Phys. Rev.* **113**, 1046–1051 (1959).
24. R. L. Rosenbaum, J. Landau and Y. Eckstein, Temperature, pressure, and concentration dependence of the thermal conductivity of very dilute solutions of ^{3}He in superfluid ^{4}He, *J. Low Temp. Phys.* **16**, 131–143 (1974).
25. R. M. Bowley and F. W. Sheard, The effect of three-phonon processes on the thermal conductivity of ^{3}He-^{4}He mixtures, *J. Phys. C* **18**, L867-L872 (1985).
26. E. R. Dobbs, *Helum Three*, International Series of Monographs on Physics **108** (Oxford Science Publications, Oxford, 2000), pp. 274–295.

Chapter 18

Computational Discovery of Hydrogen Storage Compounds

Eric H. Majzoub
University of Missouri–St. Louis
Majzoube@umsl.edu

Structure prediction of crystalline compounds using unbiased search methods remains a challenging task at the forefront of materials science. An ability to predict ground state structures as well as temperature and pressure dependent polymorphs without experimental input is not generally possible in all classes of materials. However, significant advances have been made in subclasses of materials where simplified Hamiltonians allow rapid searching of the configurational potential energy surface without the need for density functional theory or more complicated electronic structure methods. We review a novel method of structure prediction for ionic compounds consisting of a collection of charge balanced cations and complex anions based on prototype electrostatic ground states (PEGS). This method, used in concert with database search methods, is shown to successfully predict many experimentally observed crystal structures, and complicated reaction pathways for materials used in advanced hydrogen storage compounds.

18.1. Introduction to Metal Hydrides

Hydrogen storage compounds that are reversible near ambient conditions are desirable for a range of applications from transportation to stationary generation and storage of energy. Safe storage of large quantities of hydrogen, ostensibly for vehicular fuel-cell applications, presents a challenge for modern materials science. It is well known that hydrogen may be stored physically as a compressed gas or a condensed liquid.

However, safety concerns surrounding highly compressed gases and the energy required for either compression or liquefaction detract from these strategies. Solid storage of hydrogen has the potential to mitigate these shortcomings and has been an active area of materials research for over four decades. Transportation applications require rather stringent conditions on the volume and weight of the material to be competitive with conventional liquid petroleum fuels [1], resulting in a recent surge in research on so-called complex hydrides that differ significantly from the traditional metallic interstitial hydrides that have been extensively studied. An example of the latter that found its way into commercial applications for both hydrogen delivery and also electrochemical battery applications is the AB_5 compound $LaNi_5$, discovered by researchers at Philips Eindhoven Labs in the late 1960's [2]. A thorough review of hydrogen storage materials development through the mid 1990's can be found in Sandrock [3].

The discovery of new materials, most notably the reversibility of $NaAlH_4$ [4], combined with improved computational methods, has reinvigorated research in metal hydrides. This chapter will introduce a novel computational method for the discovery and study of the thermodynamic behavior of these materials. Following a brief overview of the classification of metal hydrides, this chapter will focus on recent developments in computational approaches to studying the complex ionic hydrides.

18.1.1. *Classification of metal-hydrogen compounds*

The most simple metal-hydrogen reaction may be written as

$$MH_p \rightarrow M + \frac{p}{2}H_2 \tag{18.1}$$

representing the desorption of hydrogen from the metal hydride MH_p to gaseous H_2 and the metal M. Metal hydrides generally possess much larger hydrogen specific density (e.g. 11 $Å^3$/H atom in TiH_2) than either gas (45 $Å^3/H_2$) or even liquid phase (38 $Å^3/H_2$).

Metal-hydrogen compounds fall into three general categories: *metallic interstitial*, *ionic*, and the so-called *complex hydrides*. Examples of metallic interstitial hydrides include many of the first-row transition metals such as Ti, Zr, and V, which form binary di-hydride compounds MH_2. This class of hydrides also includes intermetallic alloys in subcategories generally labeled AB, AB_2, and AB_5, where A and B represent different metal species. Metallic interstitial hydrides are distinguished by the 'tunability' of their thermodynamic properties through alloying on either the A or B sites. For example, substitutions of Ca for Li, or Al for Ni in $LaNi_5$ allow for tuning the equilibrium hydrogen pressure over a range of 0.5–20 bar [3]. Excepting some rare earth elements that undergo a metal–insulator transition on hydriding [5], many interstitial hydrides remain metallic. The hydrogen atoms release their electron to join the band structure of the metal. The hopping rate of the proton is generally quite high between interstitial sites and may be measured using nuclear magnetic resonance techniques [6]. In addition, the metal atoms may retain the same crystal structure in both the absorbed and desorbed state, and the material may naively be considered a hydrogen sponge. Well-studied and engineered, these materials possess easily tunable thermodynamics in the range of 1 bar H_2 equilibrium pressure near room temperature. However, the relative weight percent of hydrogen in these compounds is rather low due to the large weight ratio of transition metals to hydrogen.

The *ionic*, or saline, hydrides consist of the alkali and alkaline earth metals that form mono- and di-hydrides respectively. Common examples include NaH, and MgH_2. As the name suggests, bonding in these compounds is strongly ionic, and the group I hydrides take the NaCl structure. The ionic hydrides have desorption temperatures well above the operating temperature of a proton exchange membrane (PEM) fuel-cell. Sodium and magnesium hydrides obtain an equilibrium pressure of 1 bar H_2 at temperatures of 400°C and 300°C, respectively. While MgH_2 storage systems have been developed for commercial transportation applications [7], it is desirable to identify lower temperature materials.

The *complex hydrides* may be ternary, quaternary, or higher compounds that possess a more complicated bonding structure, with hydrogen atoms covalently bound to one or more elements, generally boron, aluminum, or nitrogen, forming an anion which is stabilized by the charge transfer from an alkali, alkaline-earth, or even transition-metal cation. They may have very large hydrogen weight ratios (e.g. $LiBH_4$ is about 18 wt.% hydrogen). Common examples include the alanates, (e.g. $NaAlH_4$, $LiAlH_4$, $KAlH_4$, Na_3AlH_6, Li_3AlH_6), bi-alkali alanates, (e.g. KLi_2AlH_6), borohydrides (e.g. $LiBH_4$, $NaBH_4$), and nitrogen containing compounds, (e.g. $LiNH_2$, Li_2NH, MgNH). Moreover, as the bi-alkali alanates suggest, the cations and anions in these compounds may, to some extent, be exchangeable, forming more complicated compounds such as $Li_2BN_2H_8$ (Li_2 BH_4 $(NH_2)_2$), and $Li_3BN_3H_{10}$ (Li_3 BH_4 $(NH_2)_3$), and $MgLi(NH_2)_2$. The discovery of low-temperatrue reversibility in $NaAlH_4$ [4] indicates that these materials may be much more attractive than the simple ionic hydrides. The combinatoric array of potential compounds formed by the space of cations and anions and the possibility of discovering a material with "acceptable" thermodynamic properties provides the impetus for current research in this field.

18.1.2. *Materials properties relevant to engineering*

The equilibrium hydrogen pressure for the decomposition reaction in Eq. (18.1) is the most important physical property of a metal hydride, and generally determines the practical application. Formation of the plateau is due to Gibbs' phase rule. It is temperature dependent, and described by the so-called Van't Hoff equation

$$R\ln\left(\frac{P_1}{P_0}\right)=\Delta H\left(\frac{1}{T_0}-\frac{1}{T_1}\right) \tag{18.2}$$

where R is the gas constant, P is the pressure, T is the temperature in Kelvin, and H is the enthalpy. A pedagogical statistical model that produces a pressure plateau may be obtained by considering the configurational entropy of n hydrogen atoms placed randomly at N interstitial positions in a "host" lattice. If formation of the hydride is exothermic, then each H ion has an energy -ε, and the free energy of the crystal may be written as $F_{solid}=-n\varepsilon-k_BTS$. A pressure plateau is obtained by setting the chemical potential of the solid equal to the chemical potential in the gas, which is proportional to the logarithm of the pressure. Experimentally, the enthalpy of the reaction is most accurately obtained by measuring the equilibrium pressure in the middle of the plateau at several different temperatures on desorption. Ideally the plateau would be flat in the two-phase region where there is a mixture of M and MH_p according to Eq. (18.1). However in the presence of lattice strain, multiple non-equivalent hydrogen sites, and other such effects, the energy becomes a function of the concentration, and the plateau is not precisely flat. In interstitial metallic hydrides, the energy of a proton may be described as function of the electron density at the interstitial site. Norskov, using effective medium theory, has shown that the embedding energy of a proton in a homogeneous electron gas has a minimum at a specific electron density [8]. The thermodynamics are therefore a function of the electronic structure of the compound for a given stoichiometry, explaining the relative ease in tuning thermodynamic properties of alloy intermetallic hydrides. In contrast, the polar-covalent hydrogen bonding that exists in the complex hydrides makes this simple approach to tuning their thermodynamic properties untenable.

Upper and lower bounds on enthalpies for practical materials may be taken from two materials under consideration for vehicular storage. These include the previously mentioned MgH_2, and aluminum hydride AlH_3, with dissociation enthalpies of approximately 75 and -8 kJ/mol H_2, respectively, indicating that MgH_2 is stable with an endothermic desorption reaction, while AlH_3 is unstable and should spontaneously decompose in the absence of kinetic barriers [9]. Lower enthalpies reduce the amount of heat extraction required from a hydride bed during refueling. PEM fuel cell operating temperature constraints place limitations on the accessible temperatures and pressures, and may be translated into constraints on the desired values of ΔH. Requiring hydrogen release at ambient pressures below 80°C leads to an upper enthalpy limit of about 45 kJ/(mol H_2). Requiring rehydrogenation at ambient temperatures and pressures of a few hundred bar or lower gives the lower enthalpy limit of about 20 kJ/(mol H_2). The computational methods described below allow one to rapidly determine approximate reaction enthalpies, within the physical accuracy provided by density functional theory (about +/- 10 kJ/mol H_2), of many complex hydrides.

18.2. Electronic Structure and Bonding in Complex Hydrides

Rapid crystal structure searching, using a simplified Hamiltonian to evaluate the potential energy surface, requires knowledge of the basic crystal structure building blocks. The conventional description of prototypical crystalline $NaAlH_4$, and Na_3AlH_6 is that of an ionic crystal composed of Na^+ cations, and AlH_4^- or AlH_6^{3-} anions. The tetrahedral structure of the AlH_4^- group suggests that the Al–H bond is formed between sp3 hybridized orbitals on Al and s orbitals on H atoms. Indeed, the valence band structure of $NaAlH_4$ arises from an overlap of molecular-type orbitals centered on the AlH_4^- units, while the sodium donates an electron to the AlH_4^- group and provides electrostatic stabilization of the lattice. A detailed study of the electronic structure of the constituent anions, and the relation to the crystalline band structure of $NaAlH_4$ and Na_3AlH_6 can be found elsewhere [10].

The molecular crystal description of complex hydrides has also been established through experimental measurements of the vibrational properties of these compounds. A comprehensive Raman study of borohydrides indicates the stability of BH_4^- anions and identifies many of their vibrational properties [11]. Polarized Raman scattering on single crystals of $NaAlH_4$ has been used to determine the symmetry properties and frequencies of the Raman-active vibrational modes up to the melting point of 180°C [12]. Softening of the vibrational frequencies of up to 6% is observed in the modes involving rigid translations of Na^+ cations and translations and librations of AlH_4^- anions up to 180°C. However, mode softening of less than 1.5% occurs for the Al–H stretching and bending modes of the AlH_4^- anion. These experimental results illustrate that the AlH_4^- anion remains a stable structural unit even near the melting point. The simplifying structural motif for this and many other complex hydrides is therefore a rigid anion, where the internal electronic structure may be ignored, and a simple electrostatic model for the ion charges is introduced.

18.3. Structure Prediction Using Prototype Electrostatic Ground States (PEGS)

A very successful approach to determining approximate reaction energetics and even discovering new complex hydride compounds is found in a Monte Carlo minimization of the configurational electrostatic energy of crystal structures. The method described here provides "prototype electrostatic ground states" (PEGS) and relies on a simplified Hamiltonian that captures the electrostatic interactions between ions in complex hydrides and dominates the cohesive energy of these compounds [13]. We introduce the method following a brief discussion of alloying in ionic materials that illustrates the need to search for stoichiometric compounds.

Tuning the reaction enthalpy, i.e., the H_2 equilibrium pressure, in the complex hydrides through alloying is difficult due to the strongly ionic character of the cohesive energy. An example is provided by the substitution of fluorine or chlorine atoms in NaH. The ionic radii of anions in the sodium salts are 1.48, 1.35, and 1.85, for H, F, and Cl, respectively. One may substitute F or Cl for the H ion in NaH, by mixing with the appropriate halide (e.g. $TiCl_3$, or TiF_3). X-ray diffraction studies on mechanically-milled materials indicate that F can substitute for H in NaH, continuously shifting the lattice parameter from that of NaH to NaF, while substituting Cl results in a nearly immediate phase separation of NaH and NaCl [14]. This indicates that the lattice will not tolerate changes in ionic radii leading to significant strain. Attempts to continuously mix cations of different size or charge to tailor the total energies of complex hydrides, and adjust the desorption enthalpies will likely be problematic. In many cases, new stoichiometric compounds form as molar ratios take on integer values. For example, one may consider compounds such as $(M1)_3(M2)_{3-x}(AlH_6)$, where A and B are monovalent alkai metals. Experimental attempts to prepare continuous compositions in these compounds results only in the formation of line compounds such as K_2LiAlH_6 [15]. The bialkali alanate K_2LiAlH_6 crystallizes in space group R-3m (no. 166) with unit cell parameters a = 5.62 and c = 27.40 Å, and is isostructural with the high-temperature polymorph of K_2LiAlF_6. One may attempt to synthesize this compound with the continuous composition $K_{3-x}Li_xAlH_6$, for $0<x<3$. However, only the stoichiometric compound with x=1 is observed in experiments. The change in total energy due to cation mixing is easily calculated using first-principles for the mixing of Li and K sites to estimate the energy penalty. The increase in electronic energy from swapping one pair of ions in the unit cell is nearly seven times larger than the lowering obtained through the entropy of mixing. This suggests that searching for compound crystal structures in the complex hydrides is safely confined to stoichiometric compounds.

The PEGS strategy considers searching for stable compounds within the configuration space of all combinations of mono- and di-valent alkali and alkaline earth metals with charge balanced anions of AlH_4^-, AlH_6^{3-}, or BH_4^-, or the nitrogen containing ions NH_2^-, and NH^{2-}.

18.3.1. *Database searching method*

A commonly employed complimentary method of determining low energy structures for use in theoretical investigations involves the use of a database such as the Inorganic Crystal Structure Database (ICSD), containing over 93 000 inorganic crystal structures [16]. Operationally, one searches the database for potential structures of chemically related compounds and calculates the *ab initio* total energies for the compound of interest in those structures. The structure obtaining the lowest energy is presumed to be the ground state. The strength of any database search method lies in the total number of potential structures available for a chosen stoichiometry or compound formula. For example, sodium alanate, $NaAlH_4$, would be searched for in the ICSD database using the formula string "*ABX*4," yielding over 1800 entries, and over 100 independent potential crystal structures. Database searching is successful in many cases and produces structures that agree with experimental determinations in, for example, $Ca(AlH_4)_2$ [17] and $Ca(BH_4)_2$ [18], where the compound type "*AB2X8*" yields 270 entries and at least 50 unique structures. If the number of structures in which the *ab initio* total energy is calculated is small, the probability of finding the correct structure is obviously significantly lowered. The need for methods other than database searching is easily illustrated by the much smaller number of trial structures for database entries of increasingly complicated compound stoichiometries such as "*A2BCX6*".

18.3.2. *The PEGS method and algorithm*

Complex ionic hydrides share several common features in their electronic structure and bonding, which constitute the physical basis for the PEGS approach. As discussed above, these compounds are typically wide gap insulators, with strongly bound anionic complexes. Second, experimental and theoretical structural studies on known compounds indicate that the anion complexes have very similar shapes and charge states in all low-energy crystal structures. Finally, Raman studies discussed above suggest that candidate ground states for complex hydrides may be found by minimizing the electrostatic energy of a collection of positively charged cations and negatively charged complex anions, with the latter treated as rigid units. This idea is analogous to the classical approach used in the classification of ionic crystal structure types based on the ionic charges and radii of the constituents. The PEGS method applies a modern computational approach, in the spirit of Pauling's rules [19], to the structure of complex ionic hydrides. New structure types are generated by minimizing the electrostatic energy subject to constraints on inter-ionic distances imposed by the ionic radii. These ground states are identified as functions of the parameters describing the ionic radii and charge distribution in the system. The resulting PEGS are used as input to first-principles density functional theory (DFT) calculations to obtain accurate total energies and structural coordinates. The energy functional used in our Monte Carlo approach is given in Eq. (18.3) below. For each complex anion, all center to hydrogen distances and all bond angles are kept fixed. Ionic repulsion at short distances is introduced via a soft sphere interaction potential with fixed values of ionic radii for all species in the system.

$$H_{\text{PEGS}} = \sum_{i<j} \left(\frac{Z_i Z_j}{r_{ij}} + \frac{1}{r_{ij}^{12}} \right). \tag{18.3}$$

The first term represents the Coulomb energy, where Z_i denote the ionic charges. The second term represents a repulsive, soft-sphere potential, taken from the Lennard-Jones 6–12 potential, which *only contributes when ion cores overlap*. Both sums extend over all ions in the system, including the center and vertex ions comprising the rigid complex anions but excluding ion pairs from the same complex. The input parameters for the Metropolis Monte Carlo (MMC) algorithm include the radii of the cations, vertex anions, and their charges. *We emphasize that the Hamiltonian in Eq. (18.3) is not meant to accurately represent all the details of the interionic interactions in complex hydrides.* The PEGS method identifies new prototype structures with exceptionally low electrostatic energies. The detailed structural parameters and energetics are obtained using state-of-the-art DFT methods. There are four important configuration changes employed in the PEGS algorithm. Detailed balance is maintained at each step to ensure the *a priori* probability for the system to return to a previous configuration.

1. Lattice parameter changes. Both cell shape and volume are allowed to change.
2. Cation or anion translations throughout the unit cell.
3. Anion rotations.
4. Swapping of anion and cation locations.

An exponential annealing schedule is employed for a total number of prescribed temperatures, *N*. The temperature *T* is calculated for given initial and final temperatures taken from an auto-adjusting algorithm described below. From the partition function in the canonical ensemble, at constant particle number *N*, volume *V*, and temperature *T*, one finds that the specific heat in terms of the energy fluctuations is given by

$$C_{NVT} = \frac{\langle U^2 \rangle - \langle U \rangle^2}{k_B T^2} \quad . \qquad (18.4)$$

We hold temperature and particle number constant in our Metropolis algorithm, allowing the volume to vary, and calculate a *pseudospecific* heat using Eq. (18.4). Temperature ranges corresponding to the liquid melt, where the system is condensed and the configurations are meaningful, were determined by investigating the mean square energy fluctuations versus temperature. We set the initial temperature at the onset of condensation, and the final temperature near the peak in the energy fluctuations. The onset of condensation corresponds to lattice parameter and translation rejection rates in the range 67.5–72.5%. The peak in energy fluctuations occurs for rejection rates of between

95% and 99.9%. Once the simulated anneal has been completed, the system is further relaxed using the downhill simplex method, and removes any remaining overlap. We found that potential energy smoothing produced dramatically better results over the standard Metropolis algorithm. We employed distance scaling method (DSM) [20] potential energy smoothing (PES). A scaling of the atom-atom distances $r'=\alpha r$, reduces the repulsive energy for $\alpha>1$, allowing atoms to overlap while the configuration changes rearrange the cations and anions subject to a reduced electrostatic interaction. The DSM method results in dramatic improvement in the structures generated as illustrated in Fig. 18.1Error! Reference source not found., which shows the initial and final (relaxed) DFT total energies of structures generated for trials of the compound $Ca(BH_4)_2$ through single run MC and DSM-MC approaches. PEGS structures generated through DSM-MC are subsequently relaxed using the DFT code VASP [21–23], yielding optimized cell volume, shape, and ionic positions. The structures are then searched for symmetry using the FINDSYM program in the ISOTROPY package [24].

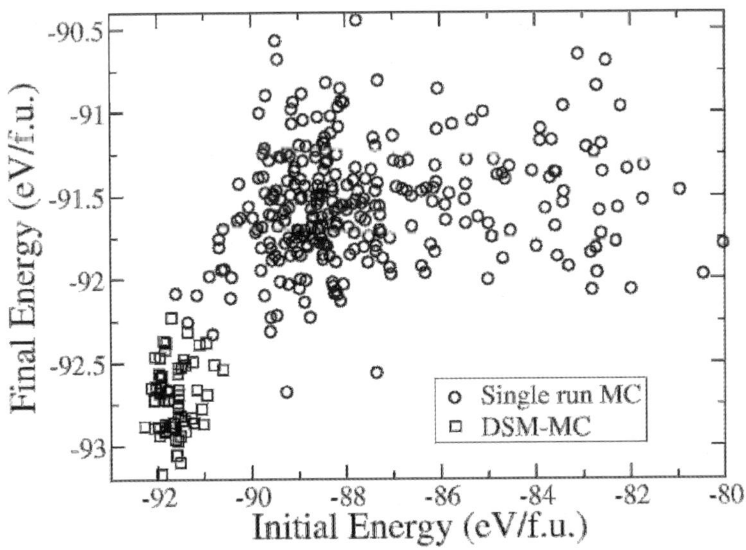

Fig. 18.1. Initial and final DFT energies for MC generated structures of 2 formula units of $Ca(BH_4)_2$. The circles correspond to single MC runs. Squares show distance scaling method DSM results. Initial DFT total energies are significantly lower using DSM, indicating superior structure generation over single run MC.

The anion coordination and bond lengths are taken from experimental diffraction data or first-principles calculations on known materials. For example, the Al–H bond lengths in four- and six-coordinated alanates are taken as 1.65 and 1.78 Å, respectively [10]. Ionic charges are obtained from first-principles calculated Born effective charges by averaging the diagonal elements for all symmetry-distinct ion types. Nominal ionic radii of the cations may be taken from standard tables for ions in their inert gas configuration [25]. Optimal hydrogen radii may be determined by examination of the *M*–H minimum distances in initial DFT-relaxed PEGS structures, where *M* is *not* the anion center atom.

18.4. Application of the PEGS Method to Complex Hydrides

18.4.1. *Confirmation of the structure of bialkali alanate K_2LiAlH_6*

In addition to structure prototypes, the PEGS method also provides complimentary confirmation that a potential crystal structure is correct, either by reproducing the experimentally determined structure, or providing prototypes with energies and interatomic distances (post-DFT relaxation) that agree with experiment. For example, the experimentally determined crystal structure of K_2LiAlH_6 in symmetry R-3m is unusual, with a c-axis of nearly 30Å, and a large aspect ratio (*c*/*a* ~ 5) [15]. In addition, the Al–H distances from the Rietveld refinement on powder X-ray data appear somewhat short due to the difficulty in locating light atom positions with X-ray data, where the atomic scattering power is proportional to Z^2. The refined Al–H distances of 1.31–1.47Å are short in comparison to other alanates, however *ab initio* relaxation of the hydrogen and metal atom positions indicate Al–H–bonding distances of 1.77–1.79Å, in agreement with the experimental and first-principles relaxed values in the Na–Al–H system for the $[AlH_6]^{3-}$ anion [10]. A PEGS search performed with two formula units using Li^+ and K^+ radii of 0.68 and 1.33Å, respectively, reproduces the R-3m K_2LiAlF_6 structure immediately, surprising considering the large aspect ratio of the unit cell. Conducting a database search for this compound produces two structures from the ICSD database that share the same lowest total energy. The first is Cs_2NaAlF_6, with symmetry *C2/m*, with four formula units (f.u.) per cell. The second structure, K_2LiAlF_6, is the high-temperature fluorite structure with symmetry R-3m containing six formula units. The energies of the remaining structures from the ICSD database search are all at least 9 kJ/mol f.u. higher in energy than the structures types of K_2LiAlF_6 and Cs_2NaAlF_6. The two models with lowest energy are assumed to be the most probable stable structures, ignoring entropy contributions. Given structures of equal energy, the unit cell with the highest symmetry is usually chosen, and the calculated Rietveld structural residuals are much lower for K_2LiAlF_6 prototype in comparison to Cs_2NaAlF_6, indicating the correct symmetry is indeed R-3m, in agreement with the PEGS and experimental structures. Had the fluorite structure not been in the database, this approach would have failed to find the correct crystal structure, in contrast to the PEGS search.

18.4.2. *The metastable bialkali borohydride $NaK(BH_4)_2$*

Structure searching with PEGS produces crystal structures that obtain an upper bound on the cohesive energy of a potential structure, i.e. there may be a lower energy structure missed due to deficiencies in the search algorithm or shortcomings of the simplified Hamiltonian. First-principles calculations utilizing PEGS prototypes for $NaK(BH_4)_2$ indicate that the decomposition reaction $NaK(BH_4)_2 \rightarrow NaBH_4 + KBH_4$ is only slightly metastable at T=0K (ignoring entropy contributions). Synthesis of this compound is possible, and it is indeed metastable as shown experimentally [26]. The PEGS search of $NaK(BH_4)_2$ prototypes produced several high symmetry crystal structures, three of which were not found in the ICSD database. The four highest symmetry crystal structures from the PEGS search for $NaK(BH_4)_2$ were, from lowest to highest symmetry, R3, R-3, P3m1, and R-3m. All structures are rhombohedral, belonging to the same basic lattice type. The total DFT energies, at T=0K, of the four structures were ordered monotonically with the crystal symmetry, with the highest symmetry R-3m having the lowest energy. The P3m1, R-3, and R3 structures obtained a T=0 K total energy of 7.6, 8, and 11 kJ/mol per formula unit above the R-3m structure, respectively. Statistical residuals via a Rietveld analysis of from powder X-ray data were used to compare the PEGS-generated structural models.

The difference plot of space group R3, shown in Fig. 18.2, gave the best fit between the observed and calculated patterns, judging from the considerably lower R_B-value and better fit between observed and calculated diffraction patterns. The predicted theoretical structure in space group R3 is most likely the experimental structure of $NaK(BH_4)_2$ based on the diffraction data obtainable as the sample spontaneously decomposes.

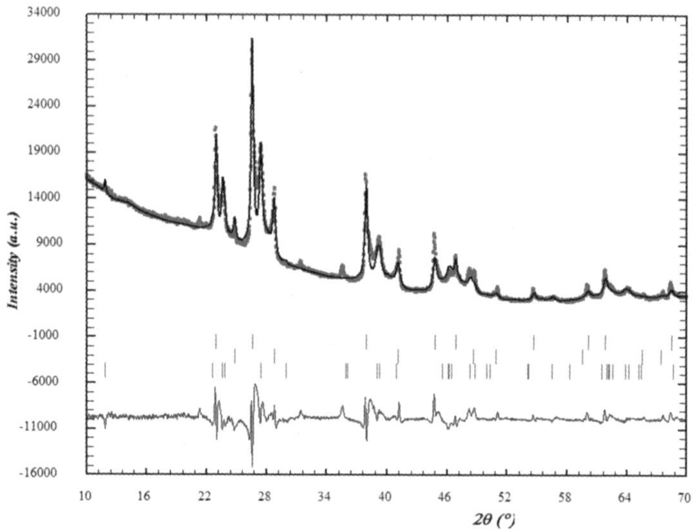

Fig. 18.2. Difference plot from Rietveld refinements of the structure for $NaK(BH_4)_2$ in space group R3. Phase markers are (from top to bottom): KBH_4, $NaBH_4$ and $NaK(BH_4)_2$.

18.4.3. *Magnesium borohydride $Mg(BH_4)_2$, alakai and alkaline earth $[B_{12}H_{12}]^{2-}$ closoborane salts and their impact on hydrogen storage in borohydrides*

The experimentally determined crystal structures for $Mg(BH_4)_2$ indicate a surprisingly complex unit cell in symmetry $P6_1$, with the low temperature structure containing 330 atoms in the primitive cell [27–29]. Such large, complicated structures raise the question of producing smaller unit cell structures with comparable total energy for computationally efficient thermodynamic assessments. The PEGS method produces such a four formula unit structure in symmetry I-4m2 that obtains a lower total energy than competing structures generated through database searching, and simulated annealing *ab initio* molecular dynamics [30]. The I-4m2 structure is also, surprisingly, lower in total energy than the experimentally observed $P6_1$, raising questions about crystal growth mechanisms when synthesized using wet-chemical methods, and has spurred further investigation into this compound [31,32].

More importantly, for hydrogen storage applications recent experimental evidence suggests that the formation of closo-borane compounds is ubiquitous in the decomposition of many borohydrides (Hwang *et al.*, 2008) [33]. The most stable closoborane salts possess the $[B_{12}H_{12}]^{2-}$ icosahedral anion. Alkali, and alkaline earth salts appear to form on decomposition of many of the corresponding borohydrides as evidenced by several magnetic resonance studies. Solid crystal structures for many of these salts were recently unknown. PEGS prototypes for these salts combined with first-principles calculations indicate that hydrogen release from $LiBH_4$ and $Mg(BH_4)_2$ is predicted to proceed via intermediate $Li_2B_{12}H_{12}$ and $MgB_{12}H_{12}$ phases, while for Ca borohydride two competing reaction pathways (into CaB_6 and CaH_2, and into $CaB_{12}H_{12}$ and CaH_2) are found to have nearly equal free energies [34]. The result is a reduction in the hydrogen storage capacity of, for example, $Mg(BH_4)_2$ of 7 wt.% with respect to the ideal decomposition reaction into MgB_2 and hydrogen. This study definitively illustrated that mitigation of the formation of the closoborane salts, or other complicated B–H species is necessary if this material is to be used as a practical storage medium.

18.4.4. *Polymorphism in calcium borohydride*

Many complex hydrides exhibit polymorphism as a function of temperature or pressure. For example, the low temperature orthorhombic *Pnma* structure of $LiBH_4$ transforms to hexagonal $P6_3mc$ at 408K [35]. At temperatures between 300–400°C, the structure transforms again to orthorhombic (Mosegaard *et al.*, 2007) [36]. Other complex hydrides behave similarly. The first structure reported for $Ca(BH_4)_2$ was in fact a theoretical determination using the database search method and found the cubic (α-phase) Fddd to have the lowest total energy [18]. However, more recent *in situ* synchrotron diffraction experiments indicate three additional $Ca(BH_4)_2$ phases as a function of temperature [37,38], indicating $Ca(BH_4)_2$ is also polymorphic, and takes on several related crystal structures as a function of temperature and pressure. An understanding of the dominant crystal structures is useful in theoretical calculations of bulk properties, surface and interface energies, and other quantities necessary for both engineering and materials properties that influence the thermodynamic pathways of desorption, and the kinetics of hydrogen sorption in this material.

Solid-state synthesis of calcium borohydride was first proposed and prepared using the hexaboride as a precursor in the reaction $CaB_6 + 2\ CaH_2 + 10\ H_2 \rightarrow 3\ Ca(BH_4)_2$. The first known solid-state synthesis required a pressure of 700 bar H_2, and temperatures between 400–450°C [39]. Wet-chemical processes may also be used to prepare $Ca(BH_4)_2$, and synchrotron experiments [37] have reported structure results for $Ca(BH_4)_2$, heated in the temperature range from 40°C – 500°C. These samples were prepared using wet-chemical synthesis technique for $Mg(BH_4)_2$ [29]. The β-phase is reported indexed on a tetragonal unit cell with lattice constants a=6.91Å, c=4.34Å, while the γ-phase is indexed on an orthorhombic cell with lattice constants a=13.10Å, b=7.52Å, and c=8.40Å. With increasing temperature, the γ-phase gradually transforms to β-phase up to 290°C. Between 290°C – 330°C, the γ-phase continues to transform, not to the β-phase, but a new phase denoted δ. Additional structure work by Buchter and coworkers [38] reports a structure solution for β-phase in space group $P4_2/m$ (#84), using a combination of Rietveld refinement with the structure solving program FOX [40]. These authors also report the γ-phase symmetry to be Pbca, but were unable to determine atomic positions.

Filinchuk and coworkers performed *in situ* synchrotron diffraction data in the temperature range 32–600°C [41]. Three structure candidates in this temperature range with symmetries F2dd, I-42d, and P-4, were suggested for the observed phases and labeled α, α' and β, respectively. The β-phase structure in P-4 was originally proposed as $P4_2nm$, and was updated to P-4, based on the results obtained in the work of Ref. [42]. The crystal structure candidates for $Ca(BH_4)_2$ predicted by database structure searching together with PEGS reveal several of the observed structures and also previously unknown candidates. The crystal structures were investigated with first-principles DFT to ensure lattice vibrations indicated dynamically stable structures. Two of these structures are clearly observed in X-ray diffraction experiments and correspond to the low temperature α-phase and high temperature β-phase. A surprising potential ground state structure for the α-phase in space group C2/c was shown to be isoenergetic with the structure in Fddd near T=0K, with larger entropy and lower total free energy at all temperatures. A complimentary database search [42] resulted in a structure for the γ-phase, which is in agreement with previous synchrotron diffraction data [37], and the Pbca symmetry proposed by Buchter *et al.* [38], which has calculated diffraction peaks in agreement with their synchrotron data.

18.4.5. *PEGS and database structures for $Ca(BH_4)_2$*

The lowest energy structure type (α-phase), found through an ICSD search is $BaMn_2O_8$–type in space group #70 (Fddd) [18]. The PEGS method finds an isoenergetic structure in the lower symmetry space group #15 (C2/c). Both Fddd and C2/c space groups can be identified from the same PEGS structure. An ICSD search also produces an identical low energy C2/c ($HfMo_2O_8$-type) candidate. The $HfMo_2O_8$-type C2/c structure generates identical diffraction to Fddd, is found to be isoenergetic with Fddd and is also dynamically stable, providing an alternative to the accepted Fddd ground state. At energies of 1.8, 1.9, and 2.1 kJ/mol f.u. above the ground state PEGS finds several structures that do not produce diffraction in agreement with experiment, and these structures are discarded. However, at 7.9, and 12.0 kJ/mol f.u. above the ground state, PEGS produces structures in space groups #12 (C2/m), and #81 (P-4), respectively. This structure may also be identified as space group #84 ($P4_2/m$). It has been shown [42] that the P-4 structure produces a better fit with the experimental diffraction spectra. We refer to the P-4 structure below interchangeably with the β-phase of $Ca(BH_4)_2$. Finally, at 14.5 kJ/mol f.u. above the ground state, PEGS find a structure in space group #61 with symmetry Pbca ($BaFe_2Br_8$-type). The lattice parameters and three lowest angle peaks appear to agree well with those previously reported [37,38]. Buchter and coworkers also identify the structure as having symmetry Pbca.

Ball and stick models of the PEGS predicted P-4, and C2/c structures are shown in Fig.18.1. In the P-4 structure, each Ca^{2+} cation is six-fold coordinated with four H ions at 2.35–2.36 Å forming an elongated rectangular plane with the Ca^{2+} ion near the geometric center, and the remaining two H ions at a distance of 2.29 Å perpendicular to this plane. In contrast the Ca-H coordination is eightfold in the Fddd structure with bond distances from 2.32 to 2.43 Å. The calculated density of the P-4 structure (1.1 g/cc) is slightly larger than Fddd (1.07 g/cc), despite the lower Ca-H coordination. The P-4 structure for the β-phase has first-principles-relaxed B–H bond lengths of 1.22Å, in agreement with many literature values for B–H bond lengths in BH_4 tetrahedra [43]. The C2/c structure is nearly identical in local order to Fddd; the Ca-H coordination is the same as Fddd, with neighbor distances in agreement within 0.05 Å out to a distance of 5 Å. Rietveld refinement residuals for *P*-4 are slightly lower than for $P4_2/m$ in Ref. [42], indicating a preference for describing the crystal structure in P-4.

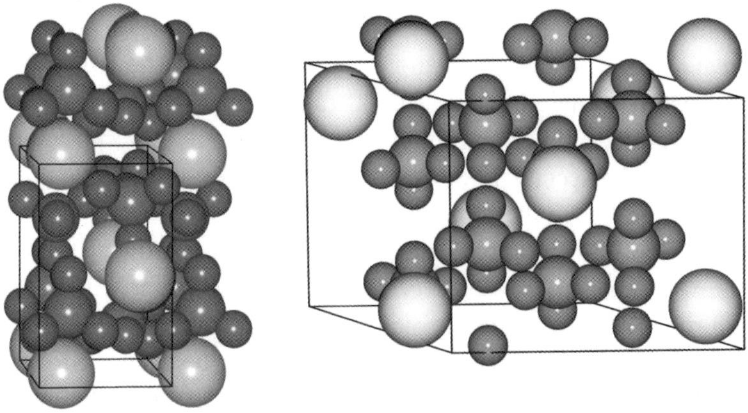

Fig. 18.3. Crystal structure of (a) β-phase $Ca(BH_4)_2$ in symmetry P-4, and (b) the ground-state structure with symmetry C2/c, which is isoenergetic with Fddd. Green boron atoms are tetrahedrally coordinated with brown hydrogen atoms in each structure. The boxes outline the conventional cell in each structure.

18.4.6. *The α- to β-phase transition in $Ca(BH_4)_2$*

The transition from α to β-phase can be understood from calculations of the total free energy of these phases as a function of temperature in the harmonic approximation using the method of frozen phonons. Calculations were performed using supercell sizes of 44 (4 f.u.) and 88 (8 f.u.) atoms for the Fddd, C2/c, and P-4 structures, and 88 (8 f.u.) atoms for Pbca. The total free energies as a function of temperature are shown in Fig. 18.4, and indicate that the larger entropy of P-4 will result in a transformation to this phase at about 350K with respect to Fddd, and at 600K with respect to C2/c. The transformation from the α to β-phase is in agreement with the experimental observations described above, while noting that the calculated transition temperature is highly sensitive to finite temperature effects, which cannot be handled to high accuracy using only the harmonic approximation.

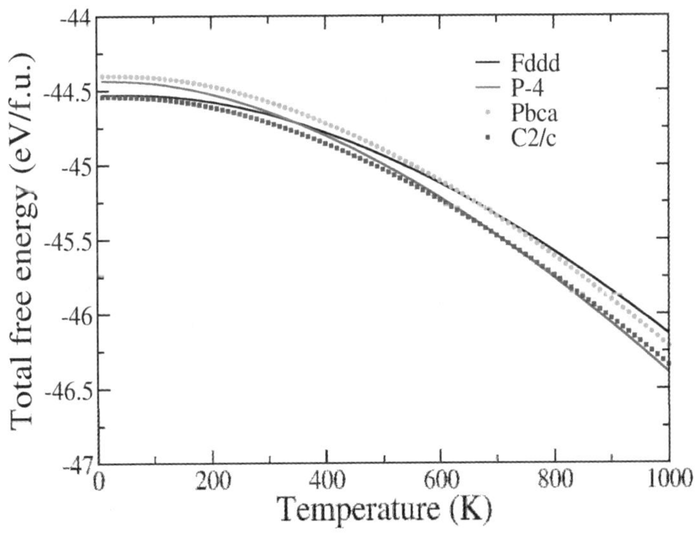

Fig. 18.4. Total free energy for four $Ca(BH_4)_2$ structures as a function of temperature in the harmonic approximation. The two lowest energy structures near 0K are Fddd (solid black) and C2/c (blue squares). The C2/c structure at T=0K is lower than Fddd in total free energy and is predicted to be the ground state. All structures are eventually lower in total free energy than Fddd, while P-4 (solid red line) attains the lowest energy above all structures at around 600 K.

18.5. Conclusions

This chapter has highlighted a few of the many successes of the application of a simple electrostatic model to the structure and thermodynamics of complex ionic hydrides. Monte Carlo simulated annealing, combined with potential energy smoothing, reproduces ground state crystal structures in complicated multi-component metal hydride systems including alanates, borohydrides, and even nitrogen containing amide and imide compounds that were not addressed here. The importance of this success lies in the fact that many technologically relevant materials are found in complex molecular crystals with strongly ionic character. Structures in this class of materials may be explored without the need for the time-consuming process of interatomic potential development for each new system.

References

1. B. Bogdanovic and G. Sandrock, Catalyzed complex metal hydrides, *Mater. Res. Bull.* **27**, 712–716 (September 2002).
2. J. J. Willems and K. H. Buschow, From permanent magnets to rechargeable hydride electrodes, *J. Less-Common Metals* **129**, 13–30 (1987).
3. G. Sandrock, State-of-the-art review of hydrogen storage in reversible metal hydrides for military fuel cell applications, Report prepared for Department of the Navy, Office of Naval Research, Ballston Tower One, 800 North Quincy Street, Arlington, VA, 22217–5660. Contract number N00014–97-M-0001, (SunaTech, Inc., 1997).
4. B. Bogdanovic, R. A. Brand, A. Marjanovic, M. Schwicardi, and J. Tolle, Metal-doped sodium aluminium hydrides as potential new hydrogen storage materials, *J. Alloys Compd.* **302**, 36–58 (2000).
5. L. Schlapbach, A. Züttel, P. Gröning, O. Gröning and P. Aebi, Hydrogen for novel materials and devices, *Appl. Phys. A* **72**, 245–253 (2001).
6. M. Conradi, M. Mendenhall, T. Ivancic, E. Carl, C. Browning, P. Notten, W. P. Kalisvaart, P. C. M. M. Magusin, R. C. Bowman, Jr., S.-J. Hwang and N. L. Adolphi, NMR to determine rates of motion and structures in metal-hydrides, *J. Alloys Comp.* **446–447**, 499–503 (2007).
7. A. Züttel, Materials for hydrogen storage, in *Fuel Cells Compendium*, eds. N. Brandon and D. Thompsett (Elsevier, San Diego, 2005), pp. 517–529.
8. L. Norskov, The hydrogen-metal interaction, *Europhysics News* **19**, 65–67 (1988).
9. J. Graetz, J. Reilly, G. Sandrock, J. Johnson, W.-M. Zhou and J. Wegrzyn, Aluminum hydride, AlH_3, as a hydrogen storage compound, Brookhaven National Laboratories, Report BNL-77336–2006 (2006).

10. V. Ozoliņš, E. H. Majzoub and T. J. Udovic, Electronic structure and Rietveld refinement parameters of Ti-doped sodium alanates, *J. Alloys Compd.* **375**, 1–10 (2004).
11. H. Hagemann, R. Gomes, G. Renaudin and K. Yvon, Raman studies of reorientation motions of $[BH_4]^-$ anions in alkali borohydrides, *J. Alloys Compd.* **363**, 129–132 (2004).
12. E. H. Majzoub, K. McCarty and V. Ozoliņš, Lattice dynamics of $NaAlH_4$ from high-temperature single-crystal Raman scattering and *ab initio* calculations: Evidence of highly stable $[AlH_4]^-$ anions," *Phys. Rev. B* **71**, 024118–1–10 (2005).
13. E. H. Majzoub and V. Ozoliņš, Prototype electrostatic ground state approach to predicting crystal structures of ionic compounds: Application to hydrogen storage materials, *Phys. Rev. B* **77**, 104115–1–13 (2008).
14. E. H. Majzoub, J. L. Herberg, R. Stumpf, S. Spangler and R. S. Maxwell, XRD and NMR investigation of Ti-compound formation in solution-doping of sodium aluminum hydrides: Solubility of Ti in $NaAlH_4$ crystals grown in THF, *J. Alloys Compd.* **394**, 265–270 (2005).
15. E. C. Rönnebro and E. H. Majzoub, Crystal structure, Raman spectroscopy, and *ab initio* calculations of a new bialkali alanate K_2LiAlH_6, *J. Phys. Chem. B* **110**, 25686–25691 (2006).
16. *Inorganic Crystal Structure Database* from http://www.fiz-karlsruhe.de/icsd.html.
17. C. Wolvertion and V. Ozoliņš, Hydrogen storage in calcium alanate: First-principles thermodynamics and crystal structures, *Phys. Rev. B* **75**, 064101–1–15 (2007).
18. K. Miwa, M. Aoki, T. Noritake, N. Ohba, Y. Nakamori, S.-I. Towata, A. Züttel and S.-I. Orimo, Thermodynamical stability of calcium borohydride $Ca(BH_4)_2$, *Phys. Rev. B* **74**, 155122–1–5 (2006).
19. L. Pauling, The principles determining the structure of complex ionic crystals, *J. Am. Chem. Soc.* **51**, 1010–1026 (1929).
20. R. J. Wawak, J. Pillardy, A. Liwo, K. D. Gibson and H. A. Scheraga, Diffusion equation and distance scaling methods of global optimization: Applications to crystal structure Prediction, *J. Phys. Chem. A* , **102**, 2904–2918 (1988)
21. G. Kresse and J. Furthmueller, Efficiency of *ab initio* total energy calculations for metals and semiconductors using a plane-wave basis set, *J. Comput. Mater. Sci.* **6**, 15–50 (1996).
22. G. Kresse and J. Furthmueller, Efficient iterative schemes for *ab initio* total-energy calculations using a plane-wave basis set, *Phys. Rev. B* **54**, 11169–11186 (1996).
23. G. Kresse and D. Joubert, From ultrasoft pseudopotentials to the projector augmented-wave method, *Phys. Rev. B* **59**, 1758–1775 (1999).
24. H. T. Stokes and D. M. Hatch, *Isotropy software package* from http://stokes.byu.edu/isotropy.html.
25. C. Kittel, C. *Introduction to Solid-State Physics*, 8th Edn. (Wiley, New York, 2005).
26. L. Seballos, J. Z. Zhang, E. C. Rönnebro, J. L. Herberg and E. H. Majzoub, Metastability and crystal structure of the bialkali complex metal borohydride $NaK(BH_4)_2$, *J. Alloys Compd.* **476**, 446–450 (2009).
27. R. Cerny, Y. Filinchuk, H. Hagemann and K. Yvon, Magnesium borohydride: Synthesis and crystal structure, *Agnew. Chem., Int. Ed.* **46**, 5765–5767 (2007).
28. J.-H. Her, P. W. Stephens, Y. Gao, G. L. Soloveichik, J. Rijssenbeek, M. Andrus and J.-C. Zhao, Structure of unsolvated magnesium borohydride $Mg(BH_4)_2$, *Acta Crystallogr. B* **63**, 561–568 (2007).

29. K. Chlopek, C. Frommen, A. Leon, O. Zabara and M. Fichtner, Synthesis and properties of magnesium tetrahydroborate, $Mg(BH_4)_2$, *J. Mater. Chem.* **17**, 3496–3503 (2007).
30. V. Ozoliņš, E. H. Majzoub and C. Wolverton, First-principles prediction of a ground state crystal structure of magnesium borohydride, *Phys. Rev. Lett.* **100**, 135501–1–4 (2008).
31. X.-F. Zhou, Q.-R. Qian, J. Zhou, B. Xu, Y. Tian and H.-T. Wang, Crystal structure and stability of magnesium borohydride from first principles, *Phys. Rev. B* **79**, 212102–1–4 (2009).
32. J. Voss, J. Hummelshoj, Z. Lodziana and T. Vegge, Structural stability and decomposition of $Mg(BH_4)_2$ isomorphs – An *ab initio* free energy study, *J. Phys.: Condens.* Matter **21**, 012203–1–7 (2009).
33. S.-J. Hwang, R. Bowman, J. Reiter, J. Rijssenbeek, G. Soloveichik, J.-C. Zhao, H. Kabbor and C. C. Ahn, NMR confirmation for formation of $[B_{12}H_{12}]^{2-}$ complexes during hydrogen desorption from metal borohydrides, *J. Phys. Chem. C* , **112**, 3164–3169 (2008).
34. V. Ozoliņš, E. H. Majzoub and C. Wolverton, First-principles prediction of thermodynamically reversible hydrogen storage reactions in the Li-Mg-Ca-B-H system, *J. Am. Chem. Soc.* **131**, 230–237 (2009).
35. J.-P. Soulie, G. Renaudin, R. Cerny and K. Yvon, Lithium boro-hydride $LiBH_4$: I. Crystal structure, *J. Alloys Compd.* **346**, 200–205 (2002).
36. L. Mosegaard, B. Moller, J.-E. Jorgensen, U. Bosenberg, M. Dornheim, J. Hanson, Y. Cerenius, G., Walker, H. J. Jakobsen, F. Besenbacher and T. Jensen, Intermediate phases observed during decomposition of $LiBH_4$, *J. Alloys Compd.* **446–447**, 301–305 (2007).
37. M. D. Riktor, M. H. Sørby, K. Chlopek, M. Fichtner, F. Bouchter, A. Züttel and B. C. Hauback, *In situ* synchrotron diffraction studies of phase transitions and thermal decomposition of $Mg(BH_4)_2$ and $Ca(BH_4)_2$, *J. Mater. Chem.* **17**, 4939–4942 (2007).
38. F. Buchter, Z. Łodziana, A. Remhof, O. Friedrichs, A. Borgschulte, P. Mauron, A. Züttel, D. Sheptyakov, G. Barkhordarian, R. Bormann, K. Chłopek, M. Fichtner, M. Sørby, M. Riktor, B. Hauback and S.-i. Orimo, Structure of $Ca(BD_4)_2$ β-Phase from combined neutron and synchrotron X-ray powder diffraction data and density functional calculations, *J. Phys. Chem. B* **112**, 8042–8048 (2008).
39. E. C. Rönnebro and E. H. Majzoub, Calcium borohydride for hydrogen storage: Catalysis and reversibility, *J. Phys. Chem. B Lett.* **111**, 12045–12047 (2007).
40. R. Cerny and V. Favre-Nicolin, FOX: A friendly tool to solve nonmolecular structures from powder diffraction, *Powder Diffr.* **4**, 359–365 (2005).
41. Y. Filinchuk, E. Rönnebro and D. Chandra, Crystal structures and phase transformations in $Ca(BH_4)_2$, *Acta Mater.* **57**, 732–738 (2009).
42. E. H. Majzoub and E. C. Rönnebro, Crystal structures of calcium borohydride: Theory and experiment, *J. Phys. Chem. C* **113**, 3352–3358 (2009).
43. G. Renaudin, S. Gomes, H. Hagemann, L. Keller and K. Yvon, Structural and spectroscopic studies on the alkali borohydrides MBH_4 (M = Na, K, Rb, Cs), *J. Alloys Compd.* **375**, 98–106 (2004).

Index

About the Editors

Thomas F. George is chancellor and professor of chemistry and physics at the University of Missouri-St. Louis. He is engaged in the St. Louis community, and in 2008 he received the (1) Distinguished Higher Education Award from the Dr. Martin Luther King, Jr., Missouri State Celebration Commission and (2) Louis North County Chamber of Commerce Citizen of the Year.

In addition to his role as campus and community leader and fund-raiser, George is an active researcher in chemistry and physics, specializing in chemical/materials/laser physics. His work has led to 725 papers, 5 authored books, 16 edited books and 210 conference abstracts. In 2004 he was elected as a foreign member of the Korean Academy of Science and Technology, and he has held the title of visiting professor of physics at Korea University in Seoul. In 2008 he received an honorary doctorate in physics (*honoris causa*) from the University of Szeged in Hungary, and in 2010 he was awarded the Medal of Honor by Gulf University for Science & Technology in Kuwait.

An accomplished jazz pianist, George has studied with faculty at the Berklee College of Music in Boston and the Eastman School of Music in Rochester, New York. He has performed extensively in public, including recent performances at the following: St. Louis Jazz & Heritage Festival, St. Louis Route 66 Festival, Art in the Park in Belleville, Illinois, Touhill Performing Arts Center, Sheldon Concert Hall, Finale and Jazz at the

Bistro in St. Louis; University of Arkansas; Nanjing University in China; and the city of Szeged in Hungary.

Born in Philadelphia, George became an Eagle Scout and received his high school diploma from Friends' Central School, where he earned varsity letters in soccer and wrestling. He received a bachelor of arts degree (*Phi Beta Kappa*) with a double major in chemistry and mathematics from Gettysburg College in Pennsylvania. He earned a master of science degree and doctor of philosophy degree in theoretical chemistry from Yale University, followed by postdoctoral appointments at MIT and the University of California at Berkeley.

Daniel Jelski, an Oregon native, spent much of his high school years in Germany. He received his undergraduate degree in chemistry from the University of Chicago and a Ph.D. degree in chemistry from Northern Illinois University. While completing his education, he worked as a Chicago cab driver and later as a computer programmer for Abbott Laboratories.

Upon finishing school, Jelski worked as a research associate in the Department of Physics and Astronomy at the State University of New York (SUNY) at Buffalo. His work involved the computational modeling of silicon clusters and of the then novel C_{60} molecule, buckminsterfullerene. Jelski was co-author on the first paper calculating the infrared and raman spectra of that molecule — a paper which has been widely cited since.

Jelski's first teaching appointment was at SUNY–Fredonia, where he started as a visiting assistant professor of chemistry. He rose through the ranks as an assistant professor and then associate professor. He continued his research in computational chemical physics, and co-authored several papers with undergraduate students.

In 1996–97, Jelski received a Fulbright Scholarship to spend a year teaching chemistry at Makerere University in Kampala, Uganda. There he had an opportunity to supervise a master's thesis, successfully graduating Mr. Godfrey Ssemakula.

Subsequently, Jelski received an appointment as professor of chemistry and head of the Department of Chemistry at Rose-Hulman Institute of Technology in Terre Haute, Indiana. He supervised a department of 12 faculty and three staff and was responsible for developing a biochemistry program.

In 2007, Jelski accepted his current position as dean of the School of Science and Engineering at SUNY–New Paltz. He also has an appointment as professor of chemistry. As dean, his efforts have emphasized student recruitment, collaboration with business and industry, and improving the quality of the academic programs. The school is initiating new programs in astronomy and biochemistry and has substantially revised curricula in biology, computer science and engineering.

Jelski is married to Valerie Castillo Jelski, and together they have two grown children, Christina and Mark. Mark is currently serving in the United States Coast Guard. In his spare time, Jelski posts occasional columns at www.newgeography.com.

Renat R. Letfullin is associate professor of physics and optical engineering department at the Rose-Hulman Institute of Technology. He has extensive academic credentials in nanotechnology and specializes in laser physics, wave and quantum optics, aerosol physics, biophotonics and nanomedicine, where he is using laser-induced explosion of absorbing nanoparticles in selective nanophotothermolysis of cancer.

Letfullin served as senior researcher at the Lebedev Physics Institute of the Russian Academy of Sciences (Samara branch), 1993–2002, and research associate at Mississippi State University, 2002–2004. His research activities have included the electrodynamics of super-high-energy pulsed chemical lasers based on photon-branched chain reactions; the auto-wave effect of photon-branched chain reactions; the effect of giant laser energy gain; optical effects of diffractive multifocal focusing of radiation; theoretical modeling of laser light interaction with biological systems containing nanostructures; development of new dynamic modes for treating cancer cells by laser-activated nanoheaters; nanoparticle optics; laser heating and evaporation of nanostructures; and optical breakdown of aerosols.

Letfullin has published over 150 articles and conference proceedings and 11 book chapters on chemical lasers, diffractive optics, ultrashort laser physics, nanotechnology, nanomedicine and aerosol physics. He owns 4 patents in laser technology and optical engineering and has presented 38 invited seminars, colloquia and talks on scientific research and higher education. He is a member of the Optical Society of America, Society of Photo-Optical Instrumentation Engineers and American Society for Engineering Education.

Letfullin earned his bachelor's degree in physics, and a master's degree in optics and spectroscopy at Russia's Samara State University. He received his doctorate in laser physics from Saratov State University in Russia, followed by postdoctoral studies in optics at Mississippi State University.

Guoping Zhang is associate professor of physics at Indiana State University and has been a Promising Scholar at Indiana State University. His research focuses on femtosecond magnetism, nano-materials engineering, first-principles density functional theory, nanostructures, fullerenes, static and time-resolved magneto-optical Kerr effect, spintronics, nonlinear optics in conjugated polymers, strongly-correlated electron systems, ultrafast spectroscopy, pump-probe and four-wave mixing signal, laser-induced ultrafast spin dynamics in nano- and low-dimensional magnets, soft x-ray spectroscopy, high-harmonic generations, mechanic properties of diamondoids, superconductivity in MgB_2, and photoisomerization in rhodopsin. Recently, his interests have extended to medical research. He has published 71 peer-reviewed research papers, with eight in *Physical Review Letters* and one in *Nature Physics*. He has co-edited three books. He has written invited articles for three journals and coauthored a review paper on femtomagnetism. He has published 8 book chapters, made more than 40 conference presentations nationally and internationally, and given 14 invited talks starting since 2000. He is a life member of both the American Physical Society and Optical Society of America.

Zhang was a co-principal investigator for the U.S. National Science Foundation grant award (2003–2006) on "Nanotechnology Undergraduate Education: Sophomore Course and Ancillaries in Nanoscience" and the U.S. Army Research Office grant award (2004–2008) on "Diamondoid Organic Nanostructures." He is the sole principal investigator for the Department of Energy grant (2006–2010) on "Laser-Induced Ultrafast Magnetization in Ferromagnets." In addition, he has been awarded over 5 million computational hours from the National Energy Research Scientific Computing Center at Lawrence Berkeley National Laboratory and the Argonne Leadership Computing Facility at Argonne National Laboratory.

He received his bachelor of science degree with a major in physics from Nanjing Normal University, where his undergraduate thesis was on chemical electrodeposition. He obtained his master of science degree in applied physics from Shanghai Jiaotong University, a top engineering school in China. He earned his doctor of philosophy degree in physics from Fudan University in Shanghai, where his thesis was on laser-induced ultrafast dynamics and pairing in C_{60}. He was a guest scientist at the Max-Planck-Institute for the Physics of Complex systems, Dresden, Germany, followed by postdoctoral appointments at the Max-Planck-Institute for Microstructure physics in Germany and the University of Tennessee at Knoxville, USA.